INTERNATIONAL REVIEW OF CELL AND MOLECULAR BIOLOGY

VOLUME THREE HUNDRED AND THIRTY FOUR

INTERNATIONAL REVIEW OF CELL AND MOLECULAR BIOLOGY

MiRNAs in Aging and Cancer

Edited by

LORENZO GALLUZZI
Department of Radiation Oncology
Weill Cornell Medical College
New York, NY

ILIO VITALE
Department of Biology
University of Rome "Tor Vergata"
Rome, Italy

ELSEVIER

ACADEMIC PRESS
An imprint of Elsevier

Academic Press is an imprint of Elsevier
50 Hampshire Street, 5th Floor, Cambridge, MA 02139, United States
525 B Street, Suite 1800, San Diego, CA 92101-4495, United States
The Boulevard, Langford Lane, Kidlington, Oxford OX5 1GB, United Kingdom
125 London Wall, London, EC2Y 5AS, United Kingdom

First edition 2017

Notices
Knowledge and best practice in this field are constantly changing. As new research and experience broaden our understanding, changes in research methods, professional practices, or medical treatment may become necessary.

Practitioners and researchers must always rely on their own experience and knowledge in evaluating and using any information, methods, compounds, or experiments described herein. In using such information or methods they should be mindful of their own safety and the safety of others, including parties for whom they have a professional responsibility.

To the fullest extent of the law, neither the Publisher nor the authors, contributors, or editors, assume any liability for any injury and/or damage to persons or property as a matter of products liability, negligence or otherwise, or from any use or operation of any methods, products, instructions, or ideas contained in the material herein.

ISBN: 978-0-12-811868-9
ISSN: 1937-6448

For information on all Academic Press publications
visit our website at https://www.elsevier.com/books-and-journals

Publisher: Zoe Kruze
Acquisition Editor: Alex White
Editorial Project Manager: Fenton Coulthurst
Production Project Manager: James Selvam
Cover Designer: Mark Rogers

Typeset by SPi Global, India

CONTENTS

CONTRIBUTORS

Kotb Abdelmohsen
Laboratory of Genetics and Genomics, National Institute on Aging, National Institutes of Health, Baltimore, MD, United States

Éric Bonneau
Maisonneuve-Rosemont Hospital, Montréal, QC, Canada

Liuxi Chen
Sir Run Run Shaw Hospital, College of Medicine, Zhejiang University, Hangzhou, China

Florence Couteau
Chromatin Structure and Cellular Senescence Research Unit, Maisonneuve-Rosemont Hospital Research Centre, Montréal, QC, Canada

Vincent De Guire
Université de Montréal; Maisonneuve-Rosemont Hospital, Montréal, QC, Canada

Tobias Engel
Royal College of Surgeons in Ireland, Dublin, Ireland

Katarzyna Goljanek-Whysall
Musculoskeletal Biology II, Centre for Integrated Research into Musculoskeletal Aging, Institute of Aging and Chronic Disease, University of Liverpool, Liverpool, United Kingdom

Myriam Gorospe
Laboratory of Genetics and Genomics, National Institute on Aging, National Institutes of Health, Baltimore, MD, United States

Ioannis Grammatikakis
Laboratory of Genetics and Genomics, National Institute on Aging, National Institutes of Health, Baltimore, MD, United States

Keisuke Ito
Ruth L. and David S. Gottesman Institute for Stem Cell and Regenerative Medicine Research, Albert Einstein Cancer Center and Einstein Diabetes Research Center, Albert Einstein College of Medicine, Bronx, NY, United States

Eva M. Jimenez-Mateos
Royal College of Surgeons in Ireland, Dublin, Ireland

Aidan Kenny
Royal College of Surgeons in Ireland, Dublin, Ireland

Frédérick A. Mallette
Chromatin Structure and Cellular Senescence Research Unit, Maisonneuve-Rosemont Hospital Research Centre; Université de Montréal, Montréal, QC, Canada

Rachel McCormick
Musculoskeletal Biology II, Centre for Integrated Research into Musculoskeletal Aging, Institute of Aging and Chronic Disease, University of Liverpool, Liverpool, United Kingdom

Miguel Medina
CIBERNED, Network Center for Biomedical Research in Neurodegenerative Diseases; CIEN Foundation, Queen Sofia Foundation Alzheimer Center, Madrid, Spain

Rachel Munk
Laboratory of Genetics and Genomics, National Institute on Aging, National Institutes of Health, Baltimore, MD, United States

Mathieu Neault
Chromatin Structure and Cellular Senescence Research Unit, Maisonneuve-Rosemont Hospital Research Centre; Université de Montréal, Montréal, QC, Canada

Hongming Pan
Sir Run Run Shaw Hospital, College of Medicine, Zhejiang University, Hangzhou, China

Amaresh C. Panda
Laboratory of Genetics and Genomics, National Institute on Aging, National Institutes of Health, Baltimore, MD, United States

Sean Quinlan
Royal College of Surgeons in Ireland, Dublin, Ireland

Blanche Schroen
Cardiovascular Research Institute Maastricht (CARIM), Maastricht University, Maastricht, The Netherlands

Xinbing Sui
Sir Run Run Shaw Hospital, College of Medicine, Zhejiang University; Zhejiang Chinese Medical University, Hangzhou, China

Qiuhua Sun
Zhejiang Chinese Medical University, Hangzhou, China

Marc van Bilsen
Cardiovascular Research Institute Maastricht (CARIM), Maastricht University, Maastricht, The Netherlands

Robin Verjans
Cardiovascular Research Institute Maastricht (CARIM), Maastricht University, Maastricht, The Netherlands

Cary N. Weiss
Ruth L. and David S. Gottesman Institute for Stem Cell and Regenerative Medicine Research, Albert Einstein Cancer Center and Einstein Diabetes Research Center, Albert Einstein College of Medicine, Bronx, NY, United States

Jichun Zhou
Sir Run Run Shaw Hospital, College of Medicine, Zhejiang University, Hangzhou, China

Yuxia Zhou
Shandong Maternal and Child Health Hospital, Ji Nan, Shandong, China

CHAPTER ONE

Regulation of Autophagy by MiRNAs and Their Emerging Roles in Tumorigenesis and Cancer Treatment

Liuxi Chen*[,2], Yuxia Zhou†[,2], Qiuhua Sun‡, Jichun Zhou*[,1], Hongming Pan*[,1], Xinbing Sui*[,‡,1]

*Sir Run Run Shaw Hospital, College of Medicine, Zhejiang University, Hangzhou, China
†Shandong Maternal and Child Health Hospital, Ji Nan, Shandong, China
‡Zhejiang Chinese Medical University, Hangzhou, China
[1]Corresponding authors: e-mail address: zjc@live.com; drpanhm@aliyun.com; hzzju@aliyun.com

Contents

Abstract

Autophagy is a conserved catabolic process for the degradation and recycling of cytosolic components or organelles through a lysosome-dependent pathway. Autophagy can be induced in response to multiple stress conditions, such as nutrient deprivation, hypoxia, energy depletion, etc. As a result, autophagy can regulate many biological processes, including cell survival, metabolism, differentiation, senescence, and cell death. MicroRNAs (MiRNAs) are small noncoding molecules that regulate gene expression by silencing mRNA targets. MiRNA dysregulation exhibits great regulatory potential during organismal development, hematopoiesis, immunity, cell proliferation and death, and autophagy. Recently, increasing studies have linked MiRNAs to autophagic regulation during cancer initiation and development. Although the relationship between

[2] These authors contributed equally to this work.

International Review of Cell and Molecular Biology, Volume 334
ISSN 1937-6448
http://dx.doi.org/10.1016/bs.ircmb.2017.03.003

MiRNAs and autophagy is quite complicated and has not been well elucidated, MiRNAs may underlie key aspects of autophagy and cancer biology. Increasing evidence shows that MiRNAs play important roles as both oncogenic MiRNAs and tumor suppressive MiRNAs in cancer initiation and development. Thus, understanding the novel relationship between MiRNAs and autophagy may allow us to develop promising cancer biomarkers and therapeutic targets.

1. INTRODUCTION

Autophagy is a catabolic process by which cellular components or organelles are sequestered into a double-membrane vesicle and subsequently delivered to the lysosome for terminal degradation and recycling (Levine and Kroemer, 2008, 2009). At present, autophagy can be divided into three major types in eukaryotic cells: macroautophagy (the best-characterized and most well-known form, often referred to as autophagy), microautophagy, and chaperone-mediated autophagy (Sui et al., 2013; Yang and Klionsky, 2009). The autophagic process involves at least the following five distinct steps: induction, nucleation, elongation and completion, docking and fusion (Feng et al., 2015; Fullgrabe et al., 2014; Levine and Klionsky, 2004; Shintani and Klionsky, 2004; Tsuchihara et al., 2009). Under the normal conditions, autophagy can function as a cellular homeostasis mechanism, which promotes the turnover of unfolded proteins and unnecessary or aged organelles (Mizushima et al., 2008). While going through the stressful conditions, such as nutrient deprivation, hypoxia, energy depletion, autophagy will be activated as a cytoprotective effect to eliminate the damaged proteins and organisms (Kroemer et al., 2010). If that situation is out of control, persistent or excessive autophagy may be resulted in cell death named as autophagic cell death, a form of cell death by autophagy (in contrast to type I programmed cell death or apoptosis) (Clarke and Puyal, 2012; Kroemer and Levine, 2008). A variety of signaling pathways or molecules have been implicated in the upregulation or downregulation of autophagy, including mammalian target of rapamycin (MTOR), AMP-responsive protein kinase (AMPK), the class III phosphatidylinositol 3-kinase (PtdIns3K), p53, BCL-2, and MiRNAs (Shintani and Klionsky, 2004).

MiRNAs are small noncoding RNAs that play essential roles in post-transcriptional protein-coding gene expression, which is regulated by MiRNA–target interactions (Bartel, 2004; Jones-Rhoades and Bartel, 2004). Currently, an increasing body of evidence suggests that MiRNAs play a double role in the regulation of autophagy, behaving both as pro-autophagy roles and antiautophagy roles. Dysregulation of autophagy occurs frequently in human cancer cells, also influenced by MiRNAs (Zhou et al., 2012). As well, MiRNAs play important roles as both oncogenic MiRNAs (oncomiRNAs) and tumor suppressive MiRNAs (tsmiRNAs) in various stages of cancer, including cancer initiation, development, metastasis, recurrence, and drug-resistance (Hanahan and Weinberg, 2011). More importantly, MiRNAs-modulated autophagy tries to explore the exact signaling pathways, networks and makes sure the exact molecules as targets for cancer diagnosis and therapy, prognosis. There is increasing support in the science, regulation of autophagy by MiRNAs may be promising for cancer diagnosis and therapy in the future. However, the relationship between MiRNAs and autophagy has not been well elucidated.

In this review, we will summarize how MiRNAs regulate autophagy and pay close attention to the impact of MiRNA-regulated autophagy on cancer initiation and development. We also classify MiRNAs and autophagy into two parts according to their functions and discuss the differentiation between different types. In addition, the scientific and clinical hurdles regarding the potential role of MiRNAs-regulated autophagy in cancer diagnosis and treatment will then be discussed.

2. MiRNAs AND ITS ROLE IN GENE REGULATION

MicroRNAs (MiRNAs) are small noncoding RNA molecules, which regulate gene expression by binding to the 3′-untranslated region (UTR) or 5′-UTR of target messenger RNAs (mRNAs). As a result, MiRNAs play essential roles in a wide variety of biological processes, including cell survival, differentiation, cell cycle arrest, senescence, and autophagy (Ambros, 2004; Calin and Croce, 2006). Currently, it has become clear that MiRNA genes are initiated by RNA polymerase II (Pol II) and generate primary MiRNA (pri-miRNA), about hundreds to thousands of nucleotides, which has a 5′-7-methylguanosine cap and a 3′-poly (A) tail (Cai et al., 2004; Lee et al., 2004). The pri-miRNA is sheared by Drosha complex to produce hairpin-structured precursor MiRNA (pre-miRNA) (Han et al., 2006). Drosha complex, also called microprocessor complex, is composed of an RNase III enzyme (DROSHA), a double-stranded RNA-binding domain protein (DiGeorge critical region 8, DGCR8) and several auxiliary factors, such as the DEAD-box RNA helicases p68 (DDX5) and p72/p82 (DDX17). DROSHA combines with DGCR8-mediated single- or double-stranded portions of pri-miRNA and releases the stem-loop pre-miRNA (Gregory et al., 2004; Han et al., 2006; Lee et al., 2003). And then, the pre-miRNA is translocated from the nucleus into the cytoplasm by exportin 5 (XPO5) via a Ran-GTP-dependent mechanism (Bohnsack et al., 2004; Lund et al., 2004; Yi et al., 2003). Subsequently, the cytoplasmic pre-miRNA is sheared by DICER1, a RNase III enzyme, which is integrated with TAR RNA-binding protein (TRBP) and kinase R-activating protein (PACT), producing the MiRNA/MiRNA* duplex, a 20–25 nucleotides mature duplex MiRNA (Chendrimada et al., 2005; Forstemann et al., 2005; Hutvagner et al., 2001; Jiang et al., 2005; Ketting et al., 2001; Lee et al., 2006; Saito et al., 2005). Next, one strand of MiRNA/MiRNA* duplex is incorporated into the RNA-induced silencing complex (RISC), containing Argonaute (AGO) family protein. The located RISC complex enables MiRNAs to recognize complementary sequences located in the 3′-UTR of target mRNAs and promotes protein translational inhibition or mRNA degradation (Bartel, 2009; Du and Zamore, 2005; Gregory et al., 2005; Hutvagner and Zamore, 2002; Jansson and Lund, 2012; Martinez and Tuschl, 2004; Schwarz et al., 2003) (Fig. 1).

It is estimated that MiRNAs can regulate more than half protein-coding genes in human (Sorel and Dewals, 2016). Therefore, MiRNAs have been shown to be crucial mediators for the posttranscriptional regulation of various cellular biological functions. What's more, MiRNAs are easily detectable not only in cells but also in various biological fluids, including serum, saliva, and urine (Izzotti et al., 2016). Both the bioinformatic approaches and the experimental identifications can be used to predict the MiRNA targets for its characteristics (Bagga et al., 2005; Sethupathy et al., 2006). Currently, MiRNAs have been implicated in different human diseases, including neurological, cardiovascular, and metabolic diseases, and various forms of cancers (Piletic and Kunej, 2016).

Increasing evidence shows that MiRNAs play important roles as both oncomiRNAs and tsmiRNAs in cancer initiation and development (Calin et al., 2004a; Lu et al., 2005). For example, MiR-155, as an oncogene, is upregulated in diverse cancers, including B-cell lymphoma, breast cancer, colon cancer, and lung cancer (Eis et al., 2005; He et al., 2005; Volinia et al., 2006). MiR-155 contributes to polyclonal B-cell malignancies with its abnormal expression, which is connected with lymphoma-genesis (Costinean et al., 2006). High expression levels of MiR-27 are associated with chemoresistance in esophageal cancer through transformation of normal fibroblasts to cancer-associated fibroblasts (Tanaka et al., 2015). While

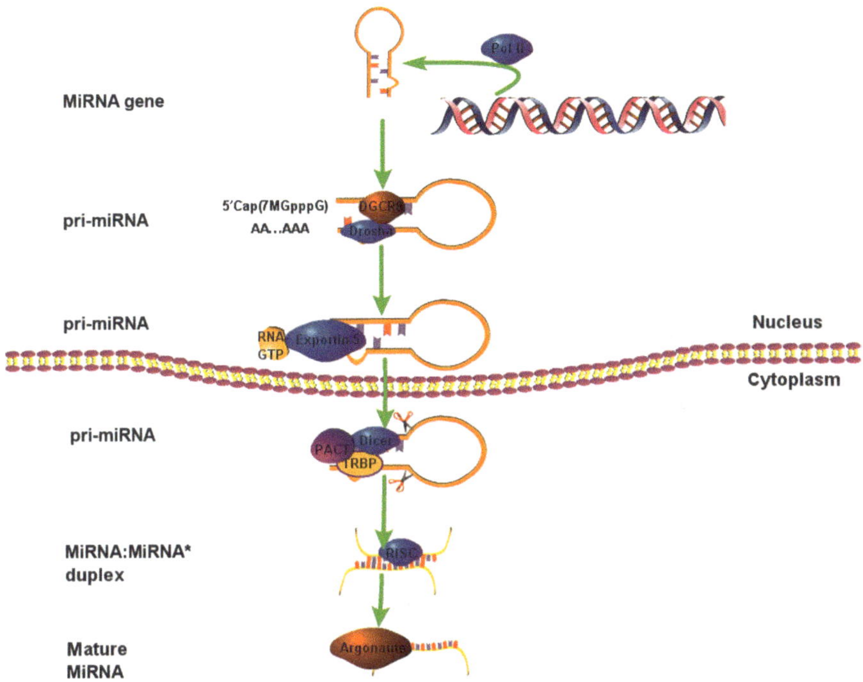

Fig. 1 MicroRNAs biogenesis and mechanism. MiRNA genes are initiated by RNA Pol II and generate pri-miRNA. The pri-miRNA is sheared by DROSHA RNase to produce about 70–100 nucleotide-long pre-miRNA, which is translocated from the nucleus into the cytoplasm by exportin 5 via a Ran-GTP-dependent mechanism. Subsequently, the cytoplasmic pre-miRNA is sheared by DICER1, a RNase III enzyme, producing the MiRNA/MiRNA* duplex incorporated into the RNA-induced silencing complex (RISC), which enables MiRNAs to recognize complementary sequences located in the 3′-UTR of target mRNAs and promotes protein translational inhibition or mRNA degradation.

MiR-15a/MiR-16, locating at chromosome 13q14, inhibits chronic lymphocytic leukemia and other cancer types through the overexpression of *BCL2*. Thus, MiR-15a/MiR-16 can act as tsmiRNA (Calin et al., 2002, 2005; Cimmino et al., 2005; Dohner et al., 2000; Dong et al., 2001). MiR-33a, as a tumor suppressor, that downregulates Pim-3 kinase *(PIM3)* expression to increase human pancreatic cancer cells sensitivity to gemcitabine via the AKT/β-catenin pathway (Liang et al., 2015). Taken together, MiRNAs play central role as tumor-associated biomarkers that are involved in not only early-stage tumors, but also advanced stage tumors, tumor recurrence, and drug sensitivities. Hence, MiRNAs can serve as diagnostic, prognostic, or even therapeutic markers for some cancers (Calin et al., 2004b; Volinia et al., 2006).

3. AUTOPHAGY AND ITS ROLE IN CANCER

Autophagy is a conserved catabolic process through which cellular unnecessary or dysfunctional proteins and organelles are transported to the lysosome for terminal degradation and recycling (Tsuchihara et al., 2009). Accumulating evidence suggests that autophagy plays a critical role in cell survival, metabolism, differentiation, senescence, and cell death (Bernard and Klionsky, 2013). When encountering multiple stresses, such as nutrient deprivation, hypoxia, energy depletion, autophagy can be constitutively activated to sustain metabolic homeostasis and protect the cells from the damages from its changing environmental conditions by promoting the turnover of damaged proteins and organelles (Inguscio et al., 2012).

However, excessive or persistent autophagy will cause cell death, termed autophagy-associated cell death or type II programmed cell death in contrast to type I programmed cell death or apoptosis (Cecconi and Levine, 2008; Glick et al., 2010). Autophagy is also shown to precede apoptosis or act in parallel with this cellular process in addition to be an alternative mechanism to cell death when apoptosis is inhibited. Autophagy related 5 (ATG5) and ATG12 have been proved to play key roles in the autophagic signaling pathway and also found to participate in the initiation of apoptosis. ATG5 induces mitochondrial apoptosis via binding to BCL2 like 1 (BCL2L1, also known as BCLXL), the prosurvival Bcl-2 family member. ATG5 translocates the N-terminal fragment to the mitochondrion by the function of calpains, which induces cytochrome *c* and causes mitochondrial apoptosis (Yousefi et al., 2006). ATG12 interacts with MCL1 and BCL2 through BH3 domain, causing damaged mitochondria, which mediates the release of cytochrome *c* (Rubinstein et al., 2011). It is seen that ATG5 and ATG12 take effect with nonconjugated forms. Vesicle nucleation is associated with the class III PI3K/Beclin 1 (BECN1) complex. While Beclin 1 could associate with BCLXL within mitochondria via a unique BH3-like motif, Beclin 1 is cleaved resulting the translocation of the C-terminal fragment to the mitochondria. Moreover, it releases cytochrome c1 /CYC1) and plays an important role in apoptotic signaling pathway (Kang et al., 2011; Noble et al., 2008). Therefore, autophagy itself can fulfill a dual role in cancer initiation and development, having tumor-promoting and tumor-suppressing properties.

Autophagic process starts with the isolation membrane (phagophore) which engulfs the cytoplasmic material and is elongated and encapsulated through the microtubule-associated protein 1 light chain 3 (MAP1LC3, also known as LC3 or ATG8) and the ATG12–ATG5–ATG16L conjugation systems. Then the autophagosome fuses with the lysosome producing the autolysosome, which degrades the engulfed components and the inner membrane by lysosomal enzymes (Feng et al., 2015; Fullgrabe et al., 2014; Levine and Klionsky, 2004; Shintani and Klionsky, 2004; Tsuchihara et al., 2009) (Fig. 2). Autophagy can be activated when cancer cells are subjected to multiple stressful conditions, including nutrient deprivation, endoplasmic reticulum stress, hypoxia, glucose/energy depletion, and other diverse stresses. A variety of signaling pathways and molecules have been implicated in autophagic process in cancer initiation and development. Among them, the central checkpoints are MTOR and AMPK signaling pathways in cancer

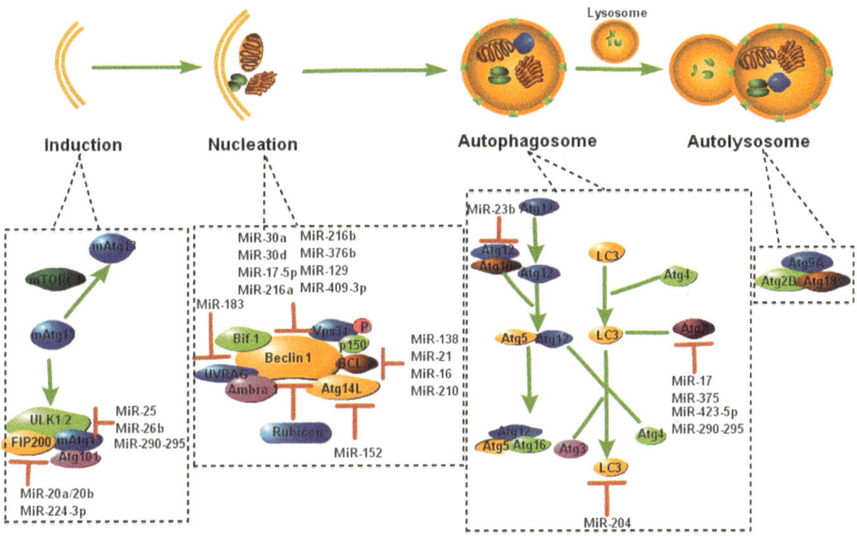

Fig. 2 The process of autophagy and MiRNAs in autophagic process. MiRNAs are involved in different autophagic stages, including induction, nucleation, elongation and completion, docking and fusion. AMPK–MTOR lies in the central of regulation of autophagy triggering by the ULK1 complex: ULK1/2 (MiR-25, MiR-26b, MiR-290–295) and FIP200 (MiR-20a/20b, MiR-224-3p). Vesicle nucleation is triggered by the class III PI3K/Beclin 1 complex and other proteins: Beclin 1 (MiR-30a, MiR-30d, MiR-17-5p, MiR-216a, MiR-216b, MiR-376b, MiR-129, MiR-409-3p), ATG14L (MiR-152), UVRAG (MiR-183), and BCL2 (MiR-138, MiR-21, MiR-16, MiR-210). Vesicle elongation and completion are referred to the two ubiquitin-like proteins, ATG12 and LC3/ATG8, which form two ubiquitin-like conjugation systems, ATG12–ATG5 and LC3/ATG8–ATG4–phosphatidylethanolamine system: LC3-II (MiR-204), ATG7 (MiR-17, MiR-375, MiR-423-5p, MiR-290–295), and ATG12 (MiR-23b). Docking and fusion of the autophagosome, targeting ATG9A, ATG2B, and ATG18, cause the maturation of the autolysosome, which can disintegrate the contents.

cells (Marx, 2006; Narita et al., 2011). MTOR may interact with two distinct cellular complexes termed MTOR complex 1 (MTORC1) and MTOR complex 2 (MTORC2). A well-established upstream regulator of MTORC1 is PtdIns3K/AKT signaling pathway. PtdIns3K binding to AKT and pyruvate dehydrogenase kinase 1 (PDK1) generates PtdIns3K-PKD1-AKT signaling pathway at the cellular plasma membrane, which inhibits the downstream target, tuberous sclerosis complex (TSC) 1/2. Loss-of-function in either TSC1 or TSC2 can cause to constitutive MTORC1 activation (Lorin et al., 2013; Shintani and Klionsky, 2004). Once activated, MTORC1 exerts a negative effect on autophagy by phosphorylating a complex of autophagy proteins (ATG13 and ULK1/2), which inhibits the downstream autophagy cascade (He and Klionsky, 2009; Sui et al., 2014). On the other hand, MTORC2 may indirectly suppress autophagy by activating MTORC1. In contrast with MTOR, AMPK can initiate autophagy by activating liver kinase B1 (LKB1) under the condition of energy depletion with the increasing AMP:ATP ratio (Singh and Cuervo, 2011). In addition, several cancer-linked genes (e.g., those encoding for p53, p62, p21, BCL2, and STAT3) and tumor-associated stressful signals (PERK/eukaryotic initiation factor 2α (eIF2α)/ATF4, MEK/ERK, and IRE1/JNK pathway) are also, respectively, stimulate or inhibit autophagy (He and Klionsky, 2009; Jiang et al., 2014; Kim et al., 2014; Maiuri et al., 2010) (Fig. 3).

4. REGULATION OF AUTOPHAGY BY MiRNAs IN CANCER

Accumulating studies show that MiRNAs are involved in different autophagic stages, including induction, nucleation, elongation and completion, docking and fusion, degradation (Jian et al., 2011; Zhu et al., 2009).

As we all know, autophagy induction is initiated by the ULK1 complex, including unc-51-like kinase 1/2 (ULK1/2), mammalian ATG13 (mATG13), RB1 inducible coiled-coil 1 (RB1CC1, also known as FIP200), and ATG101, which is inhibited by MTOR, the central checkpoint for autophagy. At the nutrient rich conditions, MTOR can phosphorylate ULK1 and mATG13, which disrupts ULK1 kinase activity. While MTOR dissociating from the ULK1 complex, the activation of ULK1/2 and the subsequent phosphorylation of FIP200 and mATG13 induce the initiation of autophagy in response to the starvation condition. Recent studies show that under the glucose condition, ULK1 complex can be phosphorylated by AMPK directly (Cheng et al., 2013; Ganley et al., 2009; Hosokawa et al., 2009; Kim et al., 2011). Both oncomiRNAs and tsmiRNAs can regulate autophagy through the ULK1/2 complex.

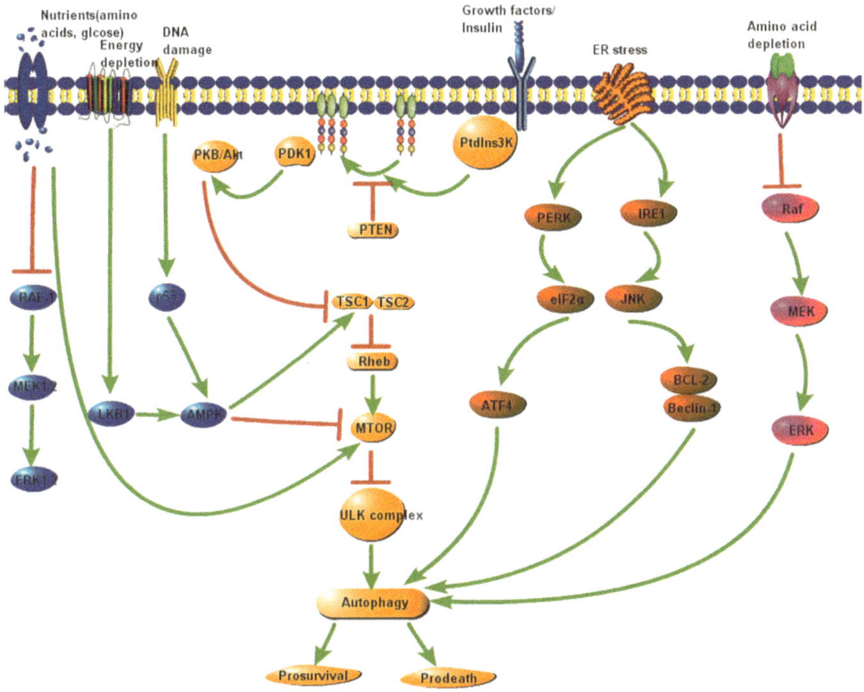

Fig. 3 Autophagy and its role in cancer. Autophagy can be activated when cancer cells are subjected to multiple stressful conditions, including nutrient deprivation, endoplasmic reticulum stress, hypoxia, glucose/energy depletion, and other diverse stresses. A variety of signaling pathways and molecules have been implicated in autophagic process in cancer initiation and development. Among them, the central checkpoints are MTOR and AMPK signaling pathways in cancer cells. A well-established upstream regulator of MTORC1 is PtdIns3K/AKT signaling pathway, which exerts a negative effect on autophagy by phosphorylating a complex of autophagy proteins (ATG13 and ULK1/2). In contrast with MTOR, AMPK can initiate autophagy by activating liver kinase B1 (LKB1) under the condition of energy depletion with the increasing AMP:ATP ratio. In addition, several cancer-linked genes (e.g., those encoding for p53, p62, p21, BCL2, and STAT3) and tumor-associated stressful signals (PERK/eukaryotic initiation factor 2α (eIF2α)/ATF4, MEK/ERK, and IRE1/JNK pathway) are also, respectively, stimulate or inhibit autophagy.

Under the condition of glucose starvation, MiR-290–295 functions as a oncogene which inhibits autophagic cell death via suppressing ULK1 and ATG 7 (Chen et al., 2012). MiR-25 has been proved to function as a novel regulator of autophagy by targeting *ULK1* and MiR-25 inhibition leads to autophagic cell death in breast cancer cells by directly increasing *ULK1* expression (Wang et al., 2014). *ULK2* is demonstrated to be a direct target of MiR-26b and MiR-26b could inhibit autophagy in prostate cancer through downregulation of *ULK2* expression (John Clotaire et al.,

2016). RB1CC1/FIP200 cellular levels are critical for autophagosome formation. Both MiR-20a and MiR-20b are found to negatively regulate autophagosome formation by targeting the expression of *RB1CC1* in the regulation of autophagosome formation in breast cancer cells and the suppression of MiR-20a or MiR-20b by specific antagomir show normal rapamycin-induced autophagic activity (Li et al., 2016). In addition to oncomiRNAs, several tsmiRNAs have been involved in the process of autophagy induction. MiR-224-3p, as a tumor suppressive role, has been shown to inhibit autophagy through downregulating the expression of ATGs and *RB1CC1* in human glioblastoma cells (Guo et al., 2015b) (Fig. 2).

Next, the vesicle nucleation is triggered by the class III PI3K/Beclin 1 complex and other proteins. Numerous binding partners of this complex, including Bax-interacting factor-1 (Bif-1), VPS34, ATG14L, UV irradiation resistance-associated gene (UVRAG), activating molecule in Beclin 1-regulated autophagy protein 1 (AMBRA 1), and RUN and cysteine rich domain containing Beclin 1 interacting protein (RUBCN, also known as RUBICON), can act as positive or negative regulators of autophagy (Fu et al., 2013; He and Levine, 2010; Kroemer et al., 2010). The first evidence about the relationship between MiRNAs and autophagy shows that MiR-30a downregulates *BECN1* expression to inhibit rapamycin-induced autophagy in cancer cells (Zhu et al., 2009). In addition, Beclin 1 is found to be a potent inducer of autophagy and plays a crucial role in cancer tumorigenesis and cancer progression. Downregulation of *BECN1* by MiR-30a attenuates the autophagic process and sensitizes the chemoresistance, including renal cell carcinoma to sorafenia, osteosarcoma to doxorubicin, and cervical cancer to hydroxycamptothecin exposure. Therefore, MiR-30a may act as a tumor suppressive role in osteosarcoma, cervical cancer, renal cell carcinoma (Xu et al., 2016; Zheng et al., 2015; Zou et al., 2012). tsmiRNA MiR-409-3p is proved to enhance the sensitivity of colon cancer cells to oxaliplatin via inhibiting Beclin 1-mediated autophagy (Tan et al., 2016). MiR-216a is another example of a Beclin 1-associated tsmiRNA which enhances the radiosensitivity of pancreatic cancer cells by inhibiting Beclin 1-mediated autophagy (Zhang et al., 2015b). MiR-30d is shown to sensitize anaplastic thyroid carcinoma cells to cisplatin or influence tumorigenesis through inhibiting autophagy via Beclin 1 and other genes, including *BINP3L*, *ATG12*, *ATG5*, and *ATG2* (Yang et al., 2013; Zhang et al., 2014). *BECN1* is found to be significantly upregulated in paclitaxel-resistant lung cancer cells, which suggests the inducting of autophagy may contribute to paclitaxel resistance. MiR-17-5p downregulation is proved to contribute to paclitaxel resistance of lung cancer cells through altering *BECN1* expression and the over-expression of MiR-17-5p resensitizes paclitaxel resistant lung cancer cells to paclitaxel (Chatterjee et al., 2014). MiR-129 plays a tumor suppressive role in human malignant glioma cells by triggering autophagy via directly regulating BECN1 (Chen et al., 2016a). MiR-216b can attenuate autophagy by downregulation of *BECN1* in nonsmall cell lung cancer and colorectal cancer (Chen and Shi, 2016; Chen et al., 2016b). MiR-376b can act as an oncomiRNA by blocking autophagy, which negatively regulates the *BECN1* expression in various cancer types, including lung cancer, renal cell carcinomas, and uterine leiomyomas (Kang et al., 2011; Korkmaz et al., 2012). In addition to Beclin 1, MiRNAs are also involved in several other autophagic regulators. BCL2, a key apoptosis regulator, is proved to inhibit autophagy by interacting with Beclin 1 (Gump and Thorburn, 2011; Thorburn, 2008). MiR-138 contributes to autophagy in temozolomide resistant glioma cells came from the downregulation of BCL2 like 11 (*BCL2L11*, best known as BIM), a proapoptotic BH3-only protein of the Bcl-2 family. In this way, BIM downregulates autophagy by binding and mislocalizing Beclin 1 (Stojcheva et al., 2016). MiR-21 is also described to inhibit tamoxifen-induced cell death in malignant glioma cells, which is correlated with Bcl-2 family members. Moreover, tamoxifen negatively regulates antiautophagic Bcl-2 family members, which causes autophagic cell death of human glioma cells (Harmalkar et al., 2015). The inhibition of MiR-16 is markedly related with paclitaxel resistant lung cancers cells via directly targeting *BCL2*. It is worth mentioning that MiR-16 along with MiR-17 can sensitive paclitaxel resistant lung cancers cells by, respectively, modulating apoptosis and autophagy. However, Chatterjee et al. have confirmed that MiR-16 and MiR-17 do not have common targets. BCL2 levels are markedly elevated through the downregulation of MiR-16, which in turn promotes the expression of *BECN1*, an autophagy regulator (Chatterjee et al., 2015). Under the condition of hypoxia, hypoxia-inducible factor-1α (HIF1A, also known as HIF-1α) facilitates the expression of MiR-210. *BCL2* is the downstream target of MiR-210, which is consistently contributed to the reduction of radiosensitivity of colon cancer cells. Furthermore, BCL2 binding to Beclin 1 may possess inappropriate inhibition of autophagy (Sun et al., 2015b). MiR-183 reduces the expression of rapamycin- or starvation-induced autophagy by targeting *UVRAG* in colorectal cancer cells (Huang et al., 2012; Huangfu et al., 2016). *ATG14*, another key component of vesicle nucleation, is a direct target of MiR-152, which suppresses cisplatin-induced autophagy in ovarian cancer (He et al., 2015b) (Fig. 2).

Subsequently, vesicle elongation and completion are referred to the two ubiquitin-like proteins, ATG12 and LC3/ATG8, which form two ubiquitin-like conjugation systems, ATG12–ATG5 and LC3/ATG8–ATG4–phosphatidylethanolamine system. Moreover, these two systems are tightly connected and also referred to the regulation of MiRNAs (Liu et al., 2010). LC3-II, a membrane-bound phosphatidylethanolamine-conjugated LC3, is processed by ATG4 and ATG7, which is differentiated from LC3-I and pro-LC3. MiR-204 may act as a tumor suppressor gene by inhibiting autophagy via directly targeting *MAP1LC3B* (Mikhaylova et al., 2012). Additionally, a tsmiRNA, MiR-375 suppresses hypoxia-induced autophagic flux through inhibiting the conversion of LC3-I to LC3-II, resulting in the decreasing cell viability in hepatocellular carcinoma (HCC) cells (Chang et al., 2012b). Similarly, MiR-17 could negatively regulate *ATG7*, which improves the sensitivity of human glioblastoma cells to temozolomide treatment (Comincini et al., 2013). MiR-423-5p can induce autophagic cell death via targeting *ATG7* and *MAP1LC3B* in hematocellular carcinoma (Stiuso et al., 2015). Conversely, MiR-290–295 as oncomiRNAs is found to inhibit autophagic cell death through downregulating the expression of *ATG7* in melanoma (Chen et al., 2012). *ATG12* is another target of MiRNAs. MiR-23b as a tsmiRNA has been proved to sensitize gastric and pancreatic cancer cells to cancer treatment via inhibiting autophagy mediated by targeting *ATG12* (An et al., 2015; Wang et al., 2013) (Fig. 2).

Docking and fusion of the autophagosome with the lysosome could cause the maturation of the autolysosome, which can disintegrate the contents (Christoffersen et al., 2010; Liu et al., 2010). So far, there is little about MiRNA-regulated autophagy in cancers in the last two steps (Fig. 2).

As a whole, MiRNAs can involve different autophagic steps, and take an effect as negative regulators (guardian) or positive autophagic modulators (executioner) (Fu et al., 2012).

4.1 The Role of Autophagy Modulated by MiRNAs in Cancer Initiation

Cancer initiation is a process described as "changing," which means that normal cells are damaged and transform into cancer cells at the cellular, genetic, and epigenetic levels (Hanahan and Weinberg, 2011). The oncomiRNA MiR-30d can impair the autophagy process by targeting multiple genes in the autophagy pathway, which is involved in tumorigenesis in various human epithelial cancer cells (Yang et al., 2013). MiR-224 is significantly upregulated in hepatitis B virus-associated HCC following by silencing its target gene SMAD family member 4 (*SMAD4*), which indicates the potential importance for autophagy inhibition in promoting tumorigenesis (Lan et al., 2014a,b). MiR-224-3p is proved to inhibit autophagy by directly suppressing the expression of *ATG5* and *RB1CC1* and overexpression of MiR224-3p in vivo inhibits inhibiting tumorigenesis in human glioblastoma cells (Guo et al., 2015b) (Table 1).

4.2 The Role of Autophagy Modulated by MiRNAs in Cancer Development

Cancer development is characterized by the growth speed, tumor differentiation, and the tumor invasiveness (Hanahan and Weinberg, 2011). Much like the role of MiRNAs-modulated autophagy in cancer initiation, the role that in cancer development is also complicated. MiR-21 downregulates autophagy through PTEN/AKT/HIF-1α feed-back loop in pancreatic cancer cells, which correlates with angiogenesis and contributes to the tumor development (Ali et al., 2010; Han et al., 2012; Liu et al., 2011). The enhancing level of MiR-224-3p is closely related to the tumor progression by inhibiting FIP200-regulated autophagy in high-risk human papillomavirus (hrHPV)-induced cervical cancer cell (Fang et al., 2016; Lin et al., 2015; Shen et al., 2013). Thus, MiR-224-3p or FIP200 can act as a diagnostic and therapeutic biomarker for hrHPV-induced cervical cancer. *BECN1* is an established target of MiR-30a and this MiRNA can negatively regulate *BECN1* expression resulting in decreased activation of autophagy induced by rapamycin. Moreover, MiR-30a as a tsmiRNA has been demonstrated to inhibit cancer development and progression in breast cancer, lung cancer, and glioma cancer through regulating autophagy (Zhu et al., 2009) (Table 1).

4.3 The Potential Role of MiRNAs-Regulated Autophagy in Cancer Diagnosis

Increasing data suggest MiRNAs play central roles as tumor-associated biomarkers which are involved in not only early-stage tumors, but also advanced stage tumors, tumor recurrence, and drug sensitivities. Therefore, MiRNAs can serve as diagnostic, prognostic, or even therapeutic markers for some cancer types (Calin et al., 2004b; Volinia et al., 2006). And how MiRNAs-regulated autophagy functions in cancer diagnosis should be further discussed. The prosurvival or prodeath function of autophagy is prone to vary according to cancer types, the stress signal, and other circumstances. MiR-372 may directly interfere with sequestosome 1 (SQSTM1) to

Table 1 The Role of Autophagy Modulated by MiRNAs in Cancer Initiation and Cancer Development

	MiRNAs	oncomiRNA/ tsmiRNA	Anti-/ Proautophagy	Target	Cancer Specificity	References
Cancer initiation	MiR-30d	oncomiRNA	Antiautophagy	*BECN1 BNIP3L, ATG12, ATG5, ATG2*	Human epithelial cancer	Yang et al. (2013)
	MiR-224	oncomiRNA	Antiautophagy	*SMAD4*	Hepatocellular carcinoma	Lan et al. (2014a,b)
	MiR-224-3p	tsmiRNA	Antiautophagy	*ATG5, RB1CC1*	Glioblastoma	Guo et al. (2015b)
Cancer development	MiR-21	oncomiRNA	Antiautophagy	*PTEN, AKT, HIF1A*	Pancreatic cancer	Ali et al. (2010), Han et al. (2012), and Liu et al. (2011b)
	MiR-224-3p	oncomiRNA	Antiautophagy	*RB1CC1*	Cervical cancer	Fang et al. (2016), Lin et al. (2015), and Shen et al. (2013)
	MiR-30a	tsmiRNA	Antiautophagy	*BECN1*	Breast cancer, lung cancer, glioma cancer	Zhu et al. (2009)

suppress autophagy activation and the growth of breast cancer, which may provide a useful histological marker for breast cancer detection (Feng et al., 2014). MiR-23b-3p inhibits autophagy by regulating *ATG12* and *HMGB2* and increases the sensitivity of multidrug resistance cells to chemotherapeutics in gastric cancer, suggesting the potential of MiR-23b-3p in drug resistance prediction and treatment (An et al., 2015). MiRNAs-regulated autophagy has not only a prosurvival role, but also plays a part in prodeath function. A recent study identified MiR-34a as a direct target of antiproliferative autophagy. MiR-34a is significantly downregulated in prostate cancer cells and transfection of MiR-34a induces autophagy, resulting in growth inhibition of tumor cells. These effects are partly responsible for clinical potential for MiR-34a as a diagnosis or therapeutic biomarker (Gaur et al., 2015) (Table 2).

4.4 Therapeutic Potential of MiRNAs-Mediated Autophagic Pathway

Currently, emerging studies have indicated that MiRNAs-mediated autophagy has a prodeath or a prosurvival role when cancer cells are subjected to metabolic or therapeutic stresses. Thus, targeting autophagy could provide a novel and potential therapeutic strategy for cancer therapy. The transfection of MiR-101 causes to inhibition of celastrol-induced autophagy and subsequently enhances the celastrol cytotoxicity in prostate cancer cells (Guo et al., 2015a). MiR-101 is also proved to be a potent inhibitor of basal, etoposide- and rapamycin-induced autophagy. Moreover, MiR-101-mediated inhibition of autophagy can sensitize breast cancer cells to 4-hydroxytamoxifen (4-OHT)-mediated cell death (Frankel et al., 2011). Sirtuin 1 (*SIRT1*) is identified as a direct target of MiR-212 in prostate cancer and loss of MiR-212 effectively induces autophagic flux and promotes cell survival. Therefore, MiR-212 may develop the therapeutic potential therapy for prostate cancer patients (Ramalinga et al., 2015). Overexpression of MiR-200c reduces irradiation-induced autophagy, and ectopic expression levels of MiR-200c inhibit irradiation-induced autophagy and enhance the therapeutic effect of radiotherapy in breast cancer (Sun et al., 2015a). MiR-22 can rescue the formation of LC3-II by targeting high-mobility group box 1 (*HMGB1*) or B-cell translocation gene 1 (*BTG1*) in osteosarcoma or colon cancer, respectively. Furthermore, the overexpressed MiR-22 blocks autophagy and promotes cell death in these cancer cell lines. These results reveal MiR-22 can be considered as a predictor of chemotherapy sensitivity and a therapeutic target for cancer patients (Li et al., 2014;

Table 2 Diagnostic or Therapeutic Potential of MiRNAs-Mediated Autophagic Pathway

	MiRNAs	The Role of Autophagy	Anti-/Proautophagy	Target	Cancer Specificity	References
Diagnosis	MiR-372	Prosurvival	Antiautophagy	*SQSTM1*	Breast cancer	Feng et al. (2014)
	MiR-23b-3p	Prosurvival	Antiautophagy	*ATG12, HMGB2*	Gastric cancer	An et al. (2015)
	MiR-34a	Prodeath	Proautophagy	*BECN1*	Prostate cancer	Gaur et al. (2015)
Therapy	MiR-101	Prosurvival	Antiautophagy	*RAB5A, STMN1, ATG4D*	Breast cancer	Frankel et al. (2011) and Guo et al. (2015a)
	MiR-212	Prosurvival	Antiautophagy	*SIRT*	Prostate cancer	Ramalinga et al. (2015)
	MiR-200c	Prosurvival	Antiautophagy	*UBQLN1*	Breast cancer	Sun et al. (2015a)
	MiR-22	Prosurvival	Antiautophagy	*HMGB1, BTG1*	Osteosarcoma, colon cancer	Li et al. (2014) and Zhang et al. (2015a)
	MiR-214	Prosurvival	Antiautophagy	*UCP2*	Breast cancer	Yu et al. (2015)
	MiR-375	Prosurvival	Antiautophagy	*ATG7*	Hepatocellular carcinoma	Chang et al. (2012a)
	MiR-30a	Prosurvival	Antiautophagy	*ATG5, RB1CC1*	Renal cell carcinoma, breast cancer, cervical cancer	Zheng et al. (2015) and Zou et al. (2012)
	MiR-140-5p	Prosurvival	Proautophagy	*IP3K2*	Osteosarcoma	Wei et al. (2016)
	MiR-21	Prodeath	Proautophagy	*PTEN*	Cervical cancer, hepatocellular cancer	He et al. (2015a), Peralta-Zaragoza et al. (2016), and Song et al. (2016)
	MiR-7	Prodeath	Proautophagy	*E2F1*	Esophageal cancer	Tazawa et al. (2012)
	MiR-100	Prodeath	Proautophagy	*MTOR, IGF1R*	Hepatocellular cancer	Ge et al. (2014)

Zhang et al., 2015a). MiR-214 could increase the sensitivity of estrogen receptor-positive (ER(+)) breast cancer cells to anticancer drugs through autophagy inhibition, indicating a potential role of MiR-214 as a novel therapeutic strategy for overcoming endocrine resistance in ER(+) breast cancers (Yu et al., 2015). MiR-375 may inhibit autophagy by reducing expression of *ATG7* and impair the viability of HCC cells under hypoxic conditions. Therefore, the direct inhibitory effect of MiR-375 on autophagy allows it to be a possible therapeutic target for cancer treatment (Chang et al., 2012a). Overexpression of MiR-30a inhibits Beclin 1 or ATG5-modulated autophagosome formation and increases tumor cell apoptosis induced by anticancer drug treatment in renal cell carcinoma, liver cancer, and cervical cancer. Thus, MiR-30a may be a potential biomarker to increase the sensitivity of chemotherapy through the abrogation of Beclin 1-dependent autophagy in cancer cells (Zheng et al., 2015; Zou et al., 2012). The increased MiR-140-5p expression can upregulate anticancer drug-induced autophagy and ameliorate the anticancer drug-induced cell proliferation in osteosarcoma cells, which suggests that MiR-140-5p may mediate drug-resistance in osteosarcoma cells by inducing autophagy (Wei et al., 2016). In addition to prosurvival autophagy, autophagic cell death can also be regulated by MiRNAs. Phosphatase and tensin homologue deleted on chromosome 10 (PTEN)-dependent autophagy is shown to be necessary for cancer cell survival. MiR-21 can increase the radioresistance in cervical cancer and sorafenib-resistance in HCC. Moreover, MiR-21 confers decreased autophagy and decreased autophagy is proved to be one of the potential mechanisms of increased radioresistance or drug-resistance. Therefore, MiR-21 may be a potential therapeutic target in gene therapy for cervical cancer and HCC (He et al., 2015a; Peralta-Zaragoza et al., 2016; Song et al., 2016). Genetically, engineered telomerase-specific oncolytic adenovirus can induce autophagic cell death through an E2F1-MiR-7-EGFR pathway in human cancer cells, which suggests the molecular mechanism of an anticancer virotherapy (Tazawa et al., 2012). MiR-100 may promote the ATG7-dependent autophagy and subsequent apoptotic cell death of HCC cells by suppressing the expression of *MTOR* or *IGF1R*, which provides a potential therapeutic target for HCC (Ge et al., 2014) (Table 2).

5. CONCLUDING REMARKS

Autophagy is a catabolic process which has been involved in biological processes, including cell survival, metabolism, differentiation, senescence, and cell death. Accumulating studies show that regulation of autophagy by MiRNAs is an essential part in tumorigenesis and cancer development, which is also a novel mechanism for anticancer drug resistance. Therefore, a thorough understanding of MiRNA-mediated autophagy is important to provide a novel and potential diagnostic or therapeutic strategy. Over the past few years, the interaction between autophagy and MiRNAs in cancer has been developed, but the exact role still remains uncertain. It is clear that MiRNAs can play as oncomiRNAs or tsmiRNAs involving in early-stage tumors, advanced stage tumors, tumor recurrence, and drug sensitivities. Further studies should pay more attention to the potential function of MiRNAs-mediated autophagy in not only physiological but also pathological conditions.

Nevertheless, there are still many problems need to be further investigated. The same MiRNA may have diverse influences on the different cancers through the dual function of autophagy, indicating that the exact role remains uncertain. And more attention should be put on how modulating MiRNAs change different cancer cells. Besides, not all MiRNAs focus on the levels of gene product, many of them only proving to affect the entire network or signaling pathways. It is conceived that recent studies get a rough idea about therapeutic potential, there is still a long way to go before applied to clinical.

ACKNOWLEDGMENTS

This study is supported by Grants from National Natural Science Foundation of China (Grant No. 81672932), Zhejiang province medical science and technology project (Grant No. 2017171227), Zhejiang province science and technology project of TCM (Grant No. 2015ZB033), and Zhengshu Medical Elite Scholarship Fund.

Conflict of Interest: The authors declare no conflict of interest.

REFERENCES

Ali, S., Ahmad, A., Banerjee, S., Padhye, S., Dominiak, K., Schaffert, J.M., Wang, Z., Philip, P.A., Sarkar, F.H., 2010. Gemcitabine sensitivity can be induced in pancreatic cancer cells through modulation of miR-200 and miR-21 expression by curcumin or its analogue CDF. Cancer Res. 70, 3606–3617. 20388782.

Ambros, V., 2004. The functions of animal microRNAs. Nature 431, 350–355. 15372042.

An, Y., Zhang, Z., Shang, Y., Jiang, X., Dong, J., Yu, P., Nie, Y., Zhao, Q., 2015. miR-23b-3p regulates the chemoresistance of gastric cancer cells by targeting ATG12 and HMGB2. Cell Death Dis. 6, e1766. 25996293.

Bagga, S., Bracht, J., Hunter, S., Massirer, K., Holtz, J., Eachus, R., Pasquinelli, A.E., 2005. Regulation by let-7 and lin-4 miRNAs results in target mRNA degradation. Cell 122, 553–563. 16122423.

Bartel, D.P., 2004. MicroRNAs: genomics, biogenesis, mechanism, and function. Cell 116, 281–297. 14744438.

Bartel, D.P., 2009. MicroRNAs: target recognition and regulatory functions. Cell 136, 215–233. 19167326.
Bernard, A., Klionsky, D.J., 2013. Autophagosome formation: tracing the source. Dev. Cell 25, 116–117. 23639440.
Bohnsack, M.T., Czaplinski, K., Gorlich, D., 2004. Exportin 5 is a RanGTP-dependent dsRNA-binding protein that mediates nuclear export of pre-miRNAs. RNA 10, 185–191. 14730017.
Cai, X., Hagedorn, C.H., Cullen, B.R., 2004. Human microRNAs are processed from capped, polyadenylated transcripts that can also function as mRNAs. RNA 10, 1957–1966. 15525708.
Calin, G.A., Croce, C.M., 2006. MicroRNA signatures in human cancers. Nat. Rev. Cancer 6, 857–866. 17060945.
Calin, G.A., Dumitru, C.D., Shimizu, M., Bichi, R., Zupo, S., Noch, E., Aldler, H., Rattan, S., Keating, M., Rai, K., Rassenti, L., Kipps, T., Negrini, M., Bullrich, F., Croce, C.M., 2002. Frequent deletions and down-regulation of micro- RNA genes miR15 and miR16 at 13q14 in chronic lymphocytic leukemia. Proc. Natl. Acad. Sci. U.S.A. 99, 15524–15529. 12434020.
Calin, G.A., Liu, C.G., Sevignani, C., Ferracin, M., Felli, N., Dumitru, C.D., Shimizu, M., Cimmino, A., Zupo, S., Dono, M., Dell'Aquila, M.L., Alder, H., Rassenti, L., Kipps, T.J., Bullrich, F., Negrini, M., Croce, C.M., 2004a. MicroRNA profiling reveals distinct signatures in B cell chronic lymphocytic leukemias. Proc. Natl. Acad. Sci. U.S.A. 101, 11755–11760. 15284443.
Calin, G.A., Sevignani, C., Dumitru, C.D., Hyslop, T., Noch, E., Yendamuri, S., Shimizu, M., Rattan, S., Bullrich, F., Negrini, M., Croce, C.M., 2004b. Human microRNA genes are frequently located at fragile sites and genomic regions involved in cancers. Proc. Natl. Acad. Sci. U.S.A. 101, 2999–3004. 14973191.
Calin, G.A., Ferracin, M., Cimmino, A., Di Leva, G., Shimizu, M., Wojcik, S.E., Iorio, M.V., Visone, R., Sever, N.I., Fabbri, M., Iuliano, R., Palumbo, T., Pichiorri, F., Roldo, C., Garzon, R., Sevignani, C., Rassenti, L., Alder, H., Volinia, S., Liu, C.G., Kipps, T.J., Negrini, M., Croce, C.M., 2005. A MicroRNA signature associated with prognosis and progression in chronic lymphocytic leukemia. N. Engl. J. Med. 353, 1793–1801. 16251535.
Cecconi, F., Levine, B., 2008. The role of autophagy in mammalian development: cell makeover rather than cell death. Dev. Cell 15, 344–357. [PubMed:WOS:000259358100007].
Chang, Y., Lin, J., Tsung, A., 2012a. Manipulation of autophagy by MIR375 generates antitumor effects in liver cancer. Autophagy 8, 1833–1834. 22929050.
Chang, Y., Yan, W., He, X., Zhang, L., Li, C., Huang, H., Nace, G., Geller, D.A., Lin, J., Tsung, A., 2012b. miR-375 inhibits autophagy and reduces viability of hepatocellular carcinoma cells under hypoxic conditions. Gastroenterology 143, 177–187, e178. 22504094.
Chatterjee, A., Chattopadhyay, D., Chakrabarti, G., 2014. miR-17-5p downregulation contributes to paclitaxel resistance of lung cancer cells through altering beclin1 expression. PLoS One 9, e95716. 24755562.
Chatterjee, A., Chattopadhyay, D., Chakrabarti, G., 2015. MiR-16 targets Bcl-2 in paclitaxel-resistant lung cancer cells and overexpression of miR-16 along with miR-17 causes unprecedented sensitivity by simultaneously modulating autophagy and apoptosis. Cell. Signal. 27, 189–203. 25435430.
Chen, K., Shi, W., 2016. Autophagy regulates resistance of non-small cell lung cancer cells to paclitaxel. Tumour Biol. 37, 10539–10544. 26852748.
Chen, Y., Liersch, R., Detmar, M., 2012. The miR-290-295 cluster suppresses autophagic cell death of melanoma cells. Sci. Rep. 2, 808. 23150779.

Chen, X., Zhang, Y., Shi, Y., Lian, H., Tu, H., Han, S., Yin, J., Peng, B., Zhou, B., He, X., Liu, W., 2016a. MiR-129 triggers autophagic flux by regulating a novel Notch-1/E2F7/Beclin-1 axis to impair the viability of human malignant glioma cells. Oncotarget 7, 9222. 26824182.

Chen, Z., Gao, S., Wang, D., Song, D., Feng, Y., 2016b. Colorectal cancer cells are resistant to anti-EGFR monoclonal antibody through adapted autophagy. Am. J. Transl. Res. 8, 1190–1196. 27158405.

Chendrimada, T.P., Gregory, R.I., Kumaraswamy, E., Norman, J., Cooch, N., Nishikura, K., Shiekhattar, R., 2005. TRBP recruits the Dicer complex to Ago2 for microRNA processing and gene silencing. Nature 436, 740–744. 15973356.

Cheng, Y., Ren, X., Hait, W.N., Yang, J.M., 2013. Therapeutic targeting of autophagy in disease: biology and pharmacology. Pharmacol. Rev. 65, 1162–1197. 23943849.

Christoffersen, N.R., Shalgi, R., Frankel, L.B., Leucci, E., Lees, M., Klausen, M., Pilpel, Y., Nielsen, F.C., Oren, M., Lund, A.H., 2010. p53-independent upregulation of miR-34a during oncogene-induced senescence represses MYC. Cell Death Differ. 17, 236–245. 19696787.

Cimmino, A., Calin, G.A., Fabbri, M., Iorio, M.V., Ferracin, M., Shimizu, M., Wojcik, S.E., Aqeilan, R.I., Zupo, S., Dono, M., Rassenti, L., Alder, H., Volinia, S., Liu, C.G., Kipps, T.J., Negrini, M., Croce, C.M., 2005. miR-15 and miR-16 induce apoptosis by targeting BCL2. Proc. Natl. Acad. Sci. U.S.A. 102, 13944–13949. 16166262.

Clarke, P.G., Puyal, J., 2012. Autophagic cell death exists. Autophagy 8, 867–869. 22652592.

Comincini, S., Allavena, G., Palumbo, S., Morini, M., Durando, F., Angeletti, F., Pirtoli, L., Miracco, C., 2013. microRNA-17 regulates the expression of ATG7 and modulates the autophagy process, improving the sensitivity to temozolomide and low-dose ionizing radiation treatments in human glioblastoma cells. Cancer Biol. Ther. 14, 574–586. 23792642.

Costinean, S., Zanesi, N., Pekarsky, Y., Tili, E., Volinia, S., Heerema, N., Croce, C.M., 2006. Pre-B cell proliferation and lymphoblastic leukemia/high-grade lymphoma in E(mu)-miR155 transgenic mice. Proc. Natl. Acad. Sci. U.S.A. 103, 7024–7029. 16641092.

Dohner, H., Stilgenbauer, S., Benner, A., Leupolt, E., Krober, A., Bullinger, L., Dohner, K., Bentz, M., Lichter, P., 2000. Genomic aberrations and survival in chronic lymphocytic leukemia. N. Engl. J. Med. 343, 1910–1916. 11136261.

Dong, J.T., Boyd, J.C., Frierson Jr., H.F., 2001. Loss of heterozygosity at 13q14 and 13q21 in high grade, high stage prostate cancer. Prostate 49, 166–171. 11746261.

Du, T., Zamore, P.D., 2005. microPrimer: the biogenesis and function of microRNA. Development 132, 4645–4652. 16224044.

Eis, P.S., Tam, W., Sun, L., Chadburn, A., Li, Z., Gomez, M.F., Lund, E., Dahlberg, J.E., 2005. Accumulation of miR-155 and BIC RNA in human B cell lymphomas. Proc. Natl. Acad. Sci. U.S.A. 102, 3627–3632. 15738415.

Fang, W., Shu, S., Yongmei, L., Endong, Z., Lirong, Y., Bei, S., 2016. miR-224-3p inhibits autophagy in cervical cancer cells by targeting FIP200. Sci. Rep. 6, 33229. 27615604.

Feng, L., Ma, Y., Sun, J., Shen, Q., Liu, L., Lu, H., Wang, F., Yue, Y., Li, J., Zhang, S., Lin, X., Chu, J., Han, W., Wang, X., Jin, H., 2014. YY1-MIR372-SQSTM1 regulatory axis in autophagy. Autophagy 10, 1442–1453. 24991827.

Feng, Y., Yao, Z., Klionsky, D.J., 2015. How to control self-digestion: transcriptional, post-transcriptional, and post-translational regulation of autophagy. Trends Cell Biol. 25, 354–363. 25759175.

Forstemann, K., Tomari, Y., Du, T., Vagin, V.V., Denli, A.M., Bratu, D.P., Klattenhoff, C., Theurkauf, W.E., Zamore, P.D., 2005. Normal microRNA maturation and germ-line

stem cell maintenance requires Loquacious, a double-stranded RNA-binding domain protein. PLoS Biol. 3, e236. 15918770.
Frankel, L.B., Wen, J., Lees, M., Hoyer-Hansen, M., Farkas, T., Krogh, A., Jaattela, M., Lund, A.H., 2011. microRNA-101 is a potent inhibitor of autophagy. EMBO J. 30, 4628–4641. 21915098.
Fu, L.L., Wen, X., Bao, J.K., Liu, B., 2012. MicroRNA-modulated autophagic signaling networks in cancer. Int. J. Biochem. Cell Biol. 44, 733–736. 22342941.
Fu, L.L., Cheng, Y., Liu, B., 2013. Beclin-1: autophagic regulator and therapeutic target in cancer. Int. J. Biochem. Cell Biol. 45, 921–924. 23420005.
Fullgrabe, J., Klionsky, D.J., Joseph, B., 2014. The return of the nucleus: transcriptional and epigenetic control of autophagy. Nat. Rev. Mol. Cell Biol. 15, 65–74. 24326622.
Ganley, I.G., Lam du, H., Wang, J., Ding, X., Chen, S., Jiang, X., 2009. ULK1.ATG13. FIP200 complex mediates mTOR signaling and is essential for autophagy. J. Biol. Chem. 284, 12297–12305. 19258318.
Gaur, S., Wen, Y., Song, J.H., Parikh, N.U., Mangala, L.S., Blessing, A.M., Ivan, C., Wu, S.Y., Varkaris, A., Shi, Y., Lopez-Berestein, G., Frigo, D.E., Sood, A.K., Gallick, G.E., 2015. Chitosan nanoparticle-mediated delivery of miRNA-34a decreases prostate tumor growth in the bone and its expression induces non-canonical autophagy. Oncotarget 6, 29161–29177. 26313360.
Ge, Y.Y., Shi, Q., Zheng, Z.Y., Gong, J., Zeng, C., Yang, J., Zhuang, S.M., 2014. MicroRNA-100 promotes the autophagy of hepatocellular carcinoma cells by inhibiting the expression of mTOR and IGF-1R. Oncotarget 5, 6218–6228. 25026290.
Glick, D., Barth, S., Macleod, K.F., 2010. Autophagy: cellular and molecular mechanisms. J. Pathol. 221, 3–12. 20225336.
Gregory, R.I., Yan, K.P., Amuthan, G., Chendrimada, T., Doratotaj, B., Cooch, N., Shiekhattar, R., 2004. The microprocessor complex mediates the genesis of microRNAs. Nature 432, 235–240. 15531877.
Gregory, R.I., Chendrimada, T.P., Cooch, N., Shiekhattar, R., 2005. Human RISC couples microRNA biogenesis and posttranscriptional gene silencing. Cell 123, 631–640. 16271387.
Gump, J.M., Thorburn, A., 2011. Autophagy and apoptosis: what is the connection? Trends Cell Biol. 21, 387–392. 21561772.
Guo, J., Huang, X., Wang, H., Yang, H., 2015a. Celastrol induces autophagy by targeting AR/miR-101 in prostate cancer cells. PLoS One 10, e0140745. 26473737.
Guo, X., Xue, H., Guo, X., Gao, X., Xu, S., Yan, S., Han, X., Li, T., Shen, J., Li, G., 2015b. MiR224-3p inhibits hypoxia-induced autophagy by targeting autophagy-related genes in human glioblastoma cells. Oncotarget 6, 41620–41637. 26536662.
Han, J., Lee, Y., Yeom, K.H., Nam, J.W., Heo, I., Rhee, J.K., Sohn, S.Y., Cho, Y., Zhang, B.T., Kim, V.N., 2006. Molecular basis for the recognition of primary microRNAs by the Drosha-DGCR8 complex. Cell 125, 887–901. 16751099.
Han, M., Wang, Y., Liu, M., Bi, X., Bao, J., Zeng, N., Zhu, Z., Mo, Z., Wu, C., Chen, X., 2012. MiR-21 regulates epithelial-mesenchymal transition phenotype and hypoxia-inducible factor-1alpha expression in third-sphere forming breast cancer stem cell-like cells. Cancer Sci. 103, 1058–1064. 22435731.
Hanahan, D., Weinberg, R.A., 2011. Hallmarks of cancer: the next generation. Cell 144, 646–674. 21376230.
Harmalkar, M., Upraity, S., Kazi, S., Shirsat, N.V., 2015. Tamoxifen-induced cell death of malignant glioma cells is brought about by oxidative-stress-mediated alterations in the expression of BCL2 family members and is enhanced on miR-21 inhibition. J. Mol. Neurosci. 57, 197–202. 26109525.
He, C., Klionsky, D.J., 2009. Regulation mechanisms and signaling pathways of autophagy. Annu. Rev. Genet. 43, 67–93. 19653858.

He, C., Levine, B., 2010. The Beclin 1 interactome. Curr. Opin. Cell Biol. 22, 140–149. 20097051.

He, H., Jazdzewski, K., Li, W., Liyanarachchi, S., Nagy, R., Volinia, S., Calin, G.A., Liu, C.G., Franssila, K., Suster, S., Kloos, R.T., Croce, C.M., de la Chapelle, A., 2005. The role of microRNA genes in papillary thyroid carcinoma. Proc. Natl. Acad. Sci. U.S.A. 102, 19075–19080. 16365291.

He, C., Dong, X., Zhai, B., Jiang, X., Dong, D., Li, B., Jiang, H., Xu, S., Sun, X., 2015a. MiR-21 mediates sorafenib resistance of hepatocellular carcinoma cells by inhibiting autophagy via the PTEN/Akt pathway. Oncotarget 6, 28867–28881. 26311740.

He, J., Yu, J.J., Xu, Q., Wang, L., Zheng, J.Z., Liu, L.Z., Jiang, B.H., 2015b. Downregulation of ATG14 by EGR1-MIR152 sensitizes ovarian cancer cells to cisplatin-induced apoptosis by inhibiting cyto-protective autophagy. Autophagy 11, 373–384. 25650716.

Hosokawa, N., Sasaki, T., Iemura, S., Natsume, T., Hara, T., Mizushima, N., 2009. Atg101, a novel mammalian autophagy protein interacting with Atg13. Autophagy 5, 973–979. 19597335.

Huang, Y., Guerrero-Preston, R., Ratovitski, E.A., 2012. Phospho-DeltaNp63alpha-dependent regulation of autophagic signaling through transcription and micro-RNA modulation. Cell Cycle 11, 1247–1259. 22356768.

Huangfu, L., Liang, H., Wang, G., Su, X., Li, L., Du, Z., Hu, M., Dong, Y., Bai, X., Liu, T., Yang, B., Shan, H., 2016. miR-183 regulates autophagy and apoptosis in colorectal cancer through targeting of UVRAG. Oncotarget 7, 4735–4745. 26717041.

Hutvagner, G., Zamore, P.D., 2002. A microRNA in a multiple-turnover RNAi enzyme complex. Science 297, 2056–2060. 12154197.

Hutvagner, G., McLachlan, J., Pasquinelli, A.E., Balint, E., Tuschl, T., Zamore, P.D., 2001. A cellular function for the RNA-interference enzyme Dicer in the maturation of the let-7 small temporal RNA. Science 293, 834–838. 11452083.

Inguscio, V., Panzarini, E., Dini, L., 2012. Autophagy contributes to the death/survival balance in cancer photodynamic therapy. Cell 1, 464–491. 24710486.

Izzotti, A., Carozzo, S., Pulliero, A., Zhabayeva, D., Ravetti, J.L., Bersimbaev, R., 2016. Extracellular MicroRNA in liquid biopsy: applicability in cancer diagnosis and prevention. Am. J. Cancer Res. 6, 1461–1493. 27508091.

Jansson, M.D., Lund, A.H., 2012. MicroRNA and cancer. Mol. Oncol. 6, 590–610. 23102669.

Jian, X., Xiao-yan, Z., Bin, H., Yu-feng, Z., Bo, K., Zhi-nong, W., Xin, N., 2011. MiR-204 regulate cardiomyocyte autophagy induced by hypoxia-reoxygenation through LC3-II. Int. J. Cardiol. 148, 110–112. 21316776.

Jiang, F., Ye, X., Liu, X., Fincher, L., McKearin, D., Liu, Q., 2005. Dicer-1 and R3D1-L catalyze microRNA maturation in Drosophila. Genes Dev. 19, 1674–1679. 15985611.

Jiang, Q., Li, F., Shi, K., Wu, P., An, J., Yang, Y., Xu, C., 2014. Involvement of p38 in signal switching from autophagy to apoptosis via the PERK/eIF2alpha/ATF4 axis in selenite-treated NB4 cells. Cell Death Dis. 5, e1270. 24874742.

John Clotaire, D.Z., Zhang, B., Wei, N., Gao, R., Zhao, F., Wang, Y., Lei, M., Huang, W., 2016. MiR-26b inhibits autophagy by targeting ULK2 in prostate cancer cells. Biochem. Biophys. Res. Commun. 472, 194–200. 26920049.

Jones-Rhoades, M.W., Bartel, D.P., 2004. Computational identification of plant MicroRNAs and their targets, including a stress-induced miRNA. Mol. Cell 14, 787–799 [PubMed:WOS:000222173800011].

Kang, R., Zeh, H.J., Lotze, M.T., Tang, D., 2011. The Beclin 1 network regulates autophagy and apoptosis. Cell Death Differ. 18, 571–580. 21311563.

Ketting, R.F., Fischer, S.E., Bernstein, E., Sijen, T., Hannon, G.J., Plasterk, R.H., 2001. Dicer functions in RNA interference and in synthesis of small RNA involved in developmental timing in C. elegans. Genes Dev. 15, 2654–2659. 11641272.

Kim, J., Kundu, M., Viollet, B., Guan, K.L., 2011. AMPK and mTOR regulate autophagy through direct phosphorylation of Ulk1. Nat. Cell Biol. 13, 132–141. 21258367.

Kim, J.H., Hong, S.K., Wu, P.K., Richards, A.L., Jackson, W.T., Park, J.I., 2014. Raf/MEK/ERK can regulate cellular levels of LC3B and SQSTM1/p62 at expression levels. Exp. Cell Res. 327, 340–352. 25128814.

Korkmaz, G., le Sage, C., Tekirdag, K.A., Agami, R., Gozuacik, D., 2012. miR-376b controls starvation and mTOR inhibition-related autophagy by targeting ATG4C and BECN1. Autophagy 8, 165–176. 22248718.

Kroemer, G., Levine, B., 2008. Autophagic cell death: the story of a misnomer. Nat. Rev. Mol. Cell Biol. 9, 1004–1010. 18971948.

Kroemer, G., Marino, G., Levine, B., 2010. Autophagy and the integrated stress response. Mol. Cell 40, 280–293. 20965422.

Lan, S.H., Wu, S.Y., Zuchini, R., Lin, X.Z., Su, I.J., Tsai, T.F., Lin, Y.J., Wu, C.T., Liu, H.S., 2014a. Autophagy suppresses tumorigenesis of hepatitis B virus-associated hepatocellular carcinoma through degradation of microRNA-224. Hepatology 59, 505–517. 23913306.

Lan, S.H., Wu, S.Y., Zuchini, R., Lin, X.Z., Su, I.J., Tsai, T.F., Lin, Y.J., Wu, C.T., Liu, H.S., 2014b. Autophagy-preferential degradation of MIR224 participates in hepatocellular carcinoma tumorigenesis. Autophagy 10, 1687–1689. 25068270.

Lee, Y., Ahn, C., Han, J., Choi, H., Kim, J., Yim, J., Lee, J., Provost, P., Radmark, O., Kim, S., Kim, V.N., 2003. The nuclear RNase III Drosha initiates microRNA processing. Nature 425, 415–419. 14508493.

Lee, Y., Kim, M., Han, J., Yeom, K.H., Lee, S., Baek, S.H., Kim, V.N., 2004. MicroRNA genes are transcribed by RNA polymerase II. EMBO J. 23, 4051–4060. 15372072.

Lee, Y., Hur, I., Park, S.Y., Kim, Y.K., Suh, M.R., Kim, V.N., 2006. The role of PACT in the RNA silencing pathway. EMBO J. 25, 522–532. 16424907.

Levine, B., Klionsky, D.J., 2004. Development by self-digestion: molecular mechanisms and biological functions of autophagy. Dev. Cell 6, 463–477. 15068787.

Levine, B., Kroemer, G., 2008. Autophagy in the pathogenesis of disease. Cell 132, 27–42. 18191218.

Levine, B., Kroemer, G., 2009. Autophagy in aging, disease and death: the true identity of a cell death impostor. Cell Death Differ. 16, 1–2. 19079285.

Li, X., Wang, S., Chen, Y., Liu, G., Yang, X., 2014. miR-22 targets the 3′ UTR of HMGB1 and inhibits the HMGB1-associated autophagy in osteosarcoma cells during chemotherapy. Tumour Biol. 35, 6021–6028. 24609901.

Li, S., Qiang, Q., Shan, H., Shi, M., Gan, G., Ma, F., Chen, B., 2016. MiR-20a and miR-20b negatively regulate autophagy by targeting RB1CC1/FIP200 in breast cancer cells. Life Sci. 147, 143–152. 26829385.

Liang, C., Yu, X.J., Guo, X.Z., Sun, M.H., Wang, Z., Song, Y., Ni, Q.X., Li, H.Y., Mukaida, N., Li, Y.Y., 2015. MicroRNA-33a-mediated downregulation of Pim-3 kinase expression renders human pancreatic cancer cells sensitivity to gemcitabine. Oncotarget 6, 14440–14455. 25971209.

Lin, F., Wang, P., Shen, Y., Xie, X., 2015. Upregulation of microRNA-224 sensitizes human cervical cells SiHa to paclitaxel. Eur. J. Gynaecol. Oncol. 36, 432–436. 26390698.

Liu, B., Cheng, Y., Liu, Q., Bao, J.K., Yang, J.M., 2010. Autophagic pathways as new targets for cancer drug development. Acta Pharmacol. Sin. 31, 1154–1164. 20694022.

Liu, L.Z., Li, C., Chen, Q., Jing, Y., Carpenter, R., Jiang, Y., Kung, H.F., Lai, L., Jiang, B.H., 2011. MiR-21 induced angiogenesis through AKT and ERK activation and HIF-1alpha expression. PLoS One 6, e19139. 21544242.
Lorin, S., Hamai, A., Mehrpour, M., Codogno, P., 2013. Autophagy regulation and its role in cancer. Semin. Cancer Biol. 23, 361–379. 23811268.
Lu, J., Getz, G., Miska, E.A., Alvarez-Saavedra, E., Lamb, J., Peck, D., Sweet-Cordero, A., Ebert, B.L., Mak, R.H., Ferrando, A.A., Downing, J.R., Jacks, T., Horvitz, H.R., Golub, T.R., 2005. MicroRNA expression profiles classify human cancers. Nature 435, 834–838. 15944708.
Lund, E., Guttinger, S., Calado, A., Dahlberg, J.E., Kutay, U., 2004. Nuclear export of microRNA precursors. Science 303, 95–98. 14631048.
Maiuri, M.C., Galluzzi, L., Morselli, E., Kepp, O., Malik, S.A., Kroemer, G., 2010. Autophagy regulation by p53. Curr. Opin. Cell Biol. 22, 181–185. 20044243.
Martinez, J., Tuschl, T., 2004. RISC is a 5′ phosphomonoester-producing RNA endonuclease. Genes Dev. 18, 975–980. 15105377.
Marx, J., 2006. Autophagy: is it cancer's friend or foe? Science 312, 1160–1161. 16728626.
Mikhaylova, O., Stratton, Y., Hall, D., Kellner, E., Ehmer, B., Drew, A.F., Gallo, C.A., Plas, D.R., Biesiada, J., Meller, J., Czyzyk-Krzeska, M.F., 2012. VHL-regulated MiR-204 suppresses tumor growth through inhibition of LC3B-mediated autophagy in renal clear cell carcinoma. Cancer Cell 21, 532–546. 22516261.
Mizushima, N., Levine, B., Cuervo, A.M., Klionsky, D.J., 2008. Autophagy fights disease through cellular self-digestion. Nature 451, 1069–1075. 18305538.
Narita, M., Young, A.R., Arakawa, S., Samarajiwa, S.A., Nakashima, T., Yoshida, S., Hong, S., Berry, L.S., Reichelt, S., Ferreira, M., Tavare, S., Inoki, K., Shimizu, S., Narita, M., 2011. Spatial coupling of mTOR and autophagy augments secretory phenotypes. Science 332, 966–970. 21512002.
Noble, C.G., Dong, J.M., Manser, E., Song, H., 2008. Bcl-xL and UVRAG cause a monomer-dimer switch in Beclin1. J. Biol. Chem. 283, 26274–26282. 18641390.
Peralta-Zaragoza, O., Deas, J., Meneses-Acosta, A., De la O-Gómez, F., Fernandez-Tilapa, G., Gomez-Ceron, C., Benitez-Boijseauneau, O., Burguete-Garcia, A., Torres-Poveda, K., Bermudez-Morales, V.H., Madrid-Marina, V., Rodriguez-Dorantes, M., Hidalgo-Miranda, A., Perez-Plasencia, C., 2016. Relevance of miR-21 in regulation of tumor suppressor gene PTEN in human cervical cancer cells. BMC Cancer 16, 215. 26975392.
Piletic, K., Kunej, T., 2016. MicroRNA epigenetic signatures in human disease. Arch. Toxicol. 90, 2405–2419. 27557899.
Ramalinga, M., Roy, A., Srivastava, A., Bhattarai, A., Harish, V., Suy, S., Collins, S., Kumar, D., 2015. MicroRNA-212 negatively regulates starvation induced autophagy in prostate cancer cells by inhibiting SIRT1 and is a modulator of angiogenesis and cellular senescence. Oncotarget 6, 34446–34457. 26439987.
Rubinstein, A.D., Eisenstein, M., Ber, Y., Bialik, S., Kimchi, A., 2011. The autophagy protein Atg12 associates with antiapoptotic Bcl-2 family members to promote mitochondrial apoptosis. Mol. Cell 44, 698–709. 22152474.
Saito, K., Ishizuka, A., Siomi, H., Siomi, M.C., 2005. Processing of pre-microRNAs by the Dicer-1-Loquacious complex in Drosophila cells. PLoS Biol. 3, e235. 15918769.
Schwarz, D.S., Hutvagner, G., Du, T., Xu, Z., Aronin, N., Zamore, P.D., 2003. Asymmetry in the assembly of the RNAi enzyme complex. Cell 115, 199–208. 14567917.
Sethupathy, P., Megraw, M., Hatzigeorgiou, A.G., 2006. A guide through present computational approaches for the identification of mammalian microRNA targets. Nat. Methods 3, 881–886. 17060911.
Shen, S.N., Wang, L.F., Jia, Y.F., Hao, Y.Q., Zhang, L., Wang, H., 2013. Upregulation of microRNA-224 is associated with aggressive progression and poor prognosis in human cervical cancer. Diagn. Pathol. 8, 69. 23631806.

Shintani, T., Klionsky, D.J., 2004. Autophagy in health and disease: a double-edged sword. Science 306, 990–995. 15528435.
Singh, R., Cuervo, A.M., 2011. Autophagy in the cellular energetic balance. Cell Metab. 13, 495–504. 21531332.
Song, L., Liu, S., Zhang, L., Yao, H., Gao, F., Xu, D., Li, Q., 2016. MiR-21 modulates radiosensitivity of cervical cancer through inhibiting autophagy via the PTEN/Akt/HIF-1alpha feedback loop and the Akt-mTOR signaling pathway. Tumour Biol. 37, 12161–12168. 27220494.
Sorel, O., Dewals, B.G., 2016. MicroRNAs in large herpesvirus DNA genomes: recent advances. Biomol. Concepts 7, 229–239. 27544723.
Stiuso, P., Potenza, N., Lombardi, A., Ferrandino, I., Monaco, A., Zappavigna, S., Vanacore, D., Mosca, N., Castiello, F., Porto, S., Addeo, R., Prete, S.D., De Vita, F., Russo, A., Caraglia, M., 2015. MicroRNA-423-5p promotes autophagy in cancer cells and is increased in serum from hepatocarcinoma patients treated with Sorafenib. Mol. Ther. Nucleic Acids 4, e233. 25782064.
Stojcheva, N., Schechtmann, G., Sass, S., Roth, P., Florea, A.M., Stefanski, A., Stuhler, K., Wolter, M., Muller, N.S., Theis, F.J., Weller, M., Reifenberger, G., Happold, C., 2016. MicroRNA-138 promotes acquired alkylator resistance in glioblastoma by targeting the Bcl-2-interacting mediator BIM. Oncotarget 7, 12937–12950. 26887050.
Sui, X., Chen, R., Wang, Z., Huang, Z., Kong, N., Zhang, M., Han, W., Lou, F., Yang, J., Zhang, Q., Wang, X., He, C., Pan, H., 2013. Autophagy and chemotherapy resistance: a promising therapeutic target for cancer treatment. Cell Death Dis. 4, e838. 24113172.
Sui, X., Kong, N., Ye, L., Han, W., Zhou, J., Zhang, Q., He, C., Pan, H., 2014. p38 and JNK MAPK pathways control the balance of apoptosis and autophagy in response to chemotherapeutic agents. Cancer Lett. 344, 174–179. 24333738.
Sun, Q., Liu, T., Yuan, Y., Guo, Z., Xie, G., Du, S., Lin, X., Xu, Z., Liu, M., Wang, W., Yuan, Q., Chen, L., 2015a. MiR-200c inhibits autophagy and enhances radiosensitivity in breast cancer cells by targeting UBQLN1. Int. J. Cancer 136, 1003–1012. 25044403.
Sun, Y., Xing, X., Liu, Q., Wang, Z., Xin, Y., Zhang, P., Hu, C., Liu, Y., 2015b. Hypoxia-induced autophagy reduces radiosensitivity by the HIF-1alpha/miR-210/Bcl-2 pathway in colon cancer cells. Int. J. Oncol. 46, 750–756. 25385144.
Tan, S., Shi, H., Ba, M., Lin, S., Tang, H., Zeng, X., Zhang, X., 2016. miR-409-3p sensitizes colon cancer cells to oxaliplatin by inhibiting Beclin-1-mediated autophagy. Int. J. Mol. Med. 37, 1030–1038. 26935807.
Tanaka, K., Miyata, H., Sugimura, K., Fukuda, S., Kanemura, T., Yamashita, K., Miyazaki, Y., Takahashi, T., Kurokawa, Y., Yamasaki, M., Wada, H., Nakajima, K., Takiguchi, S., Mori, M., Doki, Y., 2015. miR-27 is associated with chemoresistance in esophageal cancer through transformation of normal fibroblasts to cancer-associated fibroblasts. Carcinogenesis 36, 894–903. 26026166.
Tazawa, H., Yano, S., Yoshida, R., Yamasaki, Y., Sasaki, T., Hashimoto, Y., Kuroda, S., Ouchi, M., Onishi, T., Uno, F., Kagawa, S., Urata, Y., Fujiwara, T., 2012. Genetically engineered oncolytic adenovirus induces autophagic cell death through an E2F1-microRNA-7-epidermal growth factor receptor axis. Int. J. Cancer 131, 2939–2950. 22492316.
Thorburn, A., 2008. Apoptosis and autophagy: regulatory connections between two supposedly different processes. Apoptosis 13, 1–9. 17990121.
Tsuchihara, K., Fujii, S., Esumi, H., 2009. Autophagy and cancer: dynamism of the metabolism of tumor cells and tissues. Cancer Lett. 278, 130–138. 19004545.
Volinia, S., Calin, G.A., Liu, C.G., Ambs, S., Cimmino, A., Petrocca, F., Visone, R., Iorio, M., Roldo, C., Ferracin, M., Prueitt, R.L., Yanaihara, N., Lanza, G., Scarpa, A., Vecchione, A., Negrini, M., Harris, C.C., Croce, C.M., 2006. A microRNA expression signature of human solid tumors defines cancer gene targets. Proc. Natl. Acad. Sci. U.S.A. 103, 2257–2261. 16461460.

Wang, P., Zhang, J., Zhang, L., Zhu, Z., Fan, J., Chen, L., Zhuang, L., Luo, J., Chen, H., Liu, L., Chen, Z., Meng, Z., 2013. MicroRNA 23b regulates autophagy associated with radioresistance of pancreatic cancer cells. Gastroenterology 145, 1133–1143, e1112. 23916944.

Wang, Z., Wang, N., Liu, P., Chen, Q., Situ, H., Xie, T., Zhang, J., Peng, C., Lin, Y., Chen, J., 2014. MicroRNA-25 regulates chemoresistance-associated autophagy in breast cancer cells, a process modulated by the natural autophagy inducer isoliquiritigenin. Oncotarget 5, 7013–7026. 25026296.

Wei, R., Cao, G., Deng, Z., Su, J., Cai, L., 2016. miR-140-5p attenuates chemotherapeutic drug-induced cell death by regulating autophagy through inositol 1,4,5-trisphosphate kinase 2 (IP3k2) in human osteosarcoma cells. Biosci. Rep. 36. 27582507.

Xu, R., Liu, S., Chen, H., Lao, L., 2016. MicroRNA-30a downregulation contributes to chemoresistance of osteosarcoma cells through activating Beclin-1-mediated autophagy. Oncol. Rep. 35, 1757–1763. 26708607.

Yang, Z., Klionsky, D.J., 2009. An overview of the molecular mechanism of autophagy. Curr. Top. Microbiol. Immunol. 335, 1–32. 19802558.

Yang, X., Zhong, X., Tanyi, J.L., Shen, J., Xu, C., Gao, P., Zheng, T.M., DeMichele, A., Zhang, L., 2013. mir-30d regulates multiple genes in the autophagy pathway and impairs autophagy process in human cancer cells. Biochem. Biophys. Res. Commun. 431, 617–622. 23274497.

Yi, R., Qin, Y., Macara, I.G., Cullen, B.R., 2003. Exportin-5 mediates the nuclear export of pre-microRNAs and short hairpin RNAs. Genes Dev. 17, 3011–3016. 14681208.

Yousefi, S., Perozzo, R., Schmid, I., Ziemiecki, A., Schaffner, T., Scapozza, L., Brunner, T., Simon, H.U., 2006. Calpain-mediated cleavage of Atg5 switches autophagy to apoptosis. Nat. Cell Biol. 8, 1124–1132. 16998475.

Yu, X., Luo, A., Liu, Y., Wang, S., Li, Y., Shi, W., Liu, Z., Qu, X., 2015. MiR-214 increases the sensitivity of breast cancer cells to tamoxifen and fulvestrant through inhibition of autophagy. Mol. Cancer 14, 208. 26666173.

Zhang, Y., Yang, W.Q., Zhu, H., Qian, Y.Y., Zhou, L., Ren, Y.J., Ren, X.C., Zhang, L., Liu, X.P., Liu, C.G., Ming, Z.J., Li, B., Chen, B., Wang, J.R., Liu, Y.B., Yang, J.M., 2014. Regulation of autophagy by miR-30d impacts sensitivity of anaplastic thyroid carcinoma to cisplatin. Biochem. Pharmacol. 87, 562–570. 24345332.

Zhang, H., Tang, J., Li, C., Kong, J., Wang, J., Wu, Y., Xu, E., Lai, M., 2015a. MiR-22 regulates 5-FU sensitivity by inhibiting autophagy and promoting apoptosis in colorectal cancer cells. Cancer Lett. 356, 781–790. 25449431.

Zhang, X., Shi, H., Lin, S., Ba, M., Cui, S., 2015b. MicroRNA-216a enhances the radiosensitivity of pancreatic cancer cells by inhibiting beclin-1-mediated autophagy. Oncol. Rep. 34, 1557–1564. 26134156.

Zheng, B., Zhu, H., Gu, D., Pan, X., Qian, L., Xue, B., Yang, D., Zhou, J., Shan, Y., 2015. MiRNA-30a-mediated autophagy inhibition sensitizes renal cell carcinoma cells to sorafenib. Biochem. Biophys. Res. Commun. 459, 234–239. 25712526.

Zhou, S., Zhao, L., Kuang, M., Zhang, B., Liang, Z., Yi, T., Wei, Y., Zhao, X., 2012. Autophagy in tumorigenesis and cancer therapy: Dr. Jekyll or Mr. Hyde? Cancer Lett. 323, 115–127. 22542808.

Zhu, H., Wu, H., Liu, X., Li, B., Chen, Y., Ren, X., Liu, C.G., Yang, J.M., 2009. Regulation of autophagy by a beclin 1-targeted microRNA, miR-30a, in cancer cells. Autophagy 5, 816–823. 19535919.

Zou, Z., Wu, L., Ding, H., Wang, Y., Zhang, Y., Chen, X., Chen, X., Zhang, C.Y., Zhang, Q., Zen, K., 2012. MicroRNA-30a sensitizes tumor cells to cis-platinum via suppressing beclin 1-mediated autophagy. J. Biol. Chem. 287, 4148–4156. 22157765.

CHAPTER TWO

Molecular Regulation of Cellular Senescence by MicroRNAs: Implications in Cancer and Age-Related Diseases

Mathieu Neault*,†, Florence Couteau*, Éric Bonneau‡, Vincent De Guire†,‡, Frédérick A. Mallette*,†,1

*Chromatin Structure and Cellular Senescence Research Unit, Maisonneuve-Rosemont Hospital Research Centre, Montréal, QC, Canada
†Université de Montréal, Montréal, QC, Canada
‡Maisonneuve-Rosemont Hospital, Montréal, QC, Canada
[1]Corresponding author: e-mail address: fa.mallette@umontreal.ca

Contents

International Review of Cell and Molecular Biology, Volume 334
ISSN 1937-6448
http://dx.doi.org/10.1016/bs.ircmb.2017.04.001

Abstract

Cellular senescence is a tumor suppressor response that acts as a barrier to cancer development and progression. In normal cells, diverse stimuli, including excessive mitogenic signaling, DNA damage or telomere shortening, trigger a senescence response characterized by stable growth arrest. Cellular senescence is orchestrated by tumor suppressor pathways, which have to be inactivated in order to impair the establishment of senescence and promote cancer. Consequently, by overcoming or bypassing this cellular response, cancer cells evade cell cycle checkpoint control leading to genomic instability and uncontrolled proliferation. MicroRNAs (MiRs) have emerged as essential factors contributing to or preventing cellular senescence. Here we detail the molecular mechanisms underlying the fine-tuning of cellular senescence signals by MiRs, and how the senescence response itself contributes to modulation of MiR expression, with a special focus on cancer and pathologies associated with aging.

1. INTRODUCTION

MicroRNAs (MiRs) are implicated in most, if not every, cellular signaling pathway. In fact these small (19–22 bp) molecules have a huge impact on cellular processes through modulation of protein expression. They are predicted to regulate over 60% of human protein coding genes (Friedman et al., 2009) and mRNAs (Bartel, 2009), and each MiR targets hundreds of evolutionarily conserved and nonconserved genes (Bentwich et al., 2005).

The function of MiRs as regulators of gene expression was first revealed in 2000, although the initial observation of their presence was made almost a decade before. Previous studies in the model organism *Caenorhabditis elegans* indicated that *lin-4*, a gene modulating the timing of larval development, produced a short noncoding RNA instead of a protein-coding mRNA (Lee et al., 1993). A back-to-back publication by another group showed that the *lin-4* noncoding RNA could regulate another gene, *lin-14*, at the post-transcriptional level (Wightman et al., 1993). The latter group realized that *lin-4* small RNAs molecules shared sequence complementarities with its *lin-14* target mRNA in the 3′ untranslated region (3′-UTR) (Wightman et al., 1993), and demonstrated the direct inhibitory effect of the *lin-4* MiR over *lin-14* mRNA using a reporter gene (Reinhart et al., 2000).

Dysregulation of MiR expression is a hallmark of human cancer (Croce, 2009; Volinia et al., 2006). Indeed defective MiR-dependent inhibition of genes regulating cell cycle progression and survival seems to promote tumorigenesis (Cheng et al., 2005). Moreover, MiR profiling contributes to the reliability of diagnosis and prognostis of human tumors by revealing the tissue of origin and tumor stage (Lu et al., 2005; Munker and Calin, 2011). In vivo models further highlighted the potential of MiRs as biomarkers as well as their importance in cancer therapy (Kasinski and Slack, 2011).

Neoplastic transformation rarely arises from a single mutagenic event but rather from multiple events promoting genomic instability and cellular survival. Deregulated growth control in cancer cells can be attributed to compromised mechanisms of self-surveillance. One such mechanism is characterized by an irreversible growth arrest termed cellular senescence. Cellular senescence can be induced in response to numerous stress stimuli including oxidative stress, telomere shortening, chemotherapeutic drugs, and oncogene activation (i.e., Ras, Myc, STAT5, E2F1 among others), ultimately limiting the clonal expansion of premalignant cells (Lowe et al., 2004). Senescent cells remain viable and their metabolic functionality is maintained (Salama et al., 2014) while undergoing an irreversible program of tumor suppression which a potential cancer cell would need to evade (Lowe et al., 2004). Importantly, activation of cellular senescence blocks tumorigenesis in vivo (Braig et al., 2005; Chen et al., 2005b; Collado et al., 2005; Lin et al., 2010; Michaloglou et al., 2005; Ventura et al., 2007; Xue et al., 2007), confirming its critical function as a barrier to cancer progression and potential as a therapeutic avenue.

Numerous MiRs have been shown to either directly induce or bypass cellular senescence. Interestingly, the senescence response itself stimulates MiR-mediated transcriptional silencing of cell cycle regulating genes (Benhamed et al., 2012). Recent studies have stimulated avid interest in the role of MiR in modulating senescence signals. The present review will describe the critical functions of MiRs in cellular senescence as well as its involvement in senescence-associated/age-related diseases.

2. OVERVIEW OF CELLULAR SENESCENCE

Cellular senescence can be thought of as a stable biological process which emerged to protect mammalian cells from malignant transformation. Early studies revealed that transduction of normal primary cells by the oncogenic simian virus (SV40) large T antigen increased lifespan in culture

(Moyer et al., 1964), but could not effect resumption of proliferation in senescent cells (Sager et al., 1983). From these results, the hypothesis was born that senescence might constitute a tumor suppressor mechanism that must be overridden or bypassed to promote carcinogenesis (Newbold et al., 1982). In mice, the ARF/p53/p21 pathway predominantly manages establishment of the senescence program, while human cells rely on three major collaborative pathways acting in parallel, i.e., the p53/p21, $p16^{INK4A}$/pRB, and the DNA damage response (DDR) pathways (Fig. 1).

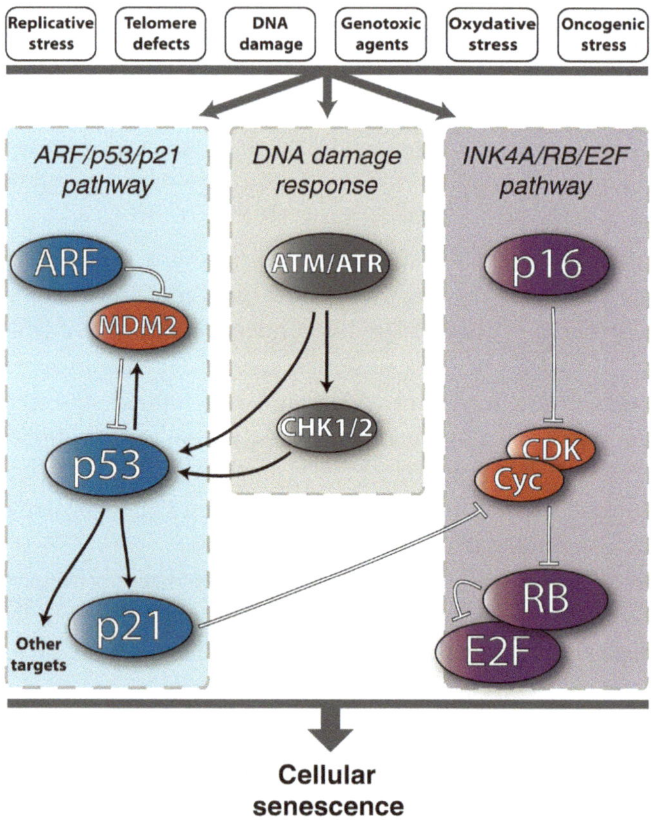

Fig. 1 Model of senescence-inducing pathways in response to cellular stresses. In human, stress signals are transmitted via the ARF/p53/p21 pathway, the $p16^{INK4A}$/pRB/E2F pathway and the DNA damage response (DDR) to promote cellular senescence. The p53 pathway is linked to the pRB network through the p21-mediated inhibition of cyclin–CDK2 complexes. The DDR contributes to the activation of p53 by ATM- and CHK2-mediated phosphorylation of p53. *Solid lines* represent activating signals, *hollow lines* represent inhibitory signals. Color legend: *blue/purple*, tumor suppressors; *red*, oncogenes; *gray*, DNA damage-related proteins.

2.1 Replicative Senescence

2.1.1 Cellular Aging and Lifespan

Before Leonard Hayflick coined the term "replicative senescence" in 1961, it was generally accepted that vertebrate fibroblasts could grow indefinitely in culture. It was wrongly assumed that the cellular growth mechanisms occurring in vivo were lost when cells were extracted and cultured in artificial growth medium (Shay and Wright, 2000). Hayflick and Moorhead—and several other independent groups—later reported that diploid cells could hardly sustain cellular division after approximately 1 year in culture. Most importantly, they identified a subtle correlation between cell ploidy and the capacity to form tumors when transplanted (Hayflick and Moorhead, 1961). Hayflick further hypothesized that normal cells possess finite replicative capabilities and subsequently estimated the replicative limit of human diploid cells at 50 ± 10 doublings. After reaching what is known as the "*Hayflick limit*," cells stop dividing while staying metabolically active; moreover, they can remain in culture for prolonged periods, if not indefinitely, provided that the culture media is periodically replenished (Matsumura et al., 1979). Hayflick also insisted to strictly limit the theory of cellular senescence to normal diploid cells and not to "heteroploid" cancer cells, proposing that the latter were most likely derived from an initially mortal cell that underwent physiological changes ultimately leading to immortality (Hayflick, 1965). In other words, replicative senescence occurs when cells reach the maximum lifespan frontier, and this reflects normal cellular aging.

2.1.2 Cellular Immortality or Irreversible Arrest: Implication for the Telomere

Chromosome ends are protected by telomeric DNA, formed by thousands of TTAGGG nucleotide repeats. Critically short telomeres coincide with the occurrence of replicative senescence of cultured cells (Olovnikov, 1996). The enzyme telomerase (comprising the catalytic subunit TERT and an RNA component, TERC) is responsible for polymerization and elongation of telomeres. Telomerase is normally inactivated in most postmitotic differentiated human cells whereas it is overexpressed and active in many immortalized cells (Greider and Blackburn, 1985; Shay and Bacchetti, 1997). Moreover, short telomeres trigger a DDR via activation of the ATM and ATR kinases, and growth arrest mediated mainly by p53 (Cong et al., 2002). Unprotected telomeric DNA caused by the loss of the telomeric protein TRF2 activates the DDR and provokes cellular senescence (Karlseder et al., 2002). Furthermore, unrepaired DNA damage in noncritically short telomeres induced by sustained exogenous genotoxic stress in human fibroblasts further supports persistent activation of the DDR (Fumagalli et al., 2012), thus contributing to the irreversibility of senescence.

2.2 Cellular Senescence and Cancer

Cellular senescence, like apoptosis, limits the clonal expansion of premalignant cells; however, in complete contrast to apoptosis which results cell death, senescent cells remain viable, and their metabolic functions are maintained (Lowe et al., 2004). Senescent cells undergo a stable program involving a network of tumor suppressors, from which a cancer cell needs to evade in order to acquire tumorigenic properties (Kuilman et al., 2010). Furthermore, senescent cells become resistant to apoptosis (Sasaki et al., 2001; Wang, 1995), possibly to maintain normal tissue homeostasis.

2.2.1 Oncogene-Induced Senescence

Activation of oncogenes such as Myc or Ras alone cannot induce cellular transformation (Stevenson and Volsky, 1986), i.e., a combination of activated oncogenes and inactivated/silenced tumor suppressors is required. In normal human and murine fibroblasts, constitutive activation of Ras provokes striking growth arrest associated with cellular senescence, a response called oncogene-induced senescence (OIS) (Palmero et al., 1998; Serrano et al., 1997). Although oncogenic Ras was able to transform immortal mouse fibroblasts (Newbold and Overell, 1983)—which were spontaneously immortalized and presumably lacking p53 (Carnero et al., 2000; Quelle et al., 1995)—expression of the telomerase hTERT alone with oncogenic Ras is insufficient to bypass senescence in normal cells (Wei and Sedivy, 1999). This suggests that activated Ras triggers a senescence pathway that is independent of telomere signaling (Ferbeyre et al., 2002). Ras activates the downstream MAPK cascade which in turn triggers the phosphorylation and activation of Raf, Mek1/2, and Erk1/2. Overexpression or activation of any of the latter MAPK effectors induces premature senescence, and these factors are necessary for proper Ras-induced senescence (RIS) signaling (Deschenes-Simard et al., 2013; Lin et al., 1998; Zhu et al., 1998). Ras also triggers a parallel pathway controlled by the stress-activated p38 MAPK via Mek1 which induces premature senescence and cyclin D1-regulated growth arrest (Ellinger-Ziegelbauer et al., 1999; Molnar et al., 1997; Wang et al., 2002). Among other oncogenes, overexpression of E2F1 (Pierce et al., 1998), BCL-2 (Feldser and Greider, 2007), AKT (Astle et al., 2012), Mos, Cdc6, cyclin E (Bartkova et al., 2006), and STAT5A (Mallette et al., 2007b, 2010) also induce premature senescence.

The induction of senescence in mouse predominantly involves the $p19^{ARF}/p53$ pathway, while in human it relies on both the p53 and $p16^{INK4A}/RB$ pathways. Functional inactivation of $p19^{ARF}$ or p53 in mouse, but not p21, is sufficient to bypass RIS (Kamijo et al., 1999; Palmero et al., 1998; Pantoja and Serrano, 1999; Serrano et al., 1997). Conversely in human cells, $p16^{INK4A}$ is the preponderant tumor suppressor activated in response to Ras (Michaloglou et al., 2005; Wei et al., 2001), where the bypass of RIS is achieved by concomitant inhibition of both p53 and $p16^{INK4A}/RB$ pathways in vitro (Hahn et al., 2002; Serrano et al., 1997; Wei et al., 2003) and in vivo (Sarkisian et al., 2007).

2.3 The Cellular Senescence Phenotype

In vitro and in vivo identification of senescent cells is possible through distinct features which, collectively, serve as biomarkers. These features can vary among different tissue types and/or experimental settings, and it is therefore advisable to rely on multiple senescence markers (Collado and Serrano, 2006).

2.3.1 Senescence-Associated Cell Cycle Arrest

The hallmark of the senescent response is stable arrest of proliferation. Senescent cells are mainly arrested at the G_1/S border (Michaloglou et al., 2005), but can also cease growth at the G_2/M transition (Wada et al., 2004). The stability—or irreversibility—of cell cycle exit seems to rely either on the levels of $p16^{INK4A}$ expression (Beausejour et al., 2003) or the level of Ras activation prior to senescence entry (Lampson et al., 2013).

2.3.2 Morphological Changes Associated With Cellular Senescence

Microscopic analysis of senescent cells in vitro reveals flat morphology with extended spindle-shaped cytoplasm (Michaloglou et al., 2005), a feature not observed in vivo (Collado and Serrano, 2010). The misshapen cytoskeleton and altered membrane organization are initiated upon cyclin-dependent kinase (CDK) 5 activation by pRB (Alexander et al., 2004; Yang et al., 2003). Ras-induced senescent cells often display a vacuolated cytoplasm as well, resulting from endoplasmic reticulum stress in response to protein misfolding (Denoyelle et al., 2006).

2.3.3 Senescence-Associated β-Galactosidase Activity

Another characteristic of senescent cells is an increase in senescence-associated β-galactosidase (SA-β-gal) activity, which can be assessed biochemically (Dimri et al., 1995). Cycling cells express β-galactosidase encoded by the *GLB1* gene which displays normal activity at pH 4. Expansion of the lysosomal compartment and increased expression of a senescence-related isoform of *GLB1* may be responsible for increased SA-β-gal activity at suboptimal pH (pH 6) (Dimri et al., 1995; Lee et al., 2006).

2.3.4 Upregulation of the p53/p21 and $p16^{INK4A}$/pRB Pathways

Cellular senescence is stimulated through activation of p53/p21 and $p16^{INK4A}$/pRB pathways (Fig. 1). Phosphorylation of p53 at serine 15 by ATM mediates the response to DNA damage resulting from oncogenic stress (Bartkova et al., 2006; Di Micco et al., 2006; Mallette et al., 2007a). Activated p53 triggers the transactivation of a multitude of target genes involved in senescence, including *CDKN1A* (p21), *MDM2*, *PAI-1*, *GADD45A*, and *PML* (Carrier et al., 1994; de Stanchina et al., 2004; Dimri et al., 1995; Ferbeyre et al., 2002; Kortlever et al., 2006; Kunz et al., 1995; Pearson et al., 2000; Zhao et al., 2000). Increased levels of hypophosphorylated pRB are also noticeable due to enhanced expression of its upstream regulators, $p16^{INK4A}$ and p21 (Otterson et al., 1994; Shapiro et al., 1995; Tanaka et al., 1998).

2.3.5 Activation of the DDR Pathway

Enhanced DNA damage signaling, resulting from either replicative- or oxidative stress is also a well-characterized hallmark of senescent cells (Fig. 1) (Bartkova et al., 2006; di Fagagna et al., 2003; Di Micco et al., 2006; Fumagalli et al., 2012; Kuilman et al., 2010; Mallette and Ferbeyre, 2007; Moiseeva et al., 2006). DNA double-strand breaks and replicative stress stimulate the recruitment and activation of the PI3K-like kinases ATM and ATR, respectively. These kinases each stimulate the DDR by phosphorylating multiple DDR proteins (many in common) including histone H2AX on serine 139 (γH2AX), CHK1, and CHK2 (Burma et al., 2001). Moreover, induction of the DDR in this manner contributes to activation of p53 and establishment of the senescence phenotype upon oncogenic stimulation (Bartkova et al., 2006; Di Micco et al., 2006; Mallette and Ferbeyre, 2007).

2.3.6 Formation of Senescence-Associated Heterochromatin Foci

Activation of Ras in normal human fibroblasts induces chromatin compaction via the $p16^{INK4A}$/pRB pathway, leading to the formation of DNA foci detectable under microscopy by DAPI staining (Narita et al., 2003). In response to Ras, these senescence-associated heterochromatin foci (SAHF) are generated in part through H3K9 trimethylation and colocalization of chromatin condensation factors, such as HP1 (Nielsen et al., 2001).

SAHF formation is a common feature of senescent cells and does not occur in reversibly arrested quiescent cells (Narita et al., 2003). SAHF formation also concurs with the recruitment of pRB and HP1 at promoter of E2F-responsive genes responsible for cell cycle progression and DNA replication, such as *CCN1A* (cyclin A) and *PCNA* (Narita et al., 2003). However, pRB-dependent chromatin condensation is dispensable to the establishment of cellular senescence (Kosar et al., 2011). Genetic studies revealed that Ras and Myc induce a KMT1A (Suv39h1)-dependent senescence response and the loss of KMT1A in Ras- or Myc-driven lymphomas accelerates tumor development (Braig et al., 2005; Reimann et al., 2010). Trimethylation of histone H3K9 by KMT1A also triggers the binding of HP1 to some E2F-responsive genes (Nielsen et al., 2001; Talluri et al., 2010), thus favoring the transcriptional silencing of those genes. These observations suggest that SAHF formation might directly promote the establishment and maintenance of senescence and could contribute to the stability of cellular senescence (Narita et al., 2003).

2.3.7 Senescence-Associated Secretory Phenotype

Senescent cells notably differ from quiescent cells by the secretion of various cytokines, matrix metalloproteases (MMP), and growth factors (Rodier and Campisi, 2011). This senescence-associated secretory phenotype (SASP) contributes to remodeling of the extracellular matrix, stimulates senescence in a paracrine fashion, and allows for immune recognition and clearance of senescent cells thus preventing their accumulation in tissues. Immune clearance of senescent cells was proven beneficial in various in vivo models such as during wound healing, following hepatocellular damage, and in liver cancer mouse models (Hoenicke and Zender, 2012; Jun and Lau, 2010;

Xue et al., 2007). Secreted proteins of the SASP include, but are not restricted, to PAI1 and IL-1-β (Ferbeyre et al., 2000), IL-6 (Kuilman et al., 2008), IL-8, IGFBP7 (Wajapeyee et al., 2008) as well as various MMP (Coppe et al., 2010).

2.3.8 Autophagy

OIS cells display enhanced autophagic activity which drives metabolic changes and protein turnover, leading to the apparition of large cytosolic vesicles (similar as lysosomal vacuoles) (Young et al., 2009). Downregulation of autophagy-related genes following Ras activation impairs the cell cycle arrest and delays cytokine production (Young et al., 2009), suggesting that autophagy is a causative agent to senescence rather than a consequence. Ras-expressing cells enter a mitotic phase prior their entry into senescence. During this transition, mTOR, which blocks early steps of autophagosome formation (Chan et al., 2007), is recruited/sequestered to TOR-autophagy spatial coupling compartments (TASCC). TASCC are amino acid-rich autolysosomal structures associated to the secretory apparatus (endoplasmic reticulum–Golgi apparatus) that facilitate secretory protein synthesis (Narita et al., 2011). Blocking mTOR recruitment to TASCC decreases SASP cytokines IL-6 and IL-8 production (Narita et al., 2011), while inhibition of mTOR by rapamycin abolishes the SASP in senescent cells (Herranz et al., 2015; Laberge et al., 2015). Therefore, mTOR recruitment to TASCC during senescence possesses dual function, such as (i) allowing phagocytosis to occur and (ii) simulating efficient synthesis of SASP proteins.

2.3.9 Autophagy-Mediated Alteration of the Nuclear Envelope

Modification of the nuclear envelope during cellular senescence is a direct outcome of increased autophagic activity. In senescent cells, cytosolic nuclear fragments (CCFs) colocalize with p62, a protein involved in the autophagosome pathway. CCFs mainly consist of degraded histones and chromatin, and result from the blebbing of nuclear chromatin through the nuclear envelope (Ivanov et al., 2013). Senescent cells displaying low nuclear γ-H2AX and H3K27me3 marks show an enrichment of these marks in CCFs, suggesting that the "leakage" of chromatin/histone complexes favors nuclear depletion of such histone marks (Ivanov et al., 2013). During replicative senescence, CCFs are also enriched for LMNB1 (lamin B1), a constituent of the nuclear lamina (Ivanov et al., 2013). LMNB1 expression is downregulated while lamin B1 proteins are degraded in Ras-expressing cells, hence contributing to the permeability of the nuclear envelope (Ivanov et al., 2013). Recently, the association between autophagy-regulating gene LC3 (Atg8) and lamin B1 was uncovered, and this interaction was shown to promote its inclusion into CCFs and subsequent degradation by autophagosomes and autolysosomes (Dou et al., 2015). Importantly, impairment of LC3–lamin B1 interaction blocks the cell cycle arrest and impaired $p16^{INK4A}$ accumulation of Ras-expressing cells, suggesting that nuclear membrane disruption following oncogenic stress may participate in tumor suppression (Dou et al., 2015).

2.4 SASP: An Inconvenient Truth?

While the senescence program blocks proliferation of potentially transformed cells, the secretion of numerous soluble factors including cytokines and proteases (Coppe et al., 2010) by senescent cells confers both beneficial (Demaria et al., 2014) and detrimental (Baker et al., 2011) effects on surrounding tissues. Proinflammatory cytokines such as IL-6 (Kuilman et al., 2008) and IL-8 (Acosta et al., 2008) as well as other cytokines including PAI-1 (Kortlever et al., 2006) and IGFBP7 (Wajapeyee et al., 2008) can either reinforce or stimulate senescence in neighboring cells (autocrine or paracrine stimulation, respectively) (Acosta et al., 2013), promoting the maintenance of senescence phenotype through a positive feedback loop within the affected tissue.

Importantly, senescent cells accumulate during aging in numerous different tissues such as liver, skin, spleen, and lung, thereby potentially contributing to tissue dysfunction (Dimri et al., 1995; Herbig et al., 2006; Jeyapalan et al., 2007; Wang et al., 2009a). Mouse models of p53 hyperactivation show accumulation of senescent cells in vivo in various tissues. These mice display a premature aging phenotype associated to degenerative changes such as decreased fertility, osteoporosis, sarcopenia, and hair growth defects (Maier et al., 2004; Tyner et al., 2002), suggesting a link between cellular senescence and organismal aging in mice. Enforcing the targeted clearance of senescent cells in vivo in a mouse model of premature aging or during normal aging deferred the appearance of age-related pathologies including cardiac hypertrophy, muscular degeneration, kidney dysfunction, and cancer (Baker et al., 2011, 2016), thus implying that cellular senescence exhibits detrimental effects during aging.

3. REGULATION OF CELLULAR SENESCENCE BY MiRs

The p53 and pRB pathways and the DDR are critical players in cellular senescence that can be negatively or positively regulated by MiRs. Herein, we describe MiRs which have been confirmed as positive or negative regulators of cellular senescence signaling (Fig. 2) with a special focus on functional senescence-associated MiRs (SA-MiRs) in vivo.

3.1 Introduction to MiR-Mediated Regulation of Gene Expression

3.1.1 MiR Biogenesis and Maturation

MiR host genes are generally located in intronic or intergenic regions, although exonic MiRs have been reported within protein-coding or long noncoding RNA genes (e.g., *MiR-985, MiR-206*) (Kim et al., 2009; Rodriguez et al., 2004). DNA polymerase II (Pol II) synthesizes primary MiR transcripts (pri-miRs) that are hundreds of nucleotides long (Lee et al., 2004). However, MiRs are often clustered at specific genomic locations such that their transcription from a common promoter can lead to polycistronic transcripts several kilobases in length (Lee et al., 2002). Pri-miRs harbor a 7-methyl guanosine cap at the 5′ end as well as a 3′ polyadenylated (poly (A)) tail (Lee et al., 2004) and form a hairpin shape with bulges due to partially complementary sequences. These primary transcripts are processed in the nucleus by the "microprocessor complex" formed by the RNase III endonuclease Drosha, in association with the RNA-binding protein DGCR8 (Kim et al., 2009). The asymmetric processing of pri-miRs results in stem loop-shaped premature-miRs (pre-miRs) displaying a 3′ overhang. Pre-miRs are then shuttled to the cytoplasm by the RanGTP-dependent DNA-binding protein, exportin 5 (Bohnsack et al., 2004; Yi et al., 2003).

Cleavage from both sides of the stem loop of pre-miRs by the DICER1/TRBP complex generates partially complementary 19–22 bp duplex RNA with 3′ overhangs. During strand selection, one strand is retained and processed by Argonaut (AGO) proteins resulting in an active, mature MiR, while the other strand may or may not be targeted for cleavage. Mature MiR from one strand of the duplex (MiR-5p) is generally more abundant than the other strand (MiR-3p), although in particular cases the balance is tilted toward the MiR-3p (Yang and Lai, 2011).

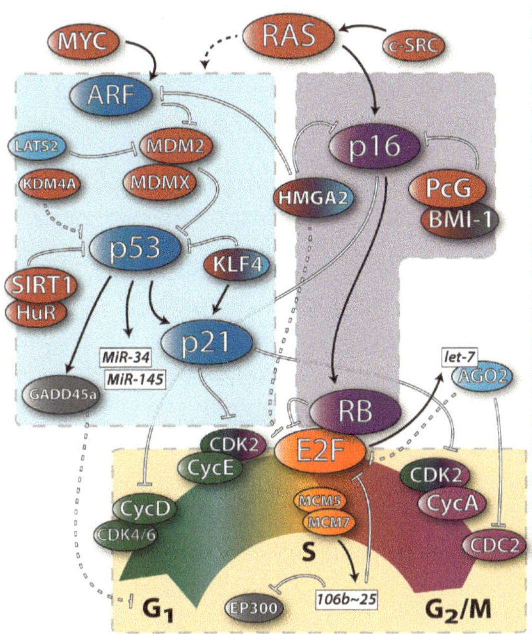

Pathway	Target	MiRs	Target	MiRs
ARF/p53/p21	p53	MiR-25, MiR-30d	SIRT1	MiR-22, MiR-34, MiR-138, MiR-181, MiR-195, MiR-203, MiR-212, MiR-217, MiR-494
	MDM2/ MDMX	MiR-34, MiR-145, MiR-192, MiR-194, MiR-215, MiR-605		
	p21	106b~25		
	KDM4A	MiR-137	HuR	MiR-519
	LATS2	MiR-372/3	Survivin	MiR-203
INK4A/RB/E2F	p16	MiR-24	BMI-1	MiR-16, MiR-128, MiR-141, MiR-300, MiR-494*
	E2F	let-7, MiR-20b, MiR-34, 106b~25, MiR-203	PcG	let-7, MiR-9, MiR-26b, MiR-181a, MiR-210, MiR-424
	HMGA2	let-7, MiR-23a, MiR-26a, MiR-30a		
	MYC	let-7, MiR-34, MiR-212, MiR-494	SCF	MiR-34c
			LRF	MiR-20a
			c-SRC	MiR-34
Cell cycle	CDK4/6	MiR-34, MiR-33, MiR-137, MiR-449a	CycE1/2	let-7, MiR-34, MiR-15a/b, MiR-16
	CycD1/2	let-7, MiR-16, MiR-34, MiR-33, MiR-449a	CDK2	let-7
			MCM5	let-7

Fig. 2 Model of senescence-associated MiRs regulating growth arrest and cellular senescence pathways. (*Left*) Stress signals such as oncogene activation (including Myc or Ras) induce cellular senescence in normal cells by activating the ARF/p53/p21 or p16^{INK4A}/pRB/E2F pathways. ARF sequesters MDM2 to stabilize and activate p53, which in turn induces p21 to promote cell cycle arrest. p21 blocks cell cycle progression by inhibition of cyclin A-CDK2 and cyclin E-CDK2 complexes. p16^{INK4A} also inhibits the cell cycle by inactivating cyclin D-CDK4/6 complexes. p21 and p16^{INK4A} prevent the CDK-mediated phosphorylation of pRB, which results in sustained repression of E2F-responsive genes required for the G_1/S-phase transition. p53 and E2F induce the expression of MiR-34/MiR-145 and *let-7*/MiR-106b~25 cluster, respectively. Color legend: *Light colored dashed squares* in the background encompass proteins involved in p53/p21 pathway (*light blue*), the p16^{INK4A}/pRB/E2F pathway (*light purple*) or genes regulating the cell cycle (*light orange*). *Blue/purple*, tumor suppressors; *red*, oncogenes; *gray*, DNA damage response/resistance-related proteins; *green/pink*, cell cycle-related proteins; *orange*, proteins involved in S-phase transition; *dual color*, context-dependent function. *Solid lines* represent activating signals, *hollow lines* represent inhibitory signals, and *dashed lines* represent indirect signals. (*Right*) Table summarizing genes involved in senescence pathways and the MiRs regulating them. SA-MiRs and their respective targets from the p53/p21 pathway (*light blue*), the p16^{INK4A}/pRB/E2F pathway (*light purple*) or the cell cycle-related genes (*light orange*). * denotes a MiR-3p.

MiRs are subclassified according to their seed sequence (nucleotides 2–8 of a MiR) which specifically recognize the 3′ untranslated region (3′-UTR) of a target mRNA (Bartel, 2004; Lin and Gregory, 2015; Melo et al., 2009). MiRs are also reported to alternatively bind to the 5′-UTR and coding sequence of some mRNAs (Miranda et al., 2006; Tay et al., 2008; Zhou et al., 2014). The mean length of 3′-UTRs is fourfold higher compared to that of 5′-UTRs which allows for increased potential MiR binding sites. Hence, much of MiR research is focused on the 3′-UTR target recognition (Miranda et al., 2006; Pesole et al., 2001).

3.1.2 Mechanisms of Gene Repression

The essential function of the mature MiR is to specifically inhibit gene expression. MiRs guide and direct Argonaut (AGO)-mediated sequestration, silencing, or decay of a target mRNA (Bartel, 2009), although the mechanisms by which mRNA repression is achieved are not fully understood. Most target mRNAs are deadenylated, decapped, and degraded by the MiR-induced silencing (miRISC) complex. However, the translation of some mRNAs can be reactivated, implying that MiR-mediated gene repression could also be due in part by a blockade of translation, without mRNA decay (Fabian and Sonenberg, 2012). MiR-mediated gene silencing appears to be ruled by a cap-dependent mechanism but, again, the precise circumstances whereby the miRISC complex will facilitate either mRNA decay or translational repression are unclear; moreover, it remains to be determined whether these two events are interdependent (Fabian and Sonenberg, 2012). Nevertheless, both modes of gene repression result in decreased protein levels. At the cellular level, MiR-mediated silencing occurs in distinct cytoplasmic structures, i.e., the processing bodies (P-bodies). Interestingly, several reports showed that P-bodies could serve as reservoir for intact mRNAs (Bhattacharyya et al., 2006; Brengues et al., 2005), enforcing the idea that some MiRs may specifically inhibit translation rather than promote mRNA decay.

3.2 MiR Regulation of the p53/p21 Pathway

Oncogenic stress activates the p53 pathway leading to cellular senescence (Serrano et al., 1997). p53 is a transcription factor regulating genes involved in apoptosis, senescence, cell cycle, and DNA repair. These genes are directly transactivated by direct recruitment of p53 at response elements (REs) within gene promoter region (Beckerman and Prives, 2010). During cellular stress, kinases including ATM and ATR phosphorylate p53 (Kruse and Gu, 2009) which in turn inhibits p53 binding to the MDM2 ubiquitin ligase. This prevents p53 proteasomal degradation, thus leading to accumulation of p53 protein (Shieh et al., 1997). Activated p53 then binds specific sequences in promoter regions of target genes to regulate their expression

(Riley et al., 2008). The functional consequences of p53-mediated gene transactivation are probably dictated by the complex interplay between other posttranslational modifications (PTMs) in addition to phosphorylation (e.g., methylation, acetylation, ubiquitination, among others) trigged by various stresses (Kruse and Gu, 2009).

TP53 is mutated in over than 50% of all human cancers (Hollstein et al., 1991; Olivier et al., 2010; Vousden and Lu, 2002). Inhibition of p53 can also be achieved by other mechanisms, including the silencing of p53 activators (Okada et al., 2014; Raman et al., 2000; Soto-Reyes and Recillas-Targa, 2010). Importantly, inactivation of p53 contributes to the bypass of cellular senescence while restoration of p53 in cancer cells can lead to cell cycle arrest and senescence in vivo (Chen et al., 2005b; Ventura et al., 2007; Xue et al., 2007). This suggests that restoration of p53 activity—or that of its effectors—might engender resumption senescence signaling in cancer cells and as such.

3.2.1 p53 and Its Downstream Effectors are Targets of oncomiRs

p53 orchestrates the senescence response by stimulating transcription of numerous genes such as *CDKN1A*, *MDM2*, *GADD45A*, *PAI-1*, and also of MiRs. Recent studies show that MiRs are also closely involved in the control of the p53 pathway, including its upstream regulators (such as MDM2, MDMX, SIRT1, and KDM4A) and downstream transcriptional targets (p21, GADD45A, and MDM2). Several MiRs are reported to directly target the *TP53* mRNA as well as several p53-dependent tumor suppressor genes. These "oncomiRs" play important roles in neoplastic transformation.

Using a dual-reporter screen, Kumar et al. identified MiR-25 and -30d as p53-targeting MiRs (Kumar et al., 2011). These two oncomiRs were revealed by their ability to (1) decrease basal p53 activity, as assessed by a firefly luciferase reporter gene under the control of p53 RE-containing promoter; (2) inhibit the expression of a *Renilla* luciferase reporter gene tagged with the WT 3′-UTR of p53; and (3) reduce p53 protein levels in 293T cells. Ectopic expression of MiR-25 and -30d following etoposide treatment reduced the expression of *GADD45a*, a p53-dependent DNA damage responsive gene. *GADD45a* has important functions both during and following OIS. Specifically, it promotes G_1/S and G_2/M cell cycle checkpoints to achieve growth arrest and premature senescence, e.g., in response to oncogenic RAS (Bulavin et al., 2003; Hildesheim et al., 2002; Jin et al., 2003; Tront et al., 2006). Furthermore, MiR-25 and -30d also impair p53-dependent apoptosis by downregulation of p21 and BAX, and decreased caspase-3/7 activity. The introduction of exogenous p53 in the p53-null H1299 lung adenocarcinoma cells could not rescue apoptosis or G_1 arrest of MiR-25- or -30d-expressing cells, suggesting that these MiRs effectively negate regulatory pathways downstream of p53. Defects in transactivation of p21 and Bax could be corroborated in HCT116 colon cancer cells when ectopic expression of MiR-25 or -30d bypassed p53-induced senescence (Kumar et al., 2011).

The *CDKN1A* locus encodes p21, a CDK inhibitor and crucial regulator of cell cycle checkpoints. p21 interacts and inhibits the activity of cyclin-CDK1/2/4/6 complexes (Harper et al., 1993; Xiong et al., 1993). The promoter region of *CDKN1A* possesses multiple p53 REs, and the induction of p21 following DNA damage was shown to be crucial to mediate p53-dependent growth arrest (el-Deiry et al., 1993, 1994). In normal human fibroblasts, restoration of p21 in the absence of p53 is sufficient to stimulate the senescence phenotype (Fang et al., 1999), confirming the importance of p21 in mediating senescence. Interestingly, many oncomiRs are organized in large, polycistronic clusters consisting of multiple MiR host genes (e.g., *MiR*-106b~25 and MiR-17–92) which simultaneously and synergistically regulate important tumor suppressor genes. *CDKN1A* is one such gene, and its inhibition consequently promotes tumorigenesis via inactivation of cell cycle checkpoints and bypass of cellular senescence.

Members of the MiR-106b~25 polycistronic cluster are notable by their ability to impede senescence signals. MiR-106b was also identified as a regulator of *CDKN1A* in a screen aimed at identifying MiRs cooperating with oncogenic Ras^{G12V} to bypass senescence in human mammary epithelial cells (HMEC). Among 328 different MiRs used in this screen, multiple members of the MiR-106b family, including MiR-93, MiR-106a, -302a/b, and -372/-373 were identified. However, only MiR-106b and -302a were able to decrease the activity of a luciferase reporter harboring the 3′-UTR of *CDKN1A* (Borgdorff et al., 2010). MiR-25, which is part the MiR-106b~25 cluster, promoted bypass of p38-induced replicative senescence (Marasa et al., 2009) as well as DNA damage-induced senescence (Zhou et al., 2014). Furthermore, the MiR-106b member of the 106b~25 cluster has been reported to inhibit *CDKN1A* expression in gastric, colon, and hepatocellular carcinoma (HCC) cells (Ivanovska et al., 2008; Li et al., 2009; Petrocca et al., 2008), while decreasing caspase-3/7 activation in prostate cancer (PC) cells (Ambs et al., 2008), suggesting redundant oncogenic roles for members of the 106b~25 cluster.

Other oncomiRs have been identified for their ability to bypass RIS, suggesting that alternate pathways can be engaged or downregulated to forestall the senescence phenotype (Borgdorff et al., 2010). For instance, MiR-372/-373-mediated inhibition of LATS2, a tumor suppressor inhibiting MDM2 ligase activity leading to p53 stabilization during the mitotic checkpoint (Aylon et al., 2006; Borgdorff et al., 2010), led to the bypass of RIS. LATS2 was identified as a direct target of MiR-372/-373 from a screen in hTERT-immortalized normal fibroblasts (Voorhoeve et al., 2006). More specifically, these MiRs could bypass RIS via interference with p53-mediated CDK inhibition. In normal cells, CDK2, a kinase participating in the G_1/S-phase transition, is inhibited by p21 following Ras activation. However, MiR-372/-373-mediated inactivation of LATS2 somehow rendered Ras-expressing cells resistant to p21-mediated inhibition of CDK2 (Voorhoeve et al., 2006).

3.2.2 MiR-34, a p53-Responsive and Tumor Suppressor MiR

Several independent groups identified three members of the MiR-34 family, MiR-34a, -34b, and -34c as potent stimulators of apoptosis, cell cycle arrest, and senescence, respectively. Of note, MiR-34 was the first MiR family identified as transcriptionally regulated by p53 via direct binding to p53 REs within promoter regions of MiR-34a and MiR-34b/c (Bommer et al., 2007; Chang et al., 2007; He et al., 2007; Raver-Shapira et al., 2007; Tarasov et al., 2007). These regions are densely hypermethylated in ovarian cancer and nonsmall cell lung cancer (NSCLC) (Corney et al., 2010; Gallardo et al., 2009). Transcriptional activation of MiR-34 during senescence is not exclusively dependent on p53. During RAF-induced senescence, MiR-34a can be induced independently of p53 to regulate Myc expression. Upon activation of B-Raf in hTERT-immortalized fibroblasts expressing a 4-hydroxytamoxifen (4-OHT)-responsive B-Raf, the *ELK1* member of the ETS transcription factor family was enriched at the MiR-34a promoter region. Furthermore, knockdown of ELK1 expression in this model abrogated the induction of MiR-34a in response to B-Raf activation (Christoffersen et al., 2010). This may suggest alternative routes to activation of MiR-34a expression in p53-deficient tumor cells. The MiR-34 family controls a vast tumor suppressive network limiting cell cycle progression, tumor invasion, EMT transition, and metabolic activity through translational repression or mRNA decay of a myriad of targets (Feng et al., 2011; Hermeking, 2012).

Recently, c-Src has been identified as a clinically relevant target of MiR-34a (Adams et al., 2016b). Oncogene c-Src is a nonreceptor tyrosine kinase involved mainly in invasion, EMT, and metastasis (Guarino, 2010). c-Src enhances focal adhesion strength in triple-negative breast cancer (TNBC) which renders this tumor subtype more sensitive to therapeutic tyrosine kinase inhibitors such as dasatinib (Lehmann et al., 2011). MiR-34a is lost in TNBC and is associated with poor prognosis (Adams et al., 2016b). In breast cancer cell lines, transfection of MiR-34a halts proliferation and stimulates senescence, while introduction of a MiR-34a-resistant c-Src cDNA reverses the effects of MiR-34a overexpression. This strongly suggests that the tumor suppressor effect of MiR-34a in TNBC cells is specifically due to its regulation of c-Src (Adams et al., 2016b). Moreover, direct delivery of MiR-34a to tumors developed from athymic mice following xenograft assays, led to significantly reduced tumor weight and decreased proliferation. Most importantly, a negative regulatory feedback loop between c-SRC and MiR-34a was observed which further sensitized TNBC cells to dasatinib (Adams et al., 2016b).

3.2.3 Tumor Suppressor MiRs Stabilize p53 Through Regulation of the ARF-MDM2 Axis

The p53 protein is under the tight control of the E3 ubiquitin ligase MDM2. K48-linked ubiquitin chains deposited by MDM2 in c-terminal lysines residues of p53 lead to its cytoplasmic localization and proteasomal degradation (Geyer et al., 2000; Honda and Yasuda, 2000; Rodriguez et al., 2000). Conversely, activation of p53 is stimulated by the gene product of an alternative reading frame (ARF) within the *CDKN2A* gene ($p14^{ARF}$ in human and $p19^{ARF}$ in mouse). ARF stabilizes p53 primarily by sequestering MDM2 to the nucleolus and inhibiting MDM2 ubiquitin ligase activity (Kruse and Gu, 2009), mediating a G_1-phase cell cycle arrest (Lowe and Sherr, 2003). In mice, oncogenic Ras induces ARF, and MEFs knocked out for ARF phenocopy p53-null MEFs (Lowe and Sherr, 2003). While ARF is a direct transcriptional target of both Myc and E2F in human (Bates et al., 1998; Zindy et al., 1998), similar induction of ARF in Ras-expressing human fibroblasts has not been observed (Itahana et al., 2004; Wei et al., 2001). Studies in mouse models and human tissues suggest that the DDR may instead support ARF-independent activation of p53 in human (Matheu et al., 2008). Nevertheless, since MDM2 cooperates with Ras during transformation (Seger et al., 2002), ARF remains a significant mediator of p53 activity. Furthermore, the fact that inactivation of both ARF and p53 rarely cooccurs in human cancers supports the important function of ARF in tumor suppression (Matheu et al., 2008).

Numerous p53-inducible MiRs have been shown to indirectly regulate p53 stability, notably through direct repression of *MDM2* (Pichiorri et al., 2010; Xiao et al., 2011). *MDM2* was revealed as a target of MiR-605 in both wild-type (WT) p53 lung and breast cancer cell lines, respectively, A549 and MCF-7. In this context, MiR-605 increased p53 transactivation activity rather than stabilizing p53 protein levels and, more specifically, ectopic expression of MiR-605 stimulated apoptosis in MCF-7, but not in p53-deficient MDA-MB-436 cells. Thus, MiR-605 is part of a p53-MDM2 regulatory loop which stimulates p53 activity (Xiao et al., 2011). In a study on novel MiRs with therapeutic potential in multiple myeloma (MM), Picchiori et al. identified MiR-192, 194, and 215 as p53-regulated MiRs. In both WT p53 and nutlin-3a-treated MM cells, MiR-194 was the only other MiR, apart from MiR-34a, to be significantly upregulated. The other two MiRs, i.e., MiR-192 which is part of a cluster with MiR-194 (MiR-194-2-192 cluster), and MiR-215 were only significantly induced upon nutlin-3a treatment. Promoter analysis revealed p53 REs at the promoter regions of these MiRs. However, p53-dependent promoter activity was observed at the MiR-194-2-192 cluster locus, but not at MiR-215, suggesting transcription-independent effect of p53 on MiR-215 levels. In addition, treatment of WT p53 MM cells with MI-219, a specific inhibitor of MDM2–p53 interaction, led to increased stabilization of p53, enhanced apoptotic signaling, and cell death. To establish the in vivo significance of the MiR-192, -194, and -215 inhibitory effects on *MDM2*, nude mice were injected subcutaneously with MM cells. Following solid tumor development, mice were treated with MI-219 plus either scrambled or pooled pre-miR-192, -194, and -215 through oral delivery. Tumors injected with pooled pre-miRs were significantly reduced in volume while scramble MiR-treated tumors did not show any sign of regression (Pichiorri et al., 2010). Since restoration of p53 can also contribute to tumor clearance by inducing senescence in vivo (Xue et al., 2007), this model demonstrates the strong potential of MiR delivery for treatment of cancers harboring a silenced, WT *TP53* allele.

MiR-34a operates in a positive feedback loop to increase p53 stability by (i) targeting Myc (Wei et al., 2008), which in turn leads to increased *ARF* expression and alleviates MDM2-dependent inactivation of p53 and also (ii) directly inhibiting *HDMX* (also known as *MDM4*) (Mandke et al., 2012; Navarro and Lieberman, 2015; Okada et al., 2014). *HDMX*, a homolog of *MDM2*, acts principally as an inhibitor of p53 transactivation (Shvarts et al., 1996). Following DNA damage, ATM and Chk2 phosphorylate HDMX to promote MDM2-mediated degradation of HDMX in order to stimulate p53 activity (Chen et al., 2005a). In addition, MDM2 controls HDMX nuclear localization in response to DNA damage (Migliorini et al., 2002), and the latter notably collaborates with oncogenic Ras to bypass senescence in normal human diploid fibroblasts (Miller et al., 2010). Several groups successfully demonstrated that *HDMX* constitutes a direct MiR-34a target, identifying MiR-34a REs in the 3′-UTR (Navarro and Lieberman, 2015; Okada et al., 2014) as well as within the coding region of *HMDX* (Mandke et al., 2012). Furthermore, HDMX protein and mRNA levels decreased independently of p53 status following genotoxic stress, consistent with the notion that the DDR could act as a surrogate pathway stimulating MiR-34a to target *HDMX* transcripts when p53 activity is impaired (Markey and Berberich, 2008).

In the K-Ras mouse model of lung cancer, the p53, MiR-34a, and HMDX positive feedback loop played a striking functional role in tumor suppression (Okada et al., 2014). *TP53* haploinsufficiency is reported to promote tumorigenesis in diverse cell types. Surprisingly, mice harboring an oncogenic *Kras* allele in a p53 heterozygous background ($Kras^{LSL\text{-}G12D/+}$; $p53^{+/-}$) did not significantly display increased lung tumor development compared to $Kras^{LSL\text{-}G12D/+}$ mouse (Okada et al., 2014). This implies that another component of the p53 pathway may salvage p53 function when its expression levels are compromised. Okada et al. suspected that MiR-34a could mediate normal p53 function in this particular model. Knockout of MiR-34a in $Kras^{LSL\text{-}G12D/+}$; $p53^{+/-}$ mice significantly promoted tumor burden and displayed high-grade clinical features. Moreover, human lung adenocarcinoma revealed increased levels of a short isoform of *MDMX*, an alternatively spliced mRNA harboring a shorter 3′-UTR resistant to MiR-34a-mediated inhibition (Okada et al., 2014).

On the other hand, homozygous deletion of the MiR-34 family did not lead to decreased lifespan nor did it promote spontaneous tumor formation with early latency, unlike the situation for $p53^{-/-}$ mice. When challenged with low dose of γ-irradiation, MiR-34-deficient newborns did not show any overt phenotype, as opposed to the previously reported short tumor-free survival time of $p53^{-/-}$ mice. However, loss of MiR-34 contributed slightly to K-Ras-mediated transformation when pRB was concomitantly inactivated by E1A (Concepcion et al., 2012). Furthermore, depletion of MiR-34a in mouse embryonic stem cells led to decreased apoptosis during differentiation (Bommer et al., 2007). MEFs from knockout mice for each member of the MiR-34 family also displayed enhanced reprogramming efficiency, similarly to $p53^{-/-}$ MEFs (Choi et al., 2011). These findings support the idea that MiR-34 possesses both p53-dependent and -independent functions, and that p53-independent roles of MiR-34 may depend on the INK4A/ARF pathway status. Nevertheless, the positive effect of MiR-34 expression on p53 activity in vivo (Okada et al., 2014; Tazawa et al., 2007) remains relevant in a clinical context.

The tumor suppressor CHD5 participates in activating the INK4A/ARF locus and is required for stimulation of p53 during OIS (Bagchi et al., 2007). The lysine demethylase KDM4A mediates transcriptional silencing of *CHD5* expression and thus hinders p53 activation during RIS (Mallette and Richard, 2012). KDM4A demethylates H3K9me2/3 and H3K36me2/3 (Whetstine et al., 2006) and is frequently upregulated in numerous cancers (Carbonneau et al., 2016). Ectopic expression of KDM4A contributes to bypass of RIS and cooperates with oncogenic Ras in transformation (Mallette and Richard, 2012). Importantly, *KDM4A* constitutes a direct target of MiR-137 (Neault et al., 2016) which is frequently downregulated in human cancers (Bemis et al., 2008; Iorio et al., 2005; Kozaki et al., 2008). Specifically, MiR-137 is induced in response to oncogenic Ras to downregulate *KDM4A* and promote p53 activation (Neault et al., 2016).

3.2.4 SIRT1, a Primary Oncogenic Target of Tumor Suppressor MiRs

SIRT1 is a NAD-dependent histone deacetylase involved in cell metabolism, DNA damage, senescence, and aging (Finkel et al., 2009; Herranz and Serrano, 2010). Importantly, SIRT1 deacetylates and negatively regulates the transcriptional activity of p53 (Luo et al., 2001; Vaziri et al., 2001). It was also reported that SIRT1 is firmly implicated in bypass of OIS through direct deacetylation of p53 (Langley et al., 2002) and extends lifespan in mice (Satoh et al., 2013). It was initially reported that MiR-34a represses *SIRT1* in colon cancer cells to induce p53-dependent apoptosis. Enhanced p53 activity following MiR-34a overexpression correlates with increased p21 and PUMA protein levels following transfection of pre-miR-34a (Yamakuchi et al., 2008). Subsequent studies provided evidence that depletion of *SIRT1* by MiR-34a induces cellular senescence in various cell lines such as in umbilical cord vein endothelial cells (HUVEC) (Ito et al., 2010) and esophageal squamous cancer cells (Ye et al., 2016). Furthermore, MiR-34a overexpression reduces cell viability and generates a senescence phenotype in mesenchymal stem cells by targeting *SIRT1* (Zhang et al., 2015).

Apart from MiR-34, silencing of *SIRT1* by other MiRs confers tumor suppression phenotypes. For example, MiR-22 directly targets *SIRT1* along with *CDK6* and is upregulated in senescent fibroblasts; moreover, it inhibits proliferation, cell motility, and invasion in human cervical cancer cells (Xu et al., 2011). In addition, overexpression of *SIRT1* and *CDK6* restored cell proliferation defects in cells transfected with MiR-22 mimics, suggesting that the senescence phenotype is due to MiR-22-mediated *SIRT1* and *CDK6* inhibition. Most significantly, intratumoral injection of MiR-22 mimics inhibited breast tumor growth and metastasis in a murine model of breast cancer development (Xu et al., 2011), further indicating the clinical potential of MiR-based therapy. MiR-195, another *SIRT1*-targeting MiR, displayed increased expression in old mouse skeletal myoblasts (SkMs), and inhibition of MiR-195 facilitated in vitro reprogramming of older SkMs and tumor formation in a xenograft model (Kondo et al., 2016). *SIRT1* is also a validated target of MiR-212, -217, and -494. Both MiR-212 and -494—which, in like *SIRT1* directly repress Myc—are downregulated in pancreatic cancer cells. These MiRs were also found to inhibit angiogenesis in HUVEC and invasion and chemoresistance in pancreatic cancer cells, while MiR-212 and -494 also engaged the senescence phenotype (Liu et al., 2015b; Ramalinga et al., 2015). MiR-217 inhibited angiogenesis and induced senescence in HUVEC, and its expression correlated with increased passage number (Menghini et al., 2009). MiR-138, -181a, and -181b are three MiRs concomitantly overexpressed in high passage primary human epidermal keratinocytes. Transfection of pre-miR-138, -181a, and -181b resulted in decreased SIRT1 levels, pRB hyphosphorylation, and stimulation of senescence. These MiRs led to reduced activity of a luciferase reporter carrying the 3′-UTR of *SIRT1*, suggesting a direct inhibitory effect over *SIRT1* (Rivetti di Val Cervo et al., 2012). In addition, ΔNp63α, an isoform of p63 supporting proliferation, stem cellness, and tumorigenesis (Guo et al., 2009; Senoo et al., 2007; Wu et al., 2003), is directly recruited to p63REs in close vicinity of MiR-138, -181a, -181b, and MiR-130b host genes (Rivetti di Val Cervo et al., 2012). ΔNp63α negatively regulates expression of these MiRs, while MiR-130b engages in a negative feedback loop by directly targeting the 3′-UTR of ΔNp63α. Finally, MiR-130b, together with MiR-138, -181a, and -181b, promoted replicative senescence in old human keratinocytes (Rivetti di Val Cervo et al., 2012).

3.2.5 Indirect Regulation of TP53 by Tumor Suppressor MiRs

While p53 levels are primarily regulated through PTM, the regulation of the *TP53* gene also participates in the activation of p53 during cellular stress.

The KLF4 transcription factor regulates p53 expression and controls a multifaceted network involving MiR-203 and p21 (Xu et al., 2016a). KLF4 acts as a context-dependent oncogene by directly repressing the promoter of *TP53* in human, while promoting p21-regulated senescence depending on the experimental context (Rowland et al., 2005). *KLF4* was thought to promote senescence in part through inhibition of Survivin, but the exact mechanism remained unclear (Zhang et al., 2009). Survivin interferes with p53 RE located in the promoter region of *CDKN1A*, thus blocking p53 transactivation of the locus. MiR-203, a putative MiR-targeting Survivin, was identified as a direct transcriptional target of KLF4. Moreover, an inhibitor of MiR-203 could partially rescue *KLF4*-induced senescence while overexpression of a pre-miR-203 precursor induced senescence in AGS gastric- and SW480 colon cancer cell lines (Xu et al., 2016a).

3.3 MiR Regulation of the p16^{INK4A}/pRB/E2F Pathway

p16^{INK4A}/pRB is a critical tumor suppressor pathway responsible for the establishment and maintenance of the senescence phenotype (Serrano et al., 1997). p16^{INK4A} accumulates during aging in mammalians, while senescence in MEFs relies on the INK4A/ARF locus (Krishnamurthy et al., 2004; Serrano et al., 1996). As such, the p16^{INK4A}/pRB pathway is a prime target for oncomiRs during tumorigenesis.

3.3.1 E2F–pRB Activity Is Modulated Mainly Through Feedback Loops

The MiR–E2F relationship is characterized by diverse complex feedback loops enforcing autoregulation of the E2F family. Hence, disregulated MiR–E2F pathways are not unexpected in neoplastic transformation. The E2F family of transcription factors regulates a subset of genes responsible for G_1/S progression. Cell cycle-dependent genes (*CCNA*, *CCNE*, *CDK2*, *E2F*, *CDC2*, and *CDC25A*) and DNA replication genes (*CDC6*, *PCNA*, and *MCM* family) are part of the E2F-dependent transcription program (Du and Pogoriler, 2006; Stevens and La Thangue, 2003). In cycling cells, entry into S-phase is monitored by pRB in complex with E2F1–3 (Giacinti and Giordano, 2006; Shirodkar et al., 1992). In response to mitogenic signals, phosphorylation of pRB by cyclin–CDK complexes in early and late G_1 inactivates pRB, which in turn can no longer sequester E2F (Chellappan et al., 1991). Moreover, cyclin–CDK complexes are critically involved in the repression of pRB-mediated senescence (Puyol et al., 2010; Rane et al., 2002). Various members of the E2F family are targeted by tumor suppressor MiRs to promote cellular senescence.

E2F2 and E2F3 are direct targets of MiR-20a, a member of the MiR-17-92 cluster. E2F1–3 mediate S-phase entry by activating the promoters of genes involved in DNA replication (DeGregori et al., 1995). E2F3 controls the G1/S transition by specifically activating numerous cell cycle-dependent genes including *MYBL2* (B-Myb) and *CCNA* (cyclin A) independently of E2F1 (Humbert et al., 2000). On the other hand, the promoter region of the MiR-17–92 cluster is transcriptionally activated by E2F1, E2F2, and E2F3, and this was shown to repress apoptosis in PC cells. The subtle tuning of E2F levels by the MiR-20a/E2F loop was suggested to be the key factor deciding whether or not cells engage apoptosis (Sylvestre et al., 2007). Of note, MiR-20a has also been shown to induce ARF-dependent senescence through direct inhibition of the leukemia/lymphoma-related factor, associated with accumulation of p16 and decreased E2F1 levels (Poliseno et al., 2008). Another feedback loop between MiR-20a and E2F1—similar to the MiR-20b member of the MiR-106a–363 cluster—appeared to directly regulate E2F1 protein expression, while being under the strict control of E2F1-dependent transcriptional activation during myogenesis (Luo et al., 2016).

E2F1 and MiR-34c are part of a negative feedback loop in colorectal cancer. E2F1 directly activated MiR-34c transcription, thereby mediating cell proliferation arrest (Yang et al., 2015). The same group previously identified stem cell factor (SCF) as a direct MiR-34c target (Yang et al., 2014), and it appeared that SCF signaling positively regulates E2F1 expression, presumably through autocrine/paracrine stimulation in colorectal cancer cells (Yang et al., 2015). E2F3 expression was also directly inhibited by MiR-449a in A549 NSCLC cells (Ren et al., 2014). MiR-449a has been identified as an E2F1-dependent target gene, while it also directly repressed E2F1–pRB activity via downregulation of *CDK6* and *CDC25A* in an complex feedback loop (Feng and Yu, 2010; Yang et al., 2009). It is worth mentioning that subcutaneous transplantation of MiR-449a-expressing A549 cells in nude mouse xenografts significantly decreased tumor growth, and that immunohistochemical analysis of these tumors displayed lower levels of *E2F3* compared to a nonspecific control MiR (Ren et al., 2014). Furthermore, expression of MiR-449a induced premature cellular senescence in lung cancer cells (Maehara et al., 2005) and triggered pRB-dependent senescence in PC cells by specifically targeting *CCND1* and *HDAC1*

(Noonan et al., 2010). These data suggest that the reduced tumor burden from the A549 xenograft model (Ren et al., 2014) could also possibly be due to the MiR-449a-dependent induction of senescence in tumor cells. MiR-449a may be considered a singular MiR since it shares sequence similarity with MiR-34. Both are coexpressed temporally in specific mammalian tissues (Song et al., 2014) and MiR-449a and MiR-34b/c share common targets in the E2F–pRB pathway during spermatogenesis in mouse (Bao et al., 2012). Consequently, it is tempting to speculate that restoration of MiR-449a expression could, like for MiR-34a, induce cellular senescence in vivo and regression of MiR-449a-depleted tumors.

The 106b~25 polycistronic cluster, which promoted bypass of RIS (Borgdorff et al., 2010), is contained within an intronic region of the *MCM7* gene. MCM7 is part of a multiprotein helicase complex required for DNA replication and G_1/S transition, and represents a marker of high proliferation rate in cancer cells (Blow and Hodgson, 2002). In gastric primary tumors, all three members of the 106b~25 cluster, MiR-106b, -93, and -25, were coregulated with *MCM7*. Furthermore, MiR-106b and -93 were shown to directly target *E2F1* mRNA, while blocking TGFβ tumor suppressor activity via direct inhibition of its downstream effector *CDKN1A* (Petrocca et al., 2008). E2F1 displays oncogenic properties through its major role as a regulator of G1/S transition; however, unlike other E2F members, its accumulation beyond a threshold triggers an intrinsic tumor suppressor activity inducing apoptosis both in vitro and in vivo (Lazzerini Denchi and Helin, 2005). These observations suggest that downregulation of E2F1 by the 106b~25 cluster participates in tumor proliferation in gastric cancer.

Lastly, MiR profiling in serum-stimulated MEF revealed several other E2F1-responsive MiR cluster, such as the *let*-7a-d, *let*-7i, MiR-15b–16-2, and MiR-106b~25 clusters. These MiRs were shown to block the G1/S transition and to protect cells from replicative stress by directly targeting the E2F1-dependent genes *CCND1/2*, *CCNE1*, *CDK2*, and *MCM5* (Bueno et al., 2010). Downregulation of *CCNE1* by MiR-15a/b and MiR-16 was also confirmed in another independent study (Ofir et al., 2011).

3.3.2 MiR-Mediated Transcriptional Regulation of E2F-Dependent Genes

MiRs were recently found to participate in gene silencing during cellular senescence. Whole genome promoter analysis of senescent WI38 fibroblasts resulted in identification of 320 putative E2F-regulated promoters (compared to 77 in control cells) which were bound by AGO2, a member of the MiRISC. Using *let-7f* as proof of concept, Benhamed et al. showed that *let-7*/AGO2 complexes were shuttled to the nucleus and directly recruited at heterochromatin during senescence, specifically to inhibit promoter activity of the E2F-dependent *CDC2* gene. Additionally, using a *CDC2*-promoter-based reporter (harboring a putative *let-7* binding site), the authors observed a synergistic function for pRB in *let-7*/AGO2-transcriptional inhibition (Benhamed et al., 2012). Knowing that decreased levels of AGO2 impair small-interfering RNA-mediated inhibition of translation (Romero-Cordoba et al., 2014; Voller et al., 2013), and that inhibition of AGO2 expression delays replicative senescence (Benhamed et al., 2012), it is tempting to speculate that the restoration of AGO2 activity could ultimately restore pRB/AGO2/MiR-mediated transcription inhibition of E2F-responsive genes and promote cellular senescence of cancer cells.

3.3.3 MiRs Regulation of INK4A-CDK Activity

Hyperstimulation of MAPK signaling by oncogenic Ras or downstream oncogenic Raf, very frequently observed in melanomas, triggers the expression of *INK4A* (Lin et al., 1998; Zhu et al., 1998). $p16^{INK4A}$ plays a pivotal role in the regulation of pRB and E2F-dependent gene transcription. $p16^{INK4A}$ interacts with CDK4/6 to prevent the latters' association with D-type cyclins necessary for G_1/S progression (Russo et al., 1998). Inhibition of CDK activity thwarts hyperphosphorylation of pRB which mediates growth arrest by repressing E2F. Although CDK4/6 has many targets other than pRB, evidence suggests that activation of pRB is the major consequence of $p16^{INK4A}$-mediated cyclin-CDK inhibition (Otterson et al., 1994; Shapiro et al., 1995; Tanaka et al., 1998).

Elevated $p16^{INK4A}$ is a hallmark of late passage, senescent normal cells (Hara et al., 1996). Microarray analysis of downregulated MiRs in senescent WI-38 fibroblasts significantly revealed MiR-24 as inversely correlated with $p16^{INK4A}$ expression. MiR-24 was shown to directly associate with *INK4A* mRNA in polysomal fractions and revealed itself as a potent inhibitor of *INK4A* expression in HeLa cells (Lal et al., 2008). However, the physiological relevance and cellular consequences of MiR-24-mediated inhibition of *INK4A* remains unknown.

Regulation of cyclins and CDKs by various MiRs also contributes to cell cycle arrest during senescence. *CDK6* and *CCND1*, two genes collaborating in G_1/S progression, are directly targeted by MiR-34 to induce G_1 arrest in A549 lung cancer cells (Sun et al., 2008). MiR-34a also targets cell cycle promoting components such as *CCNE2* and *CDK4* (He et al., 2007), while MiR-33 regulates *CDK6* and *CCND1* (Cirera-Salinas et al., 2012; Xu et al., 2016b). The E2F2-targeting MiR-449a also inhibits *CCND1* expression and promotes senescence of pancreatic cancer cells PC-3 (Bonci et al., 2008; Kitadate et al., 2016).

In addition to its ability to modulate p53 activity, MiR-137 also induces senescence in p53-depleted fibroblasts, as well as in PANC-1 cancer cells (Neault et al., 2016) which are deficient for $p16^{INK4A}$ expression and harbor a p53 heterozygous mutation (Deer et al., 2010). MiR-137 is known to target antagonists of the INK4A/pRB pathway, including *CDK2*, *CDK6*, *CDC42*, and *E2F6* (Chen et al., 2012; Kozaki et al., 2008; Liu et al., 2011), and MiR-137-mediated depletion of these targets may be responsible for activation of the $p16^{INK4A}$/pRB pathway. Nonetheless, restoration of MiR-137 expression inhibits proliferation and promotes senescence of pancreatic cancer cells (Neault et al., 2016) suggesting that modulating levels of MiR-137 may be of clinical importance for triggering in vivo senescence in pancreatic cancer.

3.3.4 Activation of the INK4A Locus by MiRs

The BMI-1 member of the polycomb group (PcG) proteins, part of the polycomb repressive complex (PRC), is a strong repressor of *INK4A* expression (Bracken et al., 2007; Gil et al., 2004; Jacobs et al., 1999). The PRC is responsible for depositing repressive methylated histone H3K27me3 marks thereby silencing the *INK4A/ARF* locus. The important influence of BMI-1 over $p16^{INK4A}$ is exemplified in *Bmi-1*-deficient mouse models, where reduced lifespan is rescued by concomitant deletion of *INK4A* (Gil et al., 2004; Jacobs et al., 1999). *BMI1* was identified as a direct target of MiR-128 in diverse cancers, including prostate and brain cancers (glioblastoma and medulloblastoma) (Godlewski et al., 2008; Jin et al., 2014; Venkataraman et al., 2010). Of note, MiR-128 expression is significantly reduced in various malignancies (Li et al., 2013). Glioma cell xenografts overexpressing MiR-128 inhibit cell proliferation, while introduction of pre-miR-128 in glioma cell lines induces growth arrest associated with induction of p21 and reduced AKT signaling (Godlewski et al., 2008). Moreover, MiR-128-mediated inhibition of *BMI1* results in increased cellular ROS, cell cycle arrest, and accumulation of $p16^{INK4A}$ (Venkataraman et al., 2010). Another SA-MiR, MiR-141, could also directly target *BMI1* in human diploid fibroblasts. Ectopic expression of MiR-141 notably impedes BMI-1 localization and H3K27me3 deposition at the *INK4A* promoter region, resulting in derepression of the INK4A locus and induction of p16^{INK4A}-dependent premature senescence (Dimri et al., 2013). Bmi-1 was further shown to positively regulate the expression of MiR-300 in mouse cardiac progenitor cells (CPC). Owing to the oncogenic nature of *BMI1*, overexpression of MiR-300 phenocopies *Bmi-1*-transduced CPC, which results in the bypass of replicative senescence and significant decrease of p16^{INK4A} expression (Cruz et al., 2015).

In a functional MiR screen, a comparison of MiRs modulating *INK4A* expression with a microarray expression profiling of SA-MiRs revealed 22 MiRs which were concomitantly represented. Using target prediction, MiR-26b, -181a, -210, and -424 were postulated to target PcG proteins CBX7, EED, EZH2, and Suz12, which were subsequently validated as direct targets via luciferase reporter assays. Chromatin immunoprecipitation assays revealed an increase of the repressive H3K27me3 mark at promoter regions of *INK4A* as well as MiR-26b, -181a, -210, and -424 loci, while H3K27me3 levels at these loci were reduced in later passage HMEC. It was therefore proposed that p16^{INK4A} acts in positive feedback loops with these SA-MiRs to inhibit PcG proteins, and vice versa (Overhoff et al., 2014). *EZH2* (and also E2F2) was further shown to be directly repressed by *let*-7b in senescent cells (Tzatsos et al., 2011). CBX7 inhibited the promoter activity of MiR-9, a MiR upregulated during replicative senescence. Moreover, ectopic expression of MiR-9 triggered premature senescence by specifically targeting *CBX7*, suggesting that MiR-9 and CBX7 participate in a negative feedback loop during cellular senescence (O'Loghlen et al., 2015). Altogether, these data support a considerable role for MiRs in mediating the oncogenic properties of the PcG proteins.

3.3.5 *let-7*-Mediated Regulation of HMGA2, a Context-Dependent Repressor of the INK4A/ARF Locus

The high-mobility group AT-hook 2 (*HMGA2*) exerts divergent functions in cellular senescence signaling (Markowski et al., 2011; Narita et al., 2006; Shi et al., 2013). *HMGA2* is a notorious *let*-7 target associated with tumorigenesis. Of note, chromosomal translocation at the *HMGA2* coding region causes loss of regulatory sequences at the 3′-UTR rendering the *HMGA2* mRNA resistant to *let*-7-mediated inhibition. Importantly, the uncoupling between *let*-7 and *HMGA2* promotes anchorage-independent growth (Lee and Dutta, 2007; Mayr et al., 2007). The expression of *HMGA2* usually occurs during organism development, but is also enriched in metastatic cancer with poor prognosis (Fedele and Fusco, 2010). Ectopic expression of *let*-7, which is downregulated in old mouse neural stem cells (NSC), contributed to the derepression of the *INK4A/ARF* locus via downregulation of HMGA2. Furthermore, *HMGA2*-deficient NSC displays poor self-renewal capabilities, while this phenotype could be partially restored to the physiological state by depletion of $p16^{INK4A}$ (Nishino et al., 2008). Together these results suggest that the alteration in *let*-7-mediated inhibition of *HMGA2* in cancer cells or old tissues might participate in tumorigenesis through decreased $p16^{INK4A}$ expression. Other members of the *let*-7 family (-7a1, -7d, and -7f) and also MiR-23a, -26a, and -30a were shown to mediate cellular senescence and regulate self-renewal in human umbilical cord blood-derived multipotent stem cells (Lee et al., 2011). *let*-7g also blocked HCC proliferation by directly inhibiting c-Myc and stabilizing $p16^{INK4A}$ (Lan et al., 2011).

On the other hand, *HMGA2* also displays tumor suppressor properties. During senescence, pRB mediates the recruitment of histone remodeling proteins which promote a condensed chromatin (heterochromatin). The formation of such SAHF, a common feature of senecent but not quiescent cells, impairs transcription of E2F-dependent genes responsible for cell cycle progression and DNA replication, such as *CCN1A* (cyclin A) and *PCNA* (Burkhart and Sage, 2008; Narita et al., 2003; Nielsen et al., 2001; Rayman et al., 2002). HMGA2 is recruited to SAHF and promotes formation of heterochromatin (Narita et al., 2006). Surprisingly, shRNA-mediated inhibition of *HMGA2* revealed a functional bias for *HMGA2* toward gene repression in Ras-induced senescent IMR90 lung fibroblasts. Conversely, the same knockdown experiment in control fibroblasts revealed no bias toward either gene activation or repression. Furthermore, xenograft of IMR90 cells transduced with oncogenic Ras and E1A led to less aggressive and penetrant tumors compared to the same cells concomitantly cotransduced with HMGA2 (Narita et al., 2006). These data support the idea that the bipolar function of HMGA2 in transformation and cancer is context dependent, and that fine-tuning of MiRs regulating the expression of *HMGA2* might be in part responsible for the context-dependent role of HMGA2.

3.3.6 SA-MiRs at the Crossroads of p53 and pRB Pathways

MiRs can engage in crosstalk between the p53 and pRB pathways. For instance, MiR-203, which promotes p21-dependent senescence (Xu et al., 2016a) (see above), also acts in the E2F–pRB pathway. In melanoma cells, ectopic expression of MiR-203 initiates a senescence phenotype—partially due to *SIRT1* depletion—but mainly through direct repression of *E2F3* (Noguchi et al., 2012).

MiR-34a, expression of which is strongly p53 dependent, halts proliferation of neuroblastoma cell lines by directly targeting *E2F3* (Welch et al., 2007). Additionally, MiR-34a induction following DNA damage promotes downregulation of *E2F1* and *E2F3*, while ectopic expression of MiR-34a alone induces senescence and growth arrest in human colon cancer cells (O'Donnell et al., 2005; Tazawa et al., 2007). MiR-125b, a brain-enriched MiR negatively affecting p53 levels in neuroblastoma (Le et al., 2009), directly represses *E2F3* to alleviate bladder cancer cell proliferation both in vitro and in vivo (Huang et al., 2011).

Interestingly, some MiR exert either a prosenescence or a proapoptotic function depending on the genetic context. For example, MiR-16 in primary cutaneous T-cell lymphoma (CTCL) triggers p53-dependent senescence in CTCL harboring a wild-type allele of *TP53*, or p53-independent apoptosis in CTCL lacking functional p53, and in both cases this occurs via direct inhibition of *BMI-1* and *CCND1* (Kitadate et al., 2016). Any possible effect via derepression of $p16^{INK4A}$ was excluded since no expression of this protein was observed; however, one cannot exclude a possible role for ARF-mediated apoptosis in this context (Nakazawa et al., 2003). These studies demonstrate the complex, multifaceted, and tissue-specific functions of MiRs in the modulation of tumor suppressor/oncogenic pathways.

3.4 MiR Regulation of the DDR

The DDR contributes to the phosphorylation and activation of p53 during senescence. During OIS, oncogene activation stimulates DDR signaling, which is required for the establishment of cellular senescence (Bartkova et al., 2006; Di Micco et al., 2006; Mallette and Ferbeyre, 2007). In addition, exogenous DNA damage in normal diploid fibroblasts leads to persistent, irreparable DNA damage in noncritically short telomeres, and to constitutive activation of DDR signaling contributing to the stability of the senescence phenotype (Fumagalli et al., 2012). The important function of the DDR in senescence is corroborated by the spontaneous bypass of replicative senescence observed in p21-null or dominant-negative p53 expressing MEFs (Bond et al., 1994; Brown et al., 1997), as well as in CHK2-inactivated human fibroblasts (Gire et al., 2004). The cooperative role of p53 and pRB in orchestrating DDR signaling is undeniable, and the study of potential DDR regulators remains a primordial aim of oncology and cancer therapy.

3.4.1 The MiR–DDR Feedback Loop

The DDR can regulate levels of multiple MiRs by directly influencing MiR processing and maturation. In response to DNA damage, activated p53 interacts with the MiR microprocessor complex to enhance processing of pri-miR-145, among others (Suzuki et al., 2009). p53-mediated upregulation of MiR-145 may enforce p53 stabilization in a feedback loop following DNA damage, since ectopic expression of MiR-145 was shown to directly inhibit *MDM2* (Zhang et al., 2013). In addition, MiR-145 demonstrates additional tumor suppressor activity as its overexpression leads to transcriptional repression of *RAS* and restores growth arrest in pancreatic cancer cells. Conversely, MiR-145 is repressed in response to oncogenic Ras, resulting in a negative feedback supporting Ras signaling (Kent et al., 2010).

3.4.2 MiR Regulation of the DDR

As mentioned above, BMI-1 blocks cellular senescence and its mRNA is targeted by many different MiRs. BMI-1 also regulates the DDR pathway by mediating homologous recombination repair of DNA double-strand breaks (Ginjala et al., 2011). MiR-494-3p, which is upregulated in head and neck cancer, is a putative inhibitor of *BMI-1* expression (Chang et al., 2015). Consistent with its important role in the DDR pathway, MiR-494-3p-mediated repression of *BMI1* increased sensitivity to ionizing radiation of the oral squamous cell carcinoma (OSCC) cell line SAS. Furthermore, ectopic expression of MiR-494-3p in these cells phenocopied the premature senescence phenotype induced by knockdown of *BMI1*. Expression of MiR-494-3p was also inversely correlated with *BMI1* levels in human OSCC, suggesting a tumor suppressor function for MiR-494-3p in this particular cancer (Weng et al., 2016).

The RNA-binding protein HuR functions as a regulator of *SIRT1* expression following DNA damage. Specifically, HuR binds and stabilizes *SIRT1* mRNA in unstressed cells, and this interaction is disrupted by direct phosphorylation of HuR by CHK2 in response to oxidative stress. Moreover, senescent WI-38 fibroblasts constitutively express low levels of HuR and SIRT1 (Abdelmohsen et al., 2007). HuR can be directly targeted by MiR-519 in multiple human cancer cell lines. Inhibition of MiR-519 using antisense oligonucleotides stimulates proliferation via upregulation of HuR translation (Abdelmohsen et al., 2008). Cancer cells often rely on active DDR signaling to endure increased genomic instability; therefore, it would be relevant to investigate whether MiR-519 expression is modulated in response to DNA damage.

The oncogenic MiR-106b~25 cluster also participates in bypass of DNA damage-induced senescence. This cluster was found to be specifically upregulated in MD60 cells, a doxorubicin- and γ-radiation-resistant cell lines derived from minimally transformed mammary epithelial breast cancer cells (MTMEC). Resistant cells also harbored characteristics of EMT and this phenotype was conferred by specific inhibition of a positive regulator of E-cadherin, EP300 (Liu et al., 2005; Zhou et al., 2014). Interestingly, treatment of MTMEC with members of the MiR-106b~25 cluster (with MiR-25 demonstrating predominent activity) did not affect proliferation in the absence of exogenous genotoxic insult. This suggests that the DNA damage induced by doxorubicin treatment could drive an EMT-associated drug resistance phenotype, and by extension the bypass of DDR-induced senescence in pretransformed cells, specifically via the upregulation of the MiR-106b~25 and subsequent decreased EP300/E-cadherin-dependent adhesion (Liu et al., 2005).

3.5 MiR Regulation of Telomerase

Normal somatic cells undergo an inevitable decrease of telomere length during their lifespan. The resulting telomere erosion ultimately exposes chromosomal DNA ends and triggers a DDR (Smith et al., 2010). Moreover, critically short telomeres and transient suppression of DDR kinases such as ATM, ATR, CHK1, and CHK2, also result in bypass of replicative senescence (di Fagagna et al., 2003; Herbig et al., 2004). Although reactivation of telomerase (hTERT) alone is not sufficient to induce cellular transformation, it does promote sustained self-renewal. Recently, *hTERT* was identified as a putative target of MiR-195 (Okada et al., 2016). The latter appeared upregulated in old mesenchymal stem cells (OMSC) compared to young ones, where the telomerase is expressed and active. OSMC displayed characteristic phenotypes associated with replicative senescence, while depletion of MiR-195 expression in OSMC resumed cellular growth and reactivated expression of *hTERT* among other proproliferative genes. Furthermore, transplantation of OMSC expressing a MiR-195 inhibitor significantly improved the regeneration of infarcted mouse heart (Okada et al., 2016). Importantly, this study implies that modulation of MiRs mediating lifespan and replicative senescence may improve stem cell-based therapy among the elderly.

Altogether, MiRs exert critical functions in tumor suppression and tumorigenesis by modulating molecular networks involved in the establishment of senescence.

4. MODES OF REGULATION OF MiR ACTIVITY IN CANCER

Cancer cells must overcome tumor suppressive mechanisms such as cellular senescence in order to achieve full transformation. Since MiRs are important regulators of cellular senescence, dysregulation of their expression, or maturation play important roles in tumorigenesis. Below we present an overview of different mechanisms underlying dysregulation of MiR expression in cancer.

4.1 Aberrant DNA and Histone Methylation

Cytosine bases within cytosine–guanine dinucleotides (CpG) located in gene promoters and intergenic regions are targeted by DNA methyltransferases (DNMTs) to regulate gene expression. DNA methylation hinders the binding of transcription factors and histone modifying enzymes, therefore blocking gene activation (Denis et al., 2011; Fazzari and Greally, 2004). DNA hypermethylation is a prevalent mechanism triggering the silencing of tumor suppressor genes in neoplastic cells (Baylin and Ohm, 2006; Robertson, 2005).

Patterns of histone methylation within nucleosomes also regulate gene expression. For instance, trimethylated H3K4 is generally linked to gene activation, whereas repressed genes are associated with di- and trimethylated H3K9 and -K27 (Barski et al., 2007; Benevolenskaya, 2007; Koch et al., 2007). Moreover, DNMT3, which methylates DNA de novo to generate 5-methylcytosine and promotes gene repression (Denis et al., 2011; Okano et al., 1999), is recruited to chromatin through interaction with methylated H3K9 (Du et al., 2015), thus highlighting the interplay between histone and DNA methylation in gene regulation.

Overall MiRs levels are generally lower in tumor cells (Lu et al., 2005), and DNA methylation and histone modifications are mainly responsible for repressing tumor suppressor MiR in cancer (Chuang and Jones, 2007; Guil and Esteller, 2009; Lehmann et al., 2008; Takamizawa et al., 2004). Comparison of MiR expression profiles between normal fibroblasts and PC cells revealed a positive correlation between activating H3K4me3 or repressive H3K27me3 histone marks and MiR transcriptional activation or inhibition, respectively (Siebig et al., 2009). Increased H3K9me2 coupled with low

H3K4me3 levels were also observed in the CpG islands of several MiR host genes frequently downregulated in acute lymphoblastic leukemia (Roman-Gomez et al., 2009).

Pharmacological inhibition of DNA methylation can alleviate the repression of tumor suppressor MiRs. For instance, drug-mediated inhibition of DNMT activity by 5-aza-2′-deoxycytidine led to derepression of MiR-124, which is embedded in a CpG island, and subsequently decreased the levels of its direct target *CDK6* in hepatocarcinoma cells (Furuta et al., 2010). Reexpression of MiR-124 also alleviated *CDK6*-mediated inactivation of pRB and led to decreased meduloblastoma proliferation (Pierson et al., 2008).

4.2 Impaired MiR Processing and Maturation

The deregulation of proteins involved in MiR maturation and processing is also an important mechanism impairing MiR functions in cancer cells (Karube et al., 2005; Kim et al., 2010; Melo et al., 2009; Suzuki et al., 2009). Altered *DICER1* expression and stability have been reported human lung cancer cell lines (Chiosea et al., 2007; Karube et al., 2005), ovarian cancer (Pampalakis et al., 2010), colorectal and gastric cancer (Melo et al., 2009), and PC (Chiosea et al., 2006). *AGO2* is also frequently deregulated in tumors (Romero-Cordoba et al., 2014). More importantly, p53 exerts transcriptional control and processing over several MiRs involved in tumor suppression and the DDR pathway (Chowdhury et al., 2013; Feng et al., 2011; Hu and Gatti, 2011). The high frequency p53 mutation or silencing in cancer adds another layer of complexity to the molecular events involved in tumorigenesis due to altered MiR processing.

4.3 Mutations and Genetic Alterations

The MiR seed sequence directs the specificity of interaction with mRNA (Bartel, 2004). Mutations within this sequence or the target mRNA MiR RE leads to impaired target recognition and erroneous MiR-mediated silencing. In human cancer, MiR seed sequence mutations rarely occur whereas SNPs or mutations of MiR RE are frequently observed (Yu et al., 2007). This is actually the case for *let-7*, which normally targets *K-RAS* mRNA (Johnson et al., 2005). A SNP located in the 3′-UTR of *K-RAS* alone significantly increases the risk of developing NSCLC (Chin et al., 2008). In addition, the shortening of 3′-UTR and poly (A) tails of mRNAs is a prevalent mechanism which inhibits MiR target recognition in cancer (Lembo et al., 2012; Mayr and Bartel, 2009). As exemplified in breast cancer cells, estrogen-dependent alternative splicing of *CDC6* results in a shortened 3′-UTR and leads to the loss of MiR regulation (Akman et al., 2012).

5. MiRs AS BIOMARKERS OF AGE-RELATED DISEASES

5.1 Cellular Senescence as a Driver of Organismal Aging

Senescent cells accumulate in organisms during aging (Herbig et al., 2006; Jeyapalan et al., 2007; Wang et al., 2009a). This most probably reflects the cumulative effects of stochastic cellular stresses that increase the proportion of cells destined to undergo the senescence program, and of reduced ability of the immune system to efficiently target and eliminate such cells (Childs et al., 2015; Hoenicke and Zender, 2012). The causative link between cellular senescence and organismal aging was initially revealed about 15 years ago by seminal studies in murine models. Specifically, p53 hyperactivity, which leads to the accumulation of senescent cells in vivo, engendered a premature aging phenotype involving degenerative alterations in various tissues and reduced life span (Maier et al., 2004; Tyner et al., 2002). It was subsequently shown that clearance of senescent cells delays onset of age-related diseases (Baker et al., 2011). Cellular senescence affects tissue homeostasis in different ways. When stem cells or progenitors are induced to senesce, their self-renewal and tissue repair capacities are impacted. Furthermore, there is an important cell nonautonomous effect on tissues via production of paracrine SASP by senescent cells, resulting in chronic but low levels of inflammation. This phenomenon is widely recognized as a marker of aging and a potential driver of age-related diseases, although little is known about unique inflammatory components and pathways that distinguish age-related pathologies from each other (Franceschi and Campisi, 2014). The exact composition of the senescent cell secretome may vary depending on the senescence trigger, and on the tissue and/or the age of the organism; indeed, such a molecular signature of SASP still awaits extensive characterization. Importantly, the specific elimination of senescent cells using so-called senolytic drugs, and the reduction of age related of SASP production are now recognized as two exciting new avenues to treat age-related diseases in humans (Childs et al., 2015).

5.2 Identification of Aging MiR Biomarkers

Hundreds of MiRs that can be used for diagnostic purposes are detectable in many different body fluids such as serum, plasma, urine, and cerebrospinal fluid (CSF) (Weber et al., 2010). The high stability of these small RNAs in blood was quite surprising at first considering the RNAse activity in body fluids (Mitchell et al., 2008). It is now well recognized that this stability is independent of their small size but rather due to their different circulating forms. For example, MiRs can circulate in plasma through exosomes, bound to proteins such as Argonaute 2 or lipoproteins (Arroyo et al., 2011; Rabinowits et al., 2009; Valadi et al., 2007; Vickers et al., 2011).

It is now established that MiRs represent powerful noninvasive biomarkers for most if not all diseases, as collectively documented by more than 13,000 peer-reviewed articles. MiRs are also biomarkers for normal aging and several groups have reported specific patterns in elderly compared to younger individuals. For example, a significant increase for 16 MiRs and decrease for 64 MiRs was observed in the elderly (ElSharawy et al., 2012). Analysis of predicted targets of the downregulated MiRs revealed significantly enriched pathways including the p53 cascade and other pathways implicated in cancer such as the citrate cycle and hedgehog signaling. It is worth mentioning that MiR-17, MiR-20a, MiR-106a, and MiR-21, which were significantly downregulated in the aforementioned analysis, are also known to manifest reduced levels in senescent cells (ElSharawy et al., 2012; Hackl et al., 2010; Marasa et al., 2010). Following qPCR validation, levels of MiR-106a, MiR-126, MiR-20a, MiR-144-3p, and MiR-18a were each shown to be significantly decreased; moreover, this MiR panel could discriminate long-lived individuals (average 95 years old) (ElSharawy et al., 2012). Another study identified MiR-103, MiR-107, MiR-128, MiR-130a, MiR-155, MiR-24, MiR-221, MiR-496, and MiR-1538 as being decreased in peripheral blood mononuclear cells in an age-dependent manner. Using ingenuity pathway analysis, cancer and cardiovascular diseases (CVDs) have been identified as the most common conditions related to these MiRs (Noren Hooten et al., 2010).

Based on the literature, it is now clear that virtually all human diseases exhibit a specific MiR signature (for a comprehensive review, see De Guire et al., 2013). In the next section, we will focus on MiRs which have been identified in the context of age-related diseases as biomarkers and/or potential therapeutic targets.

5.3 SA-MiR in Age-Related Diseases

5.3.1 Neurodegenerative Diseases

Neurodegenerative diseases (ND) are characterized by degeneration of the structure and function of neurons present in the central nervous system. Age represents an aggravating factor in all ND and correlates with progressive loss of self-renewal potential of neuronal stem cells. Furthermore, several types of nonneuronal mitotic cells in the brain can enter a senescent state. Accumulating evidence now suggests that senescent cells in the aging brain induce chronic inflammation in the central nervous system by releasing SASP factors, in turn disrupting metabolic homeostasis thus contributing to the impaired function of the aged brain (Chinta et al., 2013). It is now believed that senescent cells are a prominent feature of neurodegenerative diseases and contribute directly to the neuropathology (Chinta et al., 2015).

Since MiRs can cross the blood brain barrier, they have much potential as noninvasive biomarkers for the diagnosis of neurodegenerative diseases (Alvarez-Erviti et al., 2011). Several studies aiming to find early MiR biomarkers in biological fluids of Alzheimer's disease (AD) patients have been conducted toward developing molecular tools for presymptomatic detection of the disease. MiRs profiling in the CSF of patients with AD revealed 60 differentially expressed MiRs in the AD group among the 242 MiRs tested (Cogswell et al., 2008). More recently, a 2.2-fold increase for MiR-29a in the CSF of AD patients was reported (Muller et al., 2016). However, the fact that an increase of this MiR was also observed when CSF was contaminated with blood cells limits the use of this potential biomarker to cell-free CSF samples (Muller et al., 2016). Using an OpenArray technology, profiling of 1178 MiRs in the CSF of 22 AD patients vs 28 controls revealed that 96% of AD patients were correctly classified by the combination of deregulated MiR-100, MiR-103, and MiR-375, indicating that this could be used as a molecular signature for AD (Denk et al., 2015).

Profiling of MiRs in the blood of AD patients using next-generation sequencing represents a less invasive strategy. Interestingly, Keller et al. combined the results of two cohorts from the United States and Germany. They sequenced MiRs in the blood of 103 AD patients and 77 controls. Investigating the specificity of their MiR panel, they also included 20 patients with mild cognitive impairment and 90 patients with multiple sclerosis. MiRs with the most AD biomarker potential in blood samples were MiR-17-3p and MiR-151a-3p (Keller et al., 2016; Leidinger et al., 2013). The differential expression of both MiR-223 and MiR-125b could discriminate probable sporadic AD patients from healthy patients (Jia and Liu, 2016). MiR-34a/c was also shown to be increased in the blood of AD patients in single studies (Bhatnagar et al., 2014; Schipper et al., 2007). However, those MiRs were not detected previously in another study that searched for plasma biomarkers of AD (Kumar et al., 2013). Remarkably, the MiR-34 family members were also found to be significantly altered in aged brains of several model organisms. For example, MiR-34a expression increases with age in drosophila, mouse, and rat (Li et al., 2011a,b; Liu et al., 2012), and in a mouse model of AD (Wang et al., 2009b). MiR-34b is also a plasma biomarker in the context of Huntington disease (Gaughwin et al., 2011), another neurodegenerative disease.

MiR-146a is another promising AD-related MiR since it is overexpressed in AD patients and promotes the regulation of important inflammatory transducers (Cui et al., 2010). let-7 is also significatively overexpressed in AD (Fig. 3) (Dimmeler and Nicotera, 2013; Lehmann et al., 2012). Independent findings suggest that it might constitute a candidate MiR involved in the progression of ND and a potential therapeutic agent since it is a regulator of NSC function during development (Nishino et al., 2008). Interestingly, *let*-7 was previously identified as an MiR triggering senescence (Tzatsos et al., 2011). Independently, a *C. elegans* study in which the *let*-7 MiR family was discovered initially, shows that this family targets APL-1 mRNA, an amyloid precursor-like protein for which human homologs represent the source of amyloid plaques in AD (Niwa et al., 2008). Additionally, the *let*-7 family member *let*-7b targets HMGA2, a repressor of the INK4A/ARF locus. Therefore, overexpression of *let*-7b, potentially diminishes the self-renewal capacity of neuronal stem cells by indirectly increasing the expression of $p16^{Ink4a}$ and ARF (Nishino et al., 2008). Interestingly, *let*-7g is also upregulated in plasma of AD patients and is relatively abundant in brain (Kumar et al., 2013) (for an extensive review of AD specific MiRs, refer to Batistela et al., 2017).

Age-related macular degeneration (AMD), the leading cause of blindness in industrialized countries, is characterized by progressive loss of the central field of vision (Ambati et al., 2003). AMD is associated with aging, as more than 25% of people over 80 years old suffer from this disease (Friedman et al., 2004; Hyman, 1987). AMD is a late-onset, multifactorial, progressive neurodegenerative disease of the human retina. Degeneration correlates with inflammation in the retina, which is dependent on the level of complement factor H (CFH), a repressor of innate immunity and inflammation, and a

Neurodegenerative	MiR	Potential targets
Alzheimer's disease	↑ *let-7*	*E2F-target genes*
	↑ *MiR-29a*	*B-MYB*
	↑ *MiR-146a*	*IRAK1*
Huntington's disease	↓ *MiR-34b*	*E2F*
Old neural stem cells	↑ *let-7b*	*E2F-target genes, HMGA2*
Macular degeneration	↑ *MiR-9*	*CBX7*
	↑ *MiR-34a*	*MDM2, E2F, MYC, CDK4/6, CCND1/2, CCNE1/2*
	↓ *MiR-106b*	*CDKN1A, E2F1, MCM7, EP300*
	↓ *MiR-125b*	*E2F3*
	↑ *MiR-146a*	*IRAK1, IL6*
	↓ *MiR-146b*	*IRAK1*

Musculoskeletal	MiR	Potential targets
Sarcopenia	↑ *MiR-29*	*B-MYB*
Old skeletal muscle	↑ *let-7b/e*	*HMGA2, PAX7?*
Osteoarthritis	↑ *MiR-22*	*SIRT1, CDK5, PPARA, BMP-7*
	↓ *MiR-140*	*ADAMTS-5*
Osteoporosis	↑ *MiR-34b/c*	*SATB2, CCND1, CDK4/6*

Cardiac and vascular	MiR	Potential targets
Cardiac ageing	↑ *MiR-22*	*OGN*
Chronic heart failure	↑ *MiR-146a*	*IRAK1, IL6*
Old cardiovascular	↑ *MiR-217*	*SIRT1*
Endothelial cells	↓↑ *MiR-146a*	*IRAK1, NOX4*

Fig. 3 Differentially expressed SA-MiRs and their potential targets in senescence-associated/age-related diseases. MiRs are either upregulated (*black arrow*) or downregulated (*white arrow*) in various age-related diseases grouped under neurodegenerative, musculoskeletal, and cardiac and vascular diseases. Presented potential targets of SA-MiRs are either validated or presumed targets in their specific disease model.

pivotal player in both AD and AMD (Alexandrov et al., 2012). Several studies have identified possible circulating MiR biomarkers in AMD patients; however, no consensus has emerged when comparing all results. This may reflect the use of different methodologies (RNAseq, qPCR), and of different cohorts in which numbers of patients, stage, and evolution of the pathology vary considerably (Ertekin et al., 2014; Grassmann et al., 2014; Menard et al., 2016; Szemraj et al., 2015; Zhou et al., 2011). However some circulating MiR candidates were identified in retina, vitreous humor, or cells cultivated from AMD patients (RPE cells), such as MiR-146a, MiR-9, MiR-106b, MiR-34a, and MiR-125b, indicating these as potential therapeutic targets. The presence of both MiR-146a and MiR-106b was precise molecular signatures of the AMD group (for a comprehensive review, see Berber et al., 2017).

Importantly, the MiR with the highest number of studies supporting an involvement in AMD and other NDs (such as AD) is MiR-146a (Alexandrov et al., 2014; Lukiw et al., 2012). MiR-146a is recognized to modulate innate immune response, inflammation, and the microglial activation state. Under control of the NF-κB transactivator, MiR-146a targets the Toll-like receptor signaling molecule IL-1 receptor-associated kinase 1 (*IRAK1*), which in turn promotes the innate immune and inflammatory responses (Iyer et al., 2012; Taganov et al., 2006). Additionally, MiR-146a also directly targets IL-6, an important proinflammatory cytokine (Bhaumik et al., 2009). MiR-146b restrains the proinflammatory response via the same molecular targets as MiR-146a (Kutty et al., 2013). While MiR-146b is downregulated in AMD (Kutty et al., 2013), there is a disease-associated increase in MiR-146a (Menard et al., 2016), suggesting a fine-tuned regulation of inflammation under disease conditions.

Many screening studies taken together have shed light on the real possibility of using a combination of MiRs as early biomarkers in ND. Most MiRs identified in these studies seem to play important roles in inflammation, by either targeting or responding to the NF-κB pathway. MiR-146a exhibited the most significant expression level changes among different ND patients (AD and AMD). The therapeutic potential of MiR-146a as a proinflammatory agent needs to be analyzed more carefully. Indeed this MiR was shown to downregulate IRAK-1 in a NF-κB-dependent manner; however, in parallel NF-κB also activates IRAK-2, a related proinflammatory protein not targeted by MiR-146a (Cui et al., 2010). This suggests that MiR-146 inhibition alone may not be sufficient to provide a beneficial antiinflammatory effect without jeopardizing the innate immune system in the brain.

5.3.2 Musculoskeletal Diseases

Sarcopenia is an age-related loss of skeletal muscle strength, and cellular senescence contributes to this process by promoting chronic inflammation and growth inhibition of muscle stem cell progenitors (Beyer et al., 2012). RNA microarray analysis of muscle tissue in old vs young rodents identified members of the MiR-29 family as significantly overexpressed in old animals. MiR-29 induces pRB-dependent senescence in human fibroblasts through inhibition of *B-Myb* (Martinez et al., 2011) (Fig. 3), a transcription factor important for cell cycle regulation (Joaquin and Watson, 2003). Interestingly, ectopic expression of MiR-29 in muscle progenitor cells also induces cellular senescence in vitro and in vivo (Hu et al., 2014), suggesting that

B-Myb could be an MiR-29 target in these cells. As such, MiR-29 inhibitors may potentially be useful as therapeutic agents in age-dependent sarcopenia.

In human, *let*-7b/e levels are elevated in skeletal muscle of older subjects, and this correlates with a lower expression of several key cell cycle genes, such as, *CDK6*, *CCND1/2*, *CDKN1A*, and *E2F*. Expression *PAX7*, a crucial gene responsible for differentiation of muscle cell progenitors (satellite cells), also negatively correlated with *let*-7b/e. However, whether or not *PAX7* is a direct target of *let*-7b/e remains unknown (Drummond et al., 2011). Skeletal muscle-specific MiR transcriptome profiling in aged vs young mice revealed MiR-181a as a potential key regulator of age-dependent muscle structure. MiR-181 is predicted to regulate several hundreds of genes including genes previously associated with cell cycle control and cellular senescence, such as SIRT1 (Soriano-Arroquia et al., 2016).

Osteoarthritis (OA) is another age-related disease characterized by the progressive destruction of the articular cartilage. Specifically, OA results from the loss cartilage homeostasy and can notably promote senescence in joint cells (Miyaki and Asahara, 2012). However, relatively little is known about the molecular pathways implicated in this disease, despite its high prevalence rate especially among obese patients. Chronic low-grade inflammation linked to metabolic imbalance is thought be an important player in OA, and it is anticipated that MiR-targeting proinflammatory molecules represent a useful therapeutic tool for managing this disease. Following coordinated analyses of MiR and their targeted protein expression in patients, a few MiR–protein pairs representing signatures of OA were identified (Iliopoulos et al., 2008). In particular, the abundance of MiR-22 inversely correlated to the level of PPARA and BMP-7, two proteins connecting inflammation to metabolism by regulating the expression of IL1B. Inhibition of MiR-22 in osteoarthritic chondrocytes decreases inflammation and restores the tissue healing process (Iliopoulos et al., 2008). MiR-140 regulates cartilage-specific gene expression in response to inflammation and is downregulated in OA (Miyaki et al., 2009). However, no direct association between MiR-140 and senescence in cartilage cells has been reported yet. Interestingly, the putative MiR-140 target ADAMTS-5, a MMP (Araldi and Schipani, 2010), was shown to be overexpressed in senescent intervertebral disk cells (Dimozi et al., 2015). Moreover treatment of intervertebral disk cells with metformin, which as been shown to block the SASP (Moiseeva et al., 2013), inhibited the expression of ADAMTS-5, and protected disks from degeneration in rats (Chen et al., 2016). From these data, it is tempting to speculate that the senescence-associated dysregulation of MiR-140 and ADAMTS-5 might contribute to the onset of OA.

MiR-34b/c family was revealed as important modulators of osteoporosis by directly inhibiting the cell cycle-related gene *Ccnd1*, as well as *Satb2*, a key factor of osteoblast differentiation (Wei et al., 2012). The effect of MiR-34b/c was however not directly correlated yet with senescence or inflammation, but rather linked to its regulatory role for key effectors of bone cell proliferation and differentiation including *Cdk4/6* (Wei et al., 2012).

5.3.3 Cardiac and Vascular Aging

CVD, such as atherosclerosis and thrombosis, is a chronic age-related condition partially driven by senescence of endothelial cells (Childs et al., 2016), reduction in regenerative capacity of the endothelium, and dysfunctions of vascular tissue (for review, see Minamino and Komuro, 2007). Like in other age-related diseases, at least two MiR-targeting SIRT1 were found to be upregulated in aged vascular cells: MiR-34a in HUVEC and MiR-217 in heart endothelial cells (Fig. 3) (Ito et al., 2010; Menghini et al., 2009). MiR-22 is also upregulated during cardiac aging, and its overexpression in cardiac fibroblasts induces cellular senescence partially through direct inhibition of OGN (Jazbutyte et al., 2013), a gene associated with heart ventricular structure and function (Petretto et al., 2008). This suggests that MiR-22 may at least in part contribute to age-related cardiac disease (Jazbutyte et al., 2013).

MiR-146a was validated as a marker of senescence-associated proinflammatory processes in vascular cells and is significantly overexpressed in several types of primary vascular cells in vitro as well as in circulating angiogenic cells of patients with chronic heart failure. Furthermore, the use of MiR antagonists (antagomiR) and MiR mimics confirmed the involvement of MiR-146a in the increase expression of β-galactosidase activity and reduced telomere length, two features of senescent cells (Olivieri et al., 2013). MiR-146a overexpression in HUVEC cells induces downregulation of IRAK1 a protein involved in cytokine production (Nahid et al., 2011), as revealed in studies using other cellular models (Taganov et al., 2006). However, discrepant data were also noted where MiR-146a was either unchanged or significantly underexpressed in senescent HUVEC (Hackl et al., 2010; Vasa-Nicotera et al., 2011). It was suggested that reduced expression of

MiR-146a, which may directly target the *NOX4* NADPH oxidase, could increase ROS levels in the vascular walls in turn promoting the induction of senescence (Vasa-Nicotera et al., 2011).

5.4 Narrowing Down the Complex Network of SA-MiRs

Numerous studies focusing on the identification of clinically relevant SA-MiR in different age-related diseases seem to converge on families where the most represented target is *SIRT1*, and include MiR-34a, MiR-217, MiR-181, and MiR-22. Remarkably, MiR-34a is involved in most if not all age-related diseases. Other molecular targets of SA-MiR include effectors of the NF-κB pathway, as indicated by altered expression of MiR-146a, MiR-9, MiR-29, and MiR-125b in diverse age-related diseases. Much more work will be required to functionally characterize these MiRs, and to identify other potential SA-MiR candidates that are deregulated during aging. The best therapeutic MiRs are probably ones exhibiting the highest tissue specificity, in order to limit potential harmful side effects.

It may be tempting to categorize the above SA-MiRs as being either a cause or a consequence of aging. The complexity of MiR regulatory networks is growing with the number of studies discovering novel targets and feedback mechanisms. Indeed, it is not yet clear if SA-MiRs contribute to age-related diseases, or participate in regulatory pathways that ensure the maintenance of cellular senescence (and the SASP) once established. For example, MiR-146a is involved in regulation of proteins required for the cytokine production, and therefore participate in the SASP, a process that is both a consequence of cellular senescence and a requirement for its maintenance in a cell nonautonomous manner. Furthermore, several MiRs are involved in negative feedback loops of senescence regulation, thus contributing to its fine-tune regulation and representing both the cause and consequence of senescence.

6. MiR-BASED THERAPEUTICS

6.1 MiR-Derived Drugs

By targeting a plethora of genes, including more than 30% of protein encoding genes MiRs can broadly influence cellular pathways, resulting in drastic and rapid cellular effects. As our knowledge of cancer advances, together with the discovery of many MiR biomarkers for those diseases, many researchers are now investigating the possibility of using synthetic molecules to either block (oncogenic) or enhance (tumor suppressor) MiRs. MiR mimics are synthetic antisense oligonucleotides used to regulate levels of specific mRNAs (Farooqi et al., 2016; Metias et al., 2009). Inversely, antagomiRs are used to suppress specific MiRs that are overexpressed in a disease (Farooqi et al., 2016; Metias et al., 2009). Burgeoning progress in RNA chemistry and MiR biology have now catapulted MiR-based therapeutics to the realm of reality. Over time, different molecules have been tested as anti-miR, all based on DNA complementarity with the MiR or its target, such as antagomiR oligonucleotides, locked nucleic acids (LNAs), peptide nucleic acid, MiR sponges, and finally MiR masking which competes with MiR at the mRNA target site. Furthermore, several delivery systems are under evaluation, such as nanoparticules, liposomes, viral vectors with local or systemic action. For a comprehensive review, see Rupaimoole and Slack (2017).

6.2 MiR-Based Experimental Trials in Animal Models of Age-Related Disease

Animal models of pathologies associated to cellular senescence represent important tools to pinpoint molecular pathways involved in the progression of age-related diseases, and to test the possibility of using MiR-associated strategies as therapeutic approaches. Several studies, mainly using mice models, have already suggested such potential for MiRs in diverse pathologies.

Brain-specific downregulation of MiR-29 using an LMA antagomiR resulted in apoptosis, a phenotype that correlates with the increase of Vdac1, a mediator of apoptosis and MiR-29 target (Bargaje et al., 2012; Roshan et al., 2014). MiR-29 cluster KO mice show dysfunctional neuronal function, affecting locomotion (Papadopoulou et al., 2015). Further studies will be required to test the effect of MiR-29 overexpression on the progression of AD, and to establish all relevant molecular targets of MiR in the context of AD.

MiR-34a is another key SA-MiR that may play a role in ND. However the precise contribution of MiR-34a to NSC maintenance and differentiation is still unclear as its overexpression led to discrepant conclusions (Aranha et al., 2011; Fineberg et al., 2012). A more recent study confirmed a prominent role for MiR-34a in AD, showing that this MiR modulates levels of NF-κB, STAT1, c-FOS, CREB, and p53, and furthermore mediates a coordinated response by altering the expression of several targets involved in synaptic plasticity and energetic metabolism in the brain of AD patients

(Sarkar et al., 2016). Interestingly, inhibition of MiR-34c restores SIRT1 levels and rescues memory impairment in an AD transgenic mouse model (Zovoilis et al., 2011).

Oxygen-induced retinopathy (OIR) is a well-established protocol for studying hyperoxia-induced retinal neovascularization and inflammation in mice, two well-known characteristics of AMD (Connor et al., 2009). Intraocular injection of MiR mimics or inhibitors led to the identification of several MiRs modulating pathological neorevascularization in the OIR model, probably through their ability to regulate angiogenesis or apoptosis (Liu et al., 2015a; Shen et al., 2008; Yan et al., 2015).

Therapeutic approaches were also attempted in mice with age-related musculoskeletal diseases. In models of antigen-induced osteoarthritis, animals overexpressing MiR-140 were protected from cartilage degradation. Conversely, MiR-140 KO mice showed premature OA-like alterations of cartilage (Miyaki et al., 2010) and other skeletal defects (Nakamura et al., 2012). An osteoporotic condition is mimicked in mice or rats by surgical ablation of the ovary (OVX model), which recapitulates postmenopausal-induced loss of bone density due to increased bone resorption (elevated proliferation of osteoclasts) and decreased bone formation (reduced proliferation and differentiation of osteoblasts) in humans. Osteoblast-specific overexpression of MiR-34c in mice promotes development of age-dependent osteoporosis, due to defective proliferation of osteoblasts (Bae et al., 2012). Similarly, transgenic mice with osteoclast-specific overexpression of MiR-34a show higher bone mass. MiR-34a knockout mice exhibit decrease bone mass, correlated with elevated bone resorption. Interestingly, nanoparticule-mediated targeted delivery of MiR-34a attenuates osteoporosis in OVX models (Krzeszinski et al., 2014). Of note, in this same study, MiR-34a overexpression in osteoclasts also attenuated bone metastasis in two cancer models.

6.3 Clinical Assays Based on MiRs

Because of the promise of MiR-based therapeutics, many pharmaceutical and biotech companies have launched MiR projects. Indeed, several clinical studies are already underway using MiRs as targets. As an example, the MiR mimic MRX34 was recently tested by MiR Therapeutics (Adams et al., 2016a; Bouchie, 2013). MRX34 was tested in a phase I clinical trial, displaying partial responses in patients with renal cell carcinoma, acral melanoma, and HCC (Adams et al., 2016a; Bouchie, 2013). Considering that

MRX34 is a mimic of the SA-MiR-34, its use in the context of age-related disease, as described in the previous section, could be of potential interest.

Santaris Pharma has developed Miravirsen (or SPC3649), an antagomiR drug candidate. Miravirsen is a LNA that contains phosphorothioate modifications and specifically targets MiR-122 which is essential in the life cycle of hepatitis C virus in the liver (Gebert et al., 2014; Lindow and Kauppinen, 2012). In phase 1 clinical trials, some patients who received high doses of miravirsen in monotherapy attained undetectable HCV RNA levels (Janssen et al., 2011; Lindow and Kauppinen, 2012). Miravirsen is currently undergoing a phase 2 trial. Another compound, RG-101 produced by Regulus Therapeutics, is also currently in phase 2 clinical trials. RG-101 is an antagomiR that also targets MiR-122 in HCV-infected hepatocytes in order to reduce levels of viral RNA (Baek et al., 2014). RG-101, like miravirsen, shows considerable efficacy, with some patients displaying undetectable HCV RNA levels (Baek et al., 2014).

The development of promising therapeutics, e.g., from MiR Therapeutics (MRX34) and Regulus Therapeutics (RG-101), demonstrates that RNA biologics are closer than ever to market. Companies are vigorously engaged in initiating clinical trials for other MiR biologics, such as MiRagen Therapeutics with MiR-29b (antifibrosis) and MiR-155 (lymphoma) and Regulus/Ionis/Genzyme with MiR-21 (Alport Syndrome). It may be only a matter of time before new clinical trials emerge involving exploitation of senescence-associated/aging-related MiRs.

7. CONCLUDING REMARKS

Cellular senescence forestalls tumorigenesis in vivo. On the other hand, it may ultimately cause decreased lifespan through impairment of cell stemness, weakening of regenerative processes, and promotion of neurodegeneration. The balance between cancer and aging is far from fully understood. Clearance of senescent cells by the immune system relies on the SASP and appears beneficial for tumor regression. However, impaired removal of cytokine-expressing senescent cells in aged organisms drives inflammation and seems detrimental for tissue homeostasis (van Deursen, 2014).

MiRs regulate cellular senescence and the SASP to a considerable degree. As such, they are at the crossroads between tumor suppression and aging. Understanding how clinically relevant MiRs modulate their targets and are themselves modulated throughout the development of age-related/senescence-associated diseases is crucial toward developing new therapeutic tools. In addition, further characterization of MiR expression signatures as prognostic and diagnostic biomarkers in cancer and aging is an extremely worthy goal.

Finally, burgeoning evidence demonstrates that MiRs act as initiators, mediators, and effectors of cellular senescence. Given the high incidence of inactivation of genes controlling senescence signaling (p53, $p16^{INK4A}$/ARF, pRB) in cancer cells, and the frequent occurrence of epigenetic silencing of tumor suppressor MiRs in malignant cells, reactivation of the cellular senescence response in tumors using MiR mimics represents a promising therapeutic avenue. Owing to various delivery methods of available, several MiRs are already under scrutiny for use in promising cancer trials (Christopher et al., 2016).

ACKNOWLEDGMENTS

We thank members of the Mallette laboratory and Dr. Elliot Drobetsky for fruitful discussions and critical reading of the manuscript. Cellular senescence-related work in the Mallette laboratory is supported by the "Fondation de l'Hôpital Maisonneuve-Rosemont" and the Canadian Institutes of Health Research (MOP-133442). F.A.M. is a Junior 1 Research Scholar of the Fonds de Recherche du Québec—Santé (FRQ-S). F.A.M. holds a Canada Research Chair in Epigenetics of Aging and Cancer.

REFERENCES

Abdelmohsen, K., Pullmann Jr., R., Lal, A., Kim, H.H., Galban, S., Yang, X., Blethrow, J.D., Walker, M., Shubert, J., Gillespie, D.A., Furneaux, H., Gorospe, M., 2007. Phosphorylation of HuR by Chk2 regulates SIRT1 expression. Mol. Cell 25, 543–557.

Abdelmohsen, K., Srikantan, S., Kuwano, Y., Gorospe, M., 2008. miR-519 reduces cell proliferation by lowering RNA-binding protein HuR levels. Proc. Natl. Acad. Sci. U.S.A. 105, 20297–20302.

Acosta, J.C., O'loghlen, A., Banito, A., Guijarro, M.V., Augert, A., Raguz, S., Fumagalli, M., Da Costa, M., Brown, C., Popov, N., Takatsu, Y., Melamed, J., D'adda Di Fagagna, F., Bernard, D., Hernando, E., Gil, J., 2008. Chemokine signaling via the CXCR2 receptor reinforces senescence. Cell 133, 1006–1018.

Acosta, J.C., Banito, A., Wuestefeld, T., Georgilis, A., Janich, P., Morton, J.P., Athineos, D., Kang, T.W., Lasitschka, F., Andrulis, M., Pascual, G., Morris, K.J., Khan, S., Jin, H., Dharmalingam, G., Snijders, A.P., Carroll, T., Capper, D., Pritchard, C., Inman, G.J., Longerich, T., Sansom, O.J., Benitah, S.A., Zender, L., Gil, J., 2013. A complex secretory program orchestrated by the inflammasome controls paracrine senescence. Nat. Cell Biol. 15, 978–990.

Adams, B.D., Parsons, C., Slack, F.J., 2016a. The tumor-suppressive and potential therapeutic functions of miR-34a in epithelial carcinomas. Expert Opin. Ther. Targets 20, 737–753.

Adams, B.D., Wali, V.B., Cheng, C.J., Inukai, S., Booth, C.J., Agarwal, S., Rimm, D.L., Gyorffy, B., Santarpia, L., Pusztai, L., Saltzman, W.M., Slack, F.J., 2016b. miR-34a silences c-SRC to attenuate tumor growth in triple-negative breast cancer. Cancer Res. 76, 927–939.

Akman, B.H., Can, T., Erson-Bensan, A.E., 2012. Estrogen-induced upregulation and 3'-UTR shortening of CDC6. Nucleic Acids Res. 40, 10679–10688.
Alexander, K., Yang, H.S., Hinds, P.W., 2004. Cellular senescence requires CDK5 repression of Rac1 activity. Mol. Cell. Biol. 24, 2808–2819.
Alexandrov, P.N., Dua, P., Hill, J.M., Bhattacharjee, S., Zhao, Y., Lukiw, W.J., 2012. microRNA (miRNA) speciation in Alzheimer's disease (AD) cerebrospinal fluid (CSF) and extracellular fluid (ECF). Int. J. Biochem. Mol. Biol. 3, 365–373.
Alexandrov, P.N., Dua, P., Lukiw, W.J., 2014. Up-regulation of miRNA-146a in progressive, age-related inflammatory neurodegenerative disorders of the human CNS. Front. Neurol. 5, 181.
Alvarez-Erviti, L., Seow, Y., Yin, H., Betts, C., Lakhal, S., Wood, M.J., 2011. Delivery of siRNA to the mouse brain by systemic injection of targeted exosomes. Nat. Biotechnol. 29, 341–345.
Ambati, J., Ambati, B.K., Yoo, S.H., Ianchulev, S., Adamis, A.P., 2003. Age-related macular degeneration: etiology, pathogenesis, and therapeutic strategies. Surv. Ophthalmol. 48, 257–293.
Ambs, S., Prueitt, R.L., Yi, M., Hudson, R.S., Howe, T.M., Petrocca, F., Wallace, T.A., Liu, C.G., Volinia, S., Calin, G.A., Yfantis, H.G., Stephens, R.M., Croce, C.M., 2008. Genomic profiling of microRNA and messenger RNA reveals deregulated microRNA expression in prostate cancer. Cancer Res. 68, 6162–6170.
Araldi, E., Schipani, E., 2010. MicroRNA-140 and the silencing of osteoarthritis. Genes Dev. 24, 1075–1080.
Aranha, M.M., Santos, D.M., Sola, S., Steer, C.J., Rodrigues, C.M., 2011. miR-34a regulates mouse neural stem cell differentiation. PLoS One 6, e21396.
Arroyo, J.D., Chevillet, J.R., Kroh, E.M., Ruf, I.K., Pritchard, C.C., Gibson, D.F., Mitchell, P.S., Bennett, C.F., Pogosova-Agadjanyan, E.L., Stirewalt, D.L., Tait, J.F., Tewari, M., 2011. Argonaute2 complexes carry a population of circulating microRNAs independent of vesicles in human plasma. Proc. Natl. Acad. Sci. U.S.A. 108, 5003–5008.
Astle, M.V., Hannan, K.M., Ng, P.Y., Lee, R.S., George, A.J., Hsu, A.K., Haupt, Y., Hannan, R.D., Pearson, R.B., 2012. AKT induces senescence in human cells via mTORC1 and p53 in the absence of DNA damage: implications for targeting mTOR during malignancy. Oncogene 31, 1949–1962.
Aylon, Y., Michael, D., Shmueli, A., Yabuta, N., Nojima, H., Oren, M., 2006. A positive feedback loop between the p53 and Lats2 tumor suppressors prevents tetraploidization. Genes Dev. 20, 2687–2700.
Bae, Y., Yang, T., Zeng, H.C., Campeau, P.M., Chen, Y., Bertin, T., Dawson, B.C., Munivez, E., Tao, J., Lee, B.H., 2012. miRNA-34c regulates Notch signaling during bone development. Hum. Mol. Genet. 21, 2991–3000.
Baek, J., Kang, S., Min, H., 2014. MicroRNA-targeting therapeutics for hepatitis C. Arch. Pharm. Res. 37, 299–305.
Bagchi, A., Papazoglu, C., Wu, Y., Capurso, D., Brodt, M., Francis, D., Bredel, M., Vogel, H., Mills, A.A., 2007. CHD5 is a tumor suppressor at human 1p36. Cell 128, 459–475.
Baker, D.J., Wijshake, T., Tchkonia, T., Lebrasseur, N.K., Childs, B.G., Van De Sluis, B., Kirkland, J.L., Van Deursen, J.M., 2011. Clearance of p16Ink4a-positive senescent cells delays ageing-associated disorders. Nature 479, 232–236.
Baker, D.J., Childs, B.G., Durik, M., Wijers, M.E., Sieben, C.J., Zhong, J., Saltness, R.A., Jeganathan, K.B., Verzosa, G.C., Pezeshki, A., Khazaie, K., Miller, J.D., Van Deursen, J.M., 2016. Naturally occurring p16(Ink4a)-positive cells shorten healthy lifespan. Nature 530, 184–189.

Bao, J., Li, D., Wang, L., Wu, J., Hu, Y., Wang, Z., Chen, Y., Cao, X., Jiang, C., Yan, W., Xu, C., 2012. MicroRNA-449 and microRNA-34b/c function redundantly in murine testes by targeting E2F transcription factor-retinoblastoma protein (E2F-pRb) pathway. J. Biol. Chem. 287, 21686–21698.

Bargaje, R., Gupta, S., Sarkeshik, A., Park, R., Xu, T., Sarkar, M., Halimani, M., Roy, S.S., Yates, J., Pillai, B., 2012. Identification of novel targets for miR-29a using miRNA proteomics. PLoS One 7, e43243.

Barski, A., Cuddapah, S., Cui, K., Roh, T.Y., Schones, D.E., Wang, Z., Wei, G., Chepelev, I., Zhao, K., 2007. High-resolution profiling of histone methylations in the human genome. Cell 129, 823–837.

Bartel, D.P., 2004. MicroRNAs: genomics, biogenesis, mechanism, and function. Cell 116, 281–297.

Bartel, D.P., 2009. MicroRNAs: target recognition and regulatory functions. Cell 136, 215–233.

Bartkova, J., Rezaei, N., Liontos, M., Karakaidos, P., Kletsas, D., Issaeva, N., Vassiliou, L.V.F., Kolettas, E., Niforou, K., Zoumpourlis, V.C., Takaoka, M., Nakagawa, H., Tort, F., Fugger, K., Johansson, F., Sehested, M., Andersen, C.L., Dyrskjot, L., Orntoft, T., Lukas, J., Kittas, C., Helleday, T., Halazonetis, T.D., Bartek, J., Gorgoulis, V.G., 2006. Oncogene-induced senescence is part of the tumorigenesis barrier imposed by DNA damage checkpoints. Nature 444, 633–637.

Bates, S., Phillips, A.C., Clark, P.A., Stott, F., Peters, G., Ludwig, R.L., Vousden, K.H., 1998. p14ARF links the tumour suppressors RB and p53. Nature 395, 124–125.

Batistela, M.S., Josviak, N.D., Sulzbach, C.D., De Souza, R.L., 2017. An overview of circulating cell-free microRNAs as putative biomarkers in Alzheimer's and Parkinson's diseases. Int. J. Neurosci. 127, 547–558.

Baylin, S.B., Ohm, J.E., 2006. Epigenetic gene silencing in cancer—a mechanism for early oncogenic pathway addiction? Nat. Rev. Cancer 6, 107–116.

Beausejour, C.M., Krtolica, A., Galimi, F., Narita, M., Lowe, S.W., Yaswen, P., Campisi, J., 2003. Reversal of human cellular senescence: roles of the p53 and p16 pathways. EMBO J. 22, 4212–4222.

Beckerman, R., Prives, C., 2010. Transcriptional regulation by p53. Cold Spring Harb. Perspect. Biol. 2, a000935.

Bemis, L.T., Chen, R., Amato, C.M., Classen, E.H., Robinson, S.E., Coffey, D.G., Erickson, P.F., Shellman, Y.G., Robinson, W.A., 2008. MicroRNA-137 targets microphthalmia-associated transcription factor in melanoma cell lines. Cancer Res. 68, 1362–1368.

Benevolenskaya, E.V., 2007. Histone H3K4 demethylases are essential in development and differentiation. Biochem. Cell Biol. 85, 435–443.

Benhamed, M., Herbig, U., Ye, T., Dejean, A., Bischof, O., 2012. Senescence is an endogenous trigger for microRNA-directed transcriptional gene silencing in human cells. Nat. Cell Biol. 14, 266–275.

Bentwich, I., Avniel, A., Karov, Y., Aharonov, R., Gilad, S., Barad, O., Barzilai, A., Einat, P., Einav, U., Meiri, E., Sharon, E., Spector, Y., Bentwich, Z., 2005. Identification of hundreds of conserved and nonconserved human microRNAs. Nat. Genet. 37, 766–770.

Berber, P., Grassmann, F., Kiel, C., Weber, B.H., 2017. An eye on age-related macular degeneration: the role of MicroRNAs in disease pathology. Mol. Diagn. Ther. 21, 31–43.

Beyer, I., Njemini, R., Bautmans, I., Demanet, C., Bergmann, P., Mets, T., 2012. Inflammation-related muscle weakness and fatigue in geriatric patients. Exp. Gerontol. 47, 52–59.

Bhatnagar, S., Chertkow, H., Schipper, H.M., Yuan, Z., Shetty, V., Jenkins, S., Jones, T., Wang, E., 2014. Increased microRNA-34c abundance in Alzheimer's disease circulating blood plasma. Front. Mol. Neurosci. 7, 2.

Bhattacharyya, S.N., Habermacher, R., Martine, U., Closs, E.I., Filipowicz, W., 2006. Relief of microRNA-mediated translational repression in human cells subjected to stress. Cell 125, 1111–1124.

Bhaumik, D., Scott, G.K., Schokrpur, S., Patil, C.K., Orjalo, A.V., Rodier, F., Lithgow, G.J., Campisi, J., 2009. MicroRNAs miR-146a/b negatively modulate the senescence-associated inflammatory mediators IL-6 and IL-8. Aging (Albany NY) 1, 402–411.

Blow, J.J., Hodgson, B., 2002. Replication licensing—defining the proliferative state? Trends Cell Biol. 12, 72–78.

Bohnsack, M.T., Czaplinski, K., Gorlich, D., 2004. Exportin 5 is a RanGTP-dependent dsRNA-binding protein that mediates nuclear export of pre-miRNAs. RNA 10, 185–191.

Bommer, G.T., Gerin, I., Feng, Y., Kaczorowski, A.J., Kuick, R., Love, R.E., Zhai, Y., Giordano, T.J., Qin, Z.S., Moore, B.B., Macdougald, O.A., Cho, K.R., Fearon, E.R., 2007. p53-mediated activation of miRNA34 candidate tumor-suppressor genes. Curr. Biol. 17, 1298–1307.

Bonci, D., Coppola, V., Musumeci, M., Addario, A., Giuffrida, R., Memeo, L., D'urso, L., Pagliuca, A., Biffoni, M., Labbaye, C., Bartucci, M., Muto, G., Peschle, C., De Maria, R., 2008. The miR-15a-miR-16-1 cluster controls prostate cancer by targeting multiple oncogenic activities. Nat. Med. 14, 1271–1277.

Bond, J.A., Wyllie, F.S., Wynfordthomas, D., 1994. Escape from senescence in human-diploid fibroblasts induced directly by mutant p53. Oncogene 9, 1885–1889.

Borgdorff, V., Lleonart, M.E., Bishop, C.L., Fessart, D., Bergin, A.H., Overhoff, M.G., Beach, D.H., 2010. Multiple microRNAs rescue from Ras-induced senescence by inhibiting p21(Waf1/Cip1). Oncogene 29, 2262–2271.

Bouchie, A., 2013. First microRNA mimic enters clinic. Nat. Biotechnol. 31, 577.

Bracken, A.P., Kleine-Kohlbrecher, D., Dietrich, N., Pasini, D., Gargiulo, G., Beekman, C., Theilgaard-Monch, K., Minucci, S., Porse, B.T., Marine, J.C., Hansen, K.H., Helin, K., 2007. The polycomb group proteins bind throughout the INK4A-ARF locus and are disassociated in senescent cells. Genes Dev. 21, 525–530.

Braig, M., Lee, S., Loddenkemper, C., Rudolph, C., Peters, A.H., Schlegelberger, B., Stein, H., Dorken, B., Jenuwein, T., Schmitt, C.A., 2005. Oncogene-induced senescence as an initial barrier in lymphoma development. Nature 436, 660–665.

Brengues, M., Teixeira, D., Parker, R., 2005. Movement of eukaryotic mRNAs between polysomes and cytoplasmic processing bodies. Science 310, 486–489.

Brown, J.P., Wei, W.Y., Sedivy, J.M., 1997. Bypass of senescence after disruption of p21 (CIP1/WAF1) gene in normal diploid human fibroblasts. Science 277, 831–834.

Bueno, M.J., Gomez De Cedron, M., Laresgoiti, U., Fernandez-Piqueras, J., Zubiaga, A.M., Malumbres, M., 2010. Multiple E2F-induced microRNAs prevent replicative stress in response to mitogenic signaling. Mol. Cell. Biol. 30, 2983–2995.

Bulavin, D.V., Kovalsky, O., Hollander, M.C., Fornace, A.J., 2003. Loss of oncogenic H-ras-induced cell cycle arrest and p38 mitogen-activated protein kinase activation by disruption of gadd45a. Mol. Cell. Biol. 23, 3859–3871.

Burkhart, D.L., Sage, J., 2008. Cellular mechanisms of tumour suppression by the retinoblastoma gene. Nat. Rev. Cancer 8, 671–682.

Burma, S., Chen, B.P., Murphy, M., Kurimasa, A., Chen, D.J., 2001. ATM phosphorylates histone H2AX in response to DNA double-strand breaks. J. Biol. Chem. 276, 42462–42467.

Carbonneau, M., M. Gagné, L., Lalonde, M.E., Germain, M.A., Motorina, A., Guiot, M.C., Secco, B., Vincent, E.E., Tumber, A., Hulea, L., Bergeman, J., Oppermann, U., Jones, R.G., Laplante, M., Topisirovic, I., Petrecca, K., Huot, M.E., Mallette, F.A., 2016. The oncometabolite 2-hydroxyglutarate activates the mTOR signalling pathway. Nat. Commun. 7, 12700.

Carnero, A., Hudson, J.D., Hannon, G.J., Beach, D.H., 2000. Loss-of-function genetics in mammalian cells: the p53 tumor suppressor model. Nucleic Acids Res. 28, 2234–2241.

Carrier, F., Smith, M.L., Bae, I., Kilpatrick, K.E., Lansing, T.J., Chen, C.Y., Engelstein, M., Friend, S.H., Henner, W.D., Gilmer, T.M., et al., 1994. Characterization of human Gadd45, a p53-regulated protein. J. Biol. Chem. 269, 32672–32677.

Chan, E.Y., Kir, S., Tooze, S.A., 2007. siRNA screening of the kinome identifies ULK1 as a multidomain modulator of autophagy. J. Biol. Chem. 282, 25464–25474.

Chang, T.C., Wentzel, E.A., Kent, O.A., Ramachandran, K., Mullendore, M., Lee, K.H., Feldmann, G., Yamakuchi, M., Ferlito, M., Lowenstein, C.J., Arking, D.E., Beer, M.A., Maitra, A., Mendell, J.T., 2007. Transactivation of miR-34a by p53 broadly influences gene expression and promotes apoptosis. Mol. Cell 26, 745–752.

Chang, Y.C., Jan, C.I., Peng, C.Y., Lai, Y.C., Hu, F.W., Yu, C.C., 2015. Activation of microRNA-494-targeting Bmi1 and ADAM10 by silibinin ablates cancer stemness and predicts favourable prognostic value in head and neck squamous cell carcinomas. Oncotarget 6, 24002–24016.

Chellappan, S.P., Hiebert, S., Mudryj, M., Horowitz, J.M., Nevins, J.R., 1991. The E2F transcription factor is a cellular target for the RB protein. Cell 65, 1053–1061.

Chen, L., Gilkes, D.M., Pan, Y., Lane, W.S., Chen, J., 2005a. ATM and Chk2-dependent phosphorylation of MDMX contribute to p53 activation after DNA damage. EMBO J. 24, 3411–3422.

Chen, Z., Trotman, L.C., Shaffer, D., Lin, H.K., Dotan, Z.A., Niki, M., Koutcher, J.A., Scher, H.I., Ludwig, T., Gerald, W., Cordon-Cardo, C., Pandolfi, P.P., 2005b. Crucial role of p53-dependent cellular senescence in suppression of Pten-deficient tumorigenesis. Nature 436, 725–730.

Chen, L., Wang, X., Wang, H., Li, Y., Yan, W., Han, L., Zhang, K., Zhang, J., Wang, Y., Feng, Y., Pu, P., Jiang, T., Kang, C., Jiang, C., 2012. miR-137 is frequently down-regulated in glioblastoma and is a negative regulator of Cox-2. Eur. J. Cancer 48, 3104–3111.

Chen, D., Xia, D., Pan, Z., Xu, D., Zhou, Y., Wu, Y., Cai, N., Tang, Q., Wang, C., Yan, M., Zhang, J.J., Zhou, K., Wang, Q., Feng, Y., Wang, X., Xu, H., Zhang, X., Tian, N., 2016. Metformin protects against apoptosis and senescence in nucleus pulposus cells and ameliorates disc degeneration in vivo. Cell Death Dis. 7, e2441.

Cheng, A.M., Byrom, M.W., Shelton, J., Ford, L.P., 2005. Antisense inhibition of human miRNAs and indications for an involvement of miRNA in cell growth and apoptosis. Nucleic Acids Res. 33, 1290–1297.

Childs, B.G., Durik, M., Baker, D.J., Van Deursen, J.M., 2015. Cellular senescence in aging and age-related disease: from mechanisms to therapy. Nat. Med. 21, 1424–1435.

Childs, B.G., Baker, D.J., Wijshake, T., Conover, C.A., Campisi, J., Van Deursen, J.M., 2016. Senescent intimal foam cells are deleterious at all stages of atherosclerosis. Science 354, 472–477.

Chin, L.J., Ratner, E., Leng, S.G., Zhai, R.H., Nallur, S., Babar, I., Muller, R.U., Straka, E., Su, L., Burki, E.A., Crowell, R.E., Patel, R., Kulkarni, T., Homer, R., Zelterman, D., Kidd, K.K., Zhu, Y., Christiani, D.C., Belinsky, S.A., Slack, F.J., Weidhaas, J.B., 2008. A SNP in a let-7 microRNA complementary site in the KRAS 3' untranslated region increases non-small cell lung cancer risk. Cancer Res. 68, 8535–8540.

Chinta, S.J., Lieu, C.A., Demaria, M., Laberge, R.M., Campisi, J., Andersen, J.K., 2013. Environmental stress, ageing and glial cell senescence: a novel mechanistic link to Parkinson's disease? J. Intern. Med. 273, 429–436.
Chinta, S.J., Woods, G., Rane, A., Demaria, M., Campisi, J., Andersen, J.K., 2015. Cellular senescence and the aging brain. Exp. Gerontol. 68, 3–7.
Chiosea, S., Jelezcova, E., Chandran, U., Acquafondata, M., Mchale, T., Sobol, R.W., Dhir, R., 2006. Up-regulation of dicer, a component of the MicroRNA machinery, in prostate adenocarcinoma. Am. J. Pathol. 169, 1812–1820.
Chiosea, S., Jelezcova, E., Chandran, U., Luo, J., Mantha, G., Sobol, R.W., Dacic, S., 2007. Overexpression of Dicer in precursor lesions of lung adenocarcinoma. Cancer Res. 67, 2345–2350.
Choi, Y.J., Lin, C.P., Ho, J.J., He, X., Okada, N., Bu, P., Zhong, Y., Kim, S.Y., Bennett, M.J., Chen, C., Ozturk, A., Hicks, G.G., Hannon, G.J., He, L., 2011. miR-34 miRNAs provide a barrier for somatic cell reprogramming. Nat. Cell Biol. 13, 1353–1360.
Chowdhury, D., Choi, Y.E., Brault, M.E., 2013. DNA damage. Opinion: charity begins at home: non-coding RNA functions in DNA repair. Nat. Rev. Mol. Cell Biol. 14, 181–189.
Christoffersen, N.R., Shalgi, R., Frankel, L.B., Leucci, E., Lees, M., Klausen, M., Pilpel, Y., Nielsen, F.C., Oren, M., Lund, A.H., 2010. p53-independent upregulation of miR-34a during oncogene-induced senescence represses MYC. Cell Death Differ. 17, 236–245.
Christopher, A.F., Kaur, R.P., Kaur, G., Kaur, A., Gupta, V., Bansal, P., 2016. MicroRNA therapeutics: discovering novel targets and developing specific therapy. Perspect. Clin. Res. 7, 68–74.
Chuang, J.C., Jones, P.A., 2007. Epigenetics and microRNAs. Pediatr. Res. 61, 24R–29R.
Cirera-Salinas, D., Pauta, M., Allen, R.M., Salerno, A.G., Ramirez, C.M., Chamorro-Jorganes, A., Wanschel, A.C., Lasuncion, M.A., Morales-Ruiz, M., Suarez, Y., Baldan, A., Esplugues, E., Fernandez-Hernando, C., 2012. Mir-33 regulates cell proliferation and cell cycle progression. Cell Cycle 11, 922–933.
Cogswell, J.P., Ward, J., Taylor, I.A., Waters, M., Shi, Y., Cannon, B., Kelnar, K., Kemppainen, J., Brown, D., Chen, C., Prinjha, R.K., Richardson, J.C., Saunders, A.M., Roses, A.D., Richards, C.A., 2008. Identification of miRNA changes in Alzheimer's disease brain and CSF yields putative biomarkers and insights into disease pathways. J. Alzheimers Dis. 14, 27–41.
Collado, M., Serrano, M., 2006. The power and the promise of oncogene-induced senescence markers. Nat. Rev. Cancer 6, 472–476.
Collado, M., Serrano, M., 2010. Senescence in tumours: evidence from mice and humans. Nat. Rev. Cancer 10, 51–57.
Collado, M., Gil, J., Efeyan, A., Guerra, C., Schuhmacher, A.J., Barradas, M., Benguria, A., Zaballos, A., Flores, J.M., Barbacid, M., Beach, D., Serrano, M., 2005. Tumour biology: senescence in premalignant tumours. Nature 436, 642.
Concepcion, C.P., Han, Y.C., Mu, P., Bonetti, C., Yao, E., D'andrea, A., Vidigal, J.A., Maughan, W.P., Ogrodowski, P., Ventura, A., 2012. Intact p53-dependent responses in miR-34-deficient mice. PLoS Genet. 8, e1002797.
Cong, Y.S., Wright, W.E., Shay, J.W., 2002. Human telomerase and its regulation. Microbiol. Mol. Biol. Rev. 66, 407–425. table of contents.
Connor, K.M., Krah, N.M., Dennison, R.J., Aderman, C.M., Chen, J., Guerin, K.I., Sapieha, P., Stahl, A., Willett, K.L., Smith, L.E., 2009. Quantification of oxygen-induced retinopathy in the mouse: a model of vessel loss, vessel regrowth and pathological angiogenesis. Nat. Protoc. 4, 1565–1573.
Coppe, J.P., Desprez, P.Y., Krtolica, A., Campisi, J., 2010. The senescence-associated secretory phenotype: the dark side of tumor suppression. Annu. Rev. Pathol. 5, 99–118.

Corney, D.C., Hwang, C.I., Matoso, A., Vogt, M., Flesken-Nikitin, A., Godwin, A.K., Kamat, A.A., Sood, A.K., Ellenson, L.H., Hermeking, H., Nikitin, A.Y., 2010. Frequent downregulation of miR-34 family in human ovarian cancers. Clin. Cancer Res. 16, 1119–1128.

Croce, C.M., 2009. Causes and consequences of microRNA dysregulation in cancer. Nat. Rev. Genet. 10, 704–714.

Cruz, F.M., Tome, M., Bernal, J.A., Bernad, A., 2015. miR-300 mediates Bmi1 function and regulates differentiation in primitive cardiac progenitors. Cell Death Dis. 6, e1953.

Cui, J.G., Li, Y.Y., Zhao, Y., Bhattacharjee, S., Lukiw, W.J., 2010. Differential regulation of interleukin-1 receptor-associated kinase-1 (IRAK-1) and IRAK-2 by microRNA-146a and NF-kappaB in stressed human astroglial cells and in Alzheimer disease. J. Biol. Chem. 285, 38951–38960.

De Guire, V., Robitaille, R., Tetreault, N., Guerin, R., Menard, C., Bambace, N., Sapieha, P., 2013. Circulating miRNAs as sensitive and specific biomarkers for the diagnosis and monitoring of human diseases: promises and challenges. Clin. Biochem. 46, 846–860.

De Stanchina, E., Querido, E., Narita, M., Davuluri, R.V., Pandolfi, P.P., Ferbeyre, G., Lowe, S.W., 2004. PML is a direct p53 target that modulates p53 effector functions. Mol. Cell 13, 523–535.

Deer, E.L., Gonzalez-Hernandez, J., Coursen, J.D., Shea, J.E., Ngatia, J., Scaife, C.L., Firpo, M.A., Mulvihill, S.J., 2010. Phenotype and genotype of pancreatic cancer cell lines. Pancreas 39, 425–435.

Degregori, J., Leone, G., Ohtani, K., Miron, A., Nevins, J.R., 1995. E2F-1 accumulation bypasses a G1 arrest resulting from the inhibition of G1 cyclin-dependent kinase activity. Genes Dev. 9, 2873–2887.

Demaria, M., Ohtani, N., Youssef, S.A., Rodier, F., Toussaint, W., Mitchell, J.R., Laberge, R.M., Vijg, J., Van Steeg, H., Dolle, M.E., Hoeijmakers, J.H., De Bruin, A., Hara, E., Campisi, J., 2014. An essential role for senescent cells in optimal wound healing through secretion of PDGF-AA. Dev. Cell 31, 722–733.

Denis, H., Ndlovu, M.N., Fuks, F., 2011. Regulation of mammalian DNA methyltransferases: a route to new mechanisms. EMBO Rep. 12, 647–656.

Denk, J., Boelmans, K., Siegismund, C., Lassner, D., Arlt, S., Jahn, H., 2015. MicroRNA profiling of CSF reveals potential biomarkers to detect Alzheimer's disease. PLoS One 10, e0126423.

Denoyelle, C., Abou-Rjaily, G., Bezrookove, V., Verhaegen, M., Johnson, T.M., Fullen, D.R., Pointer, J.N., Gruber, S.B., Su, L.D., Nikiforov, M.A., Kaufman, R.J., Bastian, B.C., Soengas, M.S., 2006. Anti-oncogenic role of the endoplasmic reticulum differentially activated by mutations in the MAPK pathway. Nat. Cell Biol. 8, 1053–1063.

Deschenes-Simard, X., Gaumont-Leclerc, M.F., Bourdeau, V., Lessard, F., Moiseeva, O., Forest, V., Igelmann, S., Mallette, F.A., Saba-El-Leil, M.K., Meloche, S., Saad, F., Mes-Masson, A.M., Ferbeyre, G., 2013. Tumor suppressor activity of the ERK/MAPK pathway by promoting selective protein degradation. Genes Dev. 27, 900–915.

Di Fagagna, F.D., Reaper, P.M., Clay-Farrace, L., Fiegler, H., Carr, P., Von Zglinicki, T., Saretzki, G., Carter, N.P., Jackson, S.P., 2003. A DNA damage checkpoint response in telomere-initiated senescence. Nature 426, 194–198.

Di Micco, R., Fumagalli, M., Cicalese, A., Piccinin, S., Gasparini, P., Luise, C., Schurra, C., Garre, M., Nuciforo, P.G., Bensimon, A., Maestro, R., Pelicci, P.G., Di Fagagna, F.D., 2006. Oncogene-induced senescence is a DNA damage response triggered by DNA hyper-replication. Nature 444, 638–642.

Dimmeler, S., Nicotera, P., 2013. MicroRNAs in age-related diseases. EMBO Mol. Med. 5, 180–190.

Dimozi, A., Mavrogonatou, E., Sklirou, A., Kletsas, D., 2015. Oxidative stress inhibits the proliferation, induces premature senescence and promotes a catabolic phenotype in human nucleus pulposus intervertebral disc cells. Eur. Cell. Mater. 30, 89–102. discussion 103.

Dimri, G.P., Lee, X., Basile, G., Acosta, M., Scott, G., Roskelley, C., Medrano, E.E., Linskens, M., Rubelj, I., Pereira-Smith, O., et al., 1995. A biomarker that identifies senescent human cells in culture and in aging skin in vivo. Proc. Natl. Acad. Sci. U.S.A. 92, 9363–9367.

Dimri, M., Carroll, J.D., Cho, J.H., Dimri, G.P., 2013. microRNA-141 regulates BMI1 expression and induces senescence in human diploid fibroblasts. Cell Cycle 12, 3537–3546.

Dou, Z., Xu, C., Donahue, G., Shimi, T., Pan, J.A., Zhu, J., Ivanov, A., Capell, B.C., Drake, A.M., Shah, P.P., Catanzaro, J.M., Ricketts, M.D., Lamark, T., Adam, S.A., Marmorstein, R., Zong, W.X., Johansen, T., Goldman, R.D., Adams, P.D., Berger, S.L., 2015. Autophagy mediates degradation of nuclear lamina. Nature 527, 105–109.

Drummond, M.J., Mccarthy, J.J., Sinha, M., Spratt, H.M., Volpi, E., Esser, K.A., Rasmussen, B.B., 2011. Aging and microRNA expression in human skeletal muscle: a microarray and bioinformatics analysis. Physiol. Genomics 43, 595–603.

Du, W., Pogoriler, J., 2006. Retinoblastoma family genes. Oncogene 25, 5190–5200.

Du, J.M., Johnson, L.M., Jacobsen, S.E., Patel, D.J., 2015. DNA methylation pathways and their crosstalk with histone methylation. Nat. Rev. Mol. Cell Biol. 16, 519–532.

El-Deiry, W.S., Tokino, T., Velculescu, V.E., Levy, D.B., Parsons, R., Trent, J.M., Lin, D., Mercer, W.E., Kinzler, K.W., Vogelstein, B., 1993. WAF1, a potential mediator of p53 tumor suppression. Cell 75, 817–825.

El-Deiry, W.S., Harper, J.W., O'connor, P.M., Velculescu, V.E., Canman, C.E., Jackman, J., Pietenpol, J.A., Burrell, M., Hill, D.E., Wang, Y., et al., 1994. WAF1/CIP1 is induced in p53-mediated G1 arrest and apoptosis. Cancer Res. 54, 1169–1174.

Ellinger-Ziegelbauer, H., Kelly, K., Siebenlist, U., 1999. Cell cycle arrest and reversion of Ras-induced transformation by a conditionally activated form of mitogen-activated protein kinase kinase kinase 3. Mol. Cell. Biol. 19, 3857–3868.

Elsharawy, A., Keller, A., Flachsbart, F., Wendschlag, A., Jacobs, G., Kefer, N., Brefort, T., Leidinger, P., Backes, C., Meese, E., Schreiber, S., Rosenstiel, P., Franke, A., Nebel, A., 2012. Genome-wide miRNA signatures of human longevity. Aging Cell 11, 607–616.

Ertekin, S., Yildirim, O., Dinc, E., Ayaz, L., Fidanci, S.B., Tamer, L., 2014. Evaluation of circulating miRNAs in wet age-related macular degeneration. Mol. Vis. 20, 1057–1066.

Fabian, M.R., Sonenberg, N., 2012. The mechanics of miRNA-mediated gene silencing: a look under the hood of miRISC. Nat. Struct. Mol. Biol. 19, 586–593.

Fang, L., Igarashi, M., Leung, J., Sugrue, M.M., Lee, S.W., Aaronson, S.A., 1999. p21Waf1/Cip1/Sdi1 induces permanent growth arrest with markers of replicative senescence in human tumor cells lacking functional p53. Oncogene 18, 2789–2797.

Farooqi, A.A., Fayyaz, S., Shatynska-Mytsyk, I., Javed, Z., Jabeen, S., Yaylim, I., Gasparri, M.L., Panici, P.B., 2016. Is miR-34a a well-equipped swordsman to conquer temple of molecular oncology? Chem. Biol. Drug Des. 87, 321–334.

Fazzari, M.J., Greally, J.M., 2004. Epigenomics: beyond CpG islands. Nat. Rev. Genet. 5, 446–455.

Fedele, M., Fusco, A., 2010. HMGA and cancer. Biochim. Biophys. Acta 1799, 48–54.

Feldser, D.M., Greider, C.W., 2007. Short telomeres limit tumor progression in vivo by inducing senescence. Cancer Cell 11, 461–469.

Feng, M., Yu, Q., 2010. miR-449 regulates CDK-Rb-E2F1 through an auto-regulatory feedback circuit. Cell Cycle 9, 213–214.
Feng, Z., Zhang, C., Wu, R., Hu, W., 2011. Tumor suppressor p53 meets microRNAs. J. Mol. Cell Biol. 3, 44–50.
Ferbeyre, G., De Stanchina, E., Querido, E., Baptiste, N., Prives, C., Lowe, S.W., 2000. PML is induced by oncogenic ras and promotes premature senescence. Genes Dev. 14, 2015–2027.
Ferbeyre, G., De Stanchina, E., Lin, A.W., Querido, E., Mccurrach, M.E., Hannon, G.J., Lowe, S.W., 2002. Oncogenic ras and p53 cooperate to induce cellular senescence. Mol. Cell. Biol. 22, 3497–3508.
Fineberg, S.K., Datta, P., Stein, C.S., Davidson, B.L., 2012. MiR-34a represses Numbl in murine neural progenitor cells and antagonizes neuronal differentiation. PLoS One 7, e38562.
Finkel, T., Deng, C.X., Mostoslavsky, R., 2009. Recent progress in the biology and physiology of sirtuins. Nature 460, 587–591.
Franceschi, C., Campisi, J., 2014. Chronic inflammation (inflammaging) and its potential contribution to age-associated diseases. J. Gerontol. A Biol. Sci. Med. Sci. 69 (Suppl. 1), S4–S9.
Friedman, D.S., O'colmain, B.J., Munoz, B., Tomany, S.C., Mccarty, C., De Jong, P.T., Nemesure, B., Mitchell, P., Kempen, J., Eye Diseases Prevalence Research, G., 2004. Prevalence of age-related macular degeneration in the United States. Arch. Ophthalmol. 122, 564–572.
Friedman, R.C., Farh, K.K., Burge, C.B., Bartel, D.P., 2009. Most mammalian mRNAs are conserved targets of microRNAs. Genome Res. 19, 92–105.
Fumagalli, M., Rosiello, F., Clerici, M., Barozzi, S., Cittaro, D., Kaplunov, J.M., Bucci, G., Dobreva, M., Matti, V., Beausejour, C.M., Herbig, U., Longhese, M.P., Di Fagagna, F.D., 2012. Telomeric DNA damage is irreparable and causes persistent DNA-damage-response activation. Nat. Cell Biol. 14, 355–365.
Furuta, M., Kozaki, K.I., Tanaka, S., Arii, S., Imoto, I., Inazawa, J., 2010. miR-124 and miR-203 are epigenetically silenced tumor-suppressive microRNAs in hepatocellular carcinoma. Carcinogenesis 31, 766–776.
Gallardo, E., Navarro, A., Vinolas, N., Marrades, R.M., Diaz, T., Gel, B., Quera, A., Bandres, E., Garcia-Foncillas, J., Ramirez, J., Monzo, M., 2009. miR-34a as a prognostic marker of relapse in surgically resected non-small-cell lung cancer. Carcinogenesis 30, 1903–1909.
Gaughwin, P.M., Ciesla, M., Lahiri, N., Tabrizi, S.J., Brundin, P., Bjorkqvist, M., 2011. Hsa-miR-34b is a plasma-stable microRNA that is elevated in pre-manifest Huntington's disease. Hum. Mol. Genet. 20, 2225–2237.
Gebert, L.F., Rebhan, M.A., Crivelli, S.E., Denzler, R., Stoffel, M., Hall, J., 2014. Miravirsen (SPC3649) can inhibit the biogenesis of miR-122. Nucleic Acids Res. 42, 609–621.
Geyer, R.K., Yu, Z.K., Maki, C.G., 2000. The MDM2 RING-finger domain is required to promote p53 nuclear export. Nat. Cell Biol. 2, 569–573.
Giacinti, C., Giordano, A., 2006. RB and cell cycle progression. Oncogene 25, 5220–5227.
Gil, J., Bernard, D., Martinez, D., Beach, D., 2004. Polycomb CBX7 has a unifying role in cellular lifespan. Nat. Cell Biol. 6, 67–72.
Ginjala, V., Nacerddine, K., Kulkarni, A., Oza, J., Hill, S.J., Yao, M., Citterio, E., Van Lohuizen, M., Ganesan, S., 2011. BMI1 is recruited to DNA breaks and contributes to DNA damage-induced H2A ubiquitination and repair. Mol. Cell. Biol. 31, 1972–1982.

Gire, W., Roux, P., Wynford-Thomas, D., Brondello, J.M., Dulic, V., 2004. DNA damage checkpoint kinase Chk2 triggers replicative senescence. EMBO J. 23, 2554–2563.
Godlewski, J., Nowicki, M.O., Bronisz, A., Williams, S., Otsuki, A., Nuovo, G., Raychaudhury, A., Newton, H.B., Chiocca, E.A., Lawler, S., 2008. Targeting of the Bmi-1 oncogene/stem cell renewal factor by microRNA-128 inhibits glioma proliferation and self-renewal. Cancer Res. 68, 9125–9130.
Grassmann, F., Schoenberger, P.G., Brandl, C., Schick, T., Hasler, D., Meister, G., Fleckenstein, M., Lindner, M., Helbig, H., Fauser, S., Weber, B.H., 2014. A circulating microrna profile is associated with late-stage neovascular age-related macular degeneration. PLoS One 9, e107461.
Greider, C.W., Blackburn, E.H., 1985. Identification of a specific telomere terminal transferase-activity in tetrahymena extracts. Cell 43, 405–413.
Guarino, M., 2010. Src signaling in cancer invasion. J. Cell. Physiol. 223, 14–26.
Guil, S., Esteller, M., 2009. DNA methylomes, histone codes and miRNAs: tying it all together. Int. J. Biochem. Cell Biol. 41, 87–95.
Guo, X., Keyes, W.M., Papazoglu, C., Zuber, J., Li, W., Lowe, S.W., Vogel, H., Mills, A.A., 2009. TAp63 induces senescence and suppresses tumorigenesis in vivo. Nat. Cell Biol. 11, 1451–1457.
Hackl, M., Brunner, S., Fortschegger, K., Schreiner, C., Micutkova, L., Muck, C., Laschober, G.T., Lepperdinger, G., Sampson, N., Berger, P., Herndler-Brandstetter, D., Wieser, M., Kuhnel, H., Strasser, A., Rinnerthaler, M., Breitenbach, M., Mildner, M., Eckhart, L., Tschachler, E., Trost, A., Bauer, J.W., Papak, C., Trajanoski, Z., Scheideler, M., Grillari-Voglauer, R., Grubeck-Loebenstein, B., Jansen-Durr, P., Grillari, J., 2010. miR-17, miR-19b, miR-20a, and miR-106a are down-regulated in human aging. Aging Cell 9, 291–296.
Hahn, W.C., Dessain, S.K., Brooks, M.W., King, J.E., Elenbaas, B., Sabatini, D.M., Decaprio, J.A., Weinberg, R.A., 2002. Enumeration of the simian virus 40 early region elements necessary for human cell transformation. Mol. Cell. Biol. 22, 2111–2123.
Hara, E., Smith, R., Parry, D., Tahara, H., Stone, S., Peters, G., 1996. Regulation of p16CDKN2 expression and its implications for cell immortalization and senescence. Mol. Cell. Biol. 16, 859–867.
Harper, J.W., Adami, G.R., Wei, N., Keyomarsi, K., Elledge, S.J., 1993. The p21 Cdk-interacting protein Cip1 is a potent inhibitor of G1 cyclin-dependent kinases. Cell 75, 805–816.
Hayflick, L., 1965. The limited in vitro lifetime of human diploid cell strains. Exp. Cell Res. 37, 614–636.
Hayflick, L., Moorhead, P.S., 1961. The serial cultivation of human diploid cell strains. Exp. Cell Res. 25, 585–621.
He, L., He, X.Y., Lim, L.P., De Stanchina, E., Xuan, Z.Y., Liang, Y., Xue, W., Zender, L., Magnus, J., Ridzon, D., Jackson, A.L., Linsley, P.S., Chen, C.F., Lowe, S.W., Cleary, M.A., Hannon, G.J., 2007. A microRNA component of the p53 tumour suppressor network. Nature 447, 1130. U16.
Herbig, U., Jobling, W.A., Chen, B.P., Chen, D.J., Sedivy, J.M., 2004. Telomere shortening triggers senescence of human cells through a pathway involving ATM, p53, and p21 (CIP1), but not p16(INK4a). Mol. Cell 14, 501–513.
Herbig, U., Ferreira, M., Condel, L., Carey, D., Sedivy, J.M., 2006. Cellular senescence in aging primates. Science 311, 1257.
Hermeking, H., 2012. MicroRNAs in the p53 network: micromanagement of tumour suppression. Nat. Rev. Cancer 12, 613–626.
Herranz, D., Serrano, M., 2010. Impact of Sirt1 on mammalian aging. Aging (Albany NY) 2, 315–316.

Herranz, N., Gallage, S., Mellone, M., Wuestefeld, T., Klotz, S., Hanley, C.J., Raguz, S., Acosta, J.C., Innes, A.J., Banito, A., Georgilis, A., Montoya, A., Wolter, K., Dharmalingam, G., Faull, P., Carroll, T., Martinez-Barbera, J.P., Cutillas, P., Reisinger, F., Heikenwalder, M., Miller, R.A., Withers, D., Zender, L., Thomas, G.J., Gil, J., 2015. mTOR regulates MAPKAPK2 translation to control the senescence-associated secretory phenotype. Nat. Cell Biol. 17, 1205–1217.

Hildesheim, J., Bulavin, D.V., Anver, M.R., Alvord, W.G., Hollander, M.C., Vardanian, L., Fornace, A.J., 2002. Gadd45a protects against UV irradiation-induced skin tumors, and promotes apoptosis and stress signaling via MAPK and p53. Cancer Res. 62, 7305–7315.

Hoenicke, L., Zender, L., 2012. Immune surveillance of senescent cells—biological significance in cancer- and non-cancer pathologies. Carcinogenesis 33, 1123–1126.

Hollstein, M., Sidransky, D., Vogelstein, B., Harris, C.C., 1991. P53 mutations in human cancers. Science 253, 49–53.

Honda, R., Yasuda, H., 2000. Activity of MDM2, a ubiquitin ligase, toward p53 or itself is dependent on the RING finger domain of the ligase. Oncogene 19, 1473–1476.

Hu, H., Gatti, R.A., 2011. MicroRNAs: new players in the DNA damage response. J. Mol. Cell Biol. 3, 151–158.

Hu, Z., Klein, J.D., Mitch, W.E., Zhang, L., Martinez, I., Wang, X.H., 2014. MicroRNA-29 induces cellular senescence in aging muscle through multiple signaling pathways. Aging (Albany NY) 6, 160–175.

Huang, L., Luo, J., Cai, Q., Pan, Q., Zeng, H., Guo, Z., Dong, W., Huang, J., Lin, T., 2011. MicroRNA-125b suppresses the development of bladder cancer by targeting E2F3. Int. J. Cancer 128, 1758–1769.

Humbert, P.O., Verona, R., Trimarchi, J.M., Rogers, C., Dandapani, S., Lees, J.A., 2000. E2f3 is critical for normal cellular proliferation. Genes Dev. 14, 690–703.

Hyman, L., 1987. Epidemiology of eye disease in the elderly. Eye (Lond.) 1 (Pt. 2), 330–341.

Iliopoulos, D., Malizos, K.N., Oikonomou, P., Tsezou, A., 2008. Integrative microRNA and proteomic approaches identify novel osteoarthritis genes and their collaborative metabolic and inflammatory networks. PLoS One 3, e3740.

Iorio, M.V., Ferracin, M., Liu, C.G., Veronese, A., Spizzo, R., Sabbioni, S., Magri, E., Pedriali, M., Fabbri, M., Campiglio, M., Menard, S., Palazzo, J.P., Rosenberg, A., Musiani, P., Volinia, S., Nenci, I., Calin, G.A., Querzoli, P., Negrini, M., Croce, C.M., 2005. MicroRNA gene expression deregulation in human breast cancer. Cancer Res. 65, 7065–7070.

Itahana, K., Campisi, J., Dimri, G.P., 2004. Mechanisms of cellular senescence in human and mouse cells. Biogerontology 5, 1–10.

Ito, T., Yagi, S., Yamakuchi, M., 2010. MicroRNA-34a regulation of endothelial senescence. Biochem. Biophys. Res. Commun. 398, 735–740.

Ivanov, A., Pawlikowski, J., Manoharan, I., Van Tuyn, J., Nelson, D.M., Rai, T.S., Shah, P.P., Hewitt, G., Korolchuk, V.I., Passos, J.F., Wu, H., Berger, S.L., Adams, P.D., 2013. Lysosome-mediated processing of chromatin in senescence. J. Cell Biol. 202, 129–143.

Ivanovska, I., Ball, A.S., Diaz, R.L., Magnus, J.F., Kibukawa, M., Schelter, J.M., Kobayashi, S.V., Lim, L., Burchard, J., Jackson, A.L., Linsley, P.S., Cleary, M.A., 2008. MicroRNAs in the miR-106b family regulate p21/CDKN1A and promote cell cycle progression. Mol. Cell. Biol. 28, 2167–2174.

Iyer, A., Zurolo, E., Prabowo, A., Fluiter, K., Spliet, W.G., Van Rijen, P.C., Gorter, J.A., Aronica, E., 2012. MicroRNA-146a: a key regulator of astrocyte-mediated inflammatory response. PLoS One 7, e44789.

Jacobs, J.J., Kieboom, K., Marino, S., Depinho, R.A., Van Lohuizen, M., 1999. The oncogene and Polycomb-group gene bmi-1 regulates cell proliferation and senescence through the ink4a locus. Nature 397, 164–168.

Janssen, H.L., Reesink, H.W., Zeuzem, S., Lawitz, E., Rodriguez-Torres, M., Chen, A., Davis, C., King, B., Levin, A.A., Hodges, M.R., 2011. A randomized, double-blind, placebo (plb) controlled safety and anti-viral proof of concept study of miravirsen (MIR), an oligonucleotide targeting miR-122, in treatment naïve patients with genotype 1 (gt1) chronic HCV infection. Hepatology 54 (4), 1430A.

Jazbutyte, V., Fiedler, J., Kneitz, S., Galuppo, P., Just, A., Holzmann, A., Bauersachs, J., Thum, T., 2013. MicroRNA-22 increases senescence and activates cardiac fibroblasts in the aging heart. Age (Dordr.) 35, 747–762.

Jeyapalan, J.C., Ferreira, M., Sedivy, J.M., Herbig, U., 2007. Accumulation of senescent cells in mitotic tissue of aging primates. Mech. Ageing Dev. 128, 36–44.

Jia, L.H., Liu, Y.N., 2016. Downregulated serum miR-223 servers as biomarker in Alzheimer's disease. Cell Biochem. Funct. 34, 233–237.

Jin, S.Q., Mazzacurati, L., Zhu, X.C., Tong, T., Song, Y.M., Shao, S.J., Petrik, K.L., Rajasekaran, B., Wu, M., Zhan, Q.M., 2003. Gadd45a contributes to p53 stabilization in response to DNA damage. Oncogene 22, 8536–8540.

Jin, M., Zhang, T., Liu, C., Badeaux, M.A., Liu, B., Liu, R., Jeter, C., Chen, X., Vlassov, A.V., Tang, D.G., 2014. miRNA-128 suppresses prostate cancer by inhibiting BMI-1 to inhibit tumor-initiating cells. Cancer Res. 74, 4183–4195.

Joaquin, M., Watson, R.J., 2003. Cell cycle regulation by the B-Myb transcription factor. Cell Mol. Life Sci. 60, 2389–2401.

Johnson, S.M., Grosshans, H., Shingara, J., Byrom, M., Jarvis, R., Cheng, A., Labourier, E., Reinert, K.L., Brown, D., Slack, F.J., 2005. RAS is regulated by the let-7 MicroRNA family. Cell 120, 635–647.

Jun, J.I., Lau, L.F., 2010. The matricellular protein CCN1 induces fibroblast senescence and restricts fibrosis in cutaneous wound healing. Nat. Cell Biol. 12, 676–685.

Kamijo, T., Van De Kamp, E., Chong, M.J., Zindy, F., Diehl, J.A., Sherr, C.J., Mckinnon, P.J., 1999. Loss of the ARF tumor suppressor reverses premature replicative arrest but not radiation hypersensitivity arising from disabled atm function. Cancer Res. 59, 2464–2469.

Karlseder, J., Smogorzewska, A., De Lange, T., 2002. Senescence induced by altered telomere state, not telomere loss. Science 295, 2446–2449.

Karube, Y., Tanaka, H., Osada, H., Tomida, S., Tatematsu, Y., Yanagisawa, K., Yatabe, Y., Takamizawa, J., Miyoshi, S., Mitsudomi, T., Takahashi, T., 2005. Reduced expression of Dicer associated with poor prognosis in lung cancer patients. Cancer Sci. 96, 111–115.

Kasinski, A.L., Slack, F.J., 2011. Epigenetics and genetics. MicroRNAs en route to the clinic: progress in validating and targeting microRNAs for cancer therapy. Nat. Rev. Cancer 11, 849–864.

Keller, A., Backes, C., Haas, J., Leidinger, P., Maetzler, W., Deuschle, C., Berg, D., Ruschil, C., Galata, V., Ruprecht, K., Stahler, C., Wurstle, M., Sickert, D., Gogol, M., Meder, B., Meese, E., 2016. Validating Alzheimer's disease micro RNAs using next-generation sequencing. Alzheimers Dement. 12, 565–576.

Kent, O.A., Chivukula, R.R., Mullendore, M., Wentzel, E.A., Feldmann, G., Lee, K.H., Liu, S., Leach, S.D., Maitra, A., Mendell, J.T., 2010. Repression of the miR-143/145 cluster by oncogenic Ras initiates a tumor-promoting feed-forward pathway. Genes Dev. 24, 2754–2759.

Kim, V.N., Han, J., Siomi, M.C., 2009. Biogenesis of small RNAs in animals. Nat. Rev. Mol. Cell Biol. 10, 126–139.

Kim, M.S., Oh, J.E., Kim, Y.R., Park, S.W., Kang, M.R., Kim, S.S., Ahn, C.H., Yoo, N.J., Lee, S.H., 2010. Somatic mutations and losses of expression of microRNA regulation-related genes AGO2 and TNRC6A in gastric and colorectal cancers. J. Pathol. 221, 139–146.

Kitadate, A., Ikeda, S., Teshima, K., Ito, M., Toyota, I., Hasunuma, N., Takahashi, N., Miyagaki, T., Sugaya, M., Tagawa, H., 2016. MicroRNA-16 mediates the regulation

of a senescence-apoptosis switch in cutaneous T-cell and other non-Hodgkin lymphomas. Oncogene 35, 3692–3704.
Koch, C.M., Andrews, R.M., Flicek, P., Dillon, S.C., Karaoz, U., Clelland, G.K., Wilcox, S., Beare, D.M., Fowler, J.C., Couttet, P., James, K.D., Lefebvre, G.C., Bruce, A.W., Dovey, O.M., Ellis, P.D., Dhami, P., Langford, C.F., Weng, Z.P., Birney, E., Carter, N.P., Vetrie, D., Dunham, I., 2007. The landscape of histone modifications across 1% of the human genome in five human cell lines. Genome Res. 17, 691–707.
Kondo, H., Kim, H.W., Wang, L., Okada, M., Paul, C., Millard, R.W., Wang, Y., 2016. Blockade of senescence-associated microRNA-195 in aged skeletal muscle cells facilitates reprogramming to produce induced pluripotent stem cells. Aging Cell 15, 56–66.
Kortlever, R.M., Higgins, P.J., Bernards, R., 2006. Plasminogen activator inhibitor-1 is a critical downstream target of p53 in the induction of replicative senescence. Nat. Cell Biol. 8, 877–884.
Kosar, M., Bartkova, J., Hubackova, S., Hodny, Z., Lukas, J., Bartek, J., 2011. Senescence-associated heterochromatin foci are dispensable for cellular senescence, occur in a cell type- and insult-dependent manner and follow expression of p16(ink4a). Cell Cycle 10, 457–468.
Kozaki, K., Imoto, I., Mogi, S., Omura, K., Inazawa, J., 2008. Exploration of tumor-suppressive microRNAs silenced by DNA hypermethylation in oral cancer. Cancer Res. 68, 2094–2105.
Krishnamurthy, J., Torrice, C., Ramsey, M.R., Kovalev, G.I., Al-Regaiey, K., Su, L., Sharpless, N.E., 2004. Ink4a/Arf expression is a biomarker of aging. J. Clin. Invest. 114, 1299–1307.
Kruse, J.P., Gu, W., 2009. Modes of p53 regulation. Cell 137, 609–622.
Krzeszinski, J.Y., Wei, W., Huynh, H., Jin, Z., Wang, X., Chang, T.C., Xie, X.J., He, L., Mangala, L.S., Lopez-Berestein, G., Sood, A.K., Mendell, J.T., Wan, Y., 2014. miR-34a blocks osteoporosis and bone metastasis by inhibiting osteoclastogenesis and Tgif2. Nature 512, 431–435.
Kuilman, T., Michaloglou, C., Vredeveld, L.C., Douma, S., Van Doorn, R., Desmet, C.J., Aarden, L.A., Mooi, W.J., Peeper, D.S., 2008. Oncogene-induced senescence relayed by an interleukin-dependent inflammatory network. Cell 133, 1019–1031.
Kuilman, T., Michaloglou, C., Mooi, W.J., Peeper, D.S., 2010. The essence of senescence. Genes Dev. 24, 2463–2479.
Kumar, M., Lu, Z., Takwi, A.A., Chen, W., Callander, N.S., Ramos, K.S., Young, K.H., Li, Y., 2011. Negative regulation of the tumor suppressor p53 gene by microRNAs. Oncogene 30, 843–853.
Kumar, P., Dezso, Z., Mackenzie, C., Oestreicher, J., Agoulnik, S., Byrne, M., Bernier, F., Yanagimachi, M., Aoshima, K., Oda, Y., 2013. Circulating miRNA biomarkers for Alzheimer's disease. PLoS One 8, e69807.
Kunz, C., Pebler, S., Otte, J., Von Der Ahe, D., 1995. Differential regulation of plasminogen activator and inhibitor gene transcription by the tumor suppressor p53. Nucleic Acids Res. 23, 3710–3717.
Kutty, R.K., Nagineni, C.N., Samuel, W., Vijayasarathy, C., Jaworski, C., Duncan, T., Cameron, J.E., Flemington, E.K., Hooks, J.J., Redmond, T.M., 2013. Differential regulation of microRNA-146a and microRNA-146b-5p in human retinal pigment epithelial cells by interleukin-1beta, tumor necrosis factor-alpha, and interferon-gamma. Mol. Vis. 19, 737–750.
Laberge, R.M., Sun, Y., Orjalo, A.V., Patil, C.K., Freund, A., Zhou, L., Curran, S.C., Davalos, A.R., Wilson-Edell, K.A., Liu, S., Limbad, C., Demaria, M., Li, P., Hubbard, G.B., Ikeno, Y., Javors, M., Desprez, P.Y., Benz, C.C., Kapahi, P., Nelson, P.S., Campisi, J., 2015. MTOR regulates the pro-tumorigenic senescence-

associated secretory phenotype by promoting IL1A translation. Nat. Cell Biol. 17, 1049–1061.
Lal, A., Kim, H.H., Abdelmohsen, K., Kuwano, Y., Pullmann Jr., R., Srikantan, S., Subrahmanyam, R., Martindale, J.L., Yang, X., Ahmed, F., Navarro, F., Dykxhoorn, D., Lieberman, J., Gorospe, M., 2008. p16(INK4a) translation suppressed by miR-24. PLoS One 3, e1864.
Lampson, B.L., Pershing, N.L., Prinz, J.A., Lacsina, J.R., Marzluff, W.F., Nicchitta, C.V., Macalpine, D.M., Counter, C.M., 2013. Rare codons regulate KRas oncogenesis. Curr. Biol. 23, 70–75.
Lan, F.F., Wang, H., Chen, Y.C., Chan, C.Y., Ng, S.S., Li, K., Xie, D., He, M.L., Lin, M.C., Kung, H.F., 2011. Hsa-let-7g inhibits proliferation of hepatocellular carcinoma cells by downregulation of c-Myc and upregulation of p16(INK4A). Int. J. Cancer 128, 319–331.
Langley, E., Pearson, M., Faretta, M., Bauer, U.M., Frye, R.A., Minucci, S., Pelicci, P.G., Kouzarides, T., 2002. Human SIR2 deacetylates p53 and antagonizes PML/p53-induced cellular senescence. EMBO J. 21, 2383–2396.
Lazzerini Denchi, E., Helin, K., 2005. E2F1 is crucial for E2F-dependent apoptosis. EMBO Rep. 6, 661–668.
Le, M.T., Teh, C., Shyh-Chang, N., Xie, H., Zhou, B., Korzh, V., Lodish, H.F., Lim, B., 2009. MicroRNA-125b is a novel negative regulator of p53. Genes Dev. 23, 862–876.
Lee, Y.S., Dutta, A., 2007. The tumor suppressor microRNA let-7 represses the HMGA2 oncogene. Genes Dev. 21, 1025–1030.
Lee, R.C., Feinbaum, R.L., Ambros, V., 1993. The C. elegans heterochronic gene lin-4 encodes small RNAs with antisense complementarity to lin-14. Cell 75, 843–854.
Lee, Y., Jeon, K., Lee, J.T., Kim, S., Kim, V.N., 2002. MicroRNA maturation: stepwise processing and subcellular localization. EMBO J. 21, 4663–4670.
Lee, Y., Kim, M., Han, J., Yeom, K.H., Lee, S., Baek, S.H., Kim, V.N., 2004. MicroRNA genes are transcribed by RNA polymerase II. EMBO J. 23, 4051–4060.
Lee, B.Y., Han, J.A., Im, J.S., Morrone, A., Johung, K., Goodwin, E.C., Kleijer, W.J., Dimaio, D., Hwang, E.S., 2006. Senescence-associated beta-galactosidase is lysosomal beta-galactosidase. Aging Cell 5, 187–195.
Lee, S., Jung, J.W., Park, S.B., Roh, K., Lee, S.Y., Kim, J.H., Kang, S.K., Kang, K.S., 2011. Histone deacetylase regulates high mobility group A2-targeting microRNAs in human cord blood-derived multipotent stem cell aging. Cell. Mol. Life Sci. 68, 325–336.
Lehmann, U., Hasemeier, B., Christgen, M., Muller, M., Romermann, D., Langer, F., Kreipe, H., 2008. Epigenetic inactivation of microRNA gene hsa-mir-9-1 in human breast cancer. J. Pathol. 214, 17–24.
Lehmann, B.D., Bauer, J.A., Chen, X., Sanders, M.E., Chakravarthy, A.B., Shyr, Y., Pietenpol, J.A., 2011. Identification of human triple-negative breast cancer subtypes and preclinical models for selection of targeted therapies. J. Clin. Invest. 121, 2750–2767.
Lehmann, S.M., Kruger, C., Park, B., Derkow, K., Rosenberger, K., Baumgart, J., Trimbuch, T., Eom, G., Hinz, M., Kaul, D., Habbel, P., Kalin, R., Franzoni, E., Rybak, A., Nguyen, D., Veh, R., Ninnemann, O., Peters, O., Nitsch, R., Heppner, F.L., Golenbock, D., Schott, E., Ploegh, H.L., Wulczyn, F.G., Lehnardt, S., 2012. An unconventional role for miRNA: let-7 activates Toll-like receptor 7 and causes neurodegeneration. Nat. Neurosci. 15, 827–835.
Leidinger, P., Backes, C., Deutscher, S., Schmitt, K., Mueller, S.C., Frese, K., Haas, J., Ruprecht, K., Paul, F., Stahler, C., Lang, C.J., Meder, B., Bartfai, T., Meese, E., Keller, A., 2013. A blood based 12-miRNA signature of Alzheimer disease patients. Genome Biol. 14, R78.
Lembo, A., Di Cunto, F., Provero, P., 2012. Shortening of 3'UTRs correlates with poor prognosis in breast and lung cancer. PLoS One 7, e31129.

Li, Y., Tan, W., Neo, T.W., Aung, M.O., Wasser, S., Lim, S.G., Tan, T.M., 2009. Role of the miR-106b-25 microRNA cluster in hepatocellular carcinoma. Cancer Sci. 100, 1234–1242.

Li, N., Bates, D.J., An, J., Terry, D.A., Wang, E., 2011a. Up-regulation of key microRNAs, and inverse down-regulation of their predicted oxidative phosphorylation target genes, during aging in mouse brain. Neurobiol. Aging 32, 944–955.

Li, X., Khanna, A., Li, N., Wang, E., 2011b. Circulatory miR34a as an RNAbased, non-invasive biomarker for brain aging. Aging (Albany NY) 3, 985–1002.

Li, M., Fu, W., Wo, L., Shu, X., Liu, F., Li, C., 2013. miR-128 and its target genes in tumorigenesis and metastasis. Exp. Cell Res. 319, 3059–3064.

Lin, S., Gregory, R.I., 2015. MicroRNA biogenesis pathways in cancer. Nat. Rev. Cancer 15, 321–333.

Lin, A.W., Barradas, M., Stone, J.C., Van Aelst, L., Serrano, M., Lowe, S.W., 1998. Premature senescence involving p53 and p16 is activated in response to constitutive MEK/MAPK mitogenic signaling. Genes Dev. 12, 3008–3019.

Lin, H.K., Chen, Z., Wang, G., Nardella, C., Lee, S.W., Chan, C.H., Yang, W.L., Wang, J., Egia, A., Nakayama, K.I., Cordon-Cardo, C., Teruya-Feldstein, J., Pandolfi, P.P., 2010. Skp2 targeting suppresses tumorigenesis by Arf-p53-independent cellular senescence. Nature 464, 374–379.

Lindow, M., Kauppinen, S., 2012. Discovering the first microRNA-targeted drug. J. Cell Biol. 199, 407–412.

Liu, Y.N., Lee, W.W., Wang, C.Y., Chao, T.H., Chen, Y., Chen, J.H., 2005. Regulatory mechanisms controlling human E-cadherin gene expression. Oncogene 24, 8277–8290.

Liu, M., Lang, N., Qiu, M., Xu, F., Li, Q., Tang, Q., Chen, J., Chen, X., Zhang, S., Liu, Z., Zhou, J., Zhu, Y., Deng, Y., Zheng, Y., Bi, F., 2011. miR-137 targets Cdc42 expression, induces cell cycle G1 arrest and inhibits invasion in colorectal cancer cells. Int. J. Cancer 128, 1269–1279.

Liu, N., Landreh, M., Cao, K., Abe, M., Hendriks, G.J., Kennerdell, J.R., Zhu, Y., Wang, L.S., Bonini, N.M., 2012. The microRNA miR-34 modulates ageing and neurodegeneration in Drosophila. Nature 482, 519–523.

Liu, C.H., Sun, Y., Li, J., Gong, Y., Tian, K.T., Evans, L.P., Morss, P.C., Fredrick, T.W., Saba, N.J., Chen, J., 2015a. Endothelial microRNA-150 is an intrinsic suppressor of pathologic ocular neovascularization. Proc. Natl. Acad. Sci. U.S.A. 112, 12163–12168.

Liu, Y., Li, X., Zhu, S., Zhang, J.G., Yang, M., Qin, Q., Deng, S.C., Wang, B., Tian, K., Liu, L., Niu, Y., Wang, C.Y., Zhao, G., 2015b. Ectopic expression of miR-494 inhibited the proliferation, invasion and chemoresistance of pancreatic cancer by regulating SIRT1 and c-Myc. Gene Ther. 22, 729–738.

Lowe, S.W., Sherr, C.J., 2003. Tumor suppression by Ink4a-Arf: progress and puzzles. Curr. Opin. Genet. Dev. 13, 77–83.

Lowe, S.W., Cepero, E., Evan, G., 2004. Intrinsic tumour suppression. Nature 432, 307–315.

Lu, J., Getz, G., Miska, E.A., Alvarez-Saavedra, E., Lamb, J., Peck, D., Sweet-Cordero, A., Ebert, B.L., Mak, R.H., Ferrando, A.A., Downing, J.R., Jacks, T., Horvitz, H.R., Golub, T.R., 2005. MicroRNA expression profiles classify human cancers. Nature 435, 834–838.

Lukiw, W.J., Andreeva, T.V., Grigorenko, A.P., Rogaev, E.I., 2012. Studying micro RNA function and dysfunction in Alzheimer's disease. Front. Genet. 3, 327.

Luo, J., Nikolaev, A.Y., Imai, S., Chen, D., Su, F., Shiloh, A., Guarente, L., Gu, W., 2001. Negative control of p53 by Sir2alpha promotes cell survival under stress. Cell 107, 137–148.

Luo, W., Li, G., Yi, Z., Nie, Q., Zhang, X., 2016. E2F1-miR-20a-5p/20b-5p auto-regulatory feedback loop involved in myoblast proliferation and differentiation. Sci. Rep. 6, 27904.

Maehara, K., Yamakoshi, K., Ohtani, N., Kubo, Y., Takahashi, A., Arase, S., Jones, N., Hara, E., 2005. Reduction of total E2F/DP activity induces senescence-like cell cycle arrest in cancer cells lacking functional pRB and p53. J. Cell Biol. 168, 553–560.

Maier, B., Gluba, W., Bernier, B., Turner, T., Mohammad, K., Guise, T., Sutherland, A., Thorner, M., Scrable, H., 2004. Modulation of mammalian life span by the short isoform of p53. Genes Dev. 18, 306–319.

Mallette, F.A., Ferbeyre, G., 2007. The DNA damage signaling pathway connects oncogenic stress to cellular senescence. Cell Cycle 6, 1831–1836.

Mallette, F.A., Richard, S., 2012. JMJD2A promotes cellular transformation by blocking cellular senescence through transcriptional repression of the tumor suppressor CHD5. Cell Rep. 2, 1233–1243.

Mallette, F.A., Gaumont-Leclerc, M.F., Ferbeyre, G., 2007a. The DNA damage signaling pathway is a critical mediator of oncogene-induced senescence. Genes Dev. 21, 43–48.

Mallette, F.A., Gaumont-Leclerc, M.F., Huot, G., Ferbeyre, G., 2007b. Myc down-regulation as a mechanism to activate the Rb pathway in STAT5A-induced senescence. J. Biol. Chem. 282, 34938–34944.

Mallette, F.A., Moiseeva, O., Calabrese, V., Mao, B., Gaumont-Leclerc, M.F., Ferbeyre, G., 2010. Transcriptome analysis and tumor suppressor requirements of STAT5-induced senescence. Ann. N. Y. Acad. Sci. 1197, 142–151.

Mandke, P., Wyatt, N., Fraser, J., Bates, B., Berberich, S.J., Markey, M.P., 2012. MicroRNA-34a modulates MDM4 expression via a target site in the open reading frame. PLoS One 7, e42034.

Marasa, B.S., Srikantan, S., Masuda, K., Abdelmohsen, K., Kuwano, Y., Yang, X., Martindale, J.L., Rinker-Schaeffer, C.W., Gorospe, M., 2009. Increased MKK4 abundance with replicative senescence is linked to the joint reduction of multiple microRNAs. Sci. Signal. 2, ra69.

Marasa, B.S., Srikantan, S., Martindale, J.L., Kim, M.M., Lee, E.K., Gorospe, M., Abdelmohsen, K., 2010. MicroRNA profiling in human diploid fibroblasts uncovers miR-519 role in replicative senescence. Aging (Albany NY) 2, 333–343.

Markey, M., Berberich, S.J., 2008. Full-length hdmX transcripts decrease following genotoxic stress. Oncogene 27, 6657–6666.

Markowski, D.N., Helmke, B.M., Belge, G., Nimzyk, R., Bartnitzke, S., Deichert, U., Bullerdiek, J., 2011. HMGA2 and p14Arf: major roles in cellular senescence of fibroids and therapeutic implications. Anticancer Res. 31, 753–761.

Martinez, I., Cazalla, D., Almstead, L.L., Steitz, J.A., DiMaio, D., 2011. miR-29 and miR-30 regulate B-Myb expression during cellular senescence. Proc. Natl. Acad. Sci. U.S.A. 108, 522–527.

Matheu, A., Maraver, A., Serrano, M., 2008. The Arf/p53 pathway in cancer and aging. Cancer Res. 68, 6031–6034.

Matsumura, T., Zerrudo, Z., Hayflick, L., 1979. Senescent human diploid cells in culture: survival, DNA synthesis and morphology. J. Gerontol. 34, 328–334.

Mayr, C., Bartel, D.P., 2009. Widespread shortening of 3' UTRs by alternative cleavage and polyadenylation activates oncogenes in cancer cells. Cell 138, 673–684.

Mayr, C., Hemann, M.T., Bartel, D.P., 2007. Disrupting the pairing between let-7 and Hmga2 enhances oncogenic transformation. Science 315, 1576–1579.

Melo, S.A., Ropero, S., Moutinho, C., Aaltonen, L.A., Yamamoto, H., Calin, G.A., Rossi, S., Fernandez, A.F., Carneiro, F., Oliveira, C., Ferreira, B., Liu, C.G., Villanueva, A., Capella, G., Schwartz Jr., S., Shiekhattar, R., Esteller, M., 2009. A TARBP2 mutation in human cancer impairs microRNA processing and DICER1 function. Nat. Genet. 41, 365–370.

Menard, C., Rezende, F.A., Miloudi, K., Wilson, A., Tetreault, N., Hardy, P., Sangiovanni, J.P., De Guire, V., Sapieha, P., 2016. MicroRNA signatures in vitreous humour and plasma of patients with exudative AMD. Oncotarget 7, 19171–19184.

Menghini, R., Casagrande, V., Cardellini, M., Martelli, E., Terrinoni, A., Amati, F., Vasa-Nicotera, M., Ippoliti, A., Novelli, G., Melino, G., Lauro, R., Federici, M., 2009. MicroRNA 217 modulates endothelial cell senescence via silent information regulator 1. Circulation 120, 1524–1532.
Metias, S.M., Lianidou, E., Yousef, G.M., 2009. MicroRNAs in clinical oncology: at the crossroads between promises and problems. J. Clin. Pathol. 62, 771–776.
Michaloglou, C., Vredeveld, L.C., Soengas, M.S., Denoyelle, C., Kuilman, T., Van Der Horst, C.M., Majoor, D.M., Shay, J.W., Mooi, W.J., Peeper, D.S., 2005. BRAFE600-associated senescence-like cell cycle arrest of human naevi. Nature 436, 720–724.
Migliorini, D., Danovi, D., Colombo, E., Carbone, R., Pelicci, P.G., Marine, J.C., 2002. Hdmx recruitment into the nucleus by Hdm2 is essential for its ability to regulate p53 stability and transactivation. J. Biol. Chem. 277, 7318–7323.
Miller, K.R., Kelley, K., Tuttle, R., Berberich, S.J., 2010. HdmX overexpression inhibits oncogene induced cellular senescence. Cell Cycle 9, 3376–3382.
Minamino, T., Komuro, I., 2007. Vascular cell senescence: contribution to atherosclerosis. Circ. Res. 100, 15–26.
Miranda, K.C., Huynh, T., Tay, Y., Ang, Y.S., Tam, W.L., Thomson, A.M., Lim, B., Rigoutsos, I., 2006. A pattern-based method for the identification of MicroRNA binding sites and their corresponding heteroduplexes. Cell 126, 1203–1217.
Mitchell, P.S., Parkin, R.K., Kroh, E.M., Fritz, B.R., Wyman, S.K., Pogosova-Agadjanyan, E.L., Peterson, A., Noteboom, J., O'briant, K.C., Allen, A., Lin, D.W., Urban, N., Drescher, C.W., Knudsen, B.S., Stirewalt, D.L., Gentleman, R., Vessella, R.L., Nelson, P.S., Martin, D.B., Tewari, M., 2008. Circulating microRNAs as stable blood-based markers for cancer detection. Proc. Natl. Acad. Sci. U.S.A. 105, 10513–10518.
Miyaki, S., Asahara, H., 2012. Macro view of microRNA function in osteoarthritis. Nat. Rev. Rheumatol. 8, 543–552.
Miyaki, S., Nakasa, T., Otsuki, S., Grogan, S.P., Higashiyama, R., Inoue, A., Kato, Y., Sato, T., Lotz, M.K., Asahara, H., 2009. MicroRNA-140 is expressed in differentiated human articular chondrocytes and modulates interleukin-1 responses. Arthritis Rheum. 60, 2723–2730.
Miyaki, S., Sato, T., Inoue, A., Otsuki, S., Ito, Y., Yokoyama, S., Kato, Y., Takemoto, F., Nakasa, T., Yamashita, S., Takada, S., Lotz, M.K., Ueno-Kudo, H., Asahara, H., 2010. MicroRNA-140 plays dual roles in both cartilage development and homeostasis. Genes Dev. 24, 1173–1185.
Moiseeva, O., Mallette, F.A., Mukhopadhyay, U.K., Moores, A., Ferbeyre, G., 2006. DNA damage signaling and p53-dependent senescence after prolonged beta-interferon stimulation. Mol. Biol. Cell 17, 1583–1592.
Moiseeva, O., Deschenes-Simard, X., St-Germain, E., Igelmann, S., Huot, G., Cadar, A.E., Bourdeau, V., Pollak, M.N., Ferbeyre, G., 2013. Metformin inhibits the senescence-associated secretory phenotype by interfering with IKK/NF-kappaB activation. Aging Cell 12, 489–498.
Molnar, A., Theodoras, A.M., Zon, L.I., Kyriakis, J.M., 1997. Cdc42Hs, but not rad, inhibits serum-stimulated cell cycle progression at G(1)/S through a mechanism requiring p38/RK. J. Biol. Chem. 272, 13229–13235.
Moyer, A.W., Wallace, R., Cox, H.R., 1964. Limited growth period of human lung cell lines transformed by simian virus 40. J. Natl. Cancer Inst. 33, 227–236.
Muller, M., Jakel, L., Bruinsma, I.B., Claassen, J.A., Kuiperij, H.B., Verbeek, M.M., 2016. MicroRNA-29a is a candidate biomarker for Alzheimer's disease in cell-free cerebrospinal fluid. Mol. Neurobiol. 53, 2894–2899.
Munker, R., Calin, G.A., 2011. MicroRNA profiling in cancer. Clin. Sci. (Lond.) 121, 141–158.

Nahid, M.A., Satoh, M., Chan, E.K., 2011. MicroRNA in TLR signaling and endotoxin tolerance. Cell. Mol. Immunol. 8, 388–403.

Nakamura, Y., He, X., Kato, H., Wakitani, S., Kobayashi, T., Watanabe, S., Iida, A., Tahara, H., Warman, M.L., Watanapokasin, R., Postlethwait, J.H., 2012. Sox9 is upstream of microRNA-140 in cartilage. Appl. Biochem. Biotechnol. 166, 64–71.

Nakazawa, Y., Kamijo, T., Koike, K., Noda, T., 2003. ARF tumor suppressor induces mitochondria-dependent apoptosis by modulation of mitochondrial Bcl-2 family proteins. J. Biol. Chem. 278, 27888–27895.

Narita, M., Nunez, S., Heard, E., Lin, A.W., Hearn, S.A., Spector, D.L., Hannon, G.J., Lowe, S.W., 2003. Rb-mediated heterochromatin formation and silencing of E2F target genes during cellular senescence. Cell 113, 703–716.

Narita, M., Krizhanovsky, V., Nunez, S., Chicas, A., Hearn, S.A., Myers, M.P., Lowe, S.W., 2006. A novel role for high-mobility group a proteins in cellular senescence and heterochromatin formation. Cell 126, 503–514.

Narita, M., Young, A.R., Arakawa, S., Samarajiwa, S.A., Nakashima, T., Yoshida, S., Hong, S., Berry, L.S., Reichelt, S., Ferreira, M., Tavare, S., Inoki, K., Shimizu, S., Narita, M., 2011. Spatial coupling of mTOR and autophagy augments secretory phenotypes. Science 332, 966–970.

Navarro, F., Lieberman, J., 2015. miR-34 and p53: new insights into a complex functional relationship. PLoS One 10, e0132767.

Neault, M., Mallette, F.A., Richard, S., 2016. miR-137 modulates a tumor suppressor network-inducing senescence in pancreatic cancer cells. Cell Rep. 14, 1966–1978.

Newbold, R.F., Overell, R.W., 1983. Fibroblast immortality is a prerequisite for transformation by Ej c-Ha-ras oncogene. Nature 304, 648–651.

Newbold, R.F., Overell, R.W., Connell, J.R., 1982. Induction of immortality is an early event in malignant transformation of mammalian-cells by carcinogens. Nature 299, 633–635.

Nielsen, S.J., Schneider, R., Bauer, U.M., Bannister, A.J., Morrison, A., O'carroll, D., Firestein, R., Cleary, M., Jenuwein, T., Herrera, R.E., Kouzarides, T., 2001. Rb targets histone H3 methylation and HP1 to promoters. Nature 412, 561–565.

Nishino, J., Kim, I., Chada, K., Morrison, S.J., 2008. Hmga2 promotes neural stem cell self-renewal in young but not old mice by reducing p16Ink4a and p19Arf expression. Cell 135, 227–239.

Niwa, R., Zhou, F., Li, C., Slack, F.J., 2008. The expression of the Alzheimer's amyloid precursor protein-like gene is regulated by developmental timing microRNAs and their targets in Caenorhabditis elegans. Dev. Biol. 315, 418–425.

Noguchi, S., Mori, T., Otsuka, Y., Yamada, N., Yasui, Y., Iwasaki, J., Kumazaki, M., Maruo, K., Akao, Y., 2012. Anti-oncogenic microRNA-203 induces senescence by targeting E2F3 protein in human melanoma cells. J. Biol. Chem. 287, 11769–11777.

Noonan, E.J., Place, R.F., Basak, S., Pookot, D., Li, L.C., 2010. miR-449a causes Rb-dependent cell cycle arrest and senescence in prostate cancer cells. Oncotarget 1, 349–358.

Noren Hooten, N., Abdelmohsen, K., Gorospe, M., Ejiogu, N., Zonderman, A.B., Evans, M.K., 2010. microRNA expression patterns reveal differential expression of target genes with age. PLoS One 5, e10724.

O'donnell, K.A., Wentzel, E.A., Zeller, K.I., Dang, C.V., Mendell, J.T., 2005. c-Myc-regulated microRNAs modulate E2F1 expression. Nature 435, 839–843.

Ofir, M., Hacohen, D., Ginsberg, D., 2011. MiR-15 and miR-16 are direct transcriptional targets of E2F1 that limit E2F-induced proliferation by targeting cyclin E. Mol. Cancer Res. 9, 440–447.

Okada, N., Lin, C.P., Ribeiro, M.C., Biton, A., Lai, G., He, X., Bu, P., Vogel, H., Jablons, D.M., Keller, A.C., Wilkinson, J.E., He, B., Speed, T.P., He, L., 2014.

A positive feedback between p53 and miR-34 miRNAs mediates tumor suppression. Genes Dev. 28, 438–450.
Okada, M., Kim, H.W., Matsu-Ura, K., Wang, Y.G., Xu, M., Ashraf, M., 2016. Abrogation of age-induced MicroRNA-195 rejuvenates the senescent mesenchymal stem cells by reactivating telomerase. Stem Cells 34, 148–159.
Okano, M., Bell, D.W., Haber, D.A., Li, E., 1999. DNA methyltransferases Dnmt3a and Dnmt3b are essential for de novo methylation and mammalian development. Cell 99, 247–257.
Olivier, M., Hollstein, M., Hainaut, P., 2010. TP53 mutations in human cancers: origins, consequences, and clinical use. Cold Spring Harb. Perspect. Biol. 2, a001008.
Olivieri, F., Lazzarini, R., Recchioni, R., Marcheselli, F., Rippo, M.R., Di Nuzzo, S., Albertini, M.C., Graciotti, L., Babini, L., Mariotti, S., Spada, G., Abbatecola, A.M., Antonicelli, R., Franceschi, C., Procopio, A.D., 2013. MiR-146a as marker of senescence-associated pro-inflammatory status in cells involved in vascular remodelling. Age (Dordr.) 35, 1157–1172.
O'loghlen, A., Brookes, S., Martin, N., Rapisarda, V., Peters, G., Gil, J., 2015. CBX7 and miR-9 are part of an autoregulatory loop controlling p16(INK) (4a). Aging Cell 14, 1113–1121.
Olovnikov, A.M., 1996. Telomeres, telomerase, and aging: origin of the theory. Exp. Gerontol. 31, 443–448.
Otterson, G.A., Kratzke, R.A., Coxon, A., Kim, Y.W., Kaye, F.J., 1994. Absence of p16INK4 protein is restricted to the subset of lung cancer lines that retains wildtype RB. Oncogene 9, 3375–3378.
Overhoff, M.G., Garbe, J.C., Koh, J., Stampfer, M.R., Beach, D.H., Bishop, C.L., 2014. Cellular senescence mediated by p16INK4A-coupled miRNA pathways. Nucleic Acids Res. 42, 1606–1618.
Palmero, I., Pantoja, C., Serrano, M., 1998. p19ARF links the tumour suppressor p53 to Ras. Nature 395, 125–126.
Pampalakis, G., Diamandis, E.P., Katsaros, D., Sotiropoulou, G., 2010. Down-regulation of dicer expression in ovarian cancer tissues. Clin. Biochem. 43, 324–327.
Pantoja, C., Serrano, M., 1999. Murine fibroblasts lacking p21 undergo senescence and are resistant to transformation by oncogenic Ras. Oncogene 18, 4974–4982.
Papadopoulou, A.S., Serneels, L., Achsel, T., Mandemakers, W., Callaerts-Vegh, Z., Dooley, J., Lau, P., Ayoubi, T., Radaelli, E., Spinazzi, M., Neumann, M., Hebert, S.S., Silahtaroglu, A., Liston, A., D'hooge, R., Glatzel, M., De Strooper, B., 2015. Deficiency of the miR-29a/b-1 cluster leads to ataxic features and cerebellar alterations in mice. Neurobiol. Dis. 73, 275–288.
Pearson, M., Carbone, R., Sebastiani, C., Cioce, M., Fagioli, M., Saito, S., Higashimoto, Y., Appella, E., Minucci, S., Pandolfi, P.P., Pelicci, P.G., 2000. PML regulates p53 acetylation and premature senescence induced by oncogenic Ras. Nature 406, 207–210.
Pesole, G., Mignone, F., Gissi, C., Grillo, G., Licciulli, F., Liuni, S., 2001. Structural and functional features of eukaryotic mRNA untranslated regions. Gene 276, 73–81.
Petretto, E., Sarwar, R., Grieve, I., Lu, H., Kumaran, M.K., Muckett, P.J., Mangion, J., Schroen, B., Benson, M., Punjabi, P.P., Prasad, S.K., Pennell, D.J., Kiesewetter, C., Tasheva, E.S., Corpuz, L.M., Webb, M.D., Conrad, G.W., Kurtz, T.W., Kren, V., Fischer, J., Hubner, N., Pinto, Y.M., Pravenec, M., Aitman, T.J., Cook, S.A., 2008. Integrated genomic approaches implicate osteoglycin (Ogn) in the regulation of left ventricular mass. Nat. Genet. 40, 546–552.
Petrocca, F., Visone, R., Onelli, M.R., Shah, M.H., Nicoloso, M.S., De Martino, I., Iliopoulos, D., Pilozzi, E., Liu, C.G., Negrini, M., Cavazzini, L., Volinia, S., Alder, H., Ruco, L.P., Baldassarre, G., Croce, C.M., Vecchione, A., 2008.

E2F1-regulated microRNAs impair TGFbeta-dependent cell-cycle arrest and apoptosis in gastric cancer. Cancer Cell 13, 272–286.

Pichiorri, F., Suh, S.S., Rocci, A., De Luca, L., Taccioli, C., Santhanam, R., Zhou, W., Benson Jr., D.M., Hofmainster, C., Alder, H., Garofalo, M., Di Leva, G., Volinia, S., Lin, H.J., Perrotti, D., Kuehl, M., Aqeilan, R.I., Palumbo, A., Croce, C.M., 2010. Downregulation of p53-inducible microRNAs 192, 194, and 215 impairs the p53/MDM2 autoregulatory loop in multiple myeloma development. Cancer Cell 18, 367–381.

Pierce, A.M., Gimenez-Conti, I.B., Schneider-Broussard, R., Martinez, L.A., Conti, C.J., Johnson, D.G., 1998. Increased E2F1 activity induces skin tumors in mice heterozygous and nullizygous for p53. Proc. Natl. Acad. Sci. U.S.A. 95, 8858–8863.

Pierson, J., Hostager, B., Fan, R., Vibhakar, R., 2008. Regulation of cyclin dependent kinase 6 by microRNA 124 in medulloblastoma. J. Neurooncol 90, 1–7.

Poliseno, L., Pitto, L., Simili, M., Mariani, L., Riccardi, L., Ciucci, A., Rizzo, M., Evangelista, M., Mercatanti, A., Pandolfi, P.P., Rainaldi, G., 2008. The proto-oncogene LRF is under post-transcriptional control of MiR-20a: implications for senescence. PLoS One 3, e2542.

Puyol, M., Martin, A., Dubus, P., Mulero, F., Pizcueta, P., Khan, G., Guerra, C., Santamaria, D., Barbacid, M., 2010. A synthetic lethal interaction between K-Ras oncogenes and Cdk4 unveils a therapeutic strategy for non-small cell lung carcinoma. Cancer Cell 18, 63–73.

Quelle, D.E., Zindy, F., Ashmun, R.A., Sherr, C.J., 1995. Alternative reading frames of the Ink4a tumor-suppressor gene encode 2 unrelated proteins capable of inducing cell-cycle arrest. Cell 83, 993–1000.

Rabinowits, G., Gercel-Taylor, C., Day, J.M., Taylor, D.D., Kloecker, G.H., 2009. Exosomal microRNA: a diagnostic marker for lung cancer. Clin. Lung Cancer 10, 42–46.

Ramalinga, M., Roy, A., Srivastava, A., Bhattarai, A., Harish, V., Suy, S., Collins, S., Kumar, D., 2015. MicroRNA-212 negatively regulates starvation induced autophagy in prostate cancer cells by inhibiting SIRT1 and is a modulator of angiogenesis and cellular senescence. Oncotarget 6, 34446–34457.

Raman, V., Martensen, S.A., Reisman, D., Evron, E., Odenwald, W.F., Jaffee, E., Marks, J., Sukumar, S., 2000. Compromised HOXA5 function can limit p53 expression in human breast tumours. Nature 405, 974–978.

Rane, S.G., Cosenza, S.C., Mettus, R.V., Reddy, E.P., 2002. Germ line transmission of the Cdk4(R24C) mutation facilitates tumorigenesis and escape from cellular senescence. Mol. Cell. Biol. 22, 644–656.

Raver-Shapira, N., Marciano, E., Meiri, E., Spector, Y., Rosenfeld, N., Moskovits, N., Bentwich, Z., Oren, M., 2007. Transcriptional activation of miR-34a contributes to p53-mediated apoptosis. Mol. Cell 26, 731–743.

Rayman, J.B., Takahashi, Y., Indjeian, V.B., Dannenberg, J.H., Catchpole, S., Watson, R.J., Te Riele, H., Dynlacht, B.D., 2002. E2F mediates cell cycle-dependent transcriptional repression in vivo by recruitment of an HDAC1/mSin3B corepressor complex. Genes Dev. 16, 933–947.

Reimann, M., Lee, S., Loddenkemper, C., Dorr, J.R., Tabor, V., Aichele, P., Stein, H., Dorken, B., Jenuwein, T., Schmitt, C.A., 2010. Tumor stroma-derived TGF-beta limits myc-driven lymphomagenesis via Suv39h1-dependent senescence. Cancer Cell 17, 262–272.

Reinhart, B.J., Slack, F.J., Basson, M., Pasquinelli, A.E., Bettinger, J.C., Rougvie, A.E., Horvitz, H.R., Ruvkun, G., 2000. The 21-nucleotide let-7 RNA regulates developmental timing in Caenorhabditis elegans. Nature 403, 901–906.

Ren, X.S., Yin, M.H., Zhang, X., Wang, Z., Feng, S.P., Wang, G.X., Luo, Y.J., Liang, P.Z., Yang, X.Q., He, J.X., Zhang, B.L., 2014. Tumor-suppressive microRNA-449a induces growth arrest and senescence by targeting E2F3 in human lung cancer cells. Cancer Lett. 344, 195–203.

Riley, T., Sontag, E., Chen, P., Levine, A., 2008. Transcriptional control of human p53-regulated genes. Nat. Rev. Mol. Cell Biol. 9, 402–412.

Rivetti Di Val Cervo, P., Lena, A.M., Nicoloso, M., Rossi, S., Mancini, M., Zhou, H., Saintigny, G., Dellambra, E., Odorisio, T., Mahe, C., Calin, G.A., Candi, E., Melino, G., 2012. p63-microRNA feedback in keratinocyte senescence. Proc. Natl. Acad. Sci. U.S.A. 109, 1133–1138.

Robertson, K.D., 2005. DNA methylation and human disease. Nat. Rev. Genet. 6, 597–610.

Rodier, F., Campisi, J., 2011. Four faces of cellular senescence. J. Cell Biol. 192, 547–556.

Rodriguez, M.S., Desterro, J.M., Lain, S., Lane, D.P., Hay, R.T., 2000. Multiple C-terminal lysine residues target p53 for ubiquitin-proteasome-mediated degradation. Mol. Cell. Biol. 20, 8458–8467.

Rodriguez, A., Griffiths-Jones, S., Ashurst, J.L., Bradley, A., 2004. Identification of mammalian microRNA host genes and transcription units. Genome Res. 14, 1902–1910.

Roman-Gomez, J., Agirre, X., Jimenez-Velasco, A., Arqueros, V., Vilas-Zornoza, A., Rodriguez-Otero, P., Martin-Subero, I., Garate, L., Cordeu, L., San Jose-Eneriz, E., Martin, V., Castillejo, J.A., Bandres, E., Calasanz, M.J., Siebert, R., Heiniger, A., Torres, A., Prosper, F., 2009. Epigenetic regulation of microRNAs in acute lymphoblastic leukemia. J. Clin. Oncol. 27, 1316–1322.

Romero-Cordoba, S.L., Salido-Guadarrama, I., Rodriguez-Dorantes, M., Hidalgo-Miranda, A., 2014. miRNA biogenesis: biological impact in the development of cancer. Cancer Biol. Ther. 15, 1444–1455.

Roshan, R., Shridhar, S., Sarangdhar, M.A., Banik, A., Chawla, M., Garg, M., Singh, V.P., Pillai, B., 2014. Brain-specific knockdown of miR-29 results in neuronal cell death and ataxia in mice. RNA 20, 1287–1297.

Rowland, B.D., Bernards, R., Peeper, D.S., 2005. The KLF4 tumour suppressor is a transcriptional repressor of p53 that acts as a context-dependent oncogene. Nat. Cell Biol. 7, 1074–1082.

Rupaimoole, R., Slack, F.J., 2017. MicroRNA therapeutics: towards a new era for the management of cancer and other diseases. Nat. Rev. Drug Discov. 16, 203–222.

Russo, A.A., Tong, L., Lee, J.O., Jeffrey, P.D., Pavletich, N.P., 1998. Structural basis for inhibition of the cyclin-dependent kinase Cdk6 by the tumour suppressor p16INK4a. Nature 395, 237–243.

Sager, R., Tanaka, K., Lau, C.C., Ebina, Y., Anisowicz, A., 1983. Resistance of human cells to tumorigenesis induced by cloned transforming genes. Proc. Natl. Acad. Sci. U.S.A. 80, 7601–7605.

Salama, R., Sadaie, M., Hoare, M., Narita, M., 2014. Cellular senescence and its effector programs. Genes Dev. 28, 99–114.

Sarkar, S., Jun, S., Rellick, S., Quintana, D.D., Cavendish, J.Z., Simpkins, J.W., 2016. Expression of microRNA-34a in Alzheimer's disease brain targets genes linked to synaptic plasticity, energy metabolism, and resting state network activity. Brain Res. 1646, 139–151.

Sarkisian, C.J., Keister, B.A., Stairs, D.B., Boxer, R.B., Moody, S.E., Chodosh, L.A., 2007. Dose-dependent oncogene-induced senescence in vivo and its evasion during mammary tumorigenesis. Nat. Cell Biol. 9, 493–505.

Sasaki, M., Kumazaki, T., Takano, H., Nishiyama, M., Mitsui, Y., 2001. Senescent cells are resistant to death despite low Bcl-2 level. Mech. Ageing Dev. 122, 1695–1706.

Satoh, A., Brace, C.S., Rensing, N., Cliften, P., Wozniak, D.F., Herzog, E.D., Yamada, K.A., Imai, S., 2013. Sirt1 extends life span and delays aging in mice through the regulation of Nk2 homeobox 1 in the DMH and LH. Cell Metab. 18, 416–430.

Schipper, H.M., Maes, O.C., Chertkow, H.M., Wang, E., 2007. MicroRNA expression in Alzheimer blood mononuclear cells. Gene Regul. Syst. Biol. 1, 263–274.
Seger, Y.R., Garcia-Cao, M., Piccinin, S., Cunsolo, C.L., Doglioni, C., Blasco, M.A., Hannon, G.J., Maestro, R., 2002. Transformation of normal human cells in the absence of telomerase activation. Cancer Cell 2, 401–413.
Senoo, M., Pinto, F., Crum, C.P., Mckeon, F., 2007. p63 Is essential for the proliferative potential of stem cells in stratified epithelia. Cell 129, 523–536.
Serrano, M., Lee, H., Chin, L., Cordon-Cardo, C., Beach, D., Depinho, R.A., 1996. Role of the INK4a locus in tumor suppression and cell mortality. Cell 85, 27–37.
Serrano, M., Lin, A.W., Mccurrach, M.E., Beach, D., Lowe, S.W., 1997. Oncogenic ras provokes premature cell senescence associated with accumulation of p53 and p16INK4a. Cell 88, 593–602.
Shapiro, G.I., Edwards, C.D., Kobzik, L., Godleski, J., Richards, W., Sugarbaker, D.J., Rollins, B.J., 1995. Reciprocal Rb inactivation and p16INK4 expression in primary lung cancers and cell lines. Cancer Res. 55, 505–509.
Shay, J.W., Bacchetti, S., 1997. A survey of telomerase activity in human cancer. Eur. J. Cancer 33, 787–791.
Shay, J.W., Wright, W.E., 2000. Hayflick, his limit, and cellular ageing. Nat. Rev. Mol. Cell Biol. 1, 72–76.
Shen, J., Yang, X., Xie, B., Chen, Y., Swaim, M., Hackett, S.F., Campochiaro, P.A., 2008. MicroRNAs regulate ocular neovascularization. Mol. Ther. 16, 1208–1216.
Shi, X., Tian, B., Liu, L., Gao, Y., Ma, C., Mwichie, N., Ma, W., Han, L., Huang, B., Lu, J., Zhang, Y., 2013. Rb protein is essential to the senescence-associated heterochromatic foci formation induced by HMGA2 in primary WI38 cells. J. Genet. Genomics 40, 391–398.
Shieh, S.Y., Ikeda, M., Taya, Y., Prives, C., 1997. DNA damage-induced phosphorylation of p53 alleviates inhibition by MDM2. Cell 91, 325–334.
Shirodkar, S., Ewen, M., Decaprio, J.A., Morgan, J., Livingston, D.M., Chittenden, T., 1992. The transcription factor E2F interacts with the retinoblastoma product and a p107-cyclin A complex in a cell cycle-regulated manner. Cell 68, 157–166.
Shvarts, A., Steegenga, W.T., Riteco, N., Van Laar, T., Dekker, P., Bazuine, M., Van Ham, R.C., Van Der Houven Van Oordt, W., Hateboer, G., Van Der Eb, A.J., Jochemsen, A.G., 1996. MDMX: a novel p53-binding protein with some functional properties of MDM2. EMBO J. 15, 5349–5357.
Siebig, S., Rockmann, F., Sabel, K., Zuber-Jerger, I., Dierkes, C., Brunnler, T., Wrede, C.E., 2009. Continuous non-invasive arterial pressure technique improves patient monitoring during interventional endoscopy. Int. J. Med. Sci. 6, 37–42.
Smith, J., Tho, L.M., Xu, N., Gillespie, D.A., 2010. The ATM-Chk2 and ATR-Chk1 pathways in DNA damage signaling and cancer. Adv. Cancer Res. 108, 73–112.
Song, R., Walentek, P., Sponer, N., Klimke, A., Lee, J.S., Dixon, G., Harland, R., Wan, Y., Lishko, P., Lize, M., Kessel, M., He, L., 2014. miR-34/449 miRNAs are required for motile ciliogenesis by repressing cp110. Nature 510, 115–120.
Soriano-Arroquia, A., House, L., Tregilgas, L., Canty-Laird, E., Goljanek-Whysall, K., 2016. The functional consequences of age-related changes in microRNA expression in skeletal muscle. Biogerontology 17, 641–654.
Soto-Reyes, E., Recillas-Targa, F., 2010. Epigenetic regulation of the human p53 gene promoter by the CTCF transcription factor in transformed cell lines. Oncogene 29, 2217–2227.
Stevens, C., La Thangue, N.B., 2003. E2F and cell cycle control: a double-edged sword. Arch. Biochem. Biophys. 412, 157–169.
Stevenson, M., Volsky, D.J., 1986. Activated v-myc and v-ras oncogenes do not transform normal human lymphocytes. Mol. Cell. Biol. 6, 3410–3417.

Sun, F., Fu, H., Liu, Q., Tie, Y., Zhu, J., Xing, R., Sun, Z., Zheng, X., 2008. Down-regulation of CCND1 and CDK6 by miR-34a induces cell cycle arrest. FEBS Lett. 582, 1564–1568.

Suzuki, H.I., Yamagata, K., Sugimoto, K., Iwamoto, T., Kato, S., Miyazono, K., 2009. Modulation of microRNA processing by p53. Nature 460, 529–533.

Sylvestre, Y., De Guire, V., Querido, E., Mukhopadhyay, U.K., Bourdeau, V., Major, F., Ferbeyre, G., Chartrand, P., 2007. An E2F/miR-20a autoregulatory feedback loop. J. Biol. Chem. 282, 2135–2143.

Szemraj, M., Bielecka-Kowalska, A., Oszajca, K., Krajewska, M., Gos, R., Jurowski, P., Kowalski, M., Szemraj, J., 2015. Serum microRNAs as potential biomarkers of AMD. Med. Sci. Monit. 21, 2734–2742.

Taganov, K.D., Boldin, M.P., Chang, K.J., Baltimore, D., 2006. NF-kappaB-dependent induction of microRNA miR-146, an inhibitor targeted to signaling proteins of innate immune responses. Proc. Natl. Acad. Sci. U.S.A. 103, 12481–12486.

Takamizawa, J., Konishi, H., Yanagisawa, K., Tomida, S., Osada, H., Endoh, H., Harano, T., Yatabe, Y., Nagino, M., Nimura, Y., Mitsudomi, T., Takahashi, T., 2004. Reduced expression of the let-7 microRNAs in human lung cancers in association with shortened postoperative survival. Cancer Res. 64, 3753–3756.

Talluri, S., Isaac, C.E., Ahmad, M., Henley, S.A., Francis, S.M., Martens, A.L., Bremner, R., Dick, F.A., 2010. A G1 checkpoint mediated by the retinoblastoma protein that is dispensable in terminal differentiation but essential for senescence. Mol. Cell. Biol. 30, 948–960.

Tanaka, H., Fujii, Y., Hirabayashi, H., Miyoshi, S., Sakaguchi, M., Yoon, H.E., Matsuda, H., 1998. Disruption of the RB pathway and cell-proliferative activity in non-small-cell lung cancers. Int. J. Cancer 79, 111–115.

Tarasov, V., Jung, P., Verdoodt, B., Lodygin, D., Epanchintsev, A., Menssen, A., Meister, G., Hermeking, H., 2007. Differential regulation of microRNAs by p53 revealed by massively parallel sequencing: miR-34a is a p53 target that induces apoptosis and G1-arrest. Cell Cycle 6, 1586–1593.

Tay, Y., Zhang, J., Thomson, A.M., Lim, B., Rigoutsos, I., 2008. MicroRNAs to Nanog, Oct4 and Sox2 coding regions modulate embryonic stem cell differentiation. Nature 455, 1124–1128.

Tazawa, H., Tsuchiya, N., Izumiya, M., Nakagama, H., 2007. Tumor-suppressive miR-34a induces senescence-like growth arrest through modulation of the E2F pathway in human colon cancer cells. Proc. Natl. Acad. Sci. U.S.A. 104, 15472–15477.

Tront, J.S., Hoffman, B., Liebermann, D.A., 2006. Gadd45a suppresses ras-driven mammary tumorigenesis by activation of c-Jun NH2-terminal kinase and p38 stress signaling resulting in apoptosis and senescence. Cancer Res. 66, 8448–8454.

Tyner, S.D., Venkatachalam, S., Choi, J., Jones, S., Ghebranious, N., Igelmann, H., Lu, X., Soron, G., Cooper, B., Brayton, C., Park, S.H., Thompson, T., Karsenty, G., Bradley, A., Donehower, L.A., 2002. p53 mutant mice that display early ageing-associated phenotypes. Nature 415, 45–53.

Tzatsos, A., Paskaleva, P., Lymperi, S., Contino, G., Stoykova, S., Chen, Z., Wong, K.K., Bardeesy, N., 2011. Lysine-specific demethylase 2B (KDM2B)-let-7-enhancer of zester homolog 2 (EZH2) pathway regulates cell cycle progression and senescence in primary cells. J. Biol. Chem. 286, 33061–33069.

Valadi, H., Ekstrom, K., Bossios, A., Sjostrand, M., Lee, J.J., Lotvall, J.O., 2007. Exosome-mediated transfer of mRNAs and microRNAs is a novel mechanism of genetic exchange between cells. Nat. Cell Biol. 9, 654–659.

Van Deursen, J.M., 2014. The role of senescent cells in ageing. Nature 509, 439–446.

Vasa-Nicotera, M., Chen, H., Tucci, P., Yang, A.L., Saintigny, G., Menghini, R., Mahe, C., Agostini, M., Knight, R.A., Melino, G., Federici, M., 2011. miR-146a is modulated in human endothelial cell with aging. Atherosclerosis 217, 326–330.

Vaziri, H., Dessain, S.K., Ng Eaton, E., Imai, S.I., Frye, R.A., Pandita, T.K., Guarente, L., Weinberg, R.A., 2001. hSIR2(SIRT1) functions as an NAD-dependent p53 deacetylase. Cell 107, 149–159.

Venkataraman, S., Alimova, I., Fan, R., Harris, P., Foreman, N., Vibhakar, R., 2010. MicroRNA 128a increases intracellular ROS level by targeting Bmi-1 and inhibits medulloblastoma cancer cell growth by promoting senescence. PLoS One 5, e10748.

Ventura, A., Kirsch, D.G., Mclaughlin, M.E., Tuveson, D.A., Grimm, J., Lintault, L., Newman, J., Reczek, E.E., Weissleder, R., Jacks, T., 2007. Restoration of p53 function leads to tumour regression in vivo. Nature 445, 661–665.

Vickers, K.C., Palmisano, B.T., Shoucri, B.M., Shamburek, R.D., Remaley, A.T., 2011. MicroRNAs are transported in plasma and delivered to recipient cells by high-density lipoproteins. Nat. Cell Biol. 13, 423–433.

Volinia, S., Calin, G.A., Liu, C.G., Ambs, S., Cimmino, A., Petrocca, F., Visone, R., Iorio, M., Roldo, C., Ferracin, M., Prueitt, R.L., Yanaihara, N., Lanza, G., Scarpa, A., Vecchione, A., Negrini, M., Harris, C.C., Croce, C.M., 2006. A microRNA expression signature of human solid tumors defines cancer gene targets. Proc. Natl. Acad. Sci. U.S.A. 103, 2257–2261.

Voller, D., Reinders, J., Meister, G., Bosserhoff, A.K., 2013. Strong reduction of AGO2 expression in melanoma and cellular consequences. Br. J. Cancer 109, 3116–3124.

Voorhoeve, P.M., Le Sage, C., Schrier, M., Gillis, A.J., Stoop, H., Nagel, R., Liu, Y.P., Van Duijse, J., Drost, J., Griekspoor, A., Zlotorynski, E., Yabuta, N., De Vita, G., Nojima, H., Looijenga, L.H., Agami, R., 2006. A genetic screen implicates miRNA-372 and miRNA-373 as oncogenes in testicular germ cell tumors. Cell 124, 1169–1181.

Vousden, K.H., Lu, X., 2002. Live or let die: the cell's response to p53. Nat. Rev. Cancer 2, 594–604.

Wada, T., Joza, N., Cheng, H.Y., Sasaki, T., Kozieradzki, I., Bachmaier, K., Katada, T., Schreiber, M., Wagner, E.F., Nishina, H., Penninger, J.M., 2004. MKK7 couples stress signalling to G2/M cell-cycle progression and cellular senescence. Nat. Cell Biol. 6, 215–226.

Wajapeyee, N., Serra, R.W., Zhu, X., Mahalingam, M., Green, M.R., 2008. Oncogenic BRAF induces senescence and apoptosis through pathways mediated by the secreted protein IGFBP7. Cell 132, 363–374.

Wang, E., 1995. Senescent human fibroblasts resist programmed cell death, and failure to suppress bcl2 is involved. Cancer Res. 55, 2284–2292.

Wang, W., Chen, J.X., Liao, R., Deng, Q., Zhou, J.J., Huang, S., Sun, P., 2002. Sequential activation of the MEK-extracellular signal-regulated kinase and MKK3/6-p38 mitogen-activated protein kinase pathways mediates oncogenic ras-induced premature senescence. Mol. Cell. Biol. 22, 3389–3403.

Wang, C., Jurk, D., Maddick, M., Nelson, G., Martin-Ruiz, C., Von Zglinicki, T., 2009a. DNA damage response and cellular senescence in tissues of aging mice. Aging Cell 8, 311–323.

Wang, X., Liu, P., Zhu, H., Xu, Y., Ma, C., Dai, X., Huang, L., Liu, Y., Zhang, L., Qin, C., 2009b. miR-34a, a microRNA up-regulated in a double transgenic mouse model of Alzheimer's disease, inhibits bcl2 translation. Brain Res. Bull. 80, 268–273.

Weber, J.A., Baxter, D.H., Zhang, S., Huang, D.Y., Huang, K.H., Lee, M.J., Galas, D.J., Wang, K., 2010. The microRNA spectrum in 12 body fluids. Clin. Chem. 56, 1733–1741.

Wei, S., Sedivy, J.M., 1999. Expression of catalytically active telomerase does not prevent premature senescence caused by overexpression of oncogenic Ha-Ras in normal human fibroblasts. Cancer Res. 59, 1539–1543.

Wei, W., Hemmer, R.M., Sedivy, J.M., 2001. Role of p14(ARF) in replicative and induced senescence of human fibroblasts. Mol. Cell. Biol. 21, 6748–6757.

Wei, W., Jobling, W.A., Chen, W., Hahn, W.C., Sedivy, J.M., 2003. Abolition of cyclin-dependent kinase inhibitor p16Ink4a and p21Cip1/Waf1 functions permits Ras-induced anchorage-independent growth in telomerase-immortalized human fibroblasts. Mol. Cell. Biol. 23, 2859–2870.
Wei, J.S., Song, Y.K., Durinck, S., Chen, Q.R., Cheuk, A.T.C., Tsang, P., Zhang, Q., Thiele, C.J., Slack, A., Shohet, J., Khan, J., 2008. The MYCN oncogene is a direct target of miR-34a. Oncogene 27, 5204–5213.
Wei, J., Shi, Y., Zheng, L., Zhou, B., Inose, H., Wang, J., Guo, X.E., Grosschedl, R., Karsenty, G., 2012. miR-34s inhibit osteoblast proliferation and differentiation in the mouse by targeting SATB2. J. Cell Biol. 197, 509–521.
Welch, C., Chen, Y., Stallings, R.L., 2007. MicroRNA-34a functions as a potential tumor suppressor by inducing apoptosis in neuroblastoma cells. Oncogene 26, 5017–5022.
Weng, J.H., Yu, C.C., Lee, Y.C., Lin, C.W., Chang, W.W., Kuo, Y.L., 2016. miR-494-3p induces cellular senescence and enhances radiosensitivity in human oral squamous carcinoma cells. Int. J. Mol. Sci. 17, 1–10, article number 1092.
Whetstine, J.R., Nottke, A., Lan, F., Huarte, M., Smolikov, S., Chen, Z., Spooner, E., Li, E., Zhang, G., Colaiacovo, M., Shi, Y., 2006. Reversal of histone lysine trimethylation by the JMJD2 family of histone demethylases. Cell 125, 467–481.
Wightman, B., Ha, I., Ruvkun, G., 1993. Posttranscriptional regulation of the heterochronic gene lin-14 by lin-4 mediates temporal pattern formation in C. elegans. Cell 75, 855–862.
Wu, G., Nomoto, S., Hoque, M.O., Dracheva, T., Osada, M., Lee, C.C., Dong, S.M., Guo, Z., Benoit, N., Cohen, Y., Rechthand, P., Califano, J., Moon, C.S., Ratovitski, E., Jen, J., Sidransky, D., Trink, B., 2003. DeltaNp63alpha and TAp63alpha regulate transcription of genes with distinct biological functions in cancer and development. Cancer Res. 63, 2351–2357.
Xiao, J., Lin, H., Luo, X., Wang, Z., 2011. miR-605 joins p53 network to form a p53:miR-605:Mdm2 positive feedback loop in response to stress. EMBO J. 30, 5021.
Xiong, Y., Hannon, G.J., Zhang, H., Casso, D., Kobayashi, R., Beach, D., 1993. p21 is a universal inhibitor of cyclin kinases. Nature 366, 701–704.
Xu, D., Takeshita, F., Hino, Y., Fukunaga, S., Kudo, Y., Tamaki, A., Matsunaga, J., Takahashi, R.U., Takata, T., Shimamoto, A., Ochiya, T., Tahara, H., 2011. miR-22 represses cancer progression by inducing cellular senescence. J. Cell Biol. 193, 409–424.
Xu, Q., Liu, M., Zhang, J., Xue, L., Zhang, G., Hu, C., Wang, Z., He, S., Chen, L., Ma, K., Liu, X., Zhao, Y., Lv, N., Liang, S., Zhu, H., Xu, N., 2016a. Overexpression of KLF4 promotes cell senescence through microRNA-203-survivin-p21 pathway. Oncotarget 7, 60290–60302.
Xu, S., Huang, H., Li, N., Zhang, B., Jia, Y., Yang, Y., Yuan, Y., Xiong, X.D., Wang, D., Zheng, H.L., Liu, X., 2016b. MicroRNA-33 promotes the replicative senescence of mouse embryonic fibroblasts by suppressing CDK6. Biochem. Biophys. Res. Commun. 473, 1064–1070.
Xue, W., Zender, L., Miething, C., Dickins, R.A., Hernando, E., Krizhanovsky, V., Cordon-Cardo, C., Lowe, S.W., 2007. Senescence and tumour clearance is triggered by p53 restoration in murine liver carcinomas. Nature 445, 656–660.
Yamakuchi, M., Ferlito, M., Lowenstein, C.J., 2008. miR-34a repression of SIRT1 regulates apoptosis. Proc. Natl. Acad. Sci. U.S.A. 105, 13421–13426.
Yan, L., Lee, S., Lazzaro, D.R., Aranda, J., Grant, M.B., Chaqour, B., 2015. Single and compound knock-outs of MicroRNA (miRNA)-155 and its angiogenic gene target CCN1 in mice alter vascular and neovascular growth in the retina via resident microglia. J. Biol. Chem. 290, 23264–23281.
Yang, J.S., Lai, E.C., 2011. Alternative miRNA biogenesis pathways and the interpretation of core miRNA pathway mutants. Mol. Cell 43, 892–903.

Yang, H.S., Alexander, K., Santiago, P., Hinds, P.W., 2003. ERM proteins and Cdk5 in cellular senescence. Cell Cycle 2, 517–520.

Yang, X., Feng, M., Jiang, X., Wu, Z., Li, Z., Aau, M., Yu, Q., 2009. miR-449a and miR-449b are direct transcriptional targets of E2F1 and negatively regulate pRb-E2F1 activity through a feedback loop by targeting CDK6 and CDC25A. Genes Dev. 23, 2388–2393.

Yang, S., Li, W.S., Dong, F., Sun, H.M., Wu, B., Tan, J., Zou, W.J., Zhou, D.S., 2014. KITLG is a novel target of miR-34c that is associated with the inhibition of growth and invasion in colorectal cancer cells. J. Cell. Mol. Med. 18, 2092–2102.

Yang, S., Wu, B., Sun, H., Ji, F., Sun, T., Zhao, Y., Zhou, D., 2015. Interrupted E2F1-miR-34c-SCF negative feedback loop by hyper-methylation promotes colorectal cancer cell proliferation. Biosci. Rep. 36, e00293.

Ye, Z., Fang, J., Dai, S., Wang, Y., Fu, Z., Feng, W., Wei, Q., Huang, P., 2016. Micro-RNA-34a induces a senescence-like change via the down-regulation of SIRT1 and up-regulation of p53 protein in human esophageal squamous cancer cells with a wild-type p53 gene background. Cancer Lett. 370, 216–221.

Yi, R., Qin, Y., Macara, I.G., Cullen, B.R., 2003. Exportin-5 mediates the nuclear export of pre-microRNAs and short hairpin RNAs. Genes Dev. 17, 3011–3016.

Young, A.R., Narita, M., Ferreira, M., Kirschner, K., Sadaie, M., Darot, J.F., Tavare, S., Arakawa, S., Shimizu, S., Watt, F.M., Narita, M., 2009. Autophagy mediates the mitotic senescence transition. Genes Dev. 23, 798–803.

Yu, Z.B., Li, Z., Jolicoeur, N., Zhang, L.H., Fortin, Y., Wang, E., Wu, M.Q., Shen, S.H., 2007. Aberrant allele frequencies of the SNPs located in microRNA target sites are potentially associated with human cancers. Nucleic Acids Res. 35, 4535–4541.

Zhang, G., Zhu, H., Wang, Y., Yang, S., Liu, M., Zhang, W., Quan, L., Bai, J., Liu, Z., Xu, N., 2009. Kruppel-like factor 4 represses transcription of the survivin gene in esophageal cancer cell lines. Biol. Chem. 390, 463–469.

Zhang, J., Sun, Q., Zhang, Z., Ge, S., Han, Z.G., Chen, W.T., 2013. Loss of microRNA-143/145 disturbs cellular growth and apoptosis of human epithelial cancers by impairing the MDM2-p53 feedback loop. Oncogene 32, 61–69.

Zhang, F., Cui, J., Liu, X., Lv, B., Xie, Z., Yu, B., 2015. Roles of microRNA-34a targeting SIRT1 in mesenchymal stem cells. Stem Cell Res. Ther. 6, 195.

Zhao, W., O'malley, Y., Wei, S., Robbins, M.E., 2000. Irradiation of rat tubule epithelial cells alters the expression of gene products associated with the synthesis and degradation of extracellular matrix. Int. J. Radiat. Biol. 76, 391–402.

Zhou, Q., Gallagher, R., Ufret-Vincenty, R., Li, X., Olson, E.N., Wang, S., 2011. Regulation of angiogenesis and choroidal neovascularization by members of microRNA-23~27~24 clusters. Proc. Natl. Acad. Sci. U.S.A. 108, 8287–8292.

Zhou, Y., Hu, Y., Yang, M., Jat, P., Li, K., Lombardo, Y., Xiong, D., Coombes, R.C., Raguz, S., Yague, E., 2014. The miR-106b~25 cluster promotes bypass of doxorubicin-induced senescence and increase in motility and invasion by targeting the E-cadherin transcriptional activator EP300. Cell Death Differ. 21, 462–474.

Zhu, J., Woods, D., Mcmahon, M., Bishop, J.M., 1998. Senescence of human fibroblasts induced by oncogenic Raf. Genes Dev. 12, 2997–3007.

Zindy, F., Eischen, C.M., Randle, D.H., Kamijo, T., Cleveland, J.L., Sherr, C.J., Roussel, M.F., 1998. Myc signaling via the ARF tumor suppressor regulates p53-dependent apoptosis and immortalization. Genes Dev. 12, 2424–2433.

Zovoilis, A., Agbemenyah, H.Y., Agis-Balboa, R.C., Stilling, R.M., Edbauer, D., Rao, P., Farinelli, L., Delalle, I., Schmitt, A., Falkai, P., Bahari-Javan, S., Burkhardt, S., Sananbenesi, F., Fischer, A., 2011. microRNA-34c is a novel target to treat dementias. EMBO J. 30, 4299–4308.

CHAPTER THREE

A Macro View of MicroRNAs: The Discovery of MicroRNAs and Their Role in Hematopoiesis and Hematologic Disease

Cary N. Weiss, Keisuke Ito[1]
Ruth L. and David S. Gottesman Institute for Stem Cell and Regenerative Medicine Research, Albert Einstein Cancer Center and Einstein Diabetes Research Center, Albert Einstein College of Medicine, Bronx, NY, United States
[1]Corresponding author: e-mail address: keisuke.ito@einstein.yu.edu

Contents

International Review of Cell and Molecular Biology, Volume 334
ISSN 1937-6448
http://dx.doi.org/10.1016/bs.ircmb.2017.03.007

Abstract

MicroRNAs (MiRNAs) are a class of endogenously encoded ~22 nucleotide, noncoding, single-stranded RNAs that contribute to development, body planning, stem cell differentiation, and tissue identity through posttranscriptional regulation and degradation of transcripts. Given their importance, it is predictable that dysregulation of MiRNAs, which target a wide variety of transcripts, can result in malignant transformation. In this review, we explore the discovery of MiRNAs, their mechanism of action, and the tools that aid in their discovery and study. Strikingly, many of the studies that have expanded our understanding of the contributions of MiRNAs to normal physiology and in the development of diseases have come from studies in the hematopoietic system and hematologic malignancies, with some of the earliest identified functions for mammalian MiRNAs coming from observations made in leukemias. So, with a special focus on the hematologic system, we will discuss how MiRNAs contribute to differentiation of stem cells and how dysregulation of MiRNAs contributes to the development of malignancy, by providing examples of specific MiRNAs that function as oncogenes or tumor suppressors, as well as of defects in MiRNA processing. Finally, we will discuss the promise of MiRNA-based therapeutics and challenges for the future study of disease-causing MiRNAs.

1. INTRODUCTION

MicroRNAs (MiRNAs) are just one in an expanding class of noncoding RNAs (ncRNAs), which contribute to diverse biological processes, and the continuous discovery of different classes of ncRNAs has disrupted the conception of commonly established roles for proteins and RNAs in the regulation of cellular activity (Cech and Steitz, 2014). ncRNAs include a diverse set of RNA transcripts that are not translated into proteins. The earliest ncRNAs identified were ribosomal RNAs, which are major constituents of the ribosome and contribute to translation as a ribozyme, and transfer RNAs (tRNAs), which are the adaptor molecules that translate the triplet codon of mRNAs into an amino acid. More recently discovered ncRNAs include small nuclear RNAs, which include splicing-associated RNAs and small nucleolar RNAs; small interfering RNAs (siRNAs); MiRNAs; PIWI-associated RNAs (Aravin et al., 2006); and long noncoding RNAs (lncRNAs), which include competing endogenous RNAs (ceRNAs) (Tay et al., 2011), circular RNAs (circRNAs) (Zaphiropoulos, 1997), and transcribed pseudogenes (Poliseno et al., 2010).

MiRNAs, which will be the focus of this review, are small, ~22 nucleotide (nt), single-stranded, endogenously encoded, ncRNAs that serve a critical role in posttranscriptional regulation of protein expression, and thus contribute to a wide range of biological processes and in the development of disease. This posttranscriptional regulation is so important to normal physiology that more than 60% of human protein-coding genes are under selective evolutionary pressure to maintain MiRNA binding sites, also called MiRNA response elements (MREs), in their 3′ untranslated regions (3′-UTRs) (Friedman et al., 2009). MiRNAs are encoded and transcribed in the nucleus and are exported to the cytoplasm, where they become incorporated into ribonucleoprotein-silencing machinery. Serving as the target recognition component of the ribonucleoprotein-silencing machinery, MiRNAs identify specific transcripts, in a sequence-specific manner, for translational repression and transcript destabilization. Thousands of putative MiRNAs have been identified in the human genome, with hundreds having been experimentally validated to effect known targets.

Soon after the discovery of MiRNAs in humans and other vertebrates, researchers began to recognize the contribution of MiRNA dysregulation in the development of cancers, initially describing the role of MiR-15 and MiR-16 loss in del(13q) chronic lymphocytic leukemia (CLL) (Calin et al., 2002). With the appreciation of the important role of MiRNAs in physiology and disease, the number of studies investigating MiRNAs has exploded, with large-scale projects, such as the Cancer Genome Atlas (TCGA), collecting next-generation sequencing data for the examination of MiRNA dysregulation in disease, and researchers developing complicated conditional models of MiRNA overexpression and knockdown to more faithfully recapitulate MiRNA dysregulation in experimental animals.

In the past decades, researchers have made tremendous strides toward understanding MiRNA biogenesis and function, the structure of MiRNA-encoding genes, and the sequence specificity of MiRNA targeting. More recently, the predominant mechanisms that MiRNAs use to silence their targeted transcripts have been elucidated, shedding light on that controversial topic (Eichhorn et al., 2014; Ricci et al., 2013). With many tools at their disposal, including target-site prediction algorithms, next-generation sequencing, and animal models, among many others, researchers are now well equipped for sophisticated studies of MiRNAs, their function in normal physiology, and the mechanisms through which they can become dysregulated and contribute to disease.

With increasing attention on the role of MiRNAs as essential regulators of cellular processes, and in the development and progression of disease, more attention has been focused on the posttranscriptional regulation of MiRNAs themselves. Recent publications have suggested mechanisms through which MiRNA activity may be modulated by changing the number of available MREs. mRNAs transcribed from pseudogenes (pseudo-mRNAs), circRNAs, and other ceRNAs have been shown to antagonize MiRNA activity by competing with mRNAs sharing MREs for available MiRNAs (Arvey et al., 2010; Cazalla et al., 2010; Hansen et al., 2013; Jeyapalan et al., 2011; Poliseno et al., 2010; Tay et al., 2011; Wang et al., 2008). While work is still being done to understand how MiRNA expression is transcriptionally regulated, the emerging role of these endogenously encoded MiRNA sponges makes understanding MiRNA-regulatory networks even more complicated. Because ceRNAs can divert MiRNAs from binding to their targets, total MiRNA may not be a sufficient measure of MiRNA efficacy; rather, MiRNA availability may need to be calculated or experimentally observed using reporters. Still, in many contexts, MiRNA number may drastically outweigh the availability of MREs on ceRNAs, so careful quantitation of MiRNAs and ceRNAs is necessary for a complete understanding of MiRNA regulation. The intricacy of MiRNA regulation highlights the importance of these small RNAs to cellular processes.

In this review, we will examine the history of the discovery of MiRNAs, including their first recognition, the experiments that elucidated their mechanisms of action in repressing translation, and the various bioinformatic tools that have facilitated the discovery of MiRNAs and their targets. We will also discuss the emerging research that has demonstrated the essential roles of MiRNAs in development and normal physiological processes, as well as the increasing studies that demonstrate the role of MiRNAs in the pathogenesis of disease. As many of the earliest MiRNA-associated diseases were leukemias and much research has been conducted in this area, we will apply specific focus to MiRNAs that contribute to the regulation of the hematopoietic system and the dysregulations that contribute to hematologic malignancies. Finally, we will discuss the burgeoning research into MiRNA-based therapeutic strategies, their potential and pitfalls, and some promising clinical data that suggest that MiRNA-based therapeutics may become increasingly prevalent in the decades to come.

2. THE MicroRNA REVOLUTION: A HISTORY

In the past few decades, the scientific community has witnessed a revolution in our understanding of gene expression and posttranscriptional regulation. Previously, genome structure and subtypes of proteins, such as transcription factors and epigenetic mediators, were thought to be the only modulators of gene expression. However, in recent years, the examination of MiRNAs and other ncRNAs, which participate in posttranscriptional regulation, has revealed additional layers of regulatory machinery that are critical to a variety of cellular functions. The discovery and study of MiRNAs and their associated cellular machinery has disrupted the way in which we think about the control of gene expression and has provided new molecular tools for scientific inquiry. Further, this revolutionary period is paradigmatic of the capacity of the scientific community to adjust its beliefs and models when confronted with new information. In this section, we will review the early history of the discovery of MiRNAs and the realization that MiRNAs contribute to posttranscriptional regulation in a variety of organisms. We will also discuss the biogenesis of MiRNAs, their mechanism of action, and tools that have been developed to aid in their discovery and in the identification of their targets.

2.1 The Discovery of MicroRNAs

Many researchers in the biological sciences will remember the discovery of a class of small ncRNAs now known as MicroRNAs or MiRNAs, almost 25 years ago (Fig. 1). In a landmark paper published in the early 1990s, Lee et al. described their findings that *lin-4*, a gene important in the regulation of nematode larval development, encodes a small 22 nt RNA and a larger RNA thought to be its precursor (Lee et al., 1993). Cloning efforts led to the isolation of the genomic encoding region of *lin-4*, where the absence of start and stop codons revealed that it was unlikely that *lin-4* was protein coding, and probes developed based on the *lin-4* sequence were used to identify the *lin-4S* and *lin-4L* small RNAs, which we now know to be the MiRNA and its precursor. Further, the Ambros group recognized that a short sequence within the lin-4 small RNA was complementary to the 3′-UTR of the heterochronic gene, *lin-14*, which had been previously identified as an important negative regulatory region for lin-14 translation (Arasu et al., 1991; Lee et al., 1993; Wightman et al., 1991).

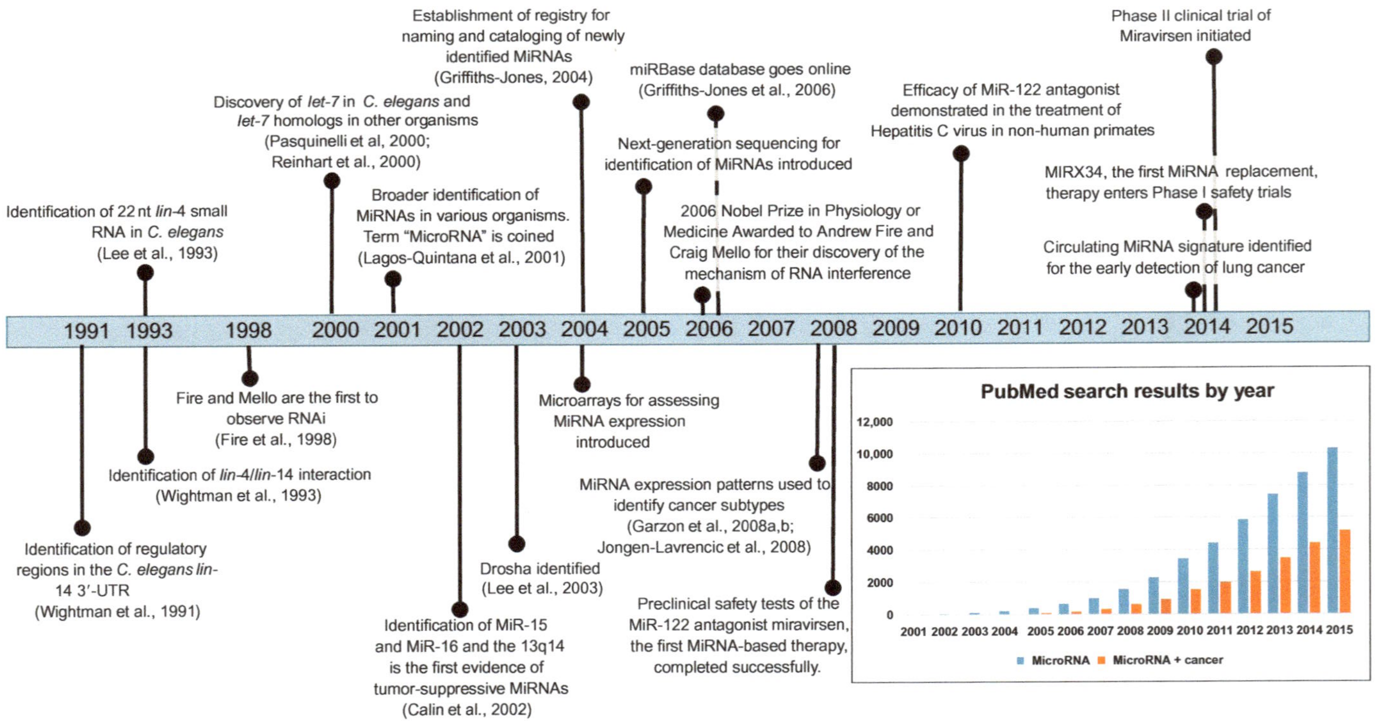

Fig. 1 Timeline for the discovery and understanding of MicroRNAs. Over the last two decades, research into MiRNAs has proceeded at a remarkable pace. This timeline takes us through the discovery of MiRNAs, to the appreciation that they represent a broadly conserved regulatory mechanism, to the appreciation of the contribution of dysregulated MiRNAs to disease, and, finally, to the initiation of clinical trials for MiRNA-based therapeutics (*main*). This MiRNA revolution can be easily tracked by examining PubMed search results for either "MicroRNA" or "MicroRNA + cancer," both of which show a year-after-year increase in the number of publications (*inset*).

This observation of complementarity between the two sequences suggested a mechanism through which the small RNA, lin-4, might exert its function—by antisense binding in the 3′-UTR. This proposed mechanism was elucidated in work published alongside the Ambros group's discovery; Wightman et al. further examined the similar phenotypes between *lin-14* gain of function and *lin-4*-null *Caenorhabditis elegans* animals, identifying seven sites in the *lin-14* 3′-UTR containing 7 nt seed sequences with complementarity to the lin-4 small RNA (Wightman et al., 1993). The Ruvkun group found that the 3′-UTR of *lin-14* was necessary and sufficient for lin-4 to exert its temporal posttranscriptional control of lin-14 expression, firmly establishing the antisense regulation of *lin-14* by lin-4, but by an unknown mechanism (Wightman et al., 1993).

While lin-4 was the earliest discovered MiRNA in animals, in the following years, with the aid of biological and sequencing tools, and with computational tools paired with the availability of sequenced genomes (*C. elegans* in 1991 (The C. elegans Sequencing Consortium, 1998), *Homo sapiens* in 2001 (McPherson et al., 2001; Venter et al., 2001), and *Mus musculus* in 2002 (Mouse Genome Sequencing Consortium et al., 2002)), an increasing number of potential MiRNAs and putatively targeted transcripts were identified, the mechanisms of translational repression by MiRNAs became further elucidated, and the importance of MiRNAs to developmental and disease processes became much clearer. No homolog for lin-4 was identified outside of *C. elegans*; so until 2000, posttranscriptional regulation by small RNAs was thought to be restricted to nematodes. In fact, it was the discovery of another heterochronic MiRNA, let-7, in *C. elegans* by the Ruvkun group (Pasquinelli et al., 2000; Reinhart et al., 2000), which does have homologs in vertebrates, including in humans and other animals, that took MiRNAs from a niche developmental regulatory mechanism specific to nematodes, to a broader phenomenon that would affect our understanding of gene regulation and reshape the central dogma of molecular biology.

With the appreciation that small regulatory RNAs represent a more widely conserved phenomenon among metazoans, additional MicroRNAs were soon identified by intensive cloning efforts, first in *Drosophila*, *C. elegans*, and humans (Lagos-Quintana et al., 2001; Lau et al., 2001; Lee and Ambros, 2001), and shortly thereafter in a variety of other organisms (Ambros et al., 2003; Aravin et al., 2003; Lagos-Quintana et al., 2003; Lim et al., 2003a,b). Variable MiRNA expression was also identified in specific tissues (Dostie et al., 2003; Houbaviy et al., 2003; Kim et al., 2004; Lagos-Quintana et al., 2002) and disease states (Michael et al., 2003). The rapid expansion of this class of regulatory molecules was facilitated by the establishment of a naming and cataloging system for MicroRNAs and their precursors (Griffiths-Jones, 2004), which prevented an overlap in naming of newly identified MiRNAs and allowed for names to be consistent for homologs across organisms. The MicroRNA Registry, still in use today under the name miRBase (Griffiths-Jones et al., 2006), also allowed depositors to identify genomic locations of a given MiRNA, to identify similar MiRNAs, and to identify MiRNAs that were derived from the different arms of the same hairpins. Further, this repository has facilitated the expansion of computational programs for MiRNA prediction.

2.2 MiRNA Biogenesis in Animals

MiRNAs are encoded in the metazoan genome in a variety of ways. MiRNA processing from primary MiRNAs (pri-miRNAs) allows for some MiRNAs to be encoded in clusters, such as the *MiR-17~92* polycistron, in which several MiRNAs are cotranscribed; thus they are coregulated. Other MiRNAs are encoded in introns, also called miRtrons, or in nonprotein-coding exons (e.g., 3′-UTRs) of host protein-coding genes where they may be cotranscribed and coregulated with the host gene (Berezikov et al., 2007; Cai et al., 2004; Rodriguez et al., 2004; Weber, 2005). The remaining MiRNAs are encoded in their own genes, where they can be individually transcriptionally regulated. Many MiRNAs are intergenic and are frequently found in the opposite orientation to neighboring genes, suggesting that MiRNA transcription is typically independently regulated (Lagos-Quintana et al., 2001; Lau et al., 2001; Lee and Ambros, 2001). Early studies led to the hypothesis that MiRNAs are processed from transcripts generated by RNA Pol II because fusion constructs generated for the study of MiRNA regulation in which protein-coding reporter genes were placed downstream of the presumed 5′-regulatory elements of MiRNA genes led to very robust protein expression (Johnson et al., 2003; Johnston and Hobert, 2003; Lagos-Quintana et al., 2002). As evidenced by the 5′-7-methylguanosine (m^7G) caps and poly(A) tails that have been demonstrated on pri-miRNA transcripts, as well as the loss of MiRNA production upon treatment with the potent RNA polymerase II (Pol II) inhibitor, α-aminitin, most MiRNA precursors are, indeed, transcribed by Pol II (Lee et al., 2004), suggesting that the structure of MiRNA host genes and promoters resembles those of protein-coding genes. The similarities between MiRNA-encoding genes and protein-coding genes have been substantiated by further studies

(Zhou et al., 2007). MiRNA transcription by RNA polymerase III (Pol III) is much rarer than Pol II transcription and occurs when MiRNA genes are near *Alu* sites, which contain a Pol III promoter (Borchert et al., 2006). As pri-miRNAs generated by Pol III undergo similar processing to Pol II generated pri-miRNAs (Faller and Guo, 2008), they do not require separate consideration.

MiRNAs that are processed from retained portions of their host transcript, such as independent MiRNA genes, MiRNA polycistrons, and MiRNAs encoded in nonprotein-coding regions of a transcript, but not miRtrons, are processed as pri-miRNAs (Lee et al., 2002), whether they are transcribed by Pol II, as is most common, or Pol III (Faller and Guo, 2008) (Fig. 2). pri-miRNAs are processed to precursor MiRNAs (pre-miRNAs), which are ~70 nt hairpin loop structures, by the Microprocessor complex, minimally composed of Drosha Ribonuclease III (DROSHA) and DiGeorge syndrome critical region 8 (DGCR8) in vertebrates, which are localized in the nucleus (Gregory et al., 2004). The hairpin loop structures that will become pre-miRNAs are recognized by double-stranded RNA (dsRNA) binding regions in the DGCR8 protein, which is the vertebrate homolog of the Pasha protein (Han et al., 2004, 2006). DROSHA, a class 2 Ribonuclease III, is responsible for cleaving the DGCR8-recognized pri-miRNA to the ~70 nt pre-miRNA at sufficient distance from the base of the hairpin, defining one end of the mature MiRNA (Lee et al., 2003). Cleavage by the Microprocessor complex can be a site of posttranscriptional MiRNA regulation, with some MiRNAs requiring specific cofactors for efficient processing and regulation, such as SMAD proteins in the processing of pri-*miR-21* and Lin-28, and RNA binding protein, in the processing of *pri-let-7* (Davis et al., 2008; Viswanathan et al., 2008). In contrast, the miRtron-processing pathway, which is better studied in invertebrates, bypasses the Microprocessor complex to generate pre-miRNAs that can be exported from the nucleus for further processing (Berezikov et al., 2007; Okamura et al., 2007; Ruby et al., 2007). Hairpin-containing lariats, the result of splicing, must be debranched, but then they behave as pre-miRNAs and are ready for nuclear export (Okamura et al., 2007; Ruby et al., 2007).

Pre-miRNAs, processed from pri-miRNAs or miRtrons, are exported from the nucleus by the EXPORTIN-5–Ran-GTP shuttle (Bohnsack et al., 2004; Lund et al., 2004; Yi et al., 2003). Knockdown of EXPORTIN-5 impairs MiRNAs from being exported from the nucleus, but does not result in their accumulation in the nucleus, suggesting that improperly processed

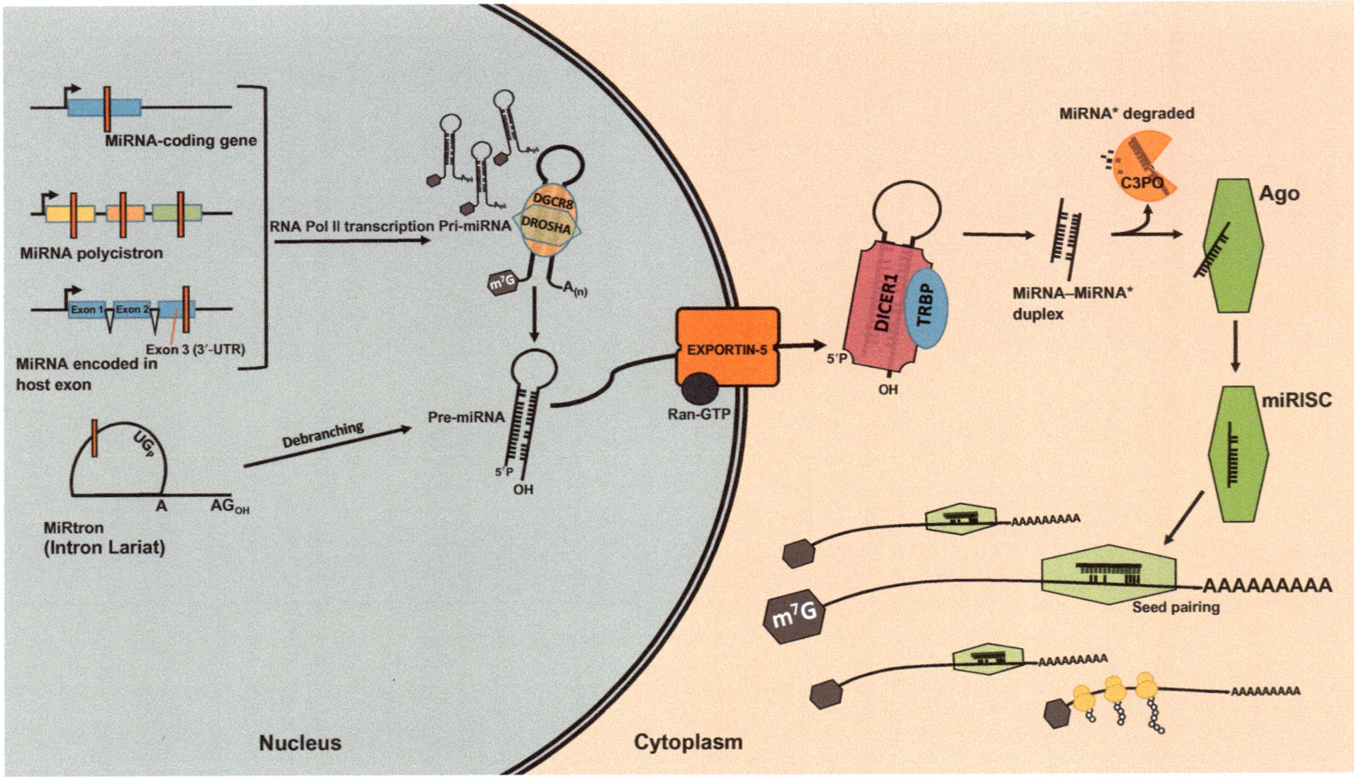

Fig. 2 MiRNA biogenesis. MiRNA biogenesis begins in the nucleus, where MiRNA-encoding genes are transcribed. For MiRNAs that are not derived from an intron, i.e., MiRNAs encoded in their own genes (*left top*), in polycistrons, or in the nonprotein-coding regions of host genes (*left middle*), hairpin primary-miRNAs (pri-miRNAs) are recognized by DROSHA, which utilizes RNAse III activity to trim the pri-miRNAs to precursor MiRNAs (pre-miRNAs). MiRNAs generated from introns and miRtrons bypass DROSHA cleavage and proceed as pre-miRNAs and normal processing (*left bottom*). Pre-miRNAs are exported from the nucleus by EXPORTIN-5 and they meet DICER in the cytoplasm. DICER further cleaves the pre-miRNA to a MiRNA duplex and selects the MiRNA with the least stable 5′-end for loading into the RNA-induced silencing complex (RISC), whose primary constituent is an Argonaute (Ago) protein (*right top*). The MiRNA-RISC (miRISC) is then capable of targeting mRNA transcripts for repression through a variety of mechanisms (*right bottom*). *C3PO*, component 3 promoter of RISC; *TRBP*, TAR RNA binding protein.

and improperly exported MiRNAs are quickly degraded (Bohnsack et al., 2004; Yi et al., 2003; Zeng and Cullen, 2004). Export is also an important checkpoint of pre-miRNA processing—only pre-miRNAs with stem loops of sufficient length and with 3′-overhangs are recognized by EXPORTIN-5 and exported from the nucleus for cytoplasmic processing (Lund and Dahlberg, 2006; Zeng and Cullen, 2004).

Pre-miRNAs are further processed in the cytoplasm, where they are cleaved to their final ~22 nt form by DICER1, a class 3 Ribonuclease III, which defines the other end of the mature MiRNA. DICER1 liberates the mature MiRNA and its opposing arm from the pre-miRNA hairpin in the form of an imperfectly complementary MiRNA:MiRNA* duplex (Grishok et al., 2001; Hutvagner et al., 2001; Ketting et al., 2001; Lee et al., 2003). DICER1 cleavage is another checkpoint site in mature MiRNA processing, with some pre-miRNAs requiring posttranscriptional base modification by adenosine deaminases for efficient cleavage by DICER1 (Kawahara et al., 2007). In summary, genomically encoded MiRNAs are transcribed by RNA polymerases to pri-miRNAs, which are processed to pre-miRNAs in the nucleus by the DCGR8–DROSHA-containing Microprocessor complex. Pre-miRNAs are exported from the nucleus by the EXPORTIN-5–Ran-GTP shuttle and are processed to mature MiRNAs, and can then exert their biological function, which we will discuss in the next section.

2.3 Mechanism of Action of MiRNAs in Animals

The biological function of MiRNAs is exerted in a way that is highly similar to the RNA interference (RNAi) mechanism described by Fire et al., work that would win Andrew Fire and Craig Mello the Nobel Prize in Physiology or Medicine in 2006 (Fire et al., 1998). Our understanding of RNAi was preceded by knowledge that injection of antisense RNA caused inhibition of protein expression in plants and *C. elegans* (Ecker and Davis, 1986; Fire et al., 1991; Guo and Kemphues, 1995; Izant and Weintraub, 1984; Nellen and Lichtenstein, 1993). But, in 1998 Fire et al. achieved much more potent RNA silencing by injecting dsRNA into *C. elegans*, whereas injection of single-stranded RNA (ssRNA) yielded only a moderate inhibitory effect (Fire et al., 1998). The observation that dsRNA yielded such a potent effect, even when diluted to only a few molecules per cell, led to the hypothesis that RNAi works by some catalytic mechanism, rather than by simple antisense binding (Fire et al., 1998). This ultimately led to the proposal of the existence of an RNA-induced silencing complex (RISC) with sequence-specific nuclease activity by Gregory Hannon's group in 2000, which demonstrated that RNAi could be carried out by *Drosophila* cell lysates (Hammond et al., 2000). Through fractionation studies of their *Drosophila* cell lysates, the Hannon group was the first to isolate the RISC, to confirm that the RISC was responsible for specific endonuclease activity, and to demonstrate that the RISC was a ribonucleoprotein complex containing the small RNA that was responsible for sequence-specific cleavage specificity (Hammond et al., 2000).

The RISC is a MiRNA ribonucleoprotein (miRNP) complex that is loaded with a MiRNA and is responsible for mediating its sequence-specific repressive effects (Schwarz and Zamore, 2002). The mammalian RISC contains components that are similar to those contained in the RISC responsible for RNAi in lower organisms. As DICER1 processes the pre-miRNA to the MiRNA:MiRNA* duplex, it unwinds the duplex and selects a strand, the strand with weaker 5′-end base pairing, for insertion into the RISC (Khvorova et al., 2003; Krol et al., 2004; Schwarz et al., 2003), with the opposite strand, the passenger strand, typically getting degraded by the C3PO (component 3 promoter of RISC) endonuclease (Liu et al., 2009; Ye et al., 2011). The primary functional constituents of the RISC are the Argonaute (AGO) family of proteins, which are responsible for mediating the function of a MiRNA against its target through sequence-specific binding.

Humans have four AGO family members, AGO1–4 (Ipsaro and Joshua-Tor, 2015). Though all four AGO family members have been demonstrated to participate in MiRNA-mediated repression (Ipsaro and Joshua-Tor, 2015), only AGO2 demonstrates the endonucleolytic activity necessary to perform siRNA-mediated slicing (Liu et al., 2004; Meister et al., 2004b), an important function that has been co-opted to a great effect by molecular biologists, but not thought to be an important mechanism of MiRNA-mediated action in mammals. AGO proteins contain four domains: N-terminal, PAZ, MID, and PIWI (Song et al., 2004). Structural analysis of AGO proteins has revealed that the 5′-end of MiRNAs, including the seed sequence, are held by the N-terminal and PIWI domains by nonspecific interactions with the 2′OH groups of the RNA sugar-phosphate backbone, making them RNA-specific binders, while the PAZ domain is responsible for holding the 3′-end of the MiRNA (Ipsaro and Joshua-Tor, 2015). Further studies have demonstrated that the tight packing of AGO and the forming RNA duplex are responsible for the perfect base-complementarity necessary in the seed sequence—bubbles formed by mismatches cannot be tolerated in such close quarters (Schirle et al., 2014). AGO undertakes a series of progressive conformational changes as the seed sequence proceeds to recognize its target through base pairing (Schirle et al., 2014). MiRNA availability affects AGO stability and degradation and, ultimately, the miRISC miRNP that is formed is highly stable and persistent in the cell (Elkayam et al., 2012; Schirle et al., 2014). The miRISC has been reported to have an exceedingly long half-life in cells, suggesting that MiRNA regulation may serve as a form of cellular memory, able to respond quickly to stimuli, bypassing responses that would require transcription or translation (Ipsaro and Joshua-Tor, 2015).

In contrast to AGO2-catalyzed mRNA cleavage, which occurs when there is perfect complementarity between a small RNA and an mRNA transcript, most human MiRNAs do not have perfect complementarity to their transcripts and translational repression must be carried out by other means. A number of mechanisms have been proposed through which MiRNAs could mediate translational repression, but recently consensus has coalesced around two methods of repression in most cellular contexts in mammals: inhibition of cap-dependent translation initiation, and mRNA destabilization and degradation (Jonas and Izaurralde, 2015), which is the dominant means of repression by MiRNAs (Eichhorn et al., 2014) (Fig. 3). Some other mechanisms of repression have been identified in other organisms and in the context of early developmental processes (Bazzini et al., 2012; Subtelny et al., 2014), but will not be discussed here.

Though the MiRNA component is responsible for mRNA target recognition and recruitment of the miRISC to the targeted transcript, it is the protein–protein interactions with the AGO component of the miRISC that are responsible for mediating translational repression. Interaction of AGO with glycine-tryptophan 182 (GW182) family members is essential for MiRNA-mediated repression, as demonstrated by studies in which GW182 is depleted or the AGO–GW182 interaction is blocked (Eulalio et al., 2008; Fabian et al., 2009; Takimoto et al., 2009; Yao et al., 2011). GW182 proteins serve as the scaffolds for the recruitment of the effector proteins of mRNA destabilization (Fabian and Sonenberg, 2012), namely the CCR4–NOT deadenylase complex (Braun et al., 2011; Chekulaeva et al., 2010; Fabian et al., 2011). Following deadenylation, further target destabilization and degradation are achieved by decapping and normal 5′-to-3′ mRNA degradation (Behm-Ansmant et al., 2006; Eulalio et al., 2007; Rehwinkel et al., 2005).

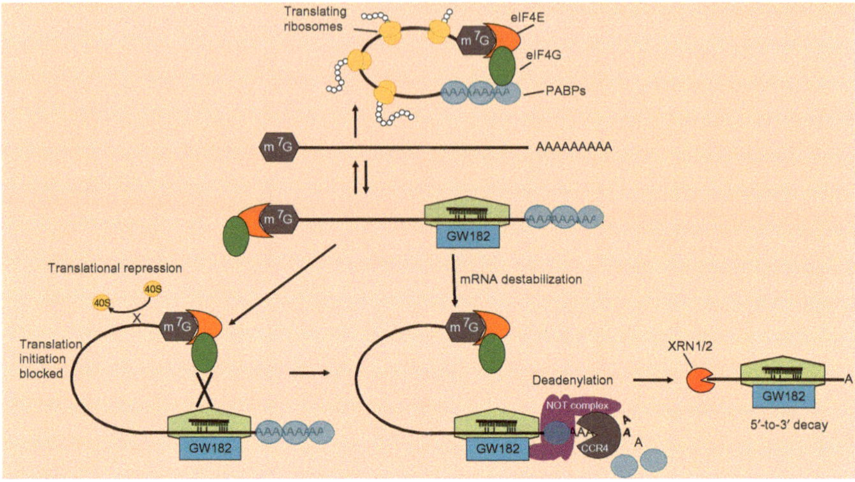

Fig. 3 Mechanism of action of MiRNA in the RISC. mRNAs are bound at the 5′-m^{7}G cap by translation initiation factors, and at the polyA tail by poly(A) binding proteins (PABPs). Interaction between the initiation factors and the PABPs is required for ribosome scanning, assembly, and translation initiation (*top*). MiRNAs disrupt translation by two mechanisms: by translational repression and through mRNA destabilization. On a short timescale, the miRISC, including GW182, blocks association of initiation factors and PABPs, disrupting ribosome assembly and translation (*left bottom*). On a longer timescale, the CCR4–NOT complex is recruited to the miRISC and GW182 and deadenylation is initiated. After deadenylation, the mRNA is degraded by 5′-to-3′ decay (*right bottom*). *eiF*, elongation initiation factor; *CCR4*, C–C motif chemokine receptor; *GW182*, glycine-tryptophan 182; *NOT*, negative on TATA; *XRN1/2*, 5′-to-3′ exoribonuclease 1/2.

Though MiRNA-mediated mRNA destabilization is the predominant mechanism of MiRNA repression (Eichhorn et al., 2014), experiments using artificial mRNAs that are resistant to deadenylation have shown that MiRNA-mediated translational repression is potent and robust (Humphreys et al., 2005; Mathonnet et al., 2007; Pillai et al., 2005). Ribosomal profiling allows for the measurement of ribosome occupancy on a given transcript—when this is compared to the total levels of that transcript, one can arrive at the transcript's "translational efficiency." The advent of effective ribosomal profiling in mammalian cells (Guo et al., 2010a) has largely settled the issue of the type of translational repression mediated by MiRNAs—dramatically decreased translational efficiencies in the presence of MiRNAs suggest that MiRNA-mediated translational repression occurs through inhibition of translation initiation, rather than some mechanism of postinitiation inhibition (Guo et al., 2010a). In short, in the presence of miRISCs, ribosomes cannot assemble on a targeted transcript; thus translation cannot initiate. In fact, more recent studies utilizing cell-free extracts and various toxins and proteases to regulate different steps of translation have demonstrated that MiRNA-mediated translational repression occurs very early, at the level of ribosome scanning, and requires cooperation from poly(A) binding proteins (PABPs) and eukaryotic translation initiation factor 4G (Ricci et al., 2013) likely through the disruption of mRNA circularization, though the precise mechanism is unclear.

The question of whether the dominant effect of MiRNAs on translation is mediated by translational repression or mRNA destabilization is a significant one to researchers studying MiRNAs—if MiRNAs act primarily by mRNA destabilization, then steady-state measures of mRNA levels, which are easily carried out by quantitative PCR, are a good approximation of the levels of MiRNA-mediated repression. In contrast, if translational repression predominates, researchers would need to assess protein levels to determine the efficacy of MiRNA repression, which proves more difficult than measuring transcript levels, especially when working with rarified cell populations. The data on translational repression and mRNA destabilization suggest a temporal order—first MiRNAs act by translational repression, while mRNA destabilization is initiated. Once mRNA destabilization begins, it quickly becomes the predominant method of MiRNA action (Eichhorn et al., 2014). Fortunately, because mRNA destabilization and degradation are the primary mechanism through which MiRNAs regulate their targeted transcripts, steady-state measurement of transcript levels yields a reasonable approximation of MiRNA effect.

We have described the transcription and synthesis of MiRNAs, their one-way transport out of the nucleus to the cytoplasm, and their role in mediating transcriptional repression by blocking translation initiation and by causing mRNA destabilization. However, a number of studies have begun to identify other regulatory functions for MiRNAs, outside of their canonical role in posttranscriptional repression. In late 2007, the Steitz group reported that MiR-369-3 was instrumental in the recruitment of AGO2 and fragile X mental retardation (FXR1) related to AU-rich elements and MREs in the 3′-UTR of tumor necrosis factor α (TNFα), a necessary signal for translation activation in the context of serum starvation and growth arrest (Vasudevan et al., 2007). This group also reported that let-7 and an artificial MiRNA mimic that they introduced contributed to translation activation upon cell cycle arrest, suggesting that MiRNA-directed translational activation might be a more broadly conserved phenomenon. Similarly, MiR-206 was found to promote translation of Kruppel-like factor 4 (Klf4) in the immortalized rat kidney epithelial cells upon reaching confluence and cell cycle arrest (Lin et al., 2011). MiRNA import into the nucleus has also been demonstrated (Hwang et al., 2007) and MiRNA-associated proteins, AGO and DICER1, have been detected in nuclear lysates (Gagnon et al., 2014; Robb et al., 2005), suggesting that MiRNAs may play an additional role as transcriptional regulators. For example, MiR-373 and MiR-205 have been associated with transcriptional activation through the recruitment of transcriptional machinery to complementary regions in gene promoters and associated activating chromatin changes (Huang et al., 2012; Majid et al., 2010; Place et al., 2008) through a mechanism similar to RNA activation (Huang et al., 2010). A number of additional functions for nuclear MiRNA have been proposed, but are poorly understood and have yet to be substantively implicated in normal physiology or the pathogenesis of disease. For the remainder of this review, we will focus on canonical MiRNA-based posttranscriptional repression, and proposed noncanonical functions of endogenous MiRNAs will not be discussed further. Noncanonical regulation by exogenously introduced small RNAs and endogenous MiRNAs is well reviewed elsewhere (Cipolla, 2014; Huang and Li, 2012; Jiao and Slack, 2014; Roberts, 2014; Vasudevan, 2012).

Armed with an understanding of MiRNA biogenesis and the mechanisms through which MiRNAs exert posttranscriptional repression, it is important to briefly discuss how the targets of MiRNAs are predicted and identified, and the tools available to researchers for assessing these putative targets.

2.4 MiRNA Target Recognition and Tools for Identifying MiRNAs and Their Targets

We have already discussed how the identification of lin-4 complementary sites in the *lin-14* 3′-UTR gave rise to the hypothesis that sequence complementarity is somehow necessary for repression of specific targets (see Section 2.1) (Lee et al., 1993; Wightman et al., 1993). In contrast to the perfect complementarity of siRNA-based silencing, the imperfect complementary of MiRNAs to their targets makes it more difficult to identify targeted transcripts with confidence, yielding a challenging computational problem for genome-wide MiRNA target prediction (Bartel, 2009), as well as the greater necessity to validate in situ MiRNA–mRNA interactions before asserting MiRNA-based effects. Poor overlap between MiRNA predictions by different groups in the same organisms, such as in humans, led to the conclusion that early prediction algorithms failed to successfully identify MiRNAs and their targeted transcripts. However, the prediction algorithms were quickly improved by adding the capability to assess sites within 3′-UTRs for evolutionary conservation, which suggests biological functionality (Lewis et al., 2005). As algorithms for MiRNA target prediction across species and utilizing different methodologies improved, three critical features emerged as essential for accurate target prediction: (1) the 5′-region of the MiRNA, specifically nucleotides 2–7, called the "seed sequence," must maintain perfect Watson–Crick complementarity to improve target prediction. The importance of the seed sequence is supported experimentally by identification of validated MiRNAs and their targets in other organisms, reporting that the 5′-end of the MiRNAs are the most highly conserved across species, and that disrupting base paring in the seed region with nucleotide substitution disrupts MiRNA function (Brennecke et al., 2005; Doench and Sharp, 2004; Lim et al., 2003b). We have already discussed the steric hindrance within the miRISC that necessitates perfect base-pair complementarity in seed region and has no tolerance for mismatches (also see Section 2.3; Schirle et al., 2014). (2) Assessment of the 7–8 nt seed sequence is sufficient for MiRNA target prediction. (3) MiRNAs with high evolutionary conservation have a high number of conserved targets, and a significant percentage of human genes are under selective pressure to maintain their 3′-UTR MiRNA target sites (Brennecke et al., 2005; Friedman et al., 2009; Krek et al., 2005; Lewis et al., 2005). Target prediction algorithms such as TargetScan (Agarwal et al., 2015), PicTar (Krek et al., 2005), and Miranda (Betel et al., 2010) have been reviewed extensively (Bartel, 2009), but suffice it to say that these tools, used alone or in combination, represent invaluable resources to MiRNA researchers.

The challenges associated with reliable target prediction highlight the importance of confirming MiRNA:target interactions experimentally. A number of experimental techniques have been used to experimentally validate these interactions, the most common technique being the luciferase reporter assay, which is applied to MiRNA biology by cloning the 3′-UTR in question, containing MRE to the end of a recombinant firefly *luciferase* gene. If the MiRNA:target interaction is genuine, luciferase activity is decreased upon overexpression of the MiRNA or is increased upon knockdown of the MiRNA. Similar reporter assays can be conducted with fluorescent proteins. High-throughput assays based on this reporter strategy, such as Sensor-seq, have revealed that many MiRNAs that can be detected in a cell by sequencing or by real-time PCR are not active in repressing transcripts (Mullokandov et al., 2012). Other large-scale methods for detecting MiRNA:target interactions have relied on cross-linking miRISCs with their targets, immunoprecipitation, and next-generation sequencing. For example, the Tuschl group reported on photoactivatable ribonucleoside-enhanced cross-linking and immunoprecipitation (PAR-CLIP), to experimentally validate MiRNA targets (Hafner et al., 2010). When modified photoactivatable nucleoside analogs are introduced to cultured cells in vitro, they can be cross-linked into their miRISCs along with the targeted transcript. Expression of FLAG-tagged AGO facilitates the pull-down of cross-linked miRISCs and their targets, which can then be processed into sequencing libraries so that MiRNA targets can be identified. PAR-CLIP is similar to high-throughput sequencing of RNA isolated by cross-linking immunoprecipitation (HITS-CLIP), which does not utilize photoactivatable nucleosides (Chi et al., 2009). Both techniques identify protein–RNA interactions and rely on the analysis of MiRNA seed sequences to predict MiRNA: mRNA interactions from two independent datasets: AGO–mRNA binding sites and AGO–MiRNA binding sites. For this reason, they fail to directly identify any noncanonical MiRNA interactions, such as with bulged seed sequences or with ncRNAs. To solve this dilemma, the Tollervey group modified their cross-linking, ligation, a sequencing of hybrids (CLASH) method to allow for the direct sequencing of MiRNA–target chimeras (Helwak et al., 2013; Kudla et al., 2011). In CLASH, cross-linked MiRNAs and targets can be ligated to each other before sequencing library preparation—the resulting library contains chimeric reads of a MiRNA and its target, allowing for direct high-resolution mapping of these interactions.

Though seed binding rules have proven very useful in predicting and identifying many MiRNA:mRNA interactions, the detailed, genome-wide mapping studies utilizing HITS-CLIP and CLASH have revealed that a significant percentage of MiRNA:mRNA interactions cannot be identified by canonical seed matching rules (Chi et al., 2012; Helwak et al., 2013). For example, careful examination of AGO binding sites in HITS-CLIP data that cannot be paired with a MiRNA or a mRNA target by seed matches, called "orphan clusters," revealed that upward of 15% of MiRNA:mRNA interactions are characterized by a bulged-G in the target sequence, corresponding to MiRNA positions 5–6 (Chi et al., 2012). This work demonstrated that bulged seed matching, which had only been observed on an individual basis, is a broader phenomenon that is conserved across nematodes and mammals (Didiano and Hobert, 2006; Ha et al., 1996; Tay et al., 2008; Vella et al., 2004). This brings attention to a number of noncanonical MiRNA:mRNA pairing interactions that have been observed. MiR-24 is, for instance, upregulated in terminal hematopoietic differentiation and has been demonstrated to inhibit cell proliferation (Lal et al., 2009). While many direct targets of MiR-24 contain canonical MiR-24 seed sequences, a number of important cell cycle genes, including *E2F2* and *MYC*, are direct targets of MiR-24, but do not contain canonical MiR-24 seed sequences. These targets were predicted to bind MiR-24 at downstream sites that tolerate mismatches, wobbles, and bulged sites. "Centered sites," which bind their targets by complementarity to MiRNA nucleotides 4–15, but without seed sequence or 3′-binding, have also been identified as functional, noncanonical MiRNA binding sites (Shin et al., 2010). As more researchers apply high-throughput methods for identifying MiRNA:mRNA interactions in a variety of cell types and organisms, it is possible that additional mechanisms of noncanonical MiRNA target recognition will be identified.

Large-scale methods for identifying MiRNA targets highlight another interesting aspect of MiRNA biology—a given MiRNA can target multiple genes and a given gene can be targeted by multiple MiRNAs, adding a level of complexity to the study of MiRNAs that can be tedious to unravel. The initial assessment of MiRNAs and their targets can be made by examining the target prediction algorithms we discussed at the beginning of this section, and we have explored various techniques, such as luciferase assays and CLASH, which can be used to validate these interactions in cell lines. For functional assessment of MiRNAs, researchers need other tools (Akbari Moqadam et al., 2013; Bernardo et al., 2012; van Rooij, 2011). Transfection of MiRNA mimics, viral overexpression, and transgenic models can all be used to study the effect of MiRNA overexpression on their targets in the relevant cellular contexts. Transfection of MiRNA antagonists, virally introduced sponge constructs, and genetic knockout models can be used to study the effect of MiRNA downregulation and loss. Conditional genetic models can be used to study the role of MiRNAs in specific tissues or during specific developmental processes. Mutation of MREs in the 3′-UTR of targets can be used to specifically study the effects of a MiRNA on a specific target. All of these models can be used in the context of other functional assays for processes such as differentiation, proliferation, apoptosis, cell migration, disease-free survival, and many others.

2.5 The Competing Endogenous MicroRNA Hypothesis and the Emerging Role of Endogenous MicroRNA Inhibition

In our introduction, we briefly presented the concept of ceRNAs. The ceRNA hypothesis is that RNA species with MREs can inhibit the activity of expressed MiRNAs by sequestering them, preventing them from exerting their activity on their protein-coding mRNA target (Salmena et al., 2011; Thomson and Dinger, 2016). Researchers have long been aware that competition of MiRNAs could inhibit their function, and the use of antisense MiRNA antagonists (e.g., antagomiRs) and MiRNA sponges has been used in experimentation and, increasingly, in the development of therapeutics, suggesting that inhibition of MiRNAs by endogenously encoded ceRNAs is similarly possible. Exogenous MiRNA antagonism and the development of MiRNA-based therapeutics will be discussed more extensively in Section 5. More recently, endogenously expressed lncRNAs, including transcribed pseudogenes and circRNAs, have been demonstrated to contribute to MiRNA inhibition as ceRNAs. Even endogenously expressed, protein-coding mRNAs, which are often highly expressed as compared to lncRNAs, have been shown to contribute to the ceRNA phenomenon. In this section, we will briefly discuss the groundbreaking works that have identified different ceRNA species and the potential for bioinformatic identification of ceRNA interactions, and address questions of the broad applicability of the ceRNA hypothesis across the highly uncharacterized lncRNA landscape.

Numerous pseudogenes have been identified as ceRNAs, including *OCT4P4* (Wang et al., 2013), *BRAFP1* (Karreth et al., 2015), *CYP4Z2P* (Zheng et al., 2015), and *Pbcas4* (Marques et al., 2012), with *PTENP1* (Johnsson et al., 2013; Poliseno et al., 2010; Yu et al., 2014) being the best characterized. The Dr. Pandolfi's group was the first to identify a noncoding role for a pseudogene in the regulation of MiRNA activity (Poliseno et al., 2010). Previous studies of *PTEN* dosage in malignancy have led researchers to examine the importance of *PTEN*-targeting MiRNAs and contributed to the hypothesis that a MiRNA-sponging pseudogene could have a significant impact on *PTEN* translation (Huse et al., 2009; Kato et al., 2009; Yang et al., 2008). The *PTENP1* pseudogene is not translated due to a mutation to its initiating methionine and retains a number of shared MREs with the *PTEN* transcript. Dysregulation of those MiRNAs, either by overexpression or by inhibition, affects *PTEN* and *PTENP1* expression similarly. Likewise, overexpression of the *PTENP1* 3′-UTR leads to the upregulation of *PTEN* and repression of the PI3K/AKT phosphorylation cascade and reduces cell proliferation and colony formation in vitro, demonstrating the role for *PTENP1* as a tumor suppressor. Consistently, focal *PTENP1* loss was found in a number of prostate cancer samples, suggesting that loss of the *PTENP1* tumor suppressor can contribute to the development of malignancy. Subsequent studies have identified a role for *PTENP1* loss in the pathogenesis of renal cell carcinoma, and for a *PTENP1* antisense transcript in stabilizing the

PTENP1 transcript for export to the cytoplasm, where it can exert its ceRNA function to sponge *PTEN*-targeting MiRNAs (Johnsson et al., 2013; Yu et al., 2014).

Other pseudogenes have also been characterized for their function as ceRNAs. Repression of the mouse *Bcas4* pseudogene, *Pbcas4*, led to changes in gene expression consistent with *Bcas4* repression (Marques et al., 2012). An earlier study identified *CYP4Z2P*, a pseudogene of *CYP4Z2*, as promoting breast cancer angiogenesis, which is similar to the role of *CYP4Z2* in breast cancer (Zheng et al., 2014). A subsequent study utilized luciferase reporter assays to demonstrate that *CYP4Z2* and *CYP4Z2P* are regulated by the same MiRNAs and that overexpression of the *CYP4Z2P* results in the downregulation of the *CYP4Z2* reporter, consistent with the ceRNA hypothesis (Zheng et al., 2015). Further, overexpression of the *CYP4Z2P* 3′-UTR promotes angiogenesis in ex vivo models and the activation of downstream pathways of *CYP4Z2*. *OCT4P4*, a pseudogene of the oncogene *OCT4*, has also been found to be frequently dysregulated in hepatocellular carcinoma (HCC) (Wang et al., 2013). The *OCT4P4* transcript serves as a sponge for MiR-145, which contributes to the derepression of *OCT4*. Consistently, *OCT4P4* and *OCT4* levels were found to be positively correlated in HCC samples. Overexpression of *OCT4P4* increases cell proliferation and colony formation in vitro, and tumorigenicity in transplanted tumor models in mice. Most strikingly, *OCT4P4* expression can significantly stratify disease-free survival and overall survival in HCC patients, with high *OCT4P4* expression being associated with poorer outcomes. Interestingly, in trying to expand their studies and examine the broader applicability of the ceRNA hypothesis, the Pandolfi group had previously recognized the conservation of the MiR-145 binding site between *OCT4* and its pseudogenes (Poliseno et al., 2010). This group further succeeded in expanding their hypothesis in their subsequent examination of the *BRAF* pseudogene, *BRAFP1*, and its contribution to the development of lymphoma in vivo (Karreth et al., 2015). In vitro, *BRAFP1* is demonstrated to regulate BRAF expression in a DICER1-dependent manner, suggesting that MiRNAs are involved in this regulation, consistent with the ceRNA hypothesis. Transgenic mice that can be made to inducibly and transiently overexpress the murine *Braf-rs1* pseudogene develop a malignancy consistent with diffuse large B-cell lymphoma (DLBCL), which regresses when *Braf-rs1* overexpression is withdrawn. Further, *BRAFP1* has been found to be upregulated in DLBCL and multiple human cancers, suggesting that it may play a role in oncogenesis.

Other RNA species have also been shown to contribute to the ceRNA phenomenon. A number of noncanonical splicing mechanisms have been proposed for the generation of circRNAs (Jeck and Sharpless, 2014). Regardless, an essential feature of circRNAs is that they lack 5′- and 3′-ends—therefore, they are resistant to exonuclease degradation and might be effective endogenous MiRNA sponges if they contain MREs. Though not many circRNAs contain MiRNA binding sites (Guo et al., 2014), several examples of circRNAs serving as ceRNAs have been identified. A circRNA derived from the *Sex-determining region Y* (*SRY*) gene, which was the first discovered circRNA (Capel et al., 1993), was found to contain 16 binding MREs for MiR-138 (Hansen et al., 2013). The efficacy in the *SRY*-generated circRNA in sequestering MiR-138 was demonstrated by luciferase assay and coimmunoprecipitation of MiR-138 and the circRNAs with tagged AGO. Another circRNA, ciRS-7 (circular RNA sponge for MiR-7, a.k.a. *CDR1-AS*), has been shown to sequester MiR-7 (Hansen et al., 2013; Memczak et al., 2013). Approximately 70 potential MiR-7 binding sites were identified in ciRS-7 (Hansen et al., 2013). Coexpression of MiR-7 and ciRS-7 results in increased immunoprecipitation of ciRS-7 with AGO than when ciRS-7 is expressed alone, demonstrating that the two RNA species interact, and expression of ciRS-7 protected MiR-7 targets from repression upon MiR-7 transfection. The Dr. Kjems' group also stressed the utility of circRNA, as compared to linear RNA, in serving as a MiRNA sponge—ciRS-7 bound by MiR-7 was resistant to degradation as compared to linear sponge constructs of the same sequence. Further characterization of ciRS-7 demonstrated that ciRS-7 and MiR-7 are coexpressed in the mammalian brain and that ciRS-7 is localized to the cytoplasm, where it can exert its effect, and introduction of ciRS-7 in zebrafish disrupts brain development comparably to MiR-7 inhibition (Memczak et al., 2013).

One ongoing criticism of the ceRNA hypothesis is that pseudogenes and circRNAs are not frequently transcribed at high enough levels to exert a significant sequestration effect on MiRNAs. In contrast, protein-coding mRNAs can be highly expressed. Recent excitement about the ceRNA hypothesis has yielded increased interest in the capacity of mRNAs to participate in MiRNA sequestration. In trying to understand how protein-coding mRNAs can regulate each other through competition over MiRNAs, we can return to *PTEN*, which has been studied by the Pandolfi group as a protein-coding mRNA that may also serve as a ceRNA. First, they sought to identify genes that are coexpressed with *PTEN* and share

MREs, and thus may serve as ceRNAs, through a computational approach that they called mutually targeted MRE enrichment (MuTaME) (Tay et al., 2011). Knockdown of potential ceRNAs against *PTEN* resulted in downregulation of *PTEN*, downregulation of PTEN protein, and activation of the PI3K/AKT pathway, in a DICER1-dependent manner, suggesting that this regulation is dependent on their competition over MiRNAs. The MuTaME approach was also used to identify ceRNAs against *PTEN* in the context of a mouse model of melanoma, driven by *B-Raf* mutation, in the context of an in vivo forward genetic screen to identify cooperating mutations (Karreth et al., 2011). This method identified *ZEB2* as a ceRNA targeting *PTEN*; when *ZEB2* expression is lost, *PTEN* expression is decreased. Low *PTEN* expression in human tumor samples is associated with low expression of *ZEB2*, consistent with the ceRNA hypothesis. A number of other mRNAs have been identified as ceRNAs, including *VCAN* in the development of HCC (Fang et al., 2013; Lee et al., 2010), and *FOXO1* in the development of breast cancer (Yang et al., 2014), among others.

When one considers that many protein-coding mRNA transcripts can compete for MiRNAs and serve as ceRNAs, it quickly becomes clear that, in most cases, focusing on the role of a single mRNA as a ceRNA might not be sufficiently informative. This has been made even clearer through a transcriptome-wide analysis of RNA–RNA interactions conducted by the Califano group, in which they revealed a broad network of MiRNA-mediated posttranscriptional regulation, much of which is mediated by sponge activity (Sumazin et al., 2011). Another example comes through the Sensor-seq approach, in which MiRNA activity, rather than abundance, was measured for a large number of MiRNAs by introducing a library of MiRNA-activity reporter constructs (Mullokandov et al., 2012). MiRNAs with a high number of expressed competing MREs were found to have low activity, as measured by the reporter, as compared to their abundance, indicating that their function was being diluted because of the large pool of targets available to them. Mathematical modeling of the ceRNA concept has suggested that derepression of a target requires near-equimolar expression of the MiRNA and ceRNA, and that ceRNA competition for abundant MiRNAs would require drastic changes in MRE availability (Ala et al., 2013; Bosia et al., 2013; Figliuzzi et al., 2013; Jens and Rajewsky, 2015). Other studies have experimentally demonstrated that changes to individual ceRNA transcripts are unlikely to affect the activity of most MiRNAs (Bosson et al., 2014; Denzler et al., 2014); thus the changing levels of ceRNAs would most dramatically be felt on lowly expressed MiRNAs.

It is clear that there is a role for RNA competition in the regulation of MiRNA activity, but studies identifying the effect of individual ceRNAs must be critically evaluated for the appropriate validation studies and assessed for the physiological relevance of performed experiments.

Equipped with an understanding of how MiRNAs are transcribed and processed, how they exert their function and how putative MiRNA targets can be identified and validated, how MiRNA function can be assessed, and the possible mechanisms through which MiRNAs may be posttranscriptionally regulated, we will spend the next several sections discussing specific MiRNAs and their roles in normal hematopoietic development, and as oncogenes and tumor suppressors whose dysregulation can contribute to malignant transformation.

3. MicroRNAs IN HEMATOPOIETIC DEVELOPMENT

Hematopoiesis is a highly regulated process of differentiation that is responsible for providing mature blood cells for the lifetime of an individual. The rigidly regulated stepwise differentiation process that characterizes hematopoiesis, as well as a sophisticated toolbox of cell type identifying surface markers and a broad variety of in vitro and in vivo assays, makes the hematopoietic system paradigmatic for assessing the mechanisms that regulate differentiation processes. At a time when there were no discovered functions for mammalian MiRNAs, the role for MiRNAs in hematopoietic differentiation became an early target of study because several identified MiRNA genes were found at sites of genomic translocation in human leukemia: *MiR-15* and *MiR-16* are located at chromosome 13q14 and are found to be downregulated or lost in a majority of B-cell CLLs (Calin et al., 2002); and *MiR-142* is located on chromosome 17 (Lagos-Quintana et al., 2002), at the junction of the t(8;17) translocation that results in an aggressive B-cell leukemia associated with upregulated MYC expression (Gauwerky et al., 1989). Preliminary studies, in which the Bartel group sought to identify differential expression of known MiRNAs in different hematopoietic cell types (fetal liver, bone marrow, spleen, and thymus), yielded MiR-181, MiR-142, and MiR-223 as differentially expressed in the various tissues (Chen et al., 2004). Differential expression of these MiRNAs suggested that they may play a role in hematopoietic differentiation, which was supported by ectopic expression studies. Overexpression of MiR-142 and MiR-223 contributed to the expansion of T cells in vitro in culture conditions supportive of lymphoid differentiation, while overexpression of MiR-181 contributed to the expansion of B cells in vitro and upon in vivo transplantation (Chen et al., 2004). This study, of course, suggested a role for MiRNAs in hematopoietic differentiation and also supported a hypothesis that gene silencing may play an important role in the decision of stem and progenitor cells to self-renew or to differentiate into a specific cell type. In this section, we will discuss several MiRNAs that have been shown to be important contributors to hematopoietic differentiation and to hematopoietic stem cell (HSC) maintenance. We will also discuss large-scale MicroRNA screening, which has been facilitated by our ability to fractionate the hematopoietic system into specific cell types based on their surface markers.

3.1 MiR-223

Following the revelation that MiR-223 is differentially expressed in the hematopoietic system, MiR-223 has been further studied for its role in differentiation, particularly granulocytic cell maturation (Fazi et al., 2005; Johnnidis et al., 2008). Soon after the Ambros group reported the expression of MiR-223 in the bone marrow (Chen et al., 2004), the competing granulopoiesis-associated transcription factors, NFI-A and C/EBPα, were shown to regulate MiR-223 expression in human granulocyte maturation, suggesting that MiR-223 is associated with granulocyte maturation or function (Fazi et al., 2005). Using reporter constructs and retinoic acid (RA)-based granulocytic differentiation in human leukemia cell lines, Fazi et al. demonstrated that NFI-A and C/EBPα compete for binding at the *MiR-223* promoter. Further, a luciferase reporter assay identified NFI-A as a direct target of MiR-223, which forms a feedback loop that reinforces granulocyte differentiation (Fazi et al., 2005). NFI-A competes for binding with C/EBPα in the promoters of granulocytic differentiation genes, including *MiR-223*. When C/EBPα binding predominates, *MiR-223* and other progranulocytic genes can be expressed. MiR-223 represses NFI-A expression, reducing its blockade on granulocytic differentiation and completing the positive feedback loop that promotes granulopoiesis. Consistently, ectopic overexpression of MiR-223 in leukemic cell lines induces their differentiation, in the absence of RA (Fazi et al., 2005). MiR-223's role is also highlighted by its upregulation in RA-induced differentiation of acute promyelocytic leukemia cell lines (Garzon et al., 2007). MiR-223 is expressed in very low levels in HSCs and steadily increased through granulocytic differentiation to bone marrow and peripheral blood neutrophils, while its expression plummets through monocyte differentiation (Johnnidis et al., 2008). Surprisingly, experiments conducted with genetically engineered mice deficient in *MiR-223* revealed an expansion of granulocytes (Johnnidis et al., 2008), suggesting that MiR-223's contribution to granulopoiesis may be more complicated than originally thought. In the context of constitutive *MiR-223* deficiency, the Camargo group revealed that the transcript of Myocyte-specific enhancer factor 2c (Mef2c), whose expression promotes myeloid progenitor expansion and enhanced granulocytic differentiation, is a specific target of MiR-223 (Johnnidis et al., 2008). The inconsistencies between MiR-223 as either a promoter or a repressor of granulopoiesis can be explained by the different experimental systems used in these publications. Whereas the Camargo group used a knockout of the *MiR-223* locus, Fazi et al. overexpressed MiR-223 in leukemic cell lines; overexpression vs knockout may each reveal different aspects of MiR-223 function. Additionally, different experimental systems may reveal temporal differences of MiR-223 function; the leukemic cell lines that were used by Fazi et al. represent an early granulocytic progenitor, while MiR-223 was disrupted throughout the hematopoietic system by the Camargo group, suggesting that MiR-223 may play different roles, and by different mechanisms, at different stages of differentiation (Fazi et al., 2005; Garzon et al., 2007).

Additional studies have identified another regulatory mechanism for MiR-223 expression—consistent with its role in hematopoietic differentiation, MiR-223 is regulated, much like other "myeloid genes," by the C/EBPs and another important myeloid transcription factor, PU.1 (Fukao et al., 2007). Fukao et al. identified a highly conserved region upstream of the *MiR-223*-encoding region, which contains two PU.1 binding sites, as well as one C/EBP binding site and one GATA-1 binding site. Chromatin immunoprecipitation (ChIP) studies have confirmed binding of PU.1, C/EBP, and GATA1 at the core promoter, and expression studies indicated that while expression of PU.1 or C/EBP alone mildly induces MiR-223 expression, they dramatically increase MiR-223 transcription when expressed in combination (Fukao et al., 2007). As expected, GATA1 expression inhibits MiR-223 expression, as is expected of a myeloid gene. The study was able to replicate the increased expression of MiR-223 upon RA treatment that was previously identified (Fazi et al., 2005) and demonstrated that this regulation is also mediated by PU.1 and C/EBP (Fukao et al., 2007). Interestingly, it has been revealed that downregulation of MiR-223 in acute myeloid leukemia (AML) can occur by an unrelated mechanism. The t(8;21) translocation is the most common karyotypic abnormality in AML and generates the *AML1–ETO* fusion oncogene, a potent epigenetic disruptor that represses transcription through interactions with histone deacetylase (HDAC) and DNA methyltransferase (DNMT) (Liu et al., 2005, 2006). The MiR-223-coding region is downregulated and packaged into heterochromatin in leukemic blasts and in cell lines harboring the t(8;21) translocation by the recruitment of AML1–ETO to an AML1 binding site near the *MiR-223* locus (Fazi et al., 2007). Ectopic overexpression of MiR-223 or RNAi against the AML1–ETO fusion protein restores MiR-223 expression and induces differentiation in leukemic cell lines (Fukao et al., 2007). Though the exact role of MiR-223 in hematopoietic differentiation remains unclear, in large part due to the contradictory findings that MiR-223 promotes granulopoiesis (Fazi et al., 2005), but that loss of *MiR-223* results in an expansion of mouse granulocytes (Johnnidis et al., 2008), it is clear that MiR-223 plays an important role in hematopoietic differentiation, and that its dysregulation can promote oncogenesis. Further studies of the role of MiR-223 in specific cell types and contexts will surely elucidate the mechanisms through which MiR-223 contributes to hematopoietic differentiation of multiple cell types.

We briefly discussed MiRNA strand selection and loading into the RISC in Section 2.3. Typically, the strand with weaker 5′-end base pairing in the MiRNA:MiRNA* duplex is loaded into the RISC and the MiRNA*, also known as the passenger strand, is degraded by C3PO (Khvorova et al., 2003; Krol et al., 2004; Liu et al., 2009; Schwarz et al., 2003; Ye et al., 2011). The advent of deep sequencing for small RNAs has made it easier to detect MiRNA* species and examine their relative abundance to MiRNAs. Examination of sequencing libraries generated from a variety of mouse and human tissues identified a number of highly abundant MiRNA* species, including MiR-223* (Kuchenbauer et al., 2011). Typically, overexpression of a MiRNA involves the overexpression of both arms of the MiRNA: MiRNA* duplex, making it difficult to untangle their separate functions. In order to study the specific function of MiR-223*, the researchers developed a retroviral construct in which the MiR-223 seed sequence was mutated so that MiR-223* activity could be studied independently. Overexpression of MiR-223* abrogates the increase in colony formation exhibited in *MiR-223*^{KO} bone marrow. The *insulin-like growth factor receptor 1* (*Igf1r*) gene is dysregulated in *MiR-223*^{KO} bone marrow and is predicted to be a target of MiR-223 and MiR-223*. Overexpression of MiR-223* in the context of *MiR-223*^{KO} restored the repression of *Igf1r*, suggesting that MiR-223* also contributes to its regulation. Overexpression of MiR-223* was also shown to repress a number of predicted MiR-223* targets, further supporting its role as an important regulatory MiRNA*. Interestingly, MiR-223* expression levels were able to stratify AML survival outcomes, with increased MiR-223* expression being associated with a superior outcome, while MiR-223 expression was not associated with any differences in survival. A number of other MiRNA* species have been identified for their contribution to physiology and malignant transformation. For example, expression of MiR-9* has also been demonstrated to be associated with a favorable outcome in AML (Nowek et al., 2016), and overexpression of MiR-27/MiR-27* has been associated with increased metastatic potential in osteosarcoma (Salah et al., 2015). Clearly, there is an emerging appreciation of the role of MiRNA* species, which will surely be the subjects of further studies.

3.2 MiR-155

MiR-155 is another MiRNA whose role in hematopoiesis was established early, due to the observation that MiR-155 is overexpressed in some B-cell lymphomas (Eis et al., 2005; Kluiver et al., 2005). Proviral integration by Avian sarcoma leukosis virus activates a gene entitled "BIC," B-cell integration cluster, which encodes *MiR-155* in a gene that generates a fully spliced and polyadenylated transcript (Tam, 2001; Tam et al., 1997). Like MiR-223, MiR-155 is only modestly expressed in HSCs, and its expression increases upon differentiation to some mature cell types, particularly upon exposure to immunogenic stimuli. MiR-155 has been shown to target activation-induced cytidine deaminase (AID), an enzyme which participates in somatic hypermutation and class-switch recombination, which is essential for B-cell maturation in germinal centers (de Yebenes et al., 2008; Teng et al., 2008). INPP5D (phosphatidylinositol-3,4,5-trisphosphate 5-phosphatase 1) is a 5′-inositol phosphatase that is a tumor suppressor in B-cell lymphoma—when *INPP5D* and *PTEN* are both deleted, it results in an aggressive B-cell lymphoma (Pedersen et al., 2009). *INPP5D* is a target of MiR-155, which has been shown to participate in a TNFα/MiR-155/INPP5D pathway in DLBCL (Pedersen et al., 2009), and MiR-155 levels can be reduced upon TNFα antagonism. MiR-155 has also been shown to contribute to the pathogenesis of DLBCL through the repression of SMAD5 (Rai et al., 2010). In the hematopoietic system, SMAD5 is responsive to transforming growth factor-beta (TGFβ) and bone morphogenetic protein (BMP) signaling and works to inhibit cell proliferation. Overexpression of MiR-155 in DLBCL cells represses SMAD5 expression, making them unresponsive to TGFβ or

BMP-signaled growth arrest, contributing to lymphomagenesis (Rai et al., 2010). Together, these data suggest that MiR-155 is an important regulator of B-cell development, and a potent oncogene in lymphoid malignancies.

3.3 MiR-181, Lin-28, and let-7

As we have previously discussed, let-7, a heterochronic MiRNA involved in developmental timing, was the second discovered MiRNA in *C. elegans* (see Section 2.1). Identification of let-7 homologs in humans and other organisms led to the realization that the MiRNA phenomenon was conserved in many species. Interestingly, *pri-let-7* is detectable in undifferentiated embryonic stem cells (ESCs) and in mature cells, whereas mature let-7 is only detectable in mature cells, suggesting that it is posttranscriptionally regulated. *Lin-28* is an RNA binding protein that is conserved from *C. elegans* to humans, and is also associated with developmental timing (Ambros and Horvitz, 1984). Lin-28 has been identified as a posttranscriptional regulator of *let-7* processing (Newman et al., 2008; Viswanathan et al., 2008). Lin-28 binds to the loop of *pre-let-7* and recruits additional factors that contribute to its degradation and the blockade of further processing. Lin-28 is highly expressed in ESCs, and the antagonism between Lin-28 and *let-7* maturation is thought to play a role in maintaining the undifferentiated state—in fact, the Thomson group included LIN-28 among their cocktail of factors used to generate induced pluripotent stem cells from human somatic cells (Yu et al., 2007). let-7 is also capable of repressing Lin-28 expression, suggesting that let-7 and Lin-28 participate in an autoregulatory loop (Rybak et al., 2008).

It was not until 2012 that a role for the let-7/LIN-28 autoregulatory loop was identified in hematopoiesis (Li et al., 2012). Overexpression of LIN-28 prevented megakaryocytic differentiation in K562 erythroleukemia cells. To demonstrate that this effect was mediated through the repression of let-7, the researchers overexpressed a *let-7* construct that was mutated to resist binding by LIN-28, which rescued megakaryocytic differentiation. Further, the researchers identified MiR-181 as the upstream regulator of LIN-28, and the switch that represses LIN-28 expression to allow for differentiation. Reporter assays demonstrated that *LIN-28* is a direct target of MiR-181, and expression studies of differentiating K562 cells and human umbilical cord blood $CD34^+$ progenitor cells revealed that MiR-181 upregulation precedes *LIN-28* downregulation and let-7 upregulation (Li et al., 2012). Additional studies have supported these findings. let-7 was found to be downregulated in a number of leukemic cell lines, including HL-60, NB4, and U937, and enforced expression of let-7 improved RA-induced granulocytic differentiation (Pelosi et al., 2013).

In another study, in which researchers were investigating factors that were lost in the postnatal transition of HSCs, Lin-28 was identified as increasing the self-renewal capacity of mouse HSCs (Copley et al., 2013). A microarray revealed that *Lin-28* was the most highly differentially expressed between fetal liver and adult bone marrow, being highly expressed in the fetal liver. Consistently, let-7 expression is low in fetal liver and high in adult bone marrow, and these findings held when examining hematopoietic stem/progenitor cells (HSPCs) and HSCs isolated from the fetal liver and adult bone marrow. let-7 binding sites were identified in a number of genes that were highly expressed in fetal liver, as compared to adult bone marrow, including in *Hmga2* (High-mobility group AT-hook 2), a known direct target of let-7 (Mayr et al., 2007), and *Hmga2* transcript levels were shown to decline through differentiation, as let-7 levels increased. Increased chimerism upon primary bone marrow transplantation, and a limiting-dilution assay of secondarily transplanted adult HSCs with enforced *Lin-28* expression, revealed that Lin-28 drastically increases the self-renewal capacity of HSCs. Similarly, enforced expression of *Hmga2* results in increased self-renewal capacity of HSCs, while fetal liver HSCs isolated from *Hmga2* knockout mice were shown to have impaired self-renewal. Taken together, these data suggest that a Lin-28/let-7/Hmga2 regulatory axis mediates HSC maintenance, with *Lin-28* downregulation upon postnatal transition resulting in a decrease in self-renewal capacity in adult HSCs (Copley et al., 2013). Other studies have also demonstrated that enforced expression of Lin-28 in adult HSCs confers broader differentiation capacity, allowing for the generation of a variety of B cells, T cells, and natural killer T cells, a capacity intrinsic to fetal liver HSCs and typical of fetal lymphopoiesis (Yuan et al., 2012). A more detailed examination of this phenomenon, using a barcoding strategy to trace the descendants of marked fetal liver HSCs upon transplantation, has confirmed that Lin-28 expression is essential for preserving the potential of fetal liver HSCs to differentiate into a variety of lymphoid cell types (Kristiansen et al., 2016). Fetal liver HSCs that still express Lin-28 can differentiate to B1-a type B cells, which can only be generated in the fetal liver, but this capacity is lost, and HSCs become B2-type restricted, as Lin-28 expression decreases. Enforced expression of Lin-28 in adult HSCs restores the capacity to differentiate B1-a type cells, the hallmark of a fetal liver HSC.

3.4 MiRNA Profiling in the Hematopoietic System

With the role of MiRNAs in hematopoietic differentiation firmly established in the earlier studies that we have discussed, a number of groups began to take advantage of the ability to fractionate human HSCs, progenitor cells, and mature cells isolated from bone marrow, mobilized peripheral blood, and umbilical cord blood for MiRNA profiling, revealing interesting patterns of differential expression between subpopulations (Bissels et al., 2011; Jin et al., 2008; Liao et al., 2008; Merkerova et al., 2010; Raghavachari et al., 2014). For example, MiRNA expression profiling was conducted on HSC-enriched $CD34^{+}CD38^{-}$ umbilical cord blood cells against total $CD34^{+}$ cells, which include committed progenitor cells (Liao et al., 2008). The study identified 9 MiRNAs that were fourfold overexpressed in $CD34^{+}CD38^{-}$ cells compared to $CD34^{+}$ cells, including MiR-520h, and 22 MiRNAs that were fourfold underexpressed, including MiR-129 (Liao et al., 2008). $CD34^{+}$ cells transfected with MiR-520h yielded increased colonies in colony-forming cell assays, suggesting that MiR-520h contributes to the maintenance of HSCs, though no target of MiR-520h responsible for the phenotype was identified in the study (Liao et al., 2008).

MiRNA profiling was also carried out on $CD133^{+}$ HSCs, $CD34^{+}CD133^{-}$ progenitors, and $CD34^{-}CD133^{-}$ differentiated cells from G-CSF (granulocyte-colony-stimulating factor) mobilized peripheral blood (Jin et al., 2008) and from isolated bone marrow (Bissels et al., 2011). Among other differentially regulated MiRNAs, both studies identified MiR-146a as being upregulated in $CD133^{+}$ HSCs.

3.5 MiR-146a

MiR-146a expression was originally identified as a target of NF-κB regulation, and a participant in a negative-feedback loop with NF-κB expression (Taganov et al., 2006). Expression of MiR-146a in various immune effector cells suggests that it plays an important role in innate immunity (Boldin et al., 2011). But following its identification as a highly expressed MiRNA in HSCs, researchers questioned whether or not it contributed to HSC maintenance and differentiation of various hematopoietic lineages. Promyelocytic zinc finger (PLZF)-overexpressing leukemic cell lines, which have a propensity for unileage megakaryocytic differentiation, were assessed for dysregulated MiRNA expression (Labbaye et al., 2008). MiR-146a was identified as significantly downregulated in PLZF-overexpressing cell lines, and its expression was further reduced upon induction of megakaryocytic differentiation in human cord blood progenitor cells. ChIP experiments confirmed that PLZF binds the promoter of MiR-146a, suggesting that it is an inhibitor of MiR-146a expression. CXCR4, the SDF1 chemokine receptor that is essential for thrombopoiesis in vivo (Avecilla et al., 2004), was identified as a putative target of MiR-146a and confirmed to be a direct target of MiR-146a by the luciferase assay (Labbaye et al., 2008). Through a series of rescue experiments with either MiR-146a overexpression or enforced expression of *CXCR4* with a mutated 3′-UTR, the researchers clearly demonstrated a PLZF/MiR-146a/CXCR4 axis for regulation of megakaryocyte differentiation, in which MiR-146a must be negatively regulated by PLZF to allow for CXCR4 expression and promotion of megakaryopoiesis (Labbaye et al., 2008).

In subsequent studies, David Baltimore's group generated constitutive *MiR-146a* knockout mice to study the effect of MiR-146a on NF-κB regulation and inflammation (Zhao et al., 2011). *MiR-146a* deficiency leads to the development of transplantable myeloid malignancies, as well as lymphomas, over a long 6- to 18-month time course (Zhao et al., 2013). Overexpression of NF-κB was responsible for the oncogenesis, suggesting that MiR-146a inhibition of NF-κB is tumor suppressive (Zhao et al., 2011). Further examination of the *MiR-146a* knockout model demonstrated that loss of *MiR-146a* resulted in chronic inflammatory stress in HSCs, and a loss in HSC number and quality over time. Reciprocal transplantation experiments, in which *MiR-146a*-deficient bone marrow is transplanted to lethally irradiated wild-type mice or wild-type bone marrow is transplanted to irradiated *MiR-146a*-deficient mice, revealed that the effect of *MiR-146a* loss is a cell intrinsic, rather than the result of a defect in the hematopoietic niche, suggesting that MiR-146a is a bona fide regulator of HSC maintenance (Zhao et al., 2013). The MiR-146a/NF-κB feedback loop responsible for this phenotype has been further elucidated—inflammatory stimuli promote the expression of interleukin-1 receptor-associated kinase 1/2 (IRAK1/2) and TRAF6, which in turn stimulates NF-κB expression (Hou et al., 2009). NF-κB induces MiR-146a expression, which is responsible for the negative feedback by inhibiting IRAK1/2 and TRAF6 expression by directly targeting their 3′-UTRs (Hou et al., 2009).

The study of the contribution of MiRNAs to hematopoietic differentiation continues, often led by investigation of a disease-causing MiRNA and how its dysregulation disrupts normal processes. While we have discussed the contribution of some physiologically important MiRNAs originally identified for their contribution to disease in previous sections, in the following sections we will examine a number of MiRNAs that have been implicated in hematologic disease, whether as oncogenes, due to their overexpression, or as tumor suppressors, due to their downregulation.

4. MicroRNAs IN HEMATOLOGIC MALIGNANCIES

By now, dysregulation of MiRNA expression has been detected in many types of cancers, with MiRNA dysfunction being implicated increasingly in the pathogenesis of certain malignancies. With MiRNAs being so important to hematopoiesis, it is no surprise that MiRNAs play a role in the development and pathogenesis of hematologic malignancies (Table 1). In fact, we have already discussed the identification of the first MiRNAs shown to play physiologically important roles in mammals, MiR-15, MiR-16, and MiR-142, which are encoded at translocation sites in lymphocytic leukemias. Further studies of the genomic locations of MiRNAs have hinted at their importance in the development hematologic disease—in a survey of six human AML and myelodysplastic syndrome (MDS) cell lines, >75% of MiRNAs were mapped to sites of genomic alterations in these cell lines, resulting in expression changes to nearly 20% of these MiRNAs (Starczynowski et al., 2011), consistent with previous studies that have mapped MiRNAs to fragile sites in the genome (Calin et al., 2004b). Other studies examining blasts from AML patients have been able to detect various disrupted MiRNA expression signatures that can be associated with AML prognosis, and known cytogenetic and karyotypic classifications (Dixon-McIver et al., 2008; Garzon et al., 2008b; Jongen-Lavrencic et al., 2008). Additionally, disruption of the MiRNA-processing machinery, such as *DICER1* deletion, can also result in the development of hematologic disease. For example, conditional deletion of *Dicer1* in mouse mesenchymal osteoprogenitor cells results in a niche defect that contributes to the development of myelodysplasia and secondary leukemia (Raaijmakers et al., 2010).

While the aforementioned studies are important for identifying MiRNAs for further investigation as mediators of disease or prognostic indicators, none of these genome-wide studies can identify specific disease-associated MiRNAs or the mechanism through which they cause malignancy. In understanding studies of dysregulation of specific MiRNAs that result in hematologic disease, it is helpful to consider a number of categories: (1) oncomiRs or MiRNAs that act as oncogenes by causing disease when they are overexpressed; (2) tumor-suppressive MiRNAs, which normally

Table 1 MiRNAs That Contribute to Hematopoiesis and Hematologic Malignancies

MicroRNA	Role in Hematopoiesis	Role in Hematologic Malignancy	Known Targets
lin-4/ MiR-125	Myelopoiesis	AML, MDS, ALL, DLBCL	BAK1, ST18, DICER1, BMF, KLF13
let-7	Megakaryopoiesis Granulopoiesis HSC differentiation		LIN-28, HMGA2, RAS
MiR-15/ MiR-16	Late erythroid differentiation	B-ALL, B-CLL, MCL, MM	BCL2, RARS, CCDN1, MCL1, CDK6
MiR-181	B- and T-cell development Megakaryopoiesis Erythropoiesis	AML, CLL	LIN-28, AID, BCL2
MiR-223	Myelopoiesis Erythropoiesis Granulopoiesis B-Cell development	CLL, ALL, AML, MALT	NFI-A, CEBPA, MEF2C, E2F1
MiR-155	B- and T-cell development	DLBCL, BCL, AML	AID, SMAD5, INPP5D, HGAL, PU.1
MiR-146a	HSC maintenance Represses megakaryopoiesis and granulopoiesis	MDS	CXCR4, TRAF6, IRAK1/2
MiR-17~92	B-Cell development	BCL, CML, CLL, MCL, MM, FL	BIM, PTEN, E2F1, CCND1
MiR-21	Myelopoiesis	CML, CLL, B-CLL, MDS, NK/T lymphoma	PTEN, PDCD4, SMAD7, MSH2, STAT3
MiR-22	HSC maintenance	MDS, AML	TET1/2/3, PTEN
MiR-203		AML, CML, ALL, CLL, TCL	ABL1
MiR-29	HSC development and maintenance	B-CLL, CLL, AML, MCL	TCL1, MCL1, SP1, DNMT3B, HBP1

This table summarizes many of the MiRNAs discussed in this review, their known roles in hematopoiesis and in hematologic malignancy, and a selection of their known targets. *ALL*, acute lymphoblastic leukemia; *AML*, acute myeloid leukemia; *BCL*, B-cell lymphoma; *CLL*, chronic lymphoblastic leukemia; *CML*, chronic myeloid leukemia; *DLBCL*, diffuse large B-cell lymphoma; *FL*, follicular lymphoma; *MALT*, mucosa-associated lymphoid tissue lymphoma; *MCL*, mantle cell lymphoma; *MDS*, myelodysplastic syndrome; *MM*, multiple myeloma; *TCL*, T-cell lymphoma.

inhibit expression of known oncogenes and result in malignancy when their expression is lost; and (3) disruption to MiRNA-processing machinery, which has more global effects on the expression of MiRNAs, but can cause malignancy depending on cellular context (Fig. 4). These divisions are important in considering how to study and evaluate disease-associated MiRNAs, but also how to approach MiRNA-based therapeutic strategies.

4.1 OncomiRs

OncomiRs represent the most expansive class of known disease-associated MiRNAs. The term "oncomiR" was first coined by Scott Hammond's group in a study in which they reported on the potent oncogenic effect of the *MiR-17~92* polycistron, which is amplified in B-cell lymphomas (He et al., 2005). As the portmanteau implies, oncogenic MiRNAs contribute to malignant transformation when they are overexpressed, typically because they downregulate the expression of tumor-suppressive proteins. Some cancers that result from the overexpression of an oncomiR have been reported to be addicted to the continued expression of the oncomiR, as is the case with MiR-21 (Medina et al., 2010), which presents the opportunity for MiRNA-based therapeutics.

4.1.1 The MiR-17~92 Polycistron

Amplification of 13q31–q32, and specifically *C13orf25*, has been identified as contributing to the transformation of malignant B-cell lymphoma (Rao et al., 1998). One of the transcripts that is found in *C13orf25* encodes seven MiRNAs in what we now know as the *MiR-17~92* polycistron (Ota et al., 2004), which is also encoded in two homologs, *MiR-106a~92* and *MiR-106b~25* (Tanzer and Stadler, 2004). The Hammond group screened lymphoma samples and discovered upregulation of the five MiRNAs encoded in the *MiR-17~92* polycistron in 65% of the isolates (He et al., 2005). When *MiR-17~92* is ectopically expressed in a transgenic model of B-cell lymphoma, in which *c-Myc* is expressed from the immunoglobulin heavy chain enhancer (*Eμ-myc*), transplantable lymphomas develop with a dramatically decreased latency, and animal succumb to disease in 65 days (He et al., 2005). The *MiR-17~92* cluster has therefore been implicated in a number of malignancies, including a variety of hematologic malignancies beyond B-cell lymphoma (Lawrie, 2013).

ChIP experiments have identified c-Myc as a positive regulator of the *MiR-17~92* cluster, suggesting that it may contribute to oncogenesis (O'Donnell et al., 2005), and a number of targets of MicroRNAs encoded in the *MiR-17~92* polycistron that may contribute to its oncogenic potential have been identified. For example, *MiR-17~92* contributes to B-cell development and B-cell maintenance through its inhibition of *Bim*, a proapoptotic protein (Ventura et al., 2008). The *MiR-17~92* has also been identified to negatively regulate a number of other important tumor suppressors, including E2F1 (Woods et al., 2007), cyclin D1 (CCND1) (Chen et al., 2008a; Deshpande et al., 2009), and PTEN protein phosphatase 2 (PP2) (Mavrakis et al., 2010).

4.1.2 MiR-125

MiR-125 has been studied for not only its role in regulation of normal HSC self-renewal but also its role in leukemogenesis, particularly in AML. We now know that the MiR-125 family is actually homologous to lin-4, in *C. elegans*, the first identified MiRNA (Lagos-Quintana et al., 2002). MiR-125 was first identified as an oncomiR in prostate cancer, where it was found to promote androgen-independent tumor growth by inhibiting the proapoptotic regulator BAK1 (Shi et al., 2007). Later, an oncogenic role for MiR-125 in the hematopoietic system was identified when overexpression studies revealed that MiR-125 inhibits terminal differentiation of human leukemia cell lines, such as HL-60 and NB4 (Bousquet et al., 2008). The t(2;11)(p21;q23) translocation is a rare abnormality that results in AML or MDS, often as a primary or single event. A number of protein-coding and MiRNA-encoding genes are located near the breakpoints, including MiR-100, let-7a-2, and MiR-125b, and MiR-125b expression was found to be dramatically upregulated in samples from affected patients (Bousquet et al., 2008). Using models of HL-60 and NB4 differentiation induced by treatment with DMSO and RA, respectively, Bousquet et al. demonstrated that transfection of MiR-125b can prevent differentiation of these cell lines to mature monocytic and granulocytic cells. Later, in vivo, studies conducted by the Lodish group demonstrated that when fetal liver cells overexpressing MiR-125b are transplanted into mice, they develop a macrocytic anemia that converts to various types of leukemia. Additionally, overexpression of MiR-125b can cooperate with the *BCR–ABL* fusion oncogene to accelerate leukemia development (Bousquet et al., 2010). MiR-125b transfection prevents differentiation of primary, patient-isolated $CD34^+$ blasts, suggesting that MiR-125b can contribute to a differentiation blockade under leukemic conditions. Overexpression of MiR-125b, which is also encoded in a second gene on chromosome 21, has also been associated with trisomy 21/ Down syndrome-associated acute megakaryoblastic leukemia (Klusmann et al., 2010). MiR-125b cooperates with *GATA1* mutation to cause increased proliferation and self-renewal of megakaryocyte progenitors. Additionally, this study identified a number of targets of MiR-125b that may contribute to its oncogenic function, including tumor suppressor, ST18, and DICER1, which will be discussed in a following section (Klusmann et al., 2010).

Building off data observed in MiR-125b overexpression in leukemia, a number of studies have identified a role for MiR-125 in normal hematopoiesis. In 2010, David Scadden's and Christopher Park's groups demonstrated that MiR-125 is highly expressed in HSCs and plays a role in regulating HSC number (Guo et al., 2010b; Ooi et al., 2010). The Scadden group observed that conditional *Dicer* deletion in the hematopoietic compartment resulted in HSPC failure due to apoptosis (Guo et al., 2010b). With the increasing evidence for the contribution of MiRNAs to hematopoiesis, this result was surely not surprising, but it was one of the earlier results to demonstrate that MiRNAs are important to HSC function, and not just in reinforcing a lineage identity during differentiation. Importantly, the Scadden group performed MiRNA profiling on the $CD34^{-}Flk2^{-}$ $c\text{-}Kit^{+}Sca1^{+}Lineage^{-}$ population, which is highly enriched for long-term HSCs (LT-HSCs), and identified MiR-99b, let-7e, and MiR-125a, all expressed from a single MiRNA cluster, as highly enriched in LT-HSCs. Ectopic overexpression of MiR-125a resulted in increased multilineage reconstitution upon bone marrow transplantation, due to a specific increase in HSC self-renewal mediated by a reduction in HSC apoptosis (Guo et al., 2010b). Reduction of apoptosis in MiR-125a-overexpressing HSCs is mediated by inhibition of the proapoptotic protein Bcl-2 (B-cell lymphoma 2) homologous antagonist/killer (Bak1), which was previously identified as a target of MiR-125 in prostate cancer. Interestingly, the expansion of HSCs upon MiR-125a overexpression must be mediated by additional unidentified targets, as well, because *Bak1* knockout was not sufficient to recapitulate the observed phenotypes (Guo et al., 2010b), highlighting a difficulty in the study of MiRNAs—in any given cellular context, MiRNAs may target multiple transcripts, to achieve unified or disparate ends.

The finding that MiR-125 contributes to HSC maintenance and expansion has been replicated and extended. Through targeting of additional antiapoptotic machinery, enforced MiR-125b expression biases HSCs toward lymphoid differentiation (Ooi et al., 2010). MiR-125b-overexpressing LT-HSCs demonstrated increased bone marrow reconstitution on primary and secondary transplantation, with a particular increase in lymphoid cells and progenitors upon secondary transplantation. Consistently, enforced MiR-125b expression led to an increase in the number of lymphoid-biased

Slamf1$^{low/neg}$ HSCs upon transplantation. Cell cycle analysis of cells over-expressing MiR-125b demonstrated that the increase in reconstitution is not the result of increased self-renewal, but a decrease in apoptosis (Ooi et al., 2010), as was previously observed (Guo et al., 2010b). Two anti-apoptotic proteins that are potential targets of MiR-125 and could contribute to the observed phenotype were also identified—*Bcl-2-modifying factor* (*Bmf*) and *Kruppel-like factor* (*Klf13*) expression was decreased upon MiR-125 overexpression and increased upon MiR-125 antagonism, though direct binding of MiR-125 to the 3′-UTRs was not observed (Ooi et al., 2010). Many additional targets of MiR-125 have been identified in other studies, in the hematopoietic system, and in other tissues (Klusmann et al., 2010; Le et al., 2009; Zhong et al., 2010). Though there is more to study, MiR-125 is clearly an important regulator of HSC maintenance and, as such, a critical oncomiR upon its dysregulation.

4.1.3 MiR-21

MiR-21 is well studied for its contribution to malignant transformation and has been found to be upregulated in nearly all types of cancers that have been analyzed (Ling et al., 2013). MiR-21's oncogenic capacity was first demonstrated in glioblastoma multiforme (GBM), where it was found to be required for survival of GBM-derived cell lines—when MiR-21 was inhibited using antisense locked nucleic acids (LNAs), the cell lines underwent apoptotic cell death, demonstrating that these cell lines were addicted to persistent MiR-21 expression (Chan et al., 2005). However, no responsible MiR-21 target gene was identified in this study. In 2010, Slack's group extended the study of MiR-21 in vivo by generating a mouse for doxycycline-controlled, tissue-specific overexpression of MiR-21, making them only the second group to demonstrate the contribution of a MiRNA to cancer development in a transgenic mouse model (Costinean et al., 2009). After birth, transgenic animals were driven to overexpress MiR-21 in the hematopoietic system, and within 3 months the animals developed a transplantable malignancy closely mimicking pre-B-cell lymphoblastic lymphoma/leukemia. Strikingly, when transgene expression was terminated in transgenic animals, full tumor regression was observed after only 7 days, suggesting that MiR-21 was sufficient for tumor initiation, but also necessary for cancer maintenance, again demonstrating oncomiR addiction for MiR-21 (Medina et al., 2010). While Medina et al. observed that tumor regression was the result of apoptosis, consistent with MiR-21's demonstrated role as an antiapoptotic factor, they did not report any specific targets of MiR-21 that may have been responsible for the dramatic phenotype they observed.

Other studies have identified targets of MiR-21, including phosphate and tensin homolog (PTEN), a known tumor suppressor, and programmed cell death 4 (PDCD4), a nuclear protein thought to contribute to apoptosis (Yamanaka et al., 2009). Treatment of K562 cells with antisense oligonucleotides against MiR-21 sensitizes the cells to apoptosis, confirming an anti-apoptotic and oncogenic role for MiR-21 (Hu et al., 2010; Li et al., 2010). MiR-21 has been studied as an indicator of poor prognosis in CLL (Rossi et al., 2010), and when qPCR for MiR-21 is added to the other diagnostic criteria for CLL, it improves prognostic risk stratification.

MiR-21 inhibition has also been investigated as a mechanism for restoring the SMAD7–TGFβ axis in MDS (Bhagat et al., 2013). TGFβ is a myelosuppressive cytokine that has been found to be upregulated in MDS, in which it is responsible for hematopoietic suppression. SMAD7, which is a negative regulator of the TGFβ receptor kinase, is downregulated in $CD34^+$ cells isolated from some MDS patients, and *SMAD7* loss results in increased TGFβ signaling, even in the absence of increased TGFβ expression (Bhagat et al., 2013). MiR-21 has also been shown to be dysregulated in samples isolated from MDS patients (Sokol et al., 2011). The Verma group demonstrated that MiR-21 binds the putative MiR-21 binding site in the 3′-UTR of *SMAD* (Bhagat et al., 2013). Overexpression of MiR-21 in $CD34^+$ cells leads to a reduction in their ability to form erythroid colonies, while MiR-21 antagonism increased the ability of $CD34^+$ cells to form erythroid colonies in response to TGFβ. In a genetic mouse model of anemia, MiR-21 antagonism improves TGFβ-induced bone marrow failure, consistent with its role in myelosuppression in MDS.

4.1.4 MiR-22

By now, many other MiRNAs have been identified as oncogenes in hematologic malignancies, some of which are reviewed elsewhere (Lawrie, 2013; Ling et al., 2013; Spizzo et al., 2009). One such oncomiR, *MiR-22*, has become a focus of Dr. Pandolfi's group (Song et al., 2013a,b). MiR-22 was identified as a potent oncogene simultaneously in metastatic breast cancer and in MDS. In both cases, MiR-22 works by causing epigenetic dysregulation in affected cells, by repressing the ten-eleven translocation (TET) family of enzymes. DNMTs are responsible for transferring methyl groups to cytosines in the genome, creating a repressive mark that blocks transcription. TET enzymes oxidize 5-methylcytosine (5-mC) to 5-hydroxymethylcytosine (5-hmC), which may serve as an epigenetic mark in its own right. More importantly, this oxidation is the first step in removing the repressive 5-mC, making the TET enzymes antagonists of the DNMTs.

In a model of metastatic breast cancer, MiR-22 overexpression results in the repression of all three Tet enzymes, Tet1/2/3 (Song et al., 2013b). Loss of this important epigenetic regulator results in the silencing of the antimetastatic *MiR-200* gene, and the promotion of breast cancer stemness, resulting in a more aggressive malignant metastatic phenotype.

MiR-22 has also been found to be overexpressed in bone marrow isolated from MDS patients. The importance of TET2 in the hematopoietic system is well established, with *TET2* mutations and inhibition of the TET2 enzyme being very frequent defects in AML and MDS (Abdel-Wahab et al., 2009; Tefferi et al., 2009). Similar to what was observed in breast cancer, conditional MiR-22 overexpression in the hematopoietic system results in downregulation of Tet2 (Song et al., 2013a). As a result, hematopoietic cells had global loss of 5-hmC and the reduced expression of known Tet2 target genes. HSCs overexpressing MiR-22 exhibit increased self-renewal in vitro and increased reconstitution in vivo upon bone marrow transplantation. Over time, mice overexpressing MiR-22 in the hematopoietic compartment develop a hematologic malignancy resembling MDS, which sometimes progresses to an AML-like disease. Further, antagonizing MiR-22 overexpression in vitro increases Tet2 expression, decreases growth in cell lines, and decreases colony-forming capacity in explanted HSPCs, suggesting that there may be potential for MiR-22 decoying in human disease (Song et al., 2013a).

Though these data clearly identify an oncogenic role for MiR-22 in MDS and in breast cancer, other studies have suggested tumor-suppressive properties for MiR-22, highlighting the importance of cellular context on MiRNA function. For example, MiR-22, which is downregulated in some lung cancer specimens, has been shown to repress the receptor tyrosine kinase, ERBB3 (Ling et al., 2012), and reduced expression of MiR-22 is associated with poor prognosis in colorectal cancer (Zhang et al., 2012). More recently, the MiR-22's role in the hematopoietic system has also grown more complex (Wurm et al., 2017). Two studies have explored a role for MiR-22 in promoting myeloid differentiation and serving as a tumor suppressor in de novo AML through the repression of known oncogenic pathways (Jiang et al., 2016; Shen et al., 2016). Further study is necessary to clarify the role of MiR-22 in the various cellular contexts of hematopoiesis and leukemogenesis, and to determine if modulation of MiR-22, either through MiR-22-replacement or antagonism, would be beneficial in the management of hematologic malignancies. In the next section, we will discuss MiRNAs with known tumor-suppressive function in hematologic disease.

4.2 Tumor-Suppressive MicroRNAs

While the class of known tumor-suppressive MiRNAs is smaller than the known oncogenic MiRNAs, and tumor-suppressive MiRNAs do not have a clever nickname, like the "oncomiRs," they are no less important in the study of human malignancies nor in the history of the discovery of disease-associated MiRNAs. MiR-15 and MiR-16, for example, were identified as lost in 13q14 deleted CLL (Calin et al., 2002), making them the first MiRNAs with a demonstrated role in human physiology or disease. It should come as no surprise that tumor-suppressive MiRNAs are responsible for repressing oncogenic genes. Tumor-suppressive MiRNAs exhibit their oncogenic function when their expression is lost or downregulated, resulting in the uninhibited expression of the normally suppressed oncogenes. For example, the *let*-7 MiRNA, whose homology between *C. elegans* and humans prompted the discovery of MiRNAs in mammals (Pasquinelli et al., 2000; Reinhart et al., 2000) and we have discussed more detail in Section 3.3, is a tumor-suppressive MiRNA that represses the known oncogenes RAS and HMGA2 (Johnson et al., 2005; Mayr et al., 2007). In this section, we will discuss tumor-suppressive MiRNAs, such as MiR-15, MiR-16, MiR-203, and MiR-29b.

4.2.1 MiR-15 and MiR-16

As previously mentioned, MiR-15 and MiR-16 became early targets of study in the hematopoietic system because of their dysregulation in B-cell chronic lymphocytic leukemia (B-CLL). These MiRNAs are encoded in the MiR-15a–MiR-16-1 cluster, which is lost with the 13q14 deletion or can be epigenetically silenced (Calin et al., 2002), suggesting that MiR-15 and MiR-16 represent tumor-suppressive MiRNAs, whose loss can result in malignancy. While Calin et al. were able to identify arginyl-tRNA synthetase (RARS) as a putative target of MiR-15/16, with RARS levels decreased in a number of samples with increased MiR-15/16 expression, they were not able to associate this defect with a pathogenic process (Calin et al., 2002). Later, however, a number of MiR-15/16 oncogenic targets were identified, helping to explain how their downregulation can result in oncogenesis. For example, MiR-15 and MiR-16 were found to repress the antiapoptotic Bcl-2 protein, which is characteristically over-expressed in CLL (Cimmino et al., 2005). Though Bcl-2 is frequently over-expressed due to a translocation in other malignancies, that is not frequently the case in CLL and the mechanism of overexpression had been previously unknown. The Croce group was able to demonstrate that when MiR-15/16 levels were high, such as in healthy $CD5^+$ lymphoid cells, Bcl-2 expression was minimal, and that Bcl2 expression was inversely correlated with

MiR-15/16 expression in cells isolated from CLL patients (Cimmino et al., 2005). Additionally, the *Bcl-2* 3′-UTR is responsive to MiR-15/16 in a luciferase assay, and overexpression of MiR-15/16 induces apoptosis in leukemia cell lines, suggesting that loss of MiR-15/16 expression can result in leukemia and that MiR-15/16 are tumor-suppressive MiRNAs in the hematopoietic system (Cimmino et al., 2005). A number of other oncogenic targets for MiR-15 and MiR-16 have been identified. Earlier studies identified a role for *CCDN1* 3′-UTR in the pathogenesis of mantle cell lymphoma (MCL)—mutations that results in the loss of the 3′-UTR result in MCL with a more aggressive clinical course (Wiestner et al., 2007). Chen et al. and Deshpande et al. recognized that stabilization of *CCDN1* occurred when a MiR-16 target site was lost along with the rest of the UTR (Chen et al., 2008a; Deshpande et al., 2009).

4.2.2 MiR-203

MiR-203 dysregulation in malignancy was first identified in studies examining a 7-Mb fragile site at chromosome 14q32, which is highly enriched in MiRNAs, and is frequently lost in T-cell malignancies (Bueno et al., 2008). The homologous region is also lost in an irradiation-induced model of T-cell lymphomas in mice. Interestingly, when a single copy of MiR-203 is lost due to deletion at the fragile site, MiR-203 expression is further reduced by epigenetic silencing of the other allele. In murine T-cell lymphoma samples and in a number of human T-cell tumor cell lines, the MiR-203 locus is dramatically hypermethylated (Bueno et al., 2008). In the context of *MiR-203* loss a number of putative MiR-203 targets are upregulated, including ABL1, a protein tyrosine kinase, and potent oncogene, which was demonstrated to be a bona fide target of MiR-203 through reporter assays. The Philadelphia chromosome, t(9;22)(q34;q11), is present in 95% of chronic myelogenous leukemia (CML) cases and results in the *BCR–ABL1* fusion oncogene. The *BCR–ABL1* transcript retains the *ABL1* 3′-UTR and is thus a target of MiR-203. Interestingly, in cell lines in which ABL1 is overexpressed or BCR–ABL1 is expressed, the MiR-203 locus is hypermethylated, suggesting a pressure to maintain low levels of MiR-203, so that ABL1 or BCR–ABL1 can exert their function. Conversely, when affected cell lines are treated with demethylating agents, such as 5′-azacytidine (Aza), expression of ABL1 and BCR–ABL1 is reduced (Bueno et al., 2008). Together, these data suggest that MiR-203 replacement in certain malignancies may function much like tyrosine kinase inhibitors, such as imatinib, in the treatment of CML. Hypermethylation of the MiR-203 tumor suppressor has now been demonstrated in a variety of hematologic malignancies and in HCC (Chim et al., 2011; Furuta et al., 2010).

4.2.3 MiR-29b

MiR-29 was first identified as tumor suppressive in B-CLL (Pekarsky et al., 2006). T-Cell leukemia/lymphoma-1 (TCL1), which is a potent activator of the PI3K/AKT prosurvival pathway, is an oncogene that is frequently overexpressed in B- and T-cell leukemias (Pekarsky et al., 2001) and is associated with aggressive B-CLL (Pekarsky et al., 2006). MiR-29 and MiR-181 were shown to be downregulated in CLL and are putative regulators of TCL1, and *MiR-29* loss has been shown to be associated with poor prognosis in CLL (Calin et al., 2005), suggesting its role as a tumor-suppressive MiRNA in CLL. MiR-29 was also demonstrated to act as a tumor suppressor in cholangiocarcinoma (Mott et al., 2007). Myeloid cell leukemia 1 protein (MCL1) is a potent antiapoptotic protein in the BCL-2 family, which has low expression in cholangiocytes. In contrast, MiR-29 is highly expressed in cholangiocytes and was shown to repress MCL1 expression. MiR-29 expression is reduced in a number of cholangiosarcoma cell lines, and enforced expression of MiR-29 reduced MCL1 expression and sensitizes cell lines to apoptosis (Mott et al., 2007).

As we discussed in the context of MiR-203 (Section 4.2.2), hypermethylation is a frequent mediator of oncogenesis. The MiR-29 family was identified as downregulated in lung cancer (Fabbri et al., 2007). In contrast, DNMT3a/b are frequently upregulated in lung cancer—upregulation of DNMT and hypermethylation is associated with poor prognosis in lung cancer. Intriguingly, the *DNMT3a/b* transcripts contain MiR-29 binding sites, and expression of DNMT3a/b and MiR-29 is inversely correlated with patient samples. Further, overexpression of MiR-29 reduces the level of hypermethylation in cell lines, as detected by mass spectrometry (MS), and upregulation of reporter genes that are frequently downregulated in lung cancer due to hypermethylation (Fabbri et al., 2007).

Aberrant DNA hypermethylation is also typical of myeloid malignancies. *DNMT3a/b* transcripts have also been demonstrated to be direct targets of MiR-29 expression in leukemic cell lines, and enforced expression of MiR-29 results in repression of DNMT expression (Garzon et al., 2009). There are no MiR-29 binding sites in the *DNMT1* 3′-UTR, suggesting that it is not a direct target of MiR-29. Interestingly, MiR-29 directly targets the zinc finger transcription factor, SP1, a positive regulator of DNMT1 expression. Taken together, MiR-29 represses the expression of all of the DNMTs. Consistently, ectopic MiR-29 overexpression in leukemic cell lines results in a reduction in global DNA methylation and partial differentiation, consistent with its role as a tumor suppressor (Garzon et al., 2009).

Further studies have placed MiR-29 downregulation within a complicated, multicomponent regulatory circuit in KIT-driven AML (Liu et al., 2010). KIT is a receptor tyrosine kinase that is expressed on hematopoietic stem and progenitor cells and is critically important to the maintenance of HSCs and normal hematopoiesis, primarily through its interaction with the stem cell factor (or the KIT ligand). SP1 and NFκB cooperate to increase KIT expression by binding at the *KIT* promoter (Liu et al., 2010). The *SP1* transcript is a direct target of MiR-29 (Garzon et al., 2009), and forced expression of MiR-29 leads to decreased KIT expression (Liu et al., 2010). Interestingly, specific chemical inhibition of SP1 or NFκB results in decreased KIT expression and increased MiR-29 expression, suggesting the existence of a feedback circuit that regulates the expression of MiR-29, SP1, and KIT. ChIP reveals that SP1/NFκB recruit HDACs to the MiR-29 locus, completing the autoregulatory loop whereby SPI/NFκB cooperate with HDACs to repress MiR-29 expression, further enhancing expression of SP1, which promotes the expression of KIT (Liu et al., 2010).

4.3 MicroRNA-Processing Mutations

We have discussed that DICER is crucial to the cytoplasmic processing of the pre-miRNA to the mature MiRNA (Section 2.2), the selection of the MiRNA strand from the passenger strand, and loading the MiRNA into the RISC so that it can affect its target transcripts. With DICER's importance to MiRNA processing, and the ability for MiRNA dysregulation to result in malignant transformation, we should expect that DICER1 dysregulation can contribute to the development of various types of cancers (Foulkes et al., 2014). We have already seen how the effect of a given MiRNA's dysregulation is cell context specific—a MiRNA that can serve as an oncomiR in some cell types can serve as a tumor suppressor in other cell types. Similarly, reduced DICER1 expression has been associated with oncogenesis in some malignancies, such as in lung, breast, skin, and ovarian cancer (Dedes et al., 2011; Karube et al., 2005; Merritt et al., 2008; Pampalakis et al., 2010; Sand et al., 2010), and DICER1 overexpression is associated with prostate cancer metastases, and in cervical and esophageal cancers (Chiosea et al., 2006; Muralidhar et al., 2007; Sugito et al., 2006). Germline DICER1 mutations have been associated with many familial and childhood cancer syndromes, including differentiated familial pulmonary blastoma (Hill et al., 2009), Wilms' tumor (Foulkes et al., 2011), ovarian sex-chord stromal tumors (Schultz et al., 2011), and many others constituting what is commonly referred to as "DICER1 syndrome" (Foulkes et al., 2014).

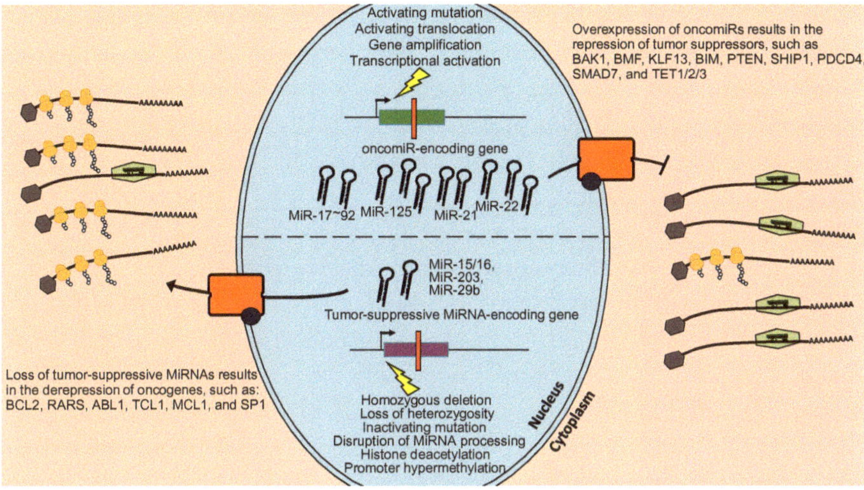

Fig. 4 MiRNAs can serve as oncogenes and tumor suppressors. MiRNAs that normally repress the expression of tumor suppressors can serve as oncogenes, also called oncomiRs (*top* and *right*). Mutations and other genetic aberrations result in the increased expression of oncomiRs, which repress their tumor-suppressive targets and contribute to the development of malignancy. On the flip side, tumor-suppressive MiRNAs normally repress the expression of oncogenes (*bottom* and *left*). When mutations or other defects result in decreased expression of tumor-suppressive MiRNAs, expression of oncogenes is increased and contributes to the development of malignancy.

In the hematopoietic system, deletion of *Dicer1* in mouse osteoprogenitor cells, driven specifically by the *Osterix-Cre Recombinase* transgene, results in the development of a myelodysplasia and AML, in an example of niche-induced leukemogenesis (Raaijmakers et al., 2010). Loss of *Dicer1* in these mice results in the failure of osteoblastic differentiation, an expansion of HSPCs, and leukopenia, anemia, thrombocytopenia, and dysplasia in cells from various lineages, consistent with a diagnosis of MDS, which converted to AML in a number of animals. Reciprocal transplantation experiments confirm that this leukemogenesis is not the result of an intrinsic defect in the HSCs, rather from defects to the HSC niche. The Scadden group identified reduced expression of the *Shwachman–Diamond–Bodian* (*Sbds*) gene, associated with stress response, and the clinical syndrome associated with MDS and AML, as mediating the observed phenotype in *Dicer1* knockout mice, though they were not able to construct a mechanism connecting Dicer1 loss and downregulation of Sbds.

Because loss of both alleles of DICER1 is necessary to result in malignancy, somatic DICER1 mutations rarely result in malignancy. However, mutations to other members of the MiRNA-processing axis have been reported to contribute to malignant transformation. Tumors exhibiting microsatellite instability due to mutations in the *Musashi 1/2* (*MSI1/2*), which is frequently mutated in hematologic malignancies and contributes to poor prognosis myeloid leukemias (Kharas et al., 2010), frequently develop somatic truncating mutations in *TARBP2*, which encodes TAR RNA binding protein, a protein essential for loading of the RISC (Melo et al., 2009). Microsatellite-unstable tumors have also exhibited mutations in XPO5, the exportin protein responsible for exporting pre-miRNAs from the nucleus (Melo et al., 2010). Msi1 has also been shown to contribute to Lin-28-mediated MiRNA processing, which we discussed in Section 3.3, in early neural differentiation (Kawahara et al., 2011). Activating mutations or overexpression of *Msi1* might contribute to oncogenesis through the repression of let-7, and the derepression of let-7's oncogenic targets, though further study of this mechanism is required.

OncomiRs and tumor suppressors both provide opportunities for therapeutic intervention. In the next section, we will discuss MiRNA-based therapeutic strategies including antagonism of oncogenic MiRNAs and replacement of lost tumor-suppressive MiRNAs.

5. THERAPEUTIC STRATEGIES

As we have discussed in Section 3, MiRNAs are important regulators of numerous physiological functions, including HSC maintenance and hematopoietic differentiation. MiRNA dysregulation, either overexpression of oncogenic MiRNAs or downregulation of tumor-suppressive MiRNAs, can result in hematologic malignancy, as we have discussed in Section 4. MiRNA-based cancer therapeutics are, in the simplest terms, a strategy of combatting pathogenesis by restoring MiRNA function of tumor-suppressive MiRNAs that are lost or downregulated, or by inhibiting oncogenic MiRNAs that are overexpressed (Li and Rana, 2014).

Other anticancer medications include highly toxic chemotherapeutic agents and highly specific protein-targeting agents, either antibody based or small molecules, which are difficult and expensive to develop. In contrast, MiRNAs operate by highly specific, Watson–Crick base pairing. In the case of oncogenic MiRNAs, base pairing can be blocked with anti-miRNA oligonucleotides, more frequently referred to as "antagomiRs," which can bind with high affinity to miRISCs, disrupting their inhibitory function. In the case of tumor-suppressive MiRNAs, MiRNA mimetics can restore MiRNA function. While strategies taking advantage of this sequence specificity are the most aggressively pursued, other therapeutic strategies include MiRNA antagonism through the expression of sponge constructs, and the development of small molecules that target the transcriptional regulation of certain MiRNAs. The primary concerns with MiRNA-based therapies are tissue-specific delivery and, in the case of systemic administration, off-target effects of the administered antagomiRs or mimetics.

5.1 Modified Nucleic Acids Are Good Candidates for MiRNA-Based Therapeutics

Unmodified RNA oligonucleotides are unsuitable for therapeutic applications because they are sensitive to serum nucleases (Hutvagner et al., 2004; Meister et al., 2004a). 2′-O-Methyl (2′-OMe)-modified nucleotides were the first modified nucleotides used in for MiRNA antagonism due to their resistance to nucleases and their ability to stably base pair with ssRNAs (Majlessi et al., 1998), such as MiRNAs in miRISCs in *Drosophila melanogaster* and in human cell lines (Hutvagner et al., 2004; Meister et al., 2004a) (Table 2). But still, 2′-OMe-modified nucleotides are not invulnerable to serum exonucleases and are therefore not suitable for human therapeutic applications. A number of other nucleotide modifications have been assessed for their resistance to nuclease degradation, their ability to permeate tissues and cell membranes, and their ability to bind and inhibit miRISCs. For example, substitution of some of the phosphodiester bonds with phosphorothioate bonds in oligonucleotides improves their absorption and makes them resistant to degradation, though oligonucleotides fully substituted with phosphorothioate-modified nucleotides do not exhibit any effect on miRISCs (Geary, 2009; Geary et al., 1997; Krutzfeldt et al., 2005; Lennox and Behlke, 2011). More recently, attempts to modify these therapeutic oligonucleotides have focused on utilizing LNAs, which utilize a bicyclic nucleic acid that is resistant to nuclease degradation, and bind antisense oligonucleotides with high strength (Singh et al., 1998). Several LNA design and modification strategies have been employed in mammalian in vivo studies, demonstrating their therapeutic potential and providing instruction for investigators who would wish to design their own (Elmen et al., 2008a,b; Lennox et al., 2013; Obad et al., 2011).

5.2 Tissue-Specific Delivery of MiRNA-Based Therapeutics Remains a Challenge

The great challenge in the application of MiRNA-based therapeutics comes in their tissue-specific delivery (Li and Rana, 2014). Oligonucleotides that achieve resistance to nuclease degradation and efficient antisense binding

Table 2 Types of MiRNA-Based Therapeutics

Therapeutic	Description	Mechanism of Action	Clinical Development	Example
MiRNA antagonism				
MiRNA sponges	MiRNA sponges are transcribed from virally introduced or transfected constructs and contain many MiRNA recognition elements (MREs)	MiRNA sponges block MiRNA activity by competing with MiRNAs, which bind their MREs. This sponge effect prevents MiRNAs from exerting their function	Preclinical studies	MiRNA sponges are frequently used in preclinical studies to antagonize MiRNA activity. Philip Sharp's lab first developed MiRNA sponges in 2007
Antisense oligonucleotides (ASOs)	ASOs are short, chemically modified, single-stranded, DNA-like, antisense oligonucleotides that are capable of binding to MiRNAs, disrupting their action, or mRNAs, disrupting their transcription. LNAs and other antagomiRs are included within the ASO category	ASOs bind their RNA targets by Watson–Crick base pairing. When ASOs target mRNAs, the resulting DNA–RNA hybrids are cleaved by RNAse H. ASOs disrupt MiRNA function by binding them in miRISCs	Preclinical studies	2′-OMe-modified oligonucleotides were used to study MiRNAs and RNAi activity in *D. melanogaster* and in human cell lines
Locked nucleic acid (LNA) antagomiRs	LNAs are locked because they are synthesized with some RNA nucleotides modified with a bridge connecting the 2′O to the 4′C. This locked conformation makes them resistant to nucleases. LNAs are usually designed to target the functional 5′-seed region of MiRNAs	LNA antagomiRs bind MiRNAs by Watson–Crick base pairing, blocking MiRNA action by binding them in miRISCs. MiRNAs are degraded and the antagomiR is recycled	Preclinical Studies Phase I trials Phase IIa trials	Miravirsen, an LNA antagomiR against MiR-122 in the treatment of HCV

Other antagomiRs	Oligonucleotides partially modified with phosphorothioate bonds can also be used to inhibit MiRNAs and are resistant to serum nucleases	Watson–Crick base pairing blocks MiRNAs in the miRISC and leads to their degradation. The antagomiR is recycled	Preclinical studies	Early studies in the targeting of MiR-122 in the liver
MiRNA replacement				
MiRNA mimics	Synthetic double-stranded oligonucleotides that are processed to single-stranded MiRNA-like molecules	MiRNA mimics are incorporated into miRISCs and act like endogenous MiRNAs. MiRNA mimics can be used to restore MiRNA levels, as in the case of lost tumor-suppressive MiRNAs	Preclinical studies Phase I trials	MRX34 for the restoration of the tumor-suppressive MiR-34 in the treatment of various malignant cancers Restoration of MiR-26a in the treatment of hepatocellular carcinoma (HCC) in mice
MiRNA expression vectors	MiRNA expression vectors can be introduced virally or by transfection	MiRNAs are transcribed from expression vectors downstream of constitutive or cell type-specific promoters	Preclinical studies	MiRNA expression vectors are commonly used in preclinical studies, but are not well suited for clinical use due to concerns with virally introduced gene therapy

exhibit limited tissue distribution—they are quickly taken up by the liver and kidneys and excreted in the urine, which necessitates administration of high doses, increasing the risk of dangerous off-target effects. A number of methods have been applied to try and improve the delivery and tissue specificity of small oligonucleotides in vivo: oligonucleotide conjugation, liposomes, nanoparticle-based methods, and antibody-based methods. The earliest conjugation method sought, simply, to improve the distribution of small oligonucleotides in tissue, generally, which was achieved by conjugating an antagomiR with cholesterol at its 3′-end (Krutzfeldt et al., 2005). Further studies demonstrated that small oligonucleotides could be assembled into either high- or low-density lipoproteins, or vitamin E for preferential accumulation at different organs, such as the liver, digestive tract, endocrine organs, and the kidneys (Nishina et al., 2008; Wolfrum et al., 2007). Additionally, small oligonucleotides conjugated to unmethylated CpG oligonucleotides can target cells expressing Toll-like receptors 9 (TLR9), such as myeloid and B cells (Kortylewski et al., 2009). Liposome-based delivery strategies have been very effective at improving delivery to the liver, greatly reducing the dose of small oligonucleotides needed to achieve an effective dose, with a library of effective lipidoids available for liposome construction (Akinc et al., 2008; Morrissey et al., 2005). Though systemic administration of liposome-based therapies leads to their accumulation in the liver, studies have demonstrated that tissue-specific administration can be achieved by localized delivery, such as to the vagina for the prevention of herpes simplex 2 infection (Palliser et al., 2006), and intracranial injection to prevent infection by the Japanese encephalitis and West Nile viruses (Kumar et al., 2006). Still, nanoparticle-based delivery methods may yield improvements over liposome-based methods because of the ability to engineer them to uniform size with more easily customizable conjugations. Strong support for nanoparticle-based delivery systems came from a Phase I clinical trial in patients suffering from Ewing's sarcoma. A systemically administered siRNA targeting the *EWS–FLI1* fusion oncogene, encapsulated in a transferrin-containing nanoparticle, meant to target tumor cells that have upregulated transferrin receptors, was able to accumulate in the solid tumors of Ewing's sarcoma and reduce the expression of the target transcript (Davis et al., 2010). More recent advances in nanoparticle assembly and size control make this a promising method for small oligonucleotide delivery, going forward (Lee et al., 2012). Of course, as it concerns tissue- and cell type specificity, the most exciting and promising delivery method is antibody based. RNA binding proteins fused to an antigen binding fragment can carry a small oligonucleotide to a given target.

For example, antibodies against gp160, the human immunodeficiency virus 1 (HIV-1) envelope protein, has been used to target siRNAs to HIV-infected cells in mice (Song et al., 2005). Similarly, antibodies against the human epidermal growth factor receptor 2 (HER2) have been able to deliver siRNAs to $HER2^+$ breast cancer cells in mice, resulting in significant therapeutic effect (Yao et al., 2012), suggesting that antibody-based oligonucleotide delivery has a lot of potential in MiRNA-based anticancer therapeutics.

5.3 Off-Target Effects of MiRNA-Based Therapeutics Remain Another Challenge

MiRNA-based therapeutics are definitely not without challenges. We have already touched upon issues related to tissue-specific delivery—this is of real concern because a MiRNA that is oncogenic in one cell type may be physiologically critical in another. Additionally, off-target effects, which can be related to oligonucleotide hybridization, or can be independent of sequence-specific hybridization, must also be considered. For the most part, the sequence specificity with which anti-miRs can be designed is a benefit that precludes significant off-target effects. However, we must consider that many MiRNAs, particularly within the same family, share similar seed sequences. Under physiological conditions, these other MiRNAs could be targeted by a synthesized antagomiR—such promiscuity has been previously reported (Li et al., 2011). One strategy that might allow for even greater specificity would be to target the pre-miRNAs, which are longer, and therefore targeting could be less dependent on the seed sequence, though this strategy has not yet been applied in mammals (Kloosterman et al., 2007).

Hybridization-independent off-target effects are largely related to the immunogenicity of oligonucleotides and their delivery systems, or any toxicity associated with them. The most ancient and evolutionary conserved arm of the innate immune system includes TLRs, some of which are capable of recognizing single-strand and double-strand RNAs, though the immunostimulatory effect can be minimized by using short, seed sequence-specific, oligonucleotides, as previous studies have demonstrated that oligos >12 nucleotides are necessary for TLR activation. Toxicity of synthetic oligonucleotides is also of some concern—phosphorothiate-modified nucleotides have been shown to transiently disrupt coagulation, phosphorothiate-modified nucleotides and LNAs can induce the complement cascade, and LNAs have been shown to have some liver toxicity (Li and Rana, 2014).

5.4 Drug Development and Clinical Trials

A number of MiRNA-targeting and MiRNA-replacement therapies are under development by pharmaceutical companies. Though not an anticancer therapy, one of the most successful antagomiR-based therapies is for the targeting of MiR-122 and deserves mention here. MiR-122 is highly expressed in the liver—observations that HCV only infects hepatocyte-derived cell lines that express MiR-122 led researchers to identify MiR-122 as an essential cofactor in HCV replication. Treatment of cell lines with 2′-OMe antisense RNA with exact complementarity to MiR-122 abrogated HCV replication (Jopling et al., 2005). With the ease with which MiRNA-based therapies can be delivered to the liver, a number of pharmaceutical companies have developed MiR-122-based therapies. One such therapy, Miravirsen, which has been developed by Santaris Pharma and Hoffman-La Roche, has entered clinical trials—it is well tolerated in non-human primates and greatly reduces HCV burden (Hildebrandt-Eriksen et al., 2012; Lanford et al., 2010). In humans, HCV RNA is undetectable after five subcutaneous doses of miravirsen, with no reported adverse effects (Lindow and Kauppinen, 2012). Trials continue for miravirsen, and it may well become the first MiRNA-based therapeutic to come to market.

As we have discussed previously, MiR-21 overexpression has been associated with the pathogenesis of several types of cancers. In vitro inhibition of MiR-21 leads to apoptotic cell death in glioblastoma, breast cancer, and liver cancer (Chan et al., 2005; Frankel et al., 2008; Meng et al., 2007). Regulus Pharmaceuticals has initiated studies for the investigation of anti-miR-21 for HCC and Alport syndrome (Wagenaar et al., 2015). Regulus Pharmaceuticals has also initiated preclinical trials into the treatment of HCC with an anti-miR-221-based therapy (Park et al., 2011).

In contrast to antagonistic MiRNA-based therapies, such as in MiR-21 and MiR-221, Mirna Therapeutics has been developing several MiRNA-replacement therapeutics for tumor-suppressive MiRNAs, namely MiR-34 and let-7. MRX34, a MiR-34-replacement therapy formulated in liposomal nanoparticles, is currently undergoing Phase 1 clinical trials for safety and is likely to be the first MiRNA-replacement therapy tested for efficacy in human cancers, particularly for its efficacy in treating melanoma. MiR-34 is a proapoptotic MiRNA that has been associated with metastatic potential in malignant metastatic melanoma (Hermeking, 2010; Lujambio et al., 2008). let-7, the MiRNA whose homology in *C. elegans* and humans led to the identification of MiRNAs in higher organisms, is frequently downregulated in human cancers (Chiu et al., 2014). Replacement therapy for let-7, as well as for MiR-215, MiR-101, and MiR-16, is also being developed by Mirna Therapeutics, but is still in preclinical investigation.

There is still much to learn about MiRNA-based therapeutics. As these drugs enter clinical trials and enter clinical practice, it is important to consider potential mechanisms of resistance against these therapies. In the case of MiRNA antagonism, targeted oncogenic MiRNAs may be upregulated, overcoming the antagonistic therapy. In the case of MiRNA replacement for tumor-suppressive MiRNAs, the seed sequence in targeted oncogenes may become mutated, registering MiRNA replacement useless. Further, upregulation of nucleases and pumps might interfere with the delivery of MiRNA-based therapies to affected tissues and cells. Of course, as identification of MiRNAs involved in disease continues, and as therapy development progresses, the need for sophisticated delivery systems for cell and tissue specification will remain an ongoing challenge. Finally, it will become increasingly important to investigate how these therapies interact with other first-line treatments, and how they can be used together to treat disease most effectively. It is likely that MiRNA-based therapies will be most effective when they are paired with other disease-specific therapies, including protein-targeting therapies.

5.5 MiRNAs as Biomarkers in Diagnosis and in Determining Prognosis

Biomarkers are measured to aid in diagnosis, and good biomarkers should be specific, sensitive, noninvasive, consistent across epidemiological groups, easily quantifiable, and cost effective. Great biomarkers should also be modifiable with the progression of the disease or treatment for easy monitoring (Faruq and Vecchione, 2015). MicroRNAs have been detected in body fluids, including blood plasma, saliva, milk, and cerebrospinal fluid (Chen et al., 2008b; Mitchell et al., 2008). MiRNAs are easily measured by microarray, next-generation sequencing, and quantitative PCR, making them ideal candidates to serve as biomarkers.

One such example is MiR-21, which we have reviewed as an oncomiR in Section 4.1.3. MiR-21, MiR-155, and MiR-210 were found to be upregulated in serum samples from 60 DLBCL patients as compared to 43 healthy controls (Lawrie et al., 2008). Moreover, elevated serum MiR-21 was associated with an improved prognosis, with extended relapse-free survival times, which is consistent with findings that elevated

MiR-21 in DLBCL tumor biopsies is associated with better prognosis (Lawrie et al., 2007).

MiR-155, which we reviewed for its role in B-cell maturation in Section 3.2, has been studied as a biomarker in B-CLL (Ferrajoli et al., 2013). Elevated MiR-155, detected in B cells and in plasma, was diagnostic for B-CLL, as compared to normal controls. Further, elevated serum MiR-155 among B-CLL patients was associated with poor response to therapy. MiR-150 and MiR-342 were detected at high levels in serum from AML patients, as compared to healthy controls (Fayyad-Kazan et al., 2013). Importantly, MiR-150 and MiR-342 levels were shown to return to normal levels in AML patients in complete remission, suggesting that these two MiRNAs are biomarkers that can be measured to monitor the efficacy of treatment. Circulating MiRNAs have also been identified as biomarkers in a number of solid tumors and other disease processes, such as cardiovascular disease (Faruq and Vecchione, 2015).

Previous and ongoing studies into MiRNA signatures in hematologic disease may aid in the development of biomarkers for the discovery and prognostic classification of leukemias. MiRNA signatures have been established from CLL—MiRNA profiling can distinguish between CLL with variable Zeta-chain-associated protein kinase 70 (ZAP70) expression, and between CLL with chromosome 13q14 deletion and expression of mutated IgV_H genes (Calin et al., 2004a, 2005). A more recent study was able to use MiRNA signatures to distinguish between five different karyotypic subtypes of CLL: normal karyotype, 11q deletion, 17p deletion, trisomy 12, and 13q deletion (Visone et al., 2009).

Though AML and acute lymphocytic leukemia (ALL) can be distinguished through clinical presentation, cell morphology, and immunohistochemistry, researchers sought after a single test that could easily distinguish between AML and ALL (Mi et al., 2007). The MiRNA signatures between ALL and AML were easily distinguished. Among the many MiRNAs with differential expression between AML and ALL, the researchers identified four MiRNAs, MiR-128a, MiR-128b, let-7b, and MiR-223, that were minimally capable of distinguishing between the two diseases with 99% accuracy, demonstrating the power of MiRNA signatures in providing diagnostic criteria that can be clinically applied. Other MiRNA signatures for AML have been established. *Nucleophosmin* (*NPM1*)-mutated AML (NPMc^{+} AML) can be distinguished from *NPM1*-unmutated AML by elevated MiR-10, let-7, and MiR-29 and downregulated MiR-204 and MiR-128 (Garzon et al., 2008a). Another group established minimal diagnostic expression signatures for several distinct karyotypes of AML, including mixed lineage leukemia 1 (MLL1)-rearranged, t(15;17), t(8;21), inv(16), and t(8;21) plus inv(16) as compared to normal controls (Li et al., 2008). Additional profiling has led to the identification of overexpression of MiR-191 and MiR-199a as prognostic of lower rates of disease-free survival in AML (Garzon et al., 2008b).

Recently, John Dick's group established MiRNA signatures to differentiate between leukemia stem cells (LSCs) and normal HSCs in AML, in which they identified a number of MiRNAs that were downregulated in LSCs or upregulated in LSCs, including MiR-155, MiR-125b, and MiR-126 (Lechman et al., 2016). The study also established a minimal LSC signature, which includes MiR-155, MiR-15b, MiR-126, and MiR-22, that is significantly prognostic for overall survival. Though the remaining experiments described in this publication deal with MiR-126, specifically, we can learn from the strategy of establishing a MiRNA signature from prospectively identified cancer stem cells. Examination of differentially expressed MiRNAs in cancer stem cells, as compared to the cancer bulk or healthy cell types, can help us to identify strongly prognostic MiRNAs, MiRNAs that may be useful in early diagnosis, or MiRNAs that can be therapeutically targeted to specifically eliminate the cancer stem cell population.

The continued development of MiRNA signatures in malignancy and in other diseases is an ongoing challenge for MiRNA researchers, but one that is likely to yield critically important results in the identification of pathology-associated MiRNAs that can be therapeutically targeted and biomarkers that can be assessed for their diagnostic and prognostic potential.

6. CONCLUDING REMARKS

Though MiRNAs were only discovered within the last 25 years, we now understand how they are encoded in the genome, how they are transcribed and processed, and the mechanisms through which they inhibit protein translation. We also have begun to appreciate the physiological importance of these posttranscriptional regulators, in large part due to the observation of malignancies that can develop when MiRNAs are dysregulated. MiRNAs have been demonstrated to play a critical role in the hematopoietic system, both in HSC maintenance and in lineage-specific differentiation. Our understanding of stepwise hematopoietic differentiation and our ability to prospectively isolate various cell types in the hematopoietic system by flow cytometry have enabled us to identify a number of specific

MiRNAs that are key regulators for certain differentiation decisions. Single MiRNAs can target many different transcripts, often within similar or related pathways. Likewise, transcripts can be targeted by multiple different MiRNAs. Studies aimed at understanding the "targetome," the interactions of MiRNAs and their targets will become increasingly important in our understanding of how MiRNAs regulate cellular processes. Techniques capable of directly identifying MiRNA:target interactions, such as PAR-CLIP, must be applied more broadly to cell lines with different tissues of origin. The continued development of animal models of MiRNA dysregulation will provide great insights into the contribution of MiRNAs to pathogenesis. Further, animal models provide a platform in which MiRNA-based therapies can be tested, which is of particular importance in the design of new tissue-targeting strategies.

While this review has discussed MiRNAs contribution to physiology and disease through their capacity to inhibit translation as part of the miRISC, other roles for MiRNAs in the regulation of various cellular processes have recently begun to emerge. MiRNAs have been implicated in modulating epigenetic effectors, such as histone modifications and DNA methylation, and in regulating alternate splicing and gene expression by transactivation. Clearly, there is more work to be done, and it is an exciting time for those researchers who choose to study the effect of MiRNAs in their model systems.

ACKNOWLEDGMENTS

We are thankful to the Ito lab members for comments and discussion on MicroRNA biology in hematopoiesis. K.I. is supported by grants from the NIH (R01DK98263 and R01DK100689) and the NYSTEM (Single cell core: C029154). C.N.W. is supported by an NIH Fellowship (F30DK108532) and an NIH MSTP training grant (T32-GM007288). Many original articles were omitted due to space limitations; for this, we apologize.

REFERENCES

Abdel-Wahab, O., Mullally, A., Hedvat, C., Garcia-Manero, G., Patel, J., Wadleigh, M., Malinge, S., Yao, J., Kilpivaara, O., Bhat, R., Huberman, K., Thomas, S., Dolgalev, I., Heguy, A., Paietta, E., Le Beau, M.M., Beran, M., Tallman, M.S., Ebert, B.L., Kantarjian, H.M., Stone, R.M., Gilliland, D.G., Crispino, J.D., Levine, R.L., 2009. Genetic characterization of TET1, TET2, and TET3 alterations in myeloid malignancies. Blood 114, 144–147.

Agarwal, V., Bell, G.W., Nam, J.W., Bartel, D.P., 2015. Predicting effective microRNA target sites in mammalian mRNAs. eLife 4, e05005.

Akbari Moqadam, F., Pieters, R., Den Boer, M.L., 2013. The hunting of targets: challenge in miRNA research. Leukemia 27, 16–23.

Akinc, A., Zumbuehl, A., Goldberg, M., Leshchiner, E.S., Busini, V., Hossain, N., Bacallado, S.A., Nguyen, D.N., Fuller, J., Alvarez, R., Borodovsky, A., Borland, T., Constien, R., De Fougerolles, A., Dorkin, J.R., Narayanannair Jayaprakash, K., Jayaraman, M., John, M., Koteliansky, V., Manoharan, M., Nechev, L., Qin, J., Racie, T., Raitcheva, D., Rajeev, K.G., Sah, D.W., Soutschek, J., Toudjarska, I., Vornlocher, H.P., Zimmermann, T.S., Langer, R., Anderson, D.G., 2008. A combinatorial library of lipid-like materials for delivery of RNAi therapeutics. Nat. Biotechnol. 26, 561–569.

Ala, U., Karreth, F.A., Bosia, C., Pagnani, A., Taulli, R., Leopold, V., Tay, Y., Provero, P., Zecchina, R., Pandolfi, P.P., 2013. Integrated transcriptional and competitive endogenous RNA networks are cross-regulated in permissive molecular environments. Proc. Natl. Acad. Sci. U.S.A. 110, 7154–7159.

Ambros, V., Horvitz, H.R., 1984. Heterochronic mutants of the nematode Caenorhabditis elegans. Science 226, 409–416.

Ambros, V., Lee, R.C., Lavanway, A., Williams, P.T., Jewell, D., 2003. MicroRNAs and other tiny endogenous RNAs in C. elegans. Curr. Biol. 13, 807–818.

Arasu, P., Wightman, B., Ruvkun, G., 1991. Temporal regulation of lin-14 by the antagonistic action of two other heterochronic genes, lin-4 and lin-28. Genes Dev. 5, 1825–1833.

Aravin, A.A., Lagos-Quintana, M., Yalcin, A., Zavolan, M., Marks, D., Snyder, B., Gaasterland, T., Meyer, J., Tuschl, T., 2003. The small RNA profile during Drosophila melanogaster development. Dev. Cell 5, 337–350.

Aravin, A., Gaidatzis, D., Pfeffer, S., Lagos-Quintana, M., Landgraf, P., Iovino, N., Morris, P., Brownstein, M.J., Kuramochi-Miyagawa, S., Nakano, T., Chien, M., Russo, J.J., Ju, J., Sheridan, R., Sander, C., Zavolan, M., Tuschl, T., 2006. A novel class of small RNAs bind to MILI protein in mouse testes. Nature 442, 203–207.

Arvey, A., Larsson, E., Sander, C., Leslie, C.S., Marks, D.S., 2010. Target mRNA abundance dilutes microRNA and siRNA activity. Mol. Syst. Biol. 6, 363.

Avecilla, S.T., Hattori, K., Heissig, B., Tejada, R., Liao, F., Shido, K., Jin, D.K., Dias, S., Zhang, F., Hartman, T.E., Hackett, N.R., Crystal, R.G., Witte, L., Hicklin, D.J., Bohlen, P., Eaton, D., Lyden, D., De Sauvage, F., Rafii, S., 2004. Chemokine-mediated interaction of hematopoietic progenitors with the bone marrow vascular niche is required for thrombopoiesis. Nat. Med. 10, 64–71.

Bartel, D.P., 2009. MicroRNAs: target recognition and regulatory functions. Cell 136, 215–233.

Bazzini, A.A., Lee, M.T., Giraldez, A.J., 2012. Ribosome profiling shows that miR-430 reduces translation before causing mRNA decay in zebrafish. Science 336, 233–237.

Behm-Ansmant, I., Rehwinkel, J., Doerks, T., Stark, A., Bork, P., Izaurralde, E., 2006. mRNA degradation by miRNAs and GW182 requires both CCR4:NOT deadenylase and DCP1:DCP2 decapping complexes. Genes Dev. 20, 1885–1898.

Berezikov, E., Chung, W.J., Willis, J., Cuppen, E., Lai, E.C., 2007. Mammalian mirtron genes. Mol. Cell 28, 328–336.

Bernardo, B.C., Charchar, F.J., Lin, R.C., Mcmullen, J.R., 2012. A microRNA guide for clinicians and basic scientists: background and experimental techniques. Heart Lung Circ. 21, 131–142.

Betel, D., Koppal, A., Agius, P., Sander, C., Leslie, C., 2010. Comprehensive modeling of microRNA targets predicts functional non-conserved and non-canonical sites. Genome Biol. 11, R90.

Bhagat, T.D., Zhou, L., Sokol, L., Kessel, R., Caceres, G., Gundabolu, K., Tamari, R., Gordon, S., Mantzaris, I., Jodlowski, T., Yu, Y., Jing, X., Polineni, R., Bhatia, K., Pellagatti, A., Boultwood, J., Kambhampati, S., Steidl, U., Stein, C., Ju, W., Liu, G., Kenny, P., List, A., Bitzer, M., Verma, A., 2013. miR-21 mediates hematopoietic suppression in MDS by activating TGF-beta signaling. Blood 121, 2875–2881.

Bissels, U., Wild, S., Tomiuk, S., Hafner, M., Scheel, H., Mihailovic, A., Choi, Y.H., Tuschl, T., Bosio, A., 2011. Combined characterization of microRNA and mRNA profiles delineates early differentiation pathways of CD133+ and CD34+ hematopoietic stem and progenitor cells. Stem Cells 29, 847–857.

Bohnsack, M.T., Czaplinski, K., Gorlich, D., 2004. Exportin 5 is a RanGTP-dependent dsRNA-binding protein that mediates nuclear export of pre-miRNAs. RNA 10, 185–191.

Boldin, M.P., Taganov, K.D., Rao, D.S., Yang, L., Zhao, J.L., Kalwani, M., Garcia-Flores, Y., Luong, M., Devrekanli, A., Xu, J., Sun, G., Tay, J., Linsley, P.S., Baltimore, D., 2011. miR-146a is a significant brake on autoimmunity, myeloproliferation, and cancer in mice. J. Exp. Med. 208, 1189–1201.

Borchert, G.M., Lanier, W., Davidson, B.L., 2006. RNA polymerase III transcribes human microRNAs. Nat. Struct. Mol. Biol. 13, 1097–1101.

Bosia, C., Pagnani, A., Zecchina, R., 2013. Modelling competing endogenous RNA networks. PLoS One 8, e66609.

Bosson, A.D., Zamudio, J.R., Sharp, P.A., 2014. Endogenous miRNA and target concentrations determine susceptibility to potential ceRNA competition. Mol. Cell 56, 347–359.

Bousquet, M., Quelen, C., Rosati, R., Mansat-De Mas, V., La Starza, R., Bastard, C., Lippert, E., Talmant, P., Lafage-Pochitaloff, M., Leroux, D., Gervais, C., Viguie, F., Lai, J.L., Terre, C., Beverlo, B., Sambani, C., Hagemeijer, A., Marynen, P., Delsol, G., Dastugue, N., Mecucci, C., Brousset, P., 2008. Myeloid cell differentiation arrest by miR-125b-1 in myelodysplastic syndrome and acute myeloid leukemia with the t(2;11)(p21;q23) translocation. J. Exp. Med. 205, 2499–2506.

Bousquet, M., Harris, M.H., Zhou, B., Lodish, H.F., 2010. MicroRNA miR-125b causes leukemia. Proc. Natl. Acad. Sci. U.S.A. 107, 21558–21563.

Braun, J.E., Huntzinger, E., Fauser, M., Izaurralde, E., 2011. GW182 proteins directly recruit cytoplasmic deadenylase complexes to miRNA targets. Mol. Cell 44, 120–133.

Brennecke, J., Stark, A., Russell, R.B., Cohen, S.M., 2005. Principles of microRNA-target recognition. PLoS Biol. 3, e85.

Bueno, M.J., Perez De Castro, I., Gomez De Cedron, M., Santos, J., Calin, G.A., Cigudosa, J.C., Croce, C.M., Fernandez-Piqueras, J., Malumbres, M., 2008. Genetic and epigenetic silencing of microRNA-203 enhances ABL1 and BCR-ABL1 oncogene expression. Cancer Cell 13, 496–506.

Cai, X., Hagedorn, C.H., Cullen, B.R., 2004. Human microRNAs are processed from capped, polyadenylated transcripts that can also function as mRNAs. RNA 10, 1957–1966.

Calin, G.A., Dumitru, C.D., Shimizu, M., Bichi, R., Zupo, S., Noch, E., Aldler, H., Rattan, S., Keating, M., Rai, K., Rassenti, L., Kipps, T., Negrini, M., Bullrich, F., Croce, C.M., 2002. Frequent deletions and down-regulation of micro-RNA genes miR15 and miR16 at 13q14 in chronic lymphocytic leukemia. Proc. Natl. Acad. Sci. U.S.A. 99, 15524–15529.

Calin, G.A., Liu, C.G., Sevignani, C., Ferracin, M., Felli, N., Dumitru, C.D., Shimizu, M., Cimmino, A., Zupo, S., Dono, M., DELL'AQUILA, M.L., Alder, H., Rassenti, L., Kipps, T.J., Bullrich, F., Negrini, M., Croce, C.M., 2004a. MicroRNA profiling reveals distinct signatures in B cell chronic lymphocytic leukemias. Proc. Natl. Acad. Sci. U.S.A. 101, 11755–11760.

Calin, G.A., Sevignani, C., Dumitru, C.D., Hyslop, T., Noch, E., Yendamuri, S., Shimizu, M., Rattan, S., Bullrich, F., Negrini, M., Croce, C.M., 2004b. Human microRNA genes are frequently located at fragile sites and genomic regions involved in cancers. Proc. Natl. Acad. Sci. U.S.A. 101, 2999–3004.

Calin, G.A., Ferracin, M., Cimmino, A., Di Leva, G., Shimizu, M., Wojcik, S.E., Iorio, M.V., Visone, R., Sever, N.I., Fabbri, M., Iuliano, R., Palumbo, T., Pichiorri, F., Roldo, C., Garzon, R., Sevignani, C., Rassenti, L., Alder, H., Volinia, S., Liu, C.G., Kipps, T.J., Negrini, M., Croce, C.M., 2005. A microRNA signature associated with prognosis and progression in chronic lymphocytic leukemia. N. Engl. J. Med. 353, 1793–1801.

Capel, B., Swain, A., Nicolis, S., Hacker, A., Walter, M., Koopman, P., Goodfellow, P., Lovell-Badge, R., 1993. Circular transcripts of the testis-determining gene Sry in adult mouse testis. Cell 73, 1019–1030.

Cazalla, D., Yario, T., Steitz, J.A., 2010. Down-regulation of a host microRNA by a Herpesvirus saimiri noncoding RNA. Science 328, 1563–1566.

Cech, T.R., Steitz, J.A., 2014. The noncoding RNA revolution-trashing old rules to forge new ones. Cell 157, 77–94.

Chan, J.A., Krichevsky, A.M., Kosik, K.S., 2005. MicroRNA-21 is an antiapoptotic factor in human glioblastoma cells. Cancer Res. 65, 6029–6033.

Chekulaeva, M., Parker, R., Filipowicz, W., 2010. The GW/WG repeats of Drosophila GW182 function as effector motifs for miRNA-mediated repression. Nucleic Acids Res. 38, 6673–6683.

Chen, C.Z., Li, L., Lodish, H.F., Bartel, D.P., 2004. MicroRNAs modulate hematopoietic lineage differentiation. Science 303, 83–86.

Chen, R.W., Bemis, L.T., Amato, C.M., Myint, H., Tran, H., Birks, D.K., Eckhardt, S.G., Robinson, W.A., 2008a. Truncation in CCND1 mRNA alters miR-16-1 regulation in mantle cell lymphoma. Blood 112, 822–829.

Chen, X., Ba, Y., Ma, L., Cai, X., Yin, Y., Wang, K., Guo, J., Zhang, Y., Chen, J., Guo, X., Li, Q., Li, X., Wang, W., Zhang, Y., Wang, J., Jiang, X., Xiang, Y., Xu, C., Zheng, P., Zhang, J., Li, R., Zhang, H., Shang, X., Gong, T., Ning, G., Wang, J., Zen, K., Zhang, J., Zhang, C.Y., 2008b. Characterization of microRNAs in serum: a novel class of biomarkers for diagnosis of cancer and other diseases. Cell Res. 18, 997–1006.

Chi, S.W., Zang, J.B., Mele, A., Darnell, R.B., 2009. Argonaute HITS-CLIP decodes microRNA-mRNA interaction maps. Nature 460, 479–486.

Chi, S.W., Hannon, G.J., Darnell, R.B., 2012. An alternative mode of microRNA target recognition. Nat. Struct. Mol. Biol. 19, 321–327.

Chim, C.S., Wong, K.Y., Leung, C.Y., Chung, L.P., Hui, P.K., Chan, S.Y., Yu, L., 2011. Epigenetic inactivation of the hsa-miR-203 in haematological malignancies. J. Cell. Mol. Med. 15, 2760–2767.

Chiosea, S., Jelezcova, E., Chandran, U., Acquafondata, M., Mchale, T., Sobol, R.W., Dhir, R., 2006. Up-regulation of dicer, a component of the microRNA machinery, in prostate adenocarcinoma. Am. J. Pathol. 169, 1812–1820.

Chiu, S.C., Chung, H.Y., Cho, D.Y., Chan, T.M., Liu, M.C., Huang, H.M., Li, T.Y., Lin, J.Y., Chou, P.C., Fu, R.H., Yang, W.K., Harn, H.J., Lin, S.Z., 2014. Therapeutic potential of microRNA let-7: tumor suppression or impeding normal stemness. Cell Transplant. 23, 459–469.

Cimmino, A., Calin, G.A., Fabbri, M., Iorio, M.V., Ferracin, M., Shimizu, M., Wojcik, S.E., Aqeilan, R.I., Zupo, S., Dono, M., Rassenti, L., Alder, H., Volinia, S., Liu, C.G., Kipps, T.J., Negrini, M., Croce, C.M., 2005. miR-15 and miR-16 induce apoptosis by targeting BCL2. Proc. Natl. Acad. Sci. U.S.A. 102, 13944–13949.

Cipolla, G.A., 2014. A non-canonical landscape of the microRNA system. Front. Genet. 5, 337.

Copley, M.R., Babovic, S., Benz, C., Knapp, D.J., Beer, P.A., Kent, D.G., Wohrer, S., Treloar, D.Q., Day, C., Rowe, K., Mader, H., Kuchenbauer, F., Humphries, R.K., Eaves, C.J., 2013. The Lin28b-let-7-Hmga2 axis determines the higher self-renewal potential of fetal haematopoietic stem cells. Nat. Cell Biol. 15, 916–925.

Costinean, S., Sandhu, S.K., Pedersen, I.M., Tili, E., Trotta, R., Perrotti, D., Ciarlariello, D., Neviani, P., Harb, J., Kauffman, L.R., Shidham, A., Croce, C.M., 2009. Src homology 2 domain-containing inositol-5-phosphatase and CCAAT enhancer-binding protein beta are targeted by miR-155 in B cells of Emicro-MiR-155 transgenic mice. Blood 114, 1374–1382.

Davis, B.N., Hilyard, A.C., Lagna, G., Hata, A., 2008. SMAD proteins control DROSHA-mediated microRNA maturation. Nature 454, 56–61.

Davis, M.E., Zuckerman, J.E., Choi, C.H., Seligson, D., Tolcher, A., Alabi, C.A., Yen, Y., Heidel, J.D., Ribas, A., 2010. Evidence of RNAi in humans from systemically administered siRNA via targeted nanoparticles. Nature 464, 1067–1070.

Dedes, K.J., Natrajan, R., Lambros, M.B., Geyer, F.C., Lopez-Garcia, M.A., Savage, K., Jones, R.L., Reis-Filho, J.S., 2011. Down-regulation of the miRNA master regulators Drosha and Dicer is associated with specific subgroups of breast cancer. Eur. J. Cancer 47, 138–150.

Denzler, R., Agarwal, V., Stefano, J., Bartel, D.P., Stoffel, M., 2014. Assessing the ceRNA hypothesis with quantitative measurements of miRNA and target abundance. Mol. Cell 54, 766–776.

Deshpande, A., Pastore, A., Deshpande, A.J., Zimmermann, Y., Hutter, G., Weinkauf, M., Buske, C., Hiddemann, W., Dreyling, M., 2009. 3'UTR mediated regulation of the cyclin D1 proto-oncogene. Cell Cycle 8, 3592–3600.

De Yebenes, V.G., Belver, L., Pisano, D.G., Gonzalez, S., Villasante, A., Croce, C., He, L., Ramiro, A.R., 2008. miR-181b negatively regulates activation-induced cytidine deaminase in B cells. J. Exp. Med. 205, 2199–2206.

Didiano, D., Hobert, O., 2006. Perfect seed pairing is not a generally reliable predictor for miRNA-target interactions. Nat. Struct. Mol. Biol. 13, 849–851.

Dixon-Mciver, A., East, P., Mein, C.A., Cazier, J.B., Molloy, G., Chaplin, T., Andrew Lister, T., Young, B.D., Debernardi, S., 2008. Distinctive patterns of microRNA expression associated with karyotype in acute myeloid leukaemia. PLoS One 3, e2141.

Doench, J.G., Sharp, P.A., 2004. Specificity of microRNA target selection in translational repression. Genes Dev. 18, 504–511.

Dostie, J., Mourelatos, Z., Yang, M., Sharma, A., Dreyfuss, G., 2003. Numerous microRNPs in neuronal cells containing novel microRNAs. RNA 9, 180–186.

Ecker, J.R., Davis, R.W., 1986. Inhibition of gene expression in plant cells by expression of antisense RNA. Proc. Natl. Acad. Sci. U.S.A. 83, 5372–5376.

Eichhorn, S.W., Guo, H., Mcgeary, S.E., Rodriguez-Mias, R.A., Shin, C., Baek, D., Hsu, S.H., Ghoshal, K., Villen, J., Bartel, D.P., 2014. mRNA destabilization is the dominant effect of mammalian microRNAs by the time substantial repression ensues. Mol. Cell 56, 104–115.

Eis, P.S., Tam, W., Sun, L., Chadburn, A., Li, Z., Gomez, M.F., Lund, E., Dahlberg, J.E., 2005. Accumulation of miR-155 and BIC RNA in human B cell lymphomas. Proc. Natl. Acad. Sci. U.S.A. 102, 3627–3632.

Elkayam, E., Kuhn, C.D., Tocilj, A., Haase, A.D., Greene, E.M., Hannon, G.J., Joshua-TOR, L., 2012. The structure of human argonaute-2 in complex with miR-20a. Cell 150, 100–110.

Elmen, J., Lindow, M., Schutz, S., Lawrence, M., Petri, A., Obad, S., Lindholm, M., Hedtjarn, M., Hansen, H.F., Berger, U., Gullans, S., Kearney, P., Sarnow, P., Straarup, E.M., Kauppinen, S., 2008a. LNA-mediated microRNA silencing in non-human primates. Nature 452, 896–899.

Elmen, J., Lindow, M., Silahtaroglu, A., Bak, M., Christensen, M., Lind-Thomsen, A., Hedtjarn, M., Hansen, J.B., Hansen, H.F., Straarup, E.M., Mccullagh, K., Kearney, P., Kauppinen, S., 2008b. Antagonism of microRNA-122 in mice by systemically administered LNA-antimiR leads to up-regulation of a large set of predicted target mRNAs in the liver. Nucleic Acids Res. 36, 1153–1162.

Eulalio, A., Rehwinkel, J., Stricker, M., Huntzinger, E., Yang, S.F., Doerks, T., Dorner, S., Bork, P., Boutros, M., Izaurralde, E., 2007. Target-specific requirements for enhancers of decapping in miRNA-mediated gene silencing. Genes Dev. 21, 2558–2570.
Eulalio, A., Huntzinger, E., Izaurralde, E., 2008. GW182 interaction with Argonaute is essential for miRNA-mediated translational repression and mRNA decay. Nat. Struct. Mol. Biol. 15, 346–353.
Fabbri, M., Garzon, R., Cimmino, A., Liu, Z., Zanesi, N., Callegari, E., Liu, S., Alder, H., Costinean, S., Fernandez-Cymering, C., Volinia, S., Guler, G., Morrison, C.D., Chan, K.K., Marcucci, G., Calin, G.A., Huebner, K., Croce, C.M., 2007. MicroRNA-29 family reverts aberrant methylation in lung cancer by targeting DNA methyltransferases 3A and 3B. Proc. Natl. Acad. Sci. U.S.A. 104, 15805–15810.
Fabian, M.R., Sonenberg, N., 2012. The mechanics of miRNA-mediated gene silencing: a look under the hood of miRISC. Nat. Struct. Mol. Biol. 19, 586–593.
Fabian, M.R., Mathonnet, G., Sundermeier, T., Mathys, H., Zipprich, J.T., Svitkin, Y.V., Rivas, F., Jinek, M., Wohlschlegel, J., Doudna, J.A., Chen, C.Y., Shyu, A.B., Yates 3rd, J.R., Hannon, G.J., Filipowicz, W., Duchaine, T.F., Sonenberg, N., 2009. Mammalian miRNA RISC recruits CAF1 and PABP to affect PABP-dependent deadenylation. Mol. Cell 35, 868–880.
Fabian, M.R., Cieplak, M.K., Frank, F., Morita, M., Green, J., Srikumar, T., Nagar, B., Yamamoto, T., Raught, B., Duchaine, T.F., Sonenberg, N., 2011. miRNA-mediated deadenylation is orchestrated by GW182 through two conserved motifs that interact with CCR4-NOT. Nat. Struct. Mol. Biol. 18, 1211–1217.
Faller, M., Guo, F., 2008. MicroRNA biogenesis: there's more than one way to skin a cat. Biochim. Biophys. Acta 1779, 663–667.
Fang, L., Du, W.W., Yang, X., Chen, K., Ghanekar, A., Levy, G., Yang, W., Yee, A.J., Lu, W.Y., Xuan, J.W., Gao, Z., Xie, F., He, C., Deng, Z., Yang, B.B., 2013. Versican 3'-untranslated region (3'-UTR) functions as a ceRNA in inducing the development of hepatocellular carcinoma by regulating miRNA activity. FASEB J. 27, 907–919.
Faruq, O., Vecchione, A., 2015. microRNA: diagnostic perspective. Front. Med. (Lausanne) 2, 51.
Fayyad-Kazan, H., Bitar, N., Najar, M., Lewalle, P., Fayyad-Kazan, M., Badran, R., Hamade, E., Daher, A., Hussein, N., Eldirani, R., Berri, F., Vanhamme, L., Burny, A., Martiat, P., Rouas, R., Badran, B., 2013. Circulating miR-150 and miR-342 in plasma are novel potential biomarkers for acute myeloid leukemia. J. Transl. Med. 11, 31.
Fazi, F., Rosa, A., Fatica, A., Gelmetti, V., De Marchis, M.L., Nervi, C., Bozzoni, I., 2005. A minicircuitry comprised of microRNA-223 and transcription factors NFI-A and C/EBPalpha regulates human granulopoiesis. Cell 123, 819–831.
Fazi, F., Racanicchi, S., Zardo, G., Starnes, L.M., Mancini, M., Travaglini, L., Diverio, D., Ammatuna, E., Cimino, G., Lo-Coco, F., Grignani, F., Nervi, C., 2007. Epigenetic silencing of the myelopoiesis regulator microRNA-223 by the AML1/ETO oncoprotein. Cancer Cell 12, 457–466.
Ferrajoli, A., Shanafelt, T.D., Ivan, C., Shimizu, M., Rabe, K.G., Nouraee, N., Ikuo, M., Ghosh, A.K., Lerner, S., Rassenti, L.Z., Xiao, L., Hu, J., Reuben, J.M., Calin, S., You, M.J., Manning, J.T., Wierda, W.G., Estrov, Z., O'Brien, S., Kipps, T.J., Keating, M.J., Kay, N.E., Calin, G.A., 2013. Prognostic value of miR-155 in individuals with monoclonal B-cell lymphocytosis and patients with B chronic lymphocytic leukemia. Blood 122, 1891–1899.
Figliuzzi, M., Marinari, E., De Martino, A., 2013. MicroRNAs as a selective channel of communication between competing RNAs: a steady-state theory. Biophys. J. 104, 1203–1213.
Fire, A., Albertson, D., Harrison, S.W., Moerman, D.G., 1991. Production of antisense RNA leads to effective and specific inhibition of gene expression in C. elegans muscle. Development 113, 503–514.

Fire, A., Xu, S., Montgomery, M.K., Kostas, S.A., Driver, S.E., Mello, C.C., 1998. Potent and specific genetic interference by double-stranded RNA in Caenorhabditis elegans. Nature 391, 806–811.
Foulkes, W.D., Bahubeshi, A., Hamel, N., Pasini, B., Asioli, S., Baynam, G., Choong, C.S., Charles, A., Frieder, R.P., Dishop, M.K., Graf, N., Ekim, M., Bouron-Dal Soglio, D., Arseneau, J., Young, R.H., Sabbaghian, N., Srivastava, A., Tischkowitz, M.D., Priest, J.R., 2011. Extending the phenotypes associated with DICER1 mutations. Hum. Mutat. 32, 1381–1384.
Foulkes, W.D., Priest, J.R., Duchaine, T.F., 2014. DICER1: mutations, microRNAs and mechanisms. Nat. Rev. Cancer 14, 662–672.
Frankel, L.B., Christoffersen, N.R., Jacobsen, A., Lindow, M., Krogh, A., Lund, A.H., 2008. Programmed cell death 4 (PDCD4) is an important functional target of the microRNA miR-21 in breast cancer cells. J. Biol. Chem. 283, 1026–1033.
Friedman, R.C., Farh, K.K., Burge, C.B., Bartel, D.P., 2009. Most mammalian mRNAs are conserved targets of microRNAs. Genome Res. 19, 92–105.
Fukao, T., Fukuda, Y., Kiga, K., Sharif, J., Hino, K., Enomoto, Y., Kawamura, A., Nakamura, K., Takeuchi, T., Tanabe, M., 2007. An evolutionarily conserved mechanism for microRNA-223 expression revealed by microRNA gene profiling. Cell 129, 617–631.
Furuta, M., Kozaki, K.I., Tanaka, S., Arii, S., Imoto, I., Inazawa, J., 2010. miR-124 and miR-203 are epigenetically silenced tumor-suppressive microRNAs in hepatocellular carcinoma. Carcinogenesis 31, 766–776.
Gagnon, K.T., Li, L., Chu, Y., Janowski, B.A., Corey, D.R., 2014. RNAi factors are present and active in human cell nuclei. Cell Rep. 6, 211–221.
Garzon, R., Pichiorri, F., Palumbo, T., Visentini, M., Aqeilan, R., Cimmino, A., Wang, H., Sun, H., Volinia, S., Alder, H., Calin, G.A., Liu, C.G., Andreeff, M., Croce, C.M., 2007. MicroRNA gene expression during retinoic acid-induced differentiation of human acute promyelocytic leukemia. Oncogene 26, 4148–4157.
Garzon, R., Garofalo, M., Martelli, M.P., Briesewitz, R., Wang, L., Fernandez-Cymering, C., Volinia, S., Liu, C.G., Schnittger, S., Haferlach, T., Liso, A., Diverio, D., Mancini, M., Meloni, G., Foa, R., Martelli, M.F., Mecucci, C., Croce, C.M., Falini, B., 2008a. Distinctive microRNA signature of acute myeloid leukemia bearing cytoplasmic mutated nucleophosmin. Proc. Natl. Acad. Sci. U.S.A. 105, 3945–3950.
Garzon, R., Volinia, S., Liu, C.G., Fernandez-Cymering, C., Palumbo, T., Pichiorri, F., Fabbri, M., Coombes, K., Alder, H., Nakamura, T., Flomenberg, N., Marcucci, G., Calin, G.A., Kornblau, S.M., Kantarjian, H., Bloomfield, C.D., Andreeff, M., Croce, C.M., 2008b. MicroRNA signatures associated with cytogenetics and prognosis in acute myeloid leukemia. Blood 111, 3183–3189.
Garzon, R., Liu, S., Fabbri, M., Liu, Z., Heaphy, C.E., Callegari, E., Schwind, S., Pang, J., Yu, J., Muthusamy, N., Havelange, V., Volinia, S., Blum, W., Rush, L.J., Perrotti, D., Andreeff, M., Bloomfield, C.D., Byrd, J.C., Chan, K., Wu, L.C., Croce, C.M., Marcucci, G., 2009. MicroRNA-29b induces global DNA hypomethylation and tumor suppressor gene reexpression in acute myeloid leukemia by targeting directly DNMT3A and 3B and indirectly DNMT1. Blood 113, 6411–6418.
Gauwerky, C.E., Huebner, K., Isobe, M., Nowell, P.C., Croce, C.M., 1989. Activation of MYC in a masked t(8;17) translocation results in an aggressive B-cell leukemia. Proc. Natl. Acad. Sci. U.S.A. 86, 8867–8871.
Geary, R.S., 2009. Antisense oligonucleotide pharmacokinetics and metabolism. Expert Opin. Drug Metab. Toxicol. 5, 381–391.
Geary, R.S., Leeds, J.M., Fitchett, J., Burckin, T., Truong, L., Spainhour, C., Creek, M., Levin, A.A., 1997. Pharmacokinetics and metabolism in mice of a phosphorothioate oligonucleotide antisense inhibitor of C-raf-1 kinase expression. Drug Metab. Dispos. 25, 1272–1281.

Gregory, R.I., Yan, K.P., Amuthan, G., Chendrimada, T., Doratotaj, B., Cooch, N., Shiekhattar, R., 2004. The Microprocessor complex mediates the genesis of microRNAs. Nature 432, 235–240.

Griffiths-Jones, S., 2004. The microRNA Registry. Nucleic Acids Res. 32, D109–D111.

Griffiths-Jones, S., Grocock, R.J., Van Dongen, S., Bateman, A., Enright, A.J., 2006. miRBase: microRNA sequences, targets and gene nomenclature. Nucleic Acids Res. 34, D140–D144.

Grishok, A., Pasquinelli, A.E., Conte, D., Li, N., Parrish, S., Ha, I., Baillie, D.L., Fire, A., Ruvkun, G., Mello, C.C., 2001. Genes and mechanisms related to RNA interference regulate expression of the small temporal RNAs that control C. elegans developmental timing. Cell 106, 23–34.

Guo, S., Kemphues, K.J., 1995. par-1, a gene required for establishing polarity in C. elegans embryos, encodes a putative Ser/Thr kinase that is asymmetrically distributed. Cell 81, 611–620.

Guo, H., Ingolia, N.T., Weissman, J.S., Bartel, D.P., 2010a. Mammalian microRNAs predominantly act to decrease target mRNA levels. Nature 466, 835–840.

Guo, S., Lu, J., Schlanger, R., Zhang, H., Wang, J.Y., Fox, M.C., Purton, L.E., Fleming, H.H., Cobb, B., Merkenschlager, M., Golub, T.R., Scadden, D.T., 2010b. MicroRNA miR-125a controls hematopoietic stem cell number. Proc. Natl. Acad. Sci. U.S.A. 107, 14229–14234.

Guo, J.U., Agarwal, V., Guo, H., Bartel, D.P., 2014. Expanded identification and characterization of mammalian circular RNAs. Genome Biol. 15, 409.

Ha, I., Wightman, B., Ruvkun, G., 1996. A bulged lin-4/lin-14 RNA duplex is sufficient for Caenorhabditis elegans lin-14 temporal gradient formation. Genes Dev. 10, 3041–3050.

Hafner, M., Landthaler, M., Burger, L., Khorshid, M., Hausser, J., Berninger, P., Rothballer, A., Ascano Jr., M., Jungkamp, A.C., Munschauer, M., Ulrich, A., Wardle, G.S., Dewell, S., Zavolan, M., Tuschl, T., 2010. Transcriptome-wide identification of RNA-binding protein and microRNA target sites by PAR-CLIP. Cell 141, 129–141.

Hammond, S.M., Bernstein, E., Beach, D., Hannon, G.J., 2000. An RNA-directed nuclease mediates post-transcriptional gene silencing in Drosophila cells. Nature 404, 293–296.

Han, J., Lee, Y., Yeom, K.H., Kim, Y.K., Jin, H., Kim, V.N., 2004. The Drosha-DGCR8 complex in primary microRNA processing. Genes Dev. 18, 3016–3027.

Han, J., Lee, Y., Yeom, K.H., Nam, J.W., Heo, I., Rhee, J.K., Sohn, S.Y., Cho, Y., Zhang, B.T., Kim, V.N., 2006. Molecular basis for the recognition of primary microRNAs by the Drosha-DGCR8 complex. Cell 125, 887–901.

Hansen, T.B., Jensen, T.I., Clausen, B.H., Bramsen, J.B., Finsen, B., Damgaard, C.K., Kjems, J., 2013. Natural RNA circles function as efficient microRNA sponges. Nature 495, 384–388.

He, L., Thomson, J.M., Hemann, M.T., Hernando-Monge, E., Mu, D., Goodson, S., Powers, S., Cordon-Cardo, C., Lowe, S.W., Hannon, G.J., Hammond, S.M., 2005. A microRNA polycistron as a potential human oncogene. Nature 435, 828–833.

Helwak, A., Kudla, G., Dudnakova, T., Tollervey, D., 2013. Mapping the human miRNA interactome by CLASH reveals frequent noncanonical binding. Cell 153, 654–665.

Hermeking, H., 2010. The miR-34 family in cancer and apoptosis. Cell Death Differ. 17, 193–199.

Hildebrandt-Eriksen, E.S., Aarup, V., Persson, R., Hansen, H.F., Munk, M.E., Orum, H., 2012. A locked nucleic acid oligonucleotide targeting microRNA 122 is well-tolerated in cynomolgus monkeys. Nucleic Acid Ther. 22, 152–161.

Hill, D.A., Ivanovich, J., Priest, J.R., Gurnett, C.A., Dehner, L.P., Desruisseau, D., Jarzembowski, J.A., Wikenheiser-Brokamp, K.A., Suarez, B.K., Whelan, A.J., Williams, G., Bracamontes, D., Messinger, Y., Goodfellow, P.J., 2009. DICER1 mutations in familial pleuropulmonary blastoma. Science 325, 965.

Hou, J., Wang, P., Lin, L., Liu, X., Ma, F., An, H., Wang, Z., Cao, X., 2009. MicroRNA-146a feedback inhibits RIG-I-dependent Type I IFN production in macrophages by targeting TRAF6, IRAK1, and IRAK2. J. Immunol. 183, 2150–2158.

Houbaviy, H.B., Murray, M.F., Sharp, P.A., 2003. Embryonic stem cell-specific Micro-RNAs. Dev. Cell 5, 351–358.

Hu, H., Li, Y., Gu, J., Zhu, X., Dong, D., Yao, J., Lin, C., Fei, J., 2010. Antisense oligonucleotide against miR-21 inhibits migration and induces apoptosis in leukemic K562 cells. Leuk. Lymphoma 51, 694–701.

Huang, V., Li, L.C., 2012. miRNA goes nuclear. RNA Biol. 9, 269–273.

Huang, V., Qin, Y., Wang, J., Wang, X., Place, R.F., Lin, G., Lue, T.F., Li, L.C., 2010. RNAa is conserved in mammalian cells. PLoS One 5, e8848.

Huang, V., Place, R.F., Portnoy, V., Wang, J., Qi, Z., Jia, Z., Yu, A., Shuman, M., Yu, J., Li, L.C., 2012. Upregulation of cyclin B1 by miRNA and its implications in cancer. Nucleic Acids Res. 40, 1695–1707.

Humphreys, D.T., Westman, B.J., Martin, D.I., Preiss, T., 2005. MicroRNAs control translation initiation by inhibiting eukaryotic initiation factor 4E/cap and poly(A) tail function. Proc. Natl. Acad. Sci. U.S.A. 102, 16961–16966.

Huse, J.T., Brennan, C., Hambardzumyan, D., Wee, B., Pena, J., Rouhanifard, S.H., Sohn-Lee, C., Le Sage, C., Agami, R., Tuschl, T., Holland, E.C., 2009. The PTEN-regulating microRNA miR-26a is amplified in high-grade glioma and facilitates gliomagenesis in vivo. Genes Dev. 23, 1327–1337.

Hutvagner, G., Mclachlan, J., Pasquinelli, A.E., Balint, E., Tuschl, T., Zamore, P.D., 2001. A cellular function for the RNA-interference enzyme Dicer in the maturation of the let-7 small temporal RNA. Science 293, 834–838.

Hutvagner, G., Simard, M.J., Mello, C.C., Zamore, P.D., 2004. Sequence-specific inhibition of small RNA function. PLoS Biol. 2, E98.

Hwang, H.W., Wentzel, E.A., Mendell, J.T., 2007. A hexanucleotide element directs microRNA nuclear import. Science 315, 97–100.

Ipsaro, J.J., Joshua-Tor, L., 2015. From guide to target: molecular insights into eukaryotic RNA-interference machinery. Nat. Struct. Mol. Biol. 22, 20–28.

Izant, J.G., Weintraub, H., 1984. Inhibition of thymidine kinase gene expression by antisense RNA: a molecular approach to genetic analysis. Cell 36, 1007–1015.

Jeck, W.R., Sharpless, N.E., 2014. Detecting and characterizing circular RNAs. Nat. Biotechnol. 32, 453–461.

Jens, M., Rajewsky, N., 2015. Competition between target sites of regulators shapes post-transcriptional gene regulation. Nat. Rev. Genet. 16, 113–126.

Jeyapalan, Z., Deng, Z., Shatseva, T., Fang, L., He, C., Yang, B.B., 2011. Expression of CD44 3'-untranslated region regulates endogenous microRNA functions in tumorigenesis and angiogenesis. Nucleic Acids Res. 39, 3026–3041.

Jiang, X., Hu, C., Arnovitz, S., Bugno, J., Yu, M., Zuo, Z., Chen, P., Huang, H., Ulrich, B., Gurbuxani, S., Weng, H., Strong, J., Wang, Y., Li, Y., Salat, J., Li, S., Elkahloun, A.G., Yang, Y., Neilly, M.B., Larson, R.A., Le Beau, M.M., Herold, T., Bohlander, S.K., Liu, P.P., Zhang, J., Li, Z., He, C., Jin, J., Hong, S., Chen, J., 2016. miR-22 has a potent anti-tumour role with therapeutic potential in acute myeloid leukaemia. Nat. Commun. 7, 11452.

Jiao, A.L., Slack, F.J., 2014. RNA-mediated gene activation. Epigenetics 9, 27–36.

Jin, P., Wang, E., Ren, J., Childs, R., Shin, J.W., Khuu, H., Marincola, F.M., Stroncek, D.F., 2008. Differentiation of two types of mobilized peripheral blood stem cells by microRNA and cDNA expression analysis. J. Transl. Med. 6, 39.

Johnnidis, J.B., Harris, M.H., Wheeler, R.T., Stehling-Sun, S., Lam, M.H., Kirak, O., Brummelkamp, T.R., Fleming, M.D., Camargo, F.D., 2008. Regulation of progenitor cell proliferation and granulocyte function by microRNA-223. Nature 451, 1125–1129.

Johnson, S.M., Lin, S.Y., Slack, F.J., 2003. The time of appearance of the C. elegans let-7 microRNA is transcriptionally controlled utilizing a temporal regulatory element in its promoter. Dev. Biol. 259, 364–379.
Johnson, S.M., Grosshans, H., Shingara, J., Byrom, M., Jarvis, R., Cheng, A., Labourier, E., Reinert, K.L., Brown, D., Slack, F.J., 2005. RAS is regulated by the let-7 microRNA family. Cell 120, 635–647.
Johnsson, P., Ackley, A., Vidarsdottir, L., Lui, W.O., Corcoran, M., Grander, D., Morris, K.V., 2013. A pseudogene long-noncoding-RNA network regulates PTEN transcription and translation in human cells. Nat. Struct. Mol. Biol. 20, 440–446.
Johnston, R.J., Hobert, O., 2003. A microRNA controlling left/right neuronal asymmetry in Caenorhabditis elegans. Nature 426, 845–849.
Jonas, S., Izaurralde, E., 2015. Towards a molecular understanding of microRNA-mediated gene silencing. Nat. Rev. Genet. 16, 421–433.
Jongen-Lavrencic, M., Sun, S.M., Dijkstra, M.K., Valk, P.J., Lowenberg, B., 2008. MicroRNA expression profiling in relation to the genetic heterogeneity of acute myeloid leukemia. Blood 111, 5078–5085.
Jopling, C.L., Yi, M., Lancaster, A.M., Lemon, S.M., Sarnow, P., 2005. Modulation of hepatitis C virus RNA abundance by a liver-specific microRNA. Science 309, 1577–1581.
Karreth, F.A., Tay, Y., Perna, D., Ala, U., Tan, S.M., Rust, A.G., Denicola, G., Webster, K.A., Weiss, D., Perez-Mancera, P.A., Krauthammer, M., Halaban, R., Provero, P., Adams, D.J., Tuveson, D.A., Pandolfi, P.P., 2011. In vivo identification of tumor- suppressive PTEN ceRNAs in an oncogenic BRAF-induced mouse model of melanoma. Cell 147, 382–395.
Karreth, F.A., Reschke, M., Ruocco, A., Ng, C., Chapuy, B., Leopold, V., Sjoberg, M., Keane, T.M., Verma, A., Ala, U., Tay, Y., Wu, D., Seitzer, N., Velasco-Herrera Mdel, C., Bothmer, A., Fung, J., Langellotto, F., Rodig, S.J., Elemento, O., Shipp, M.A., Adams, D.J., Chiarle, R., Pandolfi, P.P., 2015. The BRAF pseudogene functions as a competitive endogenous RNA and induces lymphoma in vivo. Cell 161, 319–332.
Karube, Y., Tanaka, H., Osada, H., Tomida, S., Tatematsu, Y., Yanagisawa, K., Yatabe, Y., Takamizawa, J., Miyoshi, S., Mitsudomi, T., Takahashi, T., 2005. Reduced expression of Dicer associated with poor prognosis in lung cancer patients. Cancer Sci. 96, 111–115.
Kato, M., Putta, S., Wang, M., Yuan, H., Lanting, L., Nair, I., Gunn, A., Nakagawa, Y., Shimano, H., Todorov, I., Rossi, J.J., Natarajan, R., 2009. TGF-beta activates Akt kinase through a microRNA-dependent amplifying circuit targeting PTEN. Nat. Cell Biol. 11, 881–889.
Kawahara, Y., Zinshteyn, B., Chendrimada, T.P., Shiekhattar, R., Nishikura, K., 2007. RNA editing of the microRNA-151 precursor blocks cleavage by the Dicer-TRBP complex. EMBO Rep. 8, 763–769.
Kawahara, H., Okada, Y., Imai, T., Iwanami, A., Mischel, P.S., Okano, H., 2011. Musashi1 cooperates in abnormal cell lineage protein 28 (Lin28)-mediated let-7 family microRNA biogenesis in early neural differentiation. J. Biol. Chem. 286, 16121–16130.
Ketting, R.F., Fischer, S.E., Bernstein, E., Sijen, T., Hannon, G.J., Plasterk, R.H., 2001. Dicer functions in RNA interference and in synthesis of small RNA involved in developmental timing in C. elegans. Genes Dev. 15, 2654–2659.
Kharas, M.G., Lengner, C.J., Al-Shahrour, F., Bullinger, L., Ball, B., Zaidi, S., Morgan, K., Tam, W., Paktinat, M., Okabe, R., Gozo, M., Einhorn, W., Lane, S.W., Scholl, C., Frohling, S., Fleming, M., Ebert, B.L., Gilliland, D.G., Jaenisch, R., Daley, G.Q., 2010. Musashi-2 regulates normal hematopoiesis and promotes aggressive myeloid leukemia. Nat. Med. 16, 903–908.
Khvorova, A., Reynolds, A., Jayasena, S.D., 2003. Functional siRNAs and miRNAs exhibit strand bias. Cell 115, 209–216.

Kim, J., Krichevsky, A., Grad, Y., Hayes, G.D., Kosik, K.S., Church, G.M., Ruvkun, G., 2004. Identification of many microRNAs that copurify with polyribosomes in mammalian neurons. Proc. Natl. Acad. Sci. U.S.A. 101, 360–365.

Kloosterman, W.P., Lagendijk, A.K., Ketting, R.F., Moulton, J.D., Plasterk, R.H., 2007. Targeted inhibition of miRNA maturation with morpholinos reveals a role for miR-375 in pancreatic islet development. PLoS Biol. 5, e203.

Kluiver, J., Poppema, S., De Jong, D., Blokzijl, T., Harms, G., Jacobs, S., Kroesen, B.J., Van Den Berg, A., 2005. BIC and miR-155 are highly expressed in Hodgkin, primary mediastinal and diffuse large B cell lymphomas. J. Pathol. 207, 243–249.

Klusmann, J.H., Li, Z., Bohmer, K., Maroz, A., Koch, M.L., Emmrich, S., Godinho, F.J., Orkin, S.H., Reinhardt, D., 2010. miR-125b-2 is a potential oncomiR on human chromosome 21 in megakaryoblastic leukemia. Genes Dev. 24, 478–490.

Kortylewski, M., Swiderski, P., Herrmann, A., Wang, L., Kowolik, C., Kujawski, M., Lee, H., Scuto, A., Liu, Y., Yang, C., Deng, J., Soifer, H.S., Raubitschek, A., Forman, S., Rossi, J.J., Pardoll, D.M., Jove, R., Yu, H., 2009. In vivo delivery of siRNA to immune cells by conjugation to a TLR9 agonist enhances antitumor immune responses. Nat. Biotechnol. 27, 925–932.

Krek, A., Grun, D., Poy, M.N., Wolf, R., Rosenberg, L., Epstein, E.J., Macmenamin, P., Da Piedade, I., Gunsalus, K.C., Stoffel, M., Rajewsky, N., 2005. Combinatorial microRNA target predictions. Nat. Genet. 37, 495–500.

Kristiansen, T.A., Jaensson Gyllenback, E., Zriwil, A., Bjorklund, T., Daniel, J.A., Sitnicka, E., Soneji, S., Bryder, D., Yuan, J., 2016. Cellular barcoding links B-1a B cell potential to a fetal hematopoietic stem cell state at the single-cell level. Immunity 45, 346–357.

Krol, J., Sobczak, K., Wilczynska, U., Drath, M., Jasinska, A., Kaczynska, D., Krzyzosiak, W.J., 2004. Structural features of microRNA (miRNA) precursors and their relevance to miRNA biogenesis and small interfering RNA/short hairpin RNA design. J. Biol. Chem. 279, 42230–42239.

Krutzfeldt, J., Rajewsky, N., Braich, R., Rajeev, K.G., Tuschl, T., Manoharan, M., Stoffel, M., 2005. Silencing of microRNAs in vivo with 'antagomirs'. Nature 438, 685–689.

Kuchenbauer, F., Mah, S.M., Heuser, M., Mcpherson, A., Ruschmann, J., Rouhi, A., Berg, T., Bullinger, L., Argiropoulos, B., Morin, R.D., Lai, D., Starczynowski, D.T., Karsan, A., Eaves, C.J., Watahiki, A., Wang, Y., Aparicio, S.A., Ganser, A., Krauter, J., Dohner, H., Dohner, K., Marra, M.A., Camargo, F.D., Palmqvist, L., Buske, C., Humphries, R.K., 2011. Comprehensive analysis of mammalian miRNA* species and their role in myeloid cells. Blood 118, 3350–3358.

Kudla, G., Granneman, S., Hahn, D., Beggs, J.D., Tollervey, D., 2011. Cross-linking, ligation, and sequencing of hybrids reveals RNA-RNA interactions in yeast. Proc. Natl. Acad. Sci. U.S.A. 108, 10010–10015.

Kumar, P., Lee, S.K., Shankar, P., Manjunath, N., 2006. A single siRNA suppresses fatal encephalitis induced by two different flaviviruses. PLoS Med. 3, e96.

Labbaye, C., Spinello, I., Quaranta, M.T., Pelosi, E., Pasquini, L., Petrucci, E., Biffoni, M., Nuzzolo, E.R., Billi, M., Foa, R., Brunetti, E., Grignani, F., Testa, U., Peschle, C., 2008. A three-step pathway comprising PLZF/miR-146a/CXCR4 controls megakaryopoiesis. Nat. Cell Biol. 10, 788–801.

Lagos-Quintana, M., Rauhut, R., Lendeckel, W., Tuschl, T., 2001. Identification of novel genes coding for small expressed RNAs. Science 294, 853–858.

Lagos-Quintana, M., Rauhut, R., Yalcin, A., Meyer, J., Lendeckel, W., Tuschl, T., 2002. Identification of tissue-specific microRNAs from mouse. Curr. Biol. 12, 735–739.

Lagos-Quintana, M., Rauhut, R., Meyer, J., Borkhardt, A., Tuschl, T., 2003. New microRNAs from mouse and human. RNA 9, 175–179.

Lal, A., Navarro, F., Maher, C.A., Maliszewski, L.E., Yan, N., O'DAY, E., Chowdhury, D., Dykxhoorn, D.M., Tsai, P., Hofmann, O., Becker, K.G., Gorospe, M., Hide, W., Lieberman, J., 2009. miR-24 inhibits cell proliferation by targeting E2F2, MYC, and other cell-cycle genes via binding to "seedless" 3'UTR microRNA recognition elements. Mol. Cell 35, 610–625.
Lanford, R.E., Hildebrandt-Eriksen, E.S., Petri, A., Persson, R., Lindow, M., Munk, M.E., Kauppinen, S., Orum, H., 2010. Therapeutic silencing of microRNA-122 in primates with chronic hepatitis C virus infection. Science 327, 198–201.
Lau, N.C., Lim, L.P., Weinstein, E.G., Bartel, D.P., 2001. An abundant class of tiny RNAs with probable regulatory roles in Caenorhabditis elegans. Science 294, 858–862.
Lawrie, C.H., 2013. MicroRNAs in hematological malignancies. Blood Rev. 27, 143–154.
Lawrie, C.H., Soneji, S., Marafioti, T., Cooper, C.D., Palazzo, S., Paterson, J.C., Cattan, H., Enver, T., Mager, R., Boultwood, J., Wainscoat, J.S., Hatton, C.S., 2007. MicroRNA expression distinguishes between germinal center B cell-like and activated B cell-like subtypes of diffuse large B cell lymphoma. Int. J. Cancer 121, 1156–1161.
Lawrie, C.H., Gal, S., Dunlop, H.M., Pushkaran, B., Liggins, A.P., Pulford, K., Banham, A.H., Pezzella, F., Boultwood, J., Wainscoat, J.S., Hatton, C.S., Harris, A.L., 2008. Detection of elevated levels of tumour-associated microRNAs in serum of patients with diffuse large B-cell lymphoma. Br. J. Haematol. 141, 672–675.
Le, M.T., Xie, H., Zhou, B., Chia, P.H., Rizk, P., Um, M., Udolph, G., Yang, H., Lim, B., Lodish, H.F., 2009. MicroRNA-125b promotes neuronal differentiation in human cells by repressing multiple targets. Mol. Cell. Biol. 29, 5290–5305.
Lechman, E.R., Gentner, B., Ng, S.W., Schoof, E.M., Van Galen, P., Kennedy, J.A., Nucera, S., Ciceri, F., Kaufmann, K.B., Takayama, N., Dobson, S.M., Trotman-Grant, A., Krivdova, G., Elzinga, J., Mitchell, A., Nilsson, B., Hermans, K.G., Eppert, K., Marke, R., Isserlin, R., Voisin, V., Bader, G.D., Zandstra, P.W., Golub, T.R., Ebert, B.L., Lu, J., Minden, M., Wang, J.C., Naldini, L., Dick, J.E., 2016. miR-126 regulates distinct self-renewal outcomes in normal and malignant hematopoietic stem cells. Cancer Cell 29, 214–228.
Lee, R.C., Ambros, V., 2001. An extensive class of small RNAs in Caenorhabditis elegans. Science 294, 862–864.
Lee, R.C., Feinbaum, R.L., Ambros, V., 1993. The C. elegans heterochronic gene lin-4 encodes small RNAs with antisense complementarity to lin-14. Cell 75, 843–854.
Lee, Y., Jeon, K., Lee, J.T., Kim, S., Kim, V.N., 2002. MicroRNA maturation: stepwise processing and subcellular localization. EMBO J. 21, 4663–4670.
Lee, Y., Ahn, C., Han, J., Choi, H., Kim, J., Yim, J., Lee, J., Provost, P., Radmark, O., Kim, S., Kim, V.N., 2003. The nuclear RNase III Drosha initiates microRNA processing. Nature 425, 415–419.
Lee, Y., Kim, M., Han, J., Yeom, K.H., Lee, S., Baek, S.H., Kim, V.N., 2004. MicroRNA genes are transcribed by RNA polymerase II. EMBO J. 23, 4051–4060.
Lee, D.Y., Jeyapalan, Z., Fang, L., Yang, J., Zhang, Y., Yee, A.Y., Li, M., Du, W.W., Shatseva, T., Yang, B.B., 2010. Expression of versican 3'-untranslated region modulates endogenous microRNA functions. PLoS One 5, e13599.
Lee, H., Lytton-Jean, A.K., Chen, Y., Love, K.T., Park, A.I., Karagiannis, E.D., Sehgal, A., Querbes, W., Zurenko, C.S., Jayaraman, M., Peng, C.G., Charisse, K., Borodovsky, A., Manoharan, M., Donahoe, J.S., Truelove, J., Nahrendorf, M., Langer, R., Anderson, D.G., 2012. Molecularly self-assembled nucleic acid nanoparticles for targeted in vivo siRNA delivery. Nat. Nanotechnol. 7, 389–393.
Lennox, K.A., Behlke, M.A., 2011. Chemical modification and design of anti-miRNA oligonucleotides. Gene Ther. 18, 1111–1120.

Lennox, K.A., Owczarzy, R., Thomas, D.M., Walder, J.A., Behlke, M.A., 2013. Improved performance of anti-miRNA oligonucleotides using a novel non-nucleotide modifier. Mol. Ther. Nucleic Acids 2, e117.
Lewis, B.P., Burge, C.B., Bartel, D.P., 2005. Conserved seed pairing, often flanked by adenosines, indicates that thousands of human genes are microRNA targets. Cell 120, 15–20.
Li, Z., Rana, T.M., 2014. Therapeutic targeting of microRNAs: current status and future challenges. Nat. Rev. Drug Discov. 13, 622–638.
Li, Z., Lu, J., Sun, M., Mi, S., Zhang, H., Luo, R.T., Chen, P., Wang, Y., Yan, M., Qian, Z., Neilly, M.B., Jin, J., Zhang, Y., Bohlander, S.K., Zhang, D.E., Larson, R.A., Le Beau, M.M., Thirman, M.J., Golub, T.R., Rowley, J.D., Chen, J., 2008. Distinct microRNA expression profiles in acute myeloid leukemia with common translocations. Proc. Natl. Acad. Sci. U.S.A. 105, 15535–15540.
Li, Y., Zhu, X., Gu, J., Dong, D., Yao, J., Lin, C., Huang, K., Fei, J., 2010. Anti-miR-21 oligonucleotide sensitizes leukemic K562 cells to arsenic trioxide by inducing apoptosis. Cancer Sci. 101, 948–954.
Li, Z., Yang, C.S., Nakashima, K., Rana, T.M., 2011. Small RNA-mediated regulation of iPS cell generation. EMBO J. 30, 823–834.
Li, X., Zhang, J., Gao, L., Mcclellan, S., Finan, M.A., Butler, T.W., Owen, L.B., Piazza, G.A., Xi, Y., 2012. MiR-181 mediates cell differentiation by interrupting the Lin28 and let-7 feedback circuit. Cell Death Differ. 19, 378–386.
Liao, R., Sun, J., Zhang, L., Lou, G., Chen, M., Zhou, D., Chen, Z., Zhang, S., 2008. MicroRNAs play a role in the development of human hematopoietic stem cells. J. Cell. Biochem. 104, 805–817.
Lim, L.P., Glasner, M.E., Yekta, S., Burge, C.B., Bartel, D.P., 2003a. Vertebrate microRNA genes. Science 299, 1540.
Lim, L.P., Lau, N.C., Weinstein, E.G., Abdelhakim, A., Yekta, S., Rhoades, M.W., Burge, C.B., Bartel, D.P., 2003b. The microRNAs of Caenorhabditis elegans. Genes Dev. 17, 991–1008.
Lin, C.C., Liu, L.Z., Addison, J.B., Wonderlin, W.F., Ivanov, A.V., Ruppert, J.M., 2011. A KLF4-miRNA-206 autoregulatory feedback loop can promote or inhibit protein translation depending upon cell context. Mol. Cell. Biol. 31, 2513–2527.
Lindow, M., Kauppinen, S., 2012. Discovering the first microRNA-targeted drug. J. Cell Biol. 199, 407–412.
Ling, B., Wang, G.X., Long, G., Qiu, J.H., Hu, Z.L., 2012. Tumor suppressor miR-22 suppresses lung cancer cell progression through post-transcriptional regulation of ErbB3. J. Cancer Res. Clin. Oncol. 138, 1355–1361.
Ling, H., Fabbri, M., Calin, G.A., 2013. MicroRNAs and other non-coding RNAs as targets for anticancer drug development. Nat. Rev. Drug Discov. 12, 847–865.
Liu, J., Carmell, M.A., Rivas, F.V., Marsden, C.G., Thomson, J.M., Song, J.J., Hammond, S.M., Joshua-Tor, L., Hannon, G.J., 2004. Argonaute2 is the catalytic engine of mammalian RNAi. Science 305, 1437–1441.
Liu, S., Shen, T., Huynh, L., Klisovic, M.I., Rush, L.J., Ford, J.L., Yu, J., Becknell, B., Li, Y., Liu, C., Vukosavljevic, T., Whitman, S.P., Chang, K.S., Byrd, J.C., Perrotti, D., Plass, C., Marcucci, G., 2005. Interplay of RUNX1/MTG8 and DNA methyltransferase 1 in acute myeloid leukemia. Cancer Res. 65, 1277–1284.
Liu, Y., Cheney, M.D., Gaudet, J.J., Chruszcz, M., Lukasik, S.M., Sugiyama, D., Lary, J., Cole, J., Dauter, Z., Minor, W., Speck, N.A., Bushweller, J.H., 2006. The tetramer structure of the Nervy homology two domain, NHR2, is critical for AML1/ETO's activity. Cancer Cell 9, 249–260.
Liu, Y., Ye, X., Jiang, F., Liang, C., Chen, D., Peng, J., Kinch, L.N., Grishin, N.V., Liu, Q., 2009. C3PO, an endoribonuclease that promotes RNAi by facilitating RISC activation. Science 325, 750–753.

Liu, S., Wu, L.C., Pang, J., Santhanam, R., Schwind, S., Wu, Y.Z., Hickey, C.J., Yu, J., Becker, H., Maharry, K., Radmacher, M.D., Li, C., Whitman, S.P., Mishra, A., Stauffer, N., Eiring, A.M., Briesewitz, R., Baiocchi, R.A., Chan, K.K., Paschka, P., Caligiuri, M.A., Byrd, J.C., Croce, C.M., Bloomfield, C.D., Perrotti, D., Garzon, R., Marcucci, G., 2010. Sp1/NFkappaB/HDAC/miR-29b regulatory network in KIT-driven myeloid leukemia. Cancer Cell 17, 333–347.

Lujambio, A., Calin, G.A., Villanueva, A., Ropero, S., Sanchez-Cespedes, M., Blanco, D., Montuenga, L.M., Rossi, S., Nicoloso, M.S., Faller, W.J., Gallagher, W.M., Eccles, S.A., Croce, C.M., Esteller, M., 2008. A microRNA DNA methylation signature for human cancer metastasis. Proc. Natl. Acad. Sci. U.S.A. 105, 13556–13561.

Lund, E., Dahlberg, J.E., 2006. Substrate selectivity of exportin 5 and Dicer in the biogenesis of microRNAs. Cold Spring Harb. Symp. Quant. Biol. 71, 59–66.

Lund, E., Guttinger, S., Calado, A., Dahlberg, J.E., Kutay, U., 2004. Nuclear export of microRNA precursors. Science 303, 95–98.

Majid, S., Dar, A.A., Saini, S., Yamamura, S., Hirata, H., Tanaka, Y., Deng, G., Dahiya, R., 2010. MicroRNA-205-directed transcriptional activation of tumor suppressor genes in prostate cancer. Cancer 116, 5637–5649.

Majlessi, M., Nelson, N.C., Becker, M.M., 1998. Advantages of 2'-O-methyl oligoribonucleotide probes for detecting RNA targets. Nucleic Acids Res. 26, 2224–2229.

Marques, A.C., Tan, J., Lee, S., Kong, L., Heger, A., Ponting, C.P., 2012. Evidence for conserved post-transcriptional roles of unitary pseudogenes and for frequent bifunctionality of mRNAs. Genome Biol. 13, R102.

Mathonnet, G., Fabian, M.R., Svitkin, Y.V., Parsyan, A., Huck, L., Murata, T., Biffo, S., Merrick, W.C., Darzynkiewicz, E., Pillai, R.S., Filipowicz, W., Duchaine, T.F., Sonenberg, N., 2007. MicroRNA inhibition of translation initiation in vitro by targeting the cap-binding complex eIF4F. Science 317, 1764–1767.

Mavrakis, K.J., Wolfe, A.L., Oricchio, E., Palomero, T., De Keersmaecker, K., Mcjunkin, K., Zuber, J., James, T., Khan, A.A., Leslie, C.S., Parker, J.S., Paddison, P.J., Tam, W., Ferrando, A., Wendel, H.G., 2010. Genome-wide RNA-mediated interference screen identifies miR-19 targets in Notch-induced T-cell acute lymphoblastic leukaemia. Nat. Cell Biol. 12, 372–379.

Mayr, C., Hemann, M.T., Bartel, D.P., 2007. Disrupting the pairing between let-7 and Hmga2 enhances oncogenic transformation. Science 315, 1576–1579.

Mcpherson, J.D., Marra, M., Hillier, L., Waterston, R.H., Chinwalla, A., Wallis, J., Sekhon, M., Wylie, K., Mardis, E.R., Wilson, R.K., Fulton, R., Kucaba, T.A., Wagner-Mcpherson, C., Barbazuk, W.B., Gregory, S.G., Humphray, S.J., French, L., Evans, R.S., Bethel, G., Whittaker, A., Holden, J.L., Mccann, O.T., Dunham, A., Soderlund, C., Scott, C.E., Bentley, D.R., Schuler, G., Chen, H.C., Jang, W., Green, E.D., Idol, J.R., Maduro, V.V., Montgomery, K.T., Lee, E., Miller, A., Emerling, S., Kucherlapati, Gibbs, R., Scherer, S., Gorrell, J.H., Sodergren, E., Clerc-Blankenburg, K., Tabor, P., Naylor, S., Garcia, D., De Jong, P.J., Catanese, J.J., Nowak, N., Osoegawa, K., Qin, S., Rowen, L., Madan, A., Dors, M., Hood, L., Trask, B., Friedman, C., Massa, H., Cheung, V.G., Kirsch, I.R., Reid, T., Yonescu, R., Weissenbach, J., Bruls, T., Heilig, R., Branscomb, E., Olsen, A., Doggett, N., Cheng, J.F., Hawkins, T., Myers, R.M., Shang, J., Ramirez, L., Schmutz, J., Velasquez, O., Dixon, K., Stone, N.E., Cox, D.R., Haussler, D., Kent, W.J., Furey, T., Rogic, S., Kennedy, S., Jones, S., Rosenthal, A., Wen, G., Schilhabel, M., Gloeckner, G., Nyakatura, G., Siebert, R., Schlegelberger, B., Korenberg, J., Chen, X.N., Fujiyama, A., Hattori, M., Toyoda, A., Yada, T., Park, H.S., Sakaki, Y., Shimizu, N., Asakawa, S., et al., 2001. A physical map of the human genome. Nature 409, 934–941.

Medina, P.P., Nolde, M., Slack, F.J., 2010. OncomiR addiction in an in vivo model of microRNA-21-induced pre-B-cell lymphoma. Nature 467, 86–90.
Meister, G., Landthaler, M., Dorsett, Y., Tuschl, T., 2004a. Sequence-specific inhibition of microRNA- and siRNA-induced RNA silencing. RNA 10, 544–550.
Meister, G., Landthaler, M., Patkaniowska, A., Dorsett, Y., Teng, G., Tuschl, T., 2004b. Human Argonaute2 mediates RNA cleavage targeted by miRNAs and siRNAs. Mol. Cell 15, 185–197.
Melo, S.A., Ropero, S., Moutinho, C., Aaltonen, L.A., Yamamoto, H., Calin, G.A., Rossi, S., Fernandez, A.F., Carneiro, F., Oliveira, C., Ferreira, B., Liu, C.G., Villanueva, A., Capella, G., Schwartz Jr., S., Shiekhattar, R., Esteller, M., 2009. A TARBP2 mutation in human cancer impairs microRNA processing and DICER1 function. Nat. Genet. 41, 365–370.
Melo, S.A., Moutinho, C., Ropero, S., Calin, G.A., Rossi, S., Spizzo, R., Fernandez, A.F., Davalos, V., Villanueva, A., Montoya, G., Yamamoto, H., Schwartz Jr., S., Esteller, M., 2010. A genetic defect in exportin-5 traps precursor microRNAs in the nucleus of cancer cells. Cancer Cell 18, 303–315.
Memczak, S., Jens, M., Elefsinioti, A., Torti, F., Krueger, J., Rybak, A., Maier, L., Mackowiak, S.D., Gregersen, L.H., Munschauer, M., Loewer, A., Ziebold, U., Landthaler, M., Kocks, C., Le Noble, F., Rajewsky, N., 2013. Circular RNAs are a large class of animal RNAs with regulatory potency. Nature 495, 333–338.
Meng, F., Henson, R., Wehbe-Janek, H., Ghoshal, K., Jacob, S.T., Patel, T., 2007. MicroRNA-21 regulates expression of the PTEN tumor suppressor gene in human hepatocellular cancer. Gastroenterology 133, 647–658.
Merkerova, M., Vasikova, A., Belickova, M., Bruchova, H., 2010. MicroRNA expression profiles in umbilical cord blood cell lineages. Stem Cells Dev. 19, 17–26.
Merritt, W.M., Lin, Y.G., Han, L.Y., Kamat, A.A., Spannuth, W.A., Schmandt, R., Urbauer, D., Pennacchio, L.A., Cheng, J.F., Nick, A.M., Deavers, M.T., Mourad-Zeidan, A., Wang, H., Mueller, P., Lenburg, M.E., Gray, J.W., Mok, S., Birrer, M.J., Lopez-Berestein, G., Coleman, R.L., Bar-Eli, M., Sood, A.K., 2008. Dicer, Drosha, and outcomes in patients with ovarian cancer. N. Engl. J. Med. 359, 2641–2650.
Mi, S., Lu, J., Sun, M., Li, Z., Zhang, H., Neilly, M.B., Wang, Y., Qian, Z., Jin, J., Zhang, Y., Bohlander, S.K., Le Beau, M.M., Larson, R.A., Golub, T.R., Rowley, J.D., Chen, J., 2007. MicroRNA expression signatures accurately discriminate acute lymphoblastic leukemia from acute myeloid leukemia. Proc. Natl. Acad. Sci. U.S.A. 104, 19971–19976.
Michael, M.Z., Sm, O.C., Van Holst Pellekaan, N.G., Young, G.P., James, R.J., 2003. Reduced accumulation of specific microRNAs in colorectal neoplasia. Mol. Cancer Res. 1, 882–891.
Mitchell, P.S., Parkin, R.K., Kroh, E.M., Fritz, B.R., Wyman, S.K., Pogosova-Agadjanyan, E.L., Peterson, A., Noteboom, J., O'BRIANT, K.C., Allen, A., Lin, D.W., Urban, N., Drescher, C.W., Knudsen, B.S., Stirewalt, D.L., Gentleman, R., Vessella, R.L., Nelson, P.S., Martin, D.B., Tewari, M., 2008. Circulating microRNAs as stable blood-based markers for cancer detection. Proc. Natl. Acad. Sci. U.S.A. 105, 10513–10518.
Morrissey, D.V., Lockridge, J.A., Shaw, L., Blanchard, K., Jensen, K., Breen, W., Hartsough, K., Machemer, L., Radka, S., Jadhav, V., Vaish, N., Zinnen, S., Vargeese, C., Bowman, K., Shaffer, C.S., Jeffs, L.B., Judge, A., Maclachlan, I., Polisky, B., 2005. Potent and persistent in vivo anti-HBV activity of chemically modified siRNAs. Nat. Biotechnol. 23, 1002–1007.
Mott, J.L., Kobayashi, S., Bronk, S.F., Gores, G.J., 2007. mir-29 regulates Mcl-1 protein expression and apoptosis. Oncogene 26, 6133–6140.
Mouse Genome Sequencing Consortium, Waterston, R.H., Lindblad-Toh, K., Birney, E., Rogers, J., Abril, J.F., Agarwal, P., Agarwala, R., Ainscough, R., Alexandersson, M.,

An, P., Antonarakis, S.E., Attwood, J., Baertsch, R., Bailey, J., Barlow, K., Beck, S., Berry, E., Birren, B., Bloom, T., Bork, P., Botcherby, M., Bray, N., Brent, M.R., Brown, D.G., Brown, S.D., Bult, C., Burton, J., Butler, J., Campbell, R.D., Carninci, P., Cawley, S., Chiaromonte, F., Chinwalla, A.T., Church, D.M., Clamp, M., Clee, C., Collins, F.S., Cook, L.L., Copley, R.R., Coulson, A., Couronne, O., Cuff, J., Curwen, V., Cutts, T., Daly, M., David, R., Davies, J., Delehaunty, K.D., Deri, J., Dermitzakis, E.T., Dewey, C., Dickens, N.J., Diekhans, M., Dodge, S., Dubchak, I., Dunn, D.M., Eddy, S.R., Elnitski, L., Emes, R.D., Eswara, P., Eyras, E., Felsenfeld, A., Fewell, G.A., Flicek, P., Foley, K., Frankel, W.N., Fulton, L.A., Fulton, R.S., Furey, T.S., Gage, D., Gibbs, R.A., Glusman, G., Gnerre, S., Goldman, N., Goodstadt, L., Grafham, D., Graves, T.A., Green, E.D., Gregory, S., Guigo, R., Guyer, M., Hardison, R.C., Haussler, D., Hayashizaki, Y., Hillier, L.W., Hinrichs, A., Hlavina, W., Holzer, T., Hsu, F., Hua, A., Hubbard, T., Hunt, A., Jackson, I., Jaffe, D.B., Johnson, L.S., Jones, M., Jones, T.A., Joy, A., Kamal, M., et al., 2002. Initial sequencing and comparative analysis of the mouse genome. Nature 420, 520–562.

Mullokandov, G., Baccarini, A., Ruzo, A., Jayaprakash, A.D., Tung, N., Israelow, B., Evans, M.J., Sachidanandam, R., Brown, B.D., 2012. High-throughput assessment of microRNA activity and function using microRNA sensor and decoy libraries. Nat. Methods 9, 840–846.

Muralidhar, B., Goldstein, L.D., Ng, G., Winder, D.M., Palmer, R.D., Gooding, E.L., Barbosa-Morais, N.L., Mukherjee, G., Thorne, N.P., Roberts, I., Pett, M.R., Coleman, N., 2007. Global microRNA profiles in cervical squamous cell carcinoma depend on Drosha expression levels. J. Pathol. 212, 368–377.

Nellen, W., Lichtenstein, C., 1993. What makes an mRNA anti-sense-itive? Trends Biochem. Sci. 18, 419–423.

Newman, M.A., Thomson, J.M., Hammond, S.M., 2008. Lin-28 interaction with the Let-7 precursor loop mediates regulated microRNA processing. RNA 14, 1539–1549.

Nishina, K., Unno, T., Uno, Y., Kubodera, T., Kanouchi, T., Mizusawa, H., Yokota, T., 2008. Efficient in vivo delivery of siRNA to the liver by conjugation of alpha-tocopherol. Mol. Ther. 16, 734–740.

Nowek, K., Sun, S.M., Dijkstra, M.K., Bullinger, L., Dohner, H., Erkeland, S.J., Lowenberg, B., Jongen-Lavrencic, M., 2016. Expression of a passenger miR-9* predicts favorable outcome in adults with acute myeloid leukemia less than 60 years of age. Leukemia 30, 303–309.

Obad, S., Dos Santos, C.O., Petri, A., Heidenblad, M., Broom, O., Ruse, C., Fu, C., Lindow, M., Stenvang, J., Straarup, E.M., Hansen, H.F., Koch, T., Pappin, D., Hannon, G.J., Kauppinen, S., 2011. Silencing of microRNA families by seed-targeting tiny LNAs. Nat. Genet. 43, 371–378.

O'Donnell, K.A., Wentzel, E.A., Zeller, K.I., Dang, C.V., Mendell, J.T., 2005. c-Myc-regulated microRNAs modulate E2F1 expression. Nature 435, 839–843.

Okamura, K., Hagen, J.W., Duan, H., Tyler, D.M., Lai, E.C., 2007. The mirtron pathway generates microRNA-class regulatory RNAs in Drosophila. Cell 130, 89–100.

Ooi, A.G., Sahoo, D., Adorno, M., Wang, Y., Weissman, I.L., Park, C.Y., 2010. MicroRNA-125b expands hematopoietic stem cells and enriches for the lymphoid-balanced and lymphoid-biased subsets. Proc. Natl. Acad. Sci. U.S.A. 107, 21505–21510.

Ota, A., Tagawa, H., Karnan, S., Tsuzuki, S., Karpas, A., Kira, S., Yoshida, Y., Seto, M., 2004. Identification and characterization of a novel gene, C13orf25, as a target for 13q31-q32 amplification in malignant lymphoma. Cancer Res. 64, 3087–3095.

Palliser, D., Chowdhury, D., Wang, Q.Y., Lee, S.J., Bronson, R.T., Knipe, D.M., Lieberman, J., 2006. An siRNA-based microbicide protects mice from lethal herpes simplex virus 2 infection. Nature 439, 89–94.

Pampalakis, G., Diamandis, E.P., Katsaros, D., Sotiropoulou, G., 2010. Down-regulation of dicer expression in ovarian cancer tissues. Clin. Biochem. 43, 324–327.

Park, J.K., Kogure, T., Nuovo, G.J., Jiang, J., He, L., Kim, J.H., Phelps, M.A., Papenfuss, T.L., Croce, C.M., Patel, T., Schmittgen, T.D., 2011. miR-221 silencing blocks hepatocellular carcinoma and promotes survival. Cancer Res. 71, 7608–7616.

Pasquinelli, A.E., Reinhart, B.J., Slack, F., Martindale, M.Q., Kuroda, M.I., Maller, B., Hayward, D.C., Ball, E.E., Degnan, B., Muller, P., Spring, J., Srinivasan, A., Fishman, M., Finnerty, J., Corbo, J., Levine, M., Leahy, P., Davidson, E., Ruvkun, G., 2000. Conservation of the sequence and temporal expression of let-7 heterochronic regulatory RNA. Nature 408, 86–89.

Pedersen, I.M., Otero, D., Kao, E., Miletic, A.V., Hother, C., Ralfkiaer, E., Rickert, R.C., Gronbaek, K., David, M., 2009. Onco-miR-155 targets SHIP1 to promote TNFalpha-dependent growth of B cell lymphomas. EMBO Mol. Med. 1, 288–295.

Pekarsky, Y., Hallas, C., Croce, C.M., 2001. The role of TCL1 in human T-cell leukemia. Oncogene 20, 5638–5643.

Pekarsky, Y., Santanam, U., Cimmino, A., Palamarchuk, A., Efanov, A., Maximov, V., Volinia, S., Alder, H., Liu, C.G., Rassenti, L., Calin, G.A., Hagan, J.P., Kipps, T., Croce, C.M., 2006. Tcl1 expression in chronic lymphocytic leukemia is regulated by miR-29 and miR-181. Cancer Res. 66, 11590–11593.

Pelosi, A., Careccia, S., Lulli, V., Romania, P., Marziali, G., Testa, U., Lavorgna, S., Lo-Coco, F., Petti, M.C., Calabretta, B., Levrero, M., Piaggio, G., Rizzo, M.G., 2013. miRNA let-7c promotes granulocytic differentiation in acute myeloid leukemia. Oncogene 32, 3648–3654.

Pillai, R.S., Bhattacharyya, S.N., Artus, C.G., Zoller, T., Cougot, N., Basyuk, E., Bertrand, E., Filipowicz, W., 2005. Inhibition of translational initiation by Let-7 Micro-RNA in human cells. Science 309, 1573–1576.

Place, R.F., Li, L.C., Pookot, D., Noonan, E.J., Dahiya, R., 2008. MicroRNA-373 induces expression of genes with complementary promoter sequences. Proc. Natl. Acad. Sci. U.S.A. 105, 1608–1613.

Poliseno, L., Salmena, L., Zhang, J., Carver, B., Haveman, W.J., Pandolfi, P.P., 2010. A coding-independent function of gene and pseudogene mRNAs regulates tumour biology. Nature 465, 1033–1038.

Raaijmakers, M.H., Mukherjee, S., Guo, S., Zhang, S., Kobayashi, T., Schoonmaker, J.A., Ebert, B.L., Al-Shahrour, F., Hasserjian, R.P., Scadden, E.O., Aung, Z., Matza, M., Merkenschlager, M., Lin, C., Rommens, J.M., Scadden, D.T., 2010. Bone progenitor dysfunction induces myelodysplasia and secondary leukaemia. Nature 464, 852–857.

Raghavachari, N., Liu, P., Barb, J.J., Yang, Y., Wang, R., Nguyen, Q.T., Munson, P.J., 2014. Integrated analysis of miRNA and mRNA during differentiation of human CD34+ cells delineates the regulatory roles of microRNA in hematopoiesis. Exp. Hematol. 42 (14–27), e1–e2.

Rai, D., Kim, S.W., Mckeller, M.R., Dahia, P.L., Aguiar, R.C., 2010. Targeting of SMAD5 links microRNA-155 to the TGF-beta pathway and lymphomagenesis. Proc. Natl. Acad. Sci. U.S.A. 107, 3111–3116.

Rao, P.H., Houldsworth, J., Dyomina, K., Parsa, N.Z., Cigudosa, J.C., Louie, D.C., Popplewell, L., Offit, K., Jhanwar, S.C., Chaganti, R.S., 1998. Chromosomal and gene amplification in diffuse large B-cell lymphoma. Blood 92, 234–240.

Rehwinkel, J., Behm-Ansmant, I., Gatfield, D., Izaurralde, E., 2005. A crucial role for GW182 and the DCP1:DCP2 decapping complex in miRNA-mediated gene silencing. RNA 11, 1640–1647.

Reinhart, B.J., Slack, F.J., Basson, M., Pasquinelli, A.E., Bettinger, J.C., Rougvie, A.E., Horvitz, H.R., Ruvkun, G., 2000. The 21-nucleotide let-7 RNA regulates developmental timing in Caenorhabditis elegans. Nature 403, 901–906.

Ricci, E.P., Limousin, T., Soto-Rifo, R., Rubilar, P.S., Decimo, D., Ohlmann, T., 2013. miRNA repression of translation in vitro takes place during 43S ribosomal scanning. Nucleic Acids Res. 41, 586–598.
Robb, G.B., Brown, K.M., Khurana, J., Rana, T.M., 2005. Specific and potent RNAi in the nucleus of human cells. Nat. Struct. Mol. Biol. 12, 133–137.
Roberts, T.C., 2014. The microRNA biology of the mammalian nucleus. Mol. Ther. Nucleic Acids 3 . e188
Rodriguez, A., Griffiths-Jones, S., Ashurst, J.L., Bradley, A., 2004. Identification of mammalian microRNA host genes and transcription units. Genome Res. 14, 1902–1910.
Rossi, S., Shimizu, M., Barbarotto, E., Nicoloso, M.S., Dimitri, F., Sampath, D., Fabbri, M., Lerner, S., Barron, L.L., Rassenti, L.Z., Jiang, L., Xiao, L., Hu, J., Secchiero, P., Zauli, G., Volinia, S., Negrini, M., Wierda, W., Kipps, T.J., Plunkett, W., Coombes, K.R., Abruzzo, L.V., Keating, M.J., Calin, G.A., 2010. microRNA fingerprinting of CLL patients with chromosome 17p deletion identify a miR-21 score that stratifies early survival. Blood 116, 945–952.
Ruby, J.G., Jan, C.H., Bartel, D.P., 2007. Intronic microRNA precursors that bypass Drosha processing. Nature 448, 83–86.
Rybak, A., Fuchs, H., Smirnova, L., Brandt, C., Pohl, E.E., Nitsch, R., Wulczyn, F.G., 2008. A feedback loop comprising lin-28 and let-7 controls pre-let-7 maturation during neural stem-cell commitment. Nat. Cell Biol. 10, 987–993.
Salah, Z., Arafeh, R., Maximov, V., Galasso, M., Khawaled, S., Abou-Sharieha, S., Volinia, S., Jones, K.B., Croce, C.M., Aqeilan, R.I., 2015. miR-27a and miR-27a* contribute to metastatic properties of osteosarcoma cells. Oncotarget 6, 4920–4935.
Salmena, L., Poliseno, L., Tay, Y., Kats, L., Pandolfi, P.P., 2011. A ceRNA hypothesis: the Rosetta Stone of a hidden RNA language? Cell 146, 353–358.
Sand, M., Gambichler, T., Skrygan, M., Sand, D., Scola, N., Altmeyer, P., Bechara, F.G., 2010. Expression levels of the microRNA processing enzymes Drosha and dicer in epithelial skin cancer. Cancer Invest. 28, 649–653.
Schirle, N.T., Sheu-Gruttadauria, J., Macrae, I.J., 2014. Structural basis for microRNA targeting. Science 346, 608–613.
Schultz, K.A., Pacheco, M.C., Yang, J., Williams, G.M., Messinger, Y., Hill, D.A., Dehner, L.P., Priest, J.R., 2011. Ovarian sex cord-stromal tumors, pleuropulmonary blastoma and DICER1 mutations: a report from the International Pleuropulmonary Blastoma Registry. Gynecol. Oncol. 122, 246–250.
Schwarz, D.S., Zamore, P.D., 2002. Why do miRNAs live in the miRNP? Genes Dev. 16, 1025–1031.
Schwarz, D.S., Hutvagner, G., Du, T., Xu, Z., Aronin, N., Zamore, P.D., 2003. Asymmetry in the assembly of the RNAi enzyme complex. Cell 115, 199–208.
Shen, C., Chen, M.T., Zhang, X.H., Yin, X.L., Ning, H.M., Su, R., Lin, H.S., Song, L., Wang, F., Ma, Y.N., Zhao, H.L., Yu, J., Zhang, J.W., 2016. The PU.1-modulated microRNA-22 is a regulator of monocyte/macrophage differentiation and acute myeloid leukemia. PLoS Genet. 12, e1006259.
Shi, X.B., Xue, L., Yang, J., Ma, A.H., Zhao, J., Xu, M., Tepper, C.G., Evans, C.P., Kung, H.J., Devere White, R.W., 2007. An androgen-regulated miRNA suppresses Bak1 expression and induces androgen-independent growth of prostate cancer cells. Proc. Natl. Acad. Sci. U.S.A. 104, 19983–19988.
Shin, C., Nam, J.W., Farh, K.K., Chiang, H.R., Shkumatava, A., Bartel, D.P., 2010. Expanding the microRNA targeting code: functional sites with centered pairing. Mol. Cell 38, 789–802.
Singh, S.K., Kumar, R., Wengel, J., 1998. Synthesis of novel bicyclo[2.2.1] ribonucleosides: 2'-amino- and 2'-thio-LNA monomeric nucleosides. J. Org. Chem. 63, 6078–6079.
Sokol, L., Caceres, G., Volinia, S., Alder, H., Nuovo, G.J., Liu, C.G., Mcgraw, K., Clark, J.A., Sigua, C.A., Chen, D.T., Moscinski, L., Croce, C.M., List, A.F., 2011.

Identification of a risk dependent microRNA expression signature in myelodysplastic syndromes. Br. J. Haematol. 153, 24–32.
Song, J.J., Smith, S.K., Hannon, G.J., Joshua-Tor, L., 2004. Crystal structure of argonaute and its implications for RISC slicer activity. Science 305, 1434–1437.
Song, E., Zhu, P., Lee, S.K., Chowdhury, D., Kussman, S., Dykxhoorn, D.M., Feng, Y., Palliser, D., Weiner, D.B., Shankar, P., Marasco, W.A., Lieberman, J., 2005. Antibody mediated in vivo delivery of small interfering RNAs via cell-surface receptors. Nat. Biotechnol. 23, 709–717.
Song, S.J., Ito, K., Ala, U., Kats, L., Webster, K., Sun, S.M., Jongen-Lavrencic, M., Manova-Todorova, K., Teruya-Feldstein, J., Avigan, D.E., Delwel, R., Pandolfi, P.P., 2013a. The oncogenic microRNA miR-22 targets the TET2 tumor suppressor to promote hematopoietic stem cell self-renewal and transformation. Cell Stem Cell 13, 87–101.
Song, S.J., Poliseno, L., Song, M.S., Ala, U., Webster, K., Ng, C., Beringer, G., Brikbak, N.J., Yuan, X., Cantley, L.C., Richardson, A.L., Pandolfi, P.P., 2013b. MicroRNA-antagonism regulates breast cancer stemness and metastasis via TET-family-dependent chromatin remodeling. Cell 154, 311–324.
Spizzo, R., Nicoloso, M.S., Croce, C.M., Calin, G.A., 2009. SnapShot: microRNAs in cancer. Cell 137, 586 e1.
Starczynowski, D.T., Morin, R., Mcpherson, A., Lam, J., Chari, R., Wegrzyn, J., Kuchenbauer, F., Hirst, M., Tohyama, K., Humphries, R.K., Lam, W.L., Marra, M., Karsan, A., 2011. Genome-wide identification of human microRNAs located in leukemia-associated genomic alterations. Blood 117, 595–607.
Subtelny, A.O., Eichhorn, S.W., Chen, G.R., Sive, H., Bartel, D.P., 2014. Poly(A)-tail profiling reveals an embryonic switch in translational control. Nature 508, 66–71.
Sugito, N., Ishiguro, H., Kuwabara, Y., Kimura, M., Mitsui, A., Kurehara, H., Ando, T., Mori, R., Takashima, N., Ogawa, R., Fujii, Y., 2006. RNASEN regulates cell proliferation and affects survival in esophageal cancer patients. Clin. Cancer Res. 12, 7322–7328.
Sumazin, P., Yang, X., Chiu, H.S., Chung, W.J., Iyer, A., Llobet-Navas, D., Rajbhandari, P., Bansal, M., Guarnieri, P., Silva, J., Califano, A., 2011. An extensive microRNA-mediated network of RNA-RNA interactions regulates established oncogenic pathways in glioblastoma. Cell 147, 370–381.
Taganov, K.D., Boldin, M.P., Chang, K.J., Baltimore, D., 2006. NF-kappaB-dependent induction of microRNA miR-146, an inhibitor targeted to signaling proteins of innate immune responses. Proc. Natl. Acad. Sci. U.S.A. 103, 12481–12486.
Takimoto, K., Wakiyama, M., Yokoyama, S., 2009. Mammalian GW182 contains multiple Argonaute-binding sites and functions in microRNA-mediated translational repression. RNA 15, 1078–1089.
Tam, W., 2001. Identification and characterization of human BIC, a gene on chromosome 21 that encodes a noncoding RNA. Gene 274, 157–167.
Tam, W., Ben-Yehuda, D., Hayward, W.S., 1997. bic, a novel gene activated by proviral insertions in avian leukosis virus-induced lymphomas, is likely to function through its noncoding RNA. Mol. Cell. Biol. 17, 1490–1502.
Tanzer, A., Stadler, P.F., 2004. Molecular evolution of a microRNA cluster. J. Mol. Biol. 339, 327–335.
Tay, Y., Zhang, J., Thomson, A.M., Lim, B., Rigoutsos, I., 2008. MicroRNAs to Nanog, Oct4 and Sox2 coding regions modulate embryonic stem cell differentiation. Nature 455, 1124–1128.
Tay, Y., Kats, L., Salmena, L., Weiss, D., Tan, S.M., Ala, U., Karreth, F., Poliseno, L., Provero, P., Di Cunto, F., Lieberman, J., Rigoutsos, I., Pandolfi, P.P., 2011. Coding-independent regulation of the tumor suppressor PTEN by competing endogenous mRNAs. Cell 147, 344–357.

Tefferi, A., Lim, K.H., Abdel-Wahab, O., Lasho, T.L., Patel, J., Patnaik, M.M., Hanson, C.A., Pardanani, A., Gilliland, D.G., Levine, R.L., 2009. Detection of mutant TET2 in myeloid malignancies other than myeloproliferative neoplasms: CMML, MDS, MDS/MPN and AML. Leukemia 23, 1343–1345.

Teng, G., Hakimpour, P., Landgraf, P., Rice, A., Tuschl, T., Casellas, R., Papavasiliou, F.N., 2008. MicroRNA-155 is a negative regulator of activation-induced cytidine deaminase. Immunity 28, 621–629.

The C. elegans Sequencing Consortium, 1998. Genome sequence of the nematode C. elegans: a platform for investigating biology. Science 282, 2012–2018.

Thomson, D.W., Dinger, M.E., 2016. Endogenous microRNA sponges: evidence and controversy. Nat. Rev. Genet. 17, 272–283.

Van Rooij, E., 2011. The art of microRNA research. Circ. Res. 108, 219–234.

Vasudevan, S., 2012. Posttranscriptional upregulation by microRNAs. Wiley Interdiscip. Rev. RNA 3, 311–330.

Vasudevan, S., Tong, Y., Steitz, J.A., 2007. Switching from repression to activation: microRNAs can up-regulate translation. Science 318, 1931–1934.

Vella, M.C., Choi, E.Y., Lin, S.Y., Reinert, K., Slack, F.J., 2004. The C. elegans microRNA let-7 binds to imperfect let-7 complementary sites from the lin-41 3'UTR. Genes Dev. 18, 132–137.

Venter, J.C., Adams, M.D., Myers, E.W., Li, P.W., Mural, R.J., Sutton, G.G., Smith, H.O., Yandell, M., Evans, C.A., Holt, R.A., Gocayne, J.D., Amanatides, P., Ballew, R.M., Huson, D.H., Wortman, J.R., Zhang, Q., Kodira, C.D., Zheng, X.H., Chen, L., Skupski, M., Subramanian, G., Thomas, P.D., Zhang, J., Gabor Miklos, G.L., Nelson, C., Broder, S., Clark, A.G., Nadeau, J., Mckusick, V.A., Zinder, N., Levine, A.J., Roberts, R.J., Simon, M., Slayman, C., Hunkapiller, M., Bolanos, R., Delcher, A., Dew, I., Fasulo, D., Flanigan, M., Florea, L., Halpern, A., Hannenhalli, S., Kravitz, S., Levy, S., Mobarry, C., Reinert, K., Remington, K., Abu-Threideh, J., Beasley, E., Biddick, K., Bonazzi, V., Brandon, R., Cargill, M., Chandramouliswaran, I., Charlab, R., Chaturvedi, K., Deng, Z., Di Francesco, V., Dunn, P., Eilbeck, K., Evangelista, C., Gabrielian, A.E., Gan, W., Ge, W., Gong, F., Gu, Z., Guan, P., Heiman, T.J., Higgins, M.E., Ji, R.R., Ke, Z., Ketchum, K.A., Lai, Z., Lei, Y., Li, Z., Li, J., Liang, Y., Lin, X., Lu, F., Merkulov, G.V., Milshina, N., Moore, H.M., Naik, A.K., Narayan, V.A., Neelam, B., Nusskern, D., Rusch, D.B., Salzberg, S., Shao, W., Shue, B., Sun, J., Wang, Z., Wang, A., Wang, X., Wang, J., Wei, M., Wides, R., Xiao, C., Yan, C., et al., 2001. The sequence of the human genome. Science 291, 1304–1351.

Ventura, A., Young, A.G., Winslow, M.M., Lintault, L., Meissner, A., Erkeland, S.J., Newman, J., Bronson, R.T., Crowley, D., Stone, J.R., Jaenisch, R., Sharp, P.A., Jacks, T., 2008. Targeted deletion reveals essential and overlapping functions of the miR-17 through 92 family of miRNA clusters. Cell 132, 875–886.

Visone, R., Rassenti, L.Z., Veronese, A., Taccioli, C., Costinean, S., Aguda, B.D., Volinia, S., Ferracin, M., Palatini, J., Balatti, V., Alder, H., Negrini, M., Kipps, T.J., Croce, C.M., 2009. Karyotype-specific microRNA signature in chronic lymphocytic leukemia. Blood 114, 3872–3879.

Viswanathan, S.R., Daley, G.Q., Gregory, R.I., 2008. Selective blockade of microRNA processing by Lin28. Science 320, 97–100.

Wagenaar, T.R., Zabludoff, S., Ahn, S.M., Allerson, C., Arlt, H., Baffa, R., Cao, H., Davis, S., Garcia-Echeverria, C., Gaur, R., Huang, S.M., Jiang, L., Kim, D., Metz-Weidmann, C., Pavlicek, A., Pollard, J., Reeves, J., Rocnik, J.L., Scheidler, S., Shi, C., Sun, F., Tolstykh, T., Weber, W., Winter, C., Yu, E., Yu, Q., Zheng, G., Wiederschain, D., 2015. Anti-miR-21 suppresses hepatocellular carcinoma growth via broad transcriptional network deregulation. Mol. Cancer Res. 13, 1009–1021.

Wang, C.H., Lee, D.Y., Deng, Z., Jeyapalan, Z., Lee, S.C., Kahai, S., Lu, W.Y., Zhang, Y., Yang, B.B., 2008. MicroRNA miR-328 regulates zonation morphogenesis by targeting CD44 expression. PLoS One 3 . e2420

Wang, L., Guo, Z.Y., Zhang, R., Xin, B., Chen, R., Zhao, J., Wang, T., Wen, W.H., Jia, L.T., Yao, L.B., Yang, A.G., 2013. Pseudogene OCT4-pg4 functions as a natural micro RNA sponge to regulate OCT4 expression by competing for miR-145 in hepatocellular carcinoma. Carcinogenesis 34, 1773–1781.

Weber, M.J., 2005. New human and mouse microRNA genes found by homology search. FEBS J. 272, 59–73.

Wiestner, A., Tehrani, M., Chiorazzi, M., Wright, G., Gibellini, F., Nakayama, K., Liu, H., Rosenwald, A., Muller-Hermelink, H.K., Ott, G., Chan, W.C., Greiner, T.C., Weisenburger, D.D., Vose, J., Armitage, J.O., Gascoyne, R.D., Connors, J.M., Campo, E., Montserrat, E., Bosch, F., Smeland, E.B., Kvaloy, S., Holte, H., Delabie, J., Fisher, R.I., Grogan, T.M., Miller, T.P., Wilson, W.H., Jaffe, E.S., Staudt, L.M., 2007. Point mutations and genomic deletions in CCND1 create stable truncated cyclin D1 mRNAs that are associated with increased proliferation rate and shorter survival. Blood 109, 4599–4606.

Wightman, B., Burglin, T.R., Gatto, J., Arasu, P., Ruvkun, G., 1991. Negative regulatory sequences in the lin-14 3'-untranslated region are necessary to generate a temporal switch during Caenorhabditis elegans development. Genes Dev. 5, 1813–1824.

Wightman, B., Ha, I., Ruvkun, G., 1993. Posttranscriptional regulation of the heterochronic gene lin-14 by lin-4 mediates temporal pattern formation in C. elegans. Cell 75, 855–862.

Wolfrum, C., Shi, S., Jayaprakash, K.N., Jayaraman, M., Wang, G., Pandey, R.K., Rajeev, K.G., Nakayama, T., Charrise, K., Ndungo, E.M., Zimmermann, T., Koteliansky, V., Manoharan, M., Stoffel, M., 2007. Mechanisms and optimization of in vivo delivery of lipophilic siRNAs. Nat. Biotechnol. 25, 1149–1157.

Woods, K., Thomson, J.M., Hammond, S.M., 2007. Direct regulation of an oncogenic micro-RNA cluster by E2F transcription factors. J. Biol. Chem. 282, 2130–2134.

Wurm, A.A., Tenen, D.G., Behre, G., 2017. The Janus-faced nature of miR-22 in hematopoiesis: is it an oncogenic tumor suppressor or rather a tumor-suppressive oncogene? PLoS Genet. 13, e1006505.

Yamanaka, Y., Tagawa, H., Takahashi, N., Watanabe, A., Guo, Y.M., Iwamoto, K., Yamashita, J., Saitoh, H., Kameoka, Y., Shimizu, N., Ichinohasama, R., Sawada, K., 2009. Aberrant overexpression of microRNAs activate AKT signaling via down-regulation of tumor suppressors in natural killer-cell lymphoma/leukemia. Blood 114, 3265–3275.

Yang, H., Kong, W., He, L., Zhao, J.J., O'Donnell, J.D., Wang, J., Wenham, R.M., Coppola, D., Kruk, P.A., Nicosia, S.V., Cheng, J.Q., 2008. MicroRNA expression profiling in human ovarian cancer: miR-214 induces cell survival and cisplatin resistance by targeting PTEN. Cancer Res. 68, 425–433.

Yang, J., Li, T., Gao, C., Lv, X., Liu, K., Song, H., Xing, Y., Xi, T., 2014. FOXO1 3'UTR functions as a ceRNA in repressing the metastases of breast cancer cells via regulating miRNA activity. FEBS Lett. 588, 3218–3224.

Yao, B., Li, S., Lian, S.L., Fritzler, M.J., Chan, E.K., 2011. Mapping of Ago2-GW182 functional interactions. Methods Mol. Biol. 725, 45–62.

Yao, Y.D., Sun, T.M., Huang, S.Y., Dou, S., Lin, L., Chen, J.N., Ruan, J.B., Mao, C.Q., Yu, F.Y., Zeng, M.S., Zang, J.Y., Liu, Q., Su, F.X., Zhang, P., Lieberman, J., Wang, J., Song, E., 2012. Targeted delivery of PLK1-siRNA by ScFv suppresses Her2+ breast cancer growth and metastasis. Sci. Transl. Med. 4, 130ra48.

Ye, X., Huang, N., Liu, Y., Paroo, Z., Huerta, C., Li, P., Chen, S., Liu, Q., Zhang, H., 2011. Structure of C3PO and mechanism of human RISC activation. Nat. Struct. Mol. Biol. 18, 650–657.

Yi, R., Qin, Y., Macara, I.G., Cullen, B.R., 2003. Exportin-5 mediates the nuclear export of pre-microRNAs and short hairpin RNAs. Genes Dev. 17, 3011–3016.

Yu, J., Vodyanik, M.A., Smuga-Otto, K., Antosiewicz-Bourget, J., Frane, J.L., Tian, S., Nie, J., Jonsdottir, G.A., Ruotti, V., Stewart, R., Slukvin, I.I., Thomson, J.A., 2007. Induced pluripotent stem cell lines derived from human somatic cells. Science 318, 1917–1920.

Yu, G., Yao, W., Gumireddy, K., Li, A., Wang, J., Xiao, W., Chen, K., Xiao, H., Li, H., Tang, K., Ye, Z., Huang, Q., Xu, H., 2014. Pseudogene PTENP1 functions as a competing endogenous RNA to suppress clear-cell renal cell carcinoma progression. Mol. Cancer Ther. 13, 3086–3097.

Yuan, J., Nguyen, C.K., Liu, X., Kanellopoulou, C., Muljo, S.A., 2012. Lin28b reprograms adult bone marrow hematopoietic progenitors to mediate fetal-like lymphopoiesis. Science 335, 1195–1200.

Zaphiropoulos, P.G., 1997. Exon skipping and circular RNA formation in transcripts of the human cytochrome P-450 2C18 gene in epidermis and of the rat androgen binding protein gene in testis. Mol. Cell. Biol. 17, 2985–2993.

Zeng, Y., Cullen, B.R., 2004. Structural requirements for pre-microRNA binding and nuclear export by Exportin 5. Nucleic Acids Res. 32, 4776–4785.

Zhang, G., Xia, S., Tian, H., Liu, Z., Zhou, T., 2012. Clinical significance of miR-22 expression in patients with colorectal cancer. Med. Oncol. 29, 3108–3112.

Zhao, J.L., Rao, D.S., Boldin, M.P., Taganov, K.D., O'Connell, R.M., Baltimore, D., 2011. NF-kappaB dysregulation in microRNA-146a-deficient mice drives the development of myeloid malignancies. Proc. Natl. Acad. Sci. U.S.A. 108, 9184–9189.

Zhao, J.L., Rao, D.S., O'Connell, R.M., Garcia-Flores, Y., Baltimore, D., 2013. MicroRNA-146a acts as a guardian of the quality and longevity of hematopoietic stem cells in mice. Elife 2 . e00537

Zheng, L., Li, X., Gu, Y., Ma, Y., Xi, T., 2014. Pseudogene CYP4Z2P 3'UTR promotes angiogenesis in breast cancer. Biochem. Biophys. Res. Commun. 453, 545–551.

Zheng, L., Li, X., Gu, Y., Lv, X., Xi, T., 2015. The 3'UTR of the pseudogene CYP4Z2P promotes tumor angiogenesis in breast cancer by acting as a ceRNA for CYP4Z1. Breast Cancer Res. Treat. 150, 105–118.

Zhong, X., Li, N., Liang, S., Huang, Q., Coukos, G., Zhang, L., 2010. Identification of microRNAs regulating reprogramming factor LIN28 in embryonic stem cells and cancer cells. J. Biol. Chem. 285, 41961–41971.

Zhou, X., Ruan, J., Wang, G., Zhang, W., 2007. Characterization and identification of microRNA core promoters in four model species. PLoS Comput. Biol. 3 . e37

CHAPTER FOUR

Senescence-Associated MicroRNAs

Rachel Munk, Amaresh C. Panda, Ioannis Grammatikakis, Myriam Gorospe, Kotb Abdelmohsen[1]
Laboratory of Genetics and Genomics, National Institute on Aging, National Institutes of Health, Baltimore, MD, United States
[1]Corresponding author: e-mail address: abdelmohsenk@mail.nih.gov

Contents

International Review of Cell and Molecular Biology, Volume 334
ISSN 1937-6448
http://dx.doi.org/10.1016/bs.ircmb.2017.03.008

Abstract

Senescent cells arise as a consequence of cellular damage and can have either a detrimental or advantageous impact on tissues and organs depending on the specific cell type and metabolic state. As senescent cells accumulate in tissues with advancing age, they have been implicated in many age-related declines and diseases. The major facets of senescence include two pathways responsible for establishing and maintaining a senescence program, p53/CDKN1A(p21) and CDKN2A(p16)/RB, as well as the senescence-associated secretory phenotype. Numerous MicroRNAs influence senescence by modulating the abundance of key senescence regulatory proteins, generally by lowering the stability and/or translation of mRNAs that encode such factors. Accordingly, understanding the molecular mechanisms by which MicroRNAs influence senescence will enable diagnostic and therapeutic opportunities directed at senescent cells. Here, we review senescence-associated (SA)-MicroRNAs and discuss their implications in senescence-relevant pathologies.

1. INTRODUCTION

1.1 Senescence

Senescence was first described by Leonard Hayflick 50 years ago as a state in which normal cells divide for a finite number of times before they cease proliferation indefinitely (Hayflick, 1965; Hayflick and Moorhead, 1961), although they remain viable and metabolically active (Gey and Seeger, 2013). Senescence has been studied extensively using primary cells cultured ex vivo until they stop dividing; this process is known as *replicative senescence* (Hayflick, 1992). Additionally, senescence can be achieved by exposure of primary cells to a range of sublethal damaging agents; this process is known as *premature senescence*. Knowledge from these two processes has been recently complemented by studies on senescence in vivo and its impact in physiology and pathology.

1.1.1 Replicative and Premature Senescence

Replicative senescence is achieved in primary cells isolated from tissues and placed in culture. Over time, their population doubling time (PDL) gradually increases until they are terminally arrested and cease to proliferate. At the same time, the telomeres that protect chromosome ends and permit DNA polymerase to complete replication become progressively shorter and eventually trigger a DNA damage response (DDR) resulting in growth cessation and senescence (Kuilman et al., 2010). The short, unprotected telomeres activate several downstream effectors including the kinases ataxia telangiectasia mutated (ATM) and ATM-related (ATR), the checkpoint kinases CHK1 and CHK2, and the tumor suppressor and transcription factor p53 (Ben-Porath and Weinberg, 2005; Herbig et al., 2004). In turn, p53 transcriptionally induces expression of the cyclin-dependent kinase (CDK) inhibitor CDKN1A/p21 and the alternate open reading frame protein ARF, enabling the p53/p21 axis that inhibits replication (Ben-Porath and Weinberg, 2005; Campisi, 2005). Replicative senescence is further established by the CDKN2A(p16)/RB axis, comprising of the retinoblastoma protein (RB) activated through the CDK inhibitor p16 (CDKN2A/INK4A) and the related proteins p15/INK4B, p18/INK4C, and p19/INK4D (Ben-Porath and Weinberg, 2005; Campisi, 2005).

Premature senescence, also known as stress-inducible senescence, can be rapidly achieved by exposing cells to stresses such as radiation, oxidants, toxins, chemotherapy, oncoprotein activation, or through tumor-suppressor inactivation. These harmful stimuli trigger senescence by activating stress signals including DDR; the ensuing growth arrest by CDK inhibitors and heterochromatin changes occur without apparent loss of telomere function (Kuilman et al., 2010). Premature senescence can be triggered by activation of oncoproteins such as KRAS(V12) and BRAF(V600E), and also by inactivation of the tumor suppressors phosphatase and tensin homologue (PTEN), von Hipple-Lindau (VHL), or neurofibromatosis (NF) 1 (Gorospe and Abdelmohsen, 2011). While ectopic expression of telomerase can restore cell proliferation during replicative senescence, it does not prevent premature senescence (Wei et al., 1999). The stresses that activate senescence also utilize the p53/p21 and p16/RB pathways, as described in a recent review (Loaiza and Demaria, 2016).

1.1.2 In Vivo Senescence

There are many models of replicative and premature senescence, mainly involving cultured primary cells. However, these senescence models remain limited, as they may not fully recapitulate the physiological consequences of senescence. Increasing numbers of studies have provided evidence that senescence occurs in tissues and organs in vivo and can be detected in physiologic and pathologic situations, with both detrimental and protective consequences.

In tissues from humans, mice, and monkeys, senescent cells have been detected showing the enhanced presence of DDR proteins and senescence-associated β-galactosidase activity, as explained later (Cristofalo et al., 2004; Dimri et al., 1995; Herbig et al., 2006). Senescent cells were found to accumulate with age in renewable tissues like epithelia, stroma, and the hematopoietic system (Choi et al., 2000; Dimri et al., 1995; Jeyapalan et al., 2007; Krishnamurthty et al., 2004; Krizhanovsky et al., 2008). Stem cells in muscle from aged mice express p16 as they switch from reversible quiescence to irreversible senescence, suggesting that the maintenance of quiescence depends upon the effective suppression of senescence pathways in vivo (Sousa-Victor et al., 2014). The reliance on p16 for the quiescence-to-senescence transition may explain, at least in part, the elevated proliferation of progenitors of neurons and islet cells in p16-deficient mice, which show improved neuronal and pancreatic function compared with WT mice (Krishnamurthy et al., 2006; Molofsky et al., 2006). Importantly, mice in which p16-expressing cells are selectively eliminated via caspase activation show delayed appearance of age-related conditions, including cataracts, sarcopenia, and loss of adipose tissue (Baker et al., 2011).

In physiologic contexts, senescence is essential for morphogenesis during development. It also plays a pivotal role during wound repair, as the wound-induced senescent fibroblasts and endothelial cells secrete platelet-derived growth factor AA (PDGF-AA) which accelerates myofibroblast differentiation, wound healing, and tissue repair (Demaria et al., 2014). Similarly, following liver damage, hepatic stellate cells acquire senescence, thereby increasing degradation of the extracellular matrix, preventing the formation of fibrotic scars, and preserving liver function (Krizhanovsky et al., 2008).

On the other hand, senescence has also been detected and implicated in various pathologies, prominently cancer. Interestingly, the same oncogenes that can trigger senescence (e.g., KRAS(V12), BRAF (V600E)) can also trigger tumorigenesis if cells are unable to mount a senescent response (Braig et al., 2005; Collado et al., 2005; Dankort et al., 2009; Michaloglou et al., 2005). The complexity between senescence and cancer is further illustrated by the senescence-associated secretory phenotype (SASP), as this phenotype can favor or suppress tumorigenesis. Favoring tumorigenesis is the secretion by senescent cells of angiogenic factors (e.g., VEGF) that promote local angiogenesis, and matrix metalloproteinases (MMPs) which make tissues permeable and facilitate both migration and metastasis (Freund et al., 2010; Ohtani et al., 2012). Suppressing tumorigenesis, particularly at certain tumor stages, are SASP factors that promote chemotaxis leading to the recognition and clearance of cancer cells by the immune system (Acosta et al., 2008; Kuilman et al., 2008; Massague, 2008).

Senescent cells can also exacerbate other pathologic situations. They have been implicated in age-related diseases such as atherosclerosis, diabetes, neurodegeneration, and cardiovascular illnesses (Childs et al., 2015; Chinta et al., 2015). For example, accumulation of senescent microglia with aging is linked to neurodegenerative diseases such as Parkinson's and Alzheimer's diseases (Luo et al., 2010), while the impaired vascularization in diabetes is linked to the enhanced senescence of endothelial progenitor cells (Rosso et al., 2006). Finally, senescence of skeletal muscle and precursor cells is linked to muscle aging and the related disease sarcopenia (Navarro et al., 2001). Collectively, there is rising recognition that in vivo senescence impacts human physiology and pathology.

1.1.3 Senescence Markers

As cells approach senescence, they undergo specific morphological and molecular changes. Morphologically, senescent cells become flat and enlarged, and accumulate stress granules and vacuoles (Campisi, 1997; Goldstein, 1990). Senescent cells normally display increased activity of the acidic senescence-associated β-galactosidase (SA-β-gal), which is widely used as a marker to identify senescent cells in tissues and cell cultures (Dimri et al., 1995). Senescent cells display significant changes in their nuclear morphology including large nuclei, irregular nuclear envelope, and condensed and disordered chromosomes (Freund et al., 2012; Mehta et al., 2007; Zhang et al., 2007).

The senescence phenotype can also be confirmed by other methods. High levels of ROS are widely used as markers of senescence; as ROS can induce DNA damage. DDR proteins including TP53, p21, γ-H2AX, p16, as well as senescence-associated heterochromatin foci (SAHF) are also major markers of senescence (Narita et al., 2003; Zhang et al., 2005). The localization of HMGB1 can indicate senescence as it enters the extracellular space to stimulate cytokine production through the Toll-like receptor (TLR) 4 pathway (Davalos et al., 2013). Senescent cells can also be identified by the SASP, a phenotype that includes heightened secretion of major chemokines (e.g., IL8, GROA/B/G, MCP2, and HCC4), cytokines (IL6, IL1A, IL1B, and IL7), MMPs (including MMP1/3/10, TIMP1/2), and additional factors (e.g., VEGF, EGF, IGFBP2, NGF, and FGF7) that influence both the immediate cellular microenvironment and distant tissues and organs (Campisi, 2005; Coppe et al., 2006; Kuilman and Peeper, 2009). Together, senescent cells display many specific changes that can serve as markers, although it is important to mention that no single marker is conclusive and thus assessment of multiple markers is needed to identify senescent cells.

1.2 MicroRNAs

MicroRNAs (MiRNAs) are small noncoding RNA molecules (~22 nucleotides long) that generally influence gene expression by binding to subsets of mRNAs with which they share partial complementarity, and suppressing the stability and/or translation of such mRNAs. They are transcribed as long hairpin structures, primary (Pri)-miRNAs, which are processed in the nucleus by Drosha/DGCR8 into precursor (Pre)-miRNAs that are exported to the cytoplasm and further processed by DICER1 into mature single-stranded MiRNAs. Mature MiRNAs are key components of the RNA-inducible silencing complex (RISC) that direct RISC to target mRNAs (Bartel, 2009; Fabian et al., 2010; Kim et al., 2009; Newman and Hammond, 2010). Through their interaction with a wide spectra of mRNAs encoding proteins with diverse functions, MiRNAs have emerged as potent posttranscriptional regulators of several cellular processes, including cell survival, death, division, differentiation, and senescence (Abdelmohsen and Gorospe, 2015; Bushati and Cohen, 2007; Catalanotto et al., 2016; Kloosterman and Plasterk, 2006; Lingor, 2010; Yao, 2016). Like senescence, MiRNAs have been implicated in several disease conditions such as cancer, cardiovascular disease, and neurodegeneration (Saeidimehr et al., 2016; Shah et al., 2016).

The activity of the p53/p21 and p16/RB senescence pathways, as well as the SASP, is governed by the relative levels of the constituents and modulators of these pathways. Among other mechanisms, the levels of the protein components of these signaling systems are strongly regulated by MiRNAs, as discussed in Sections 2–4.

2. SA-MicroRNAs IN THE P53/P21 PATHWAY

In addition to the finding that suppression of global MiRNA biogenesis induces p53/p21 signaling and senescence, individual MiRNAs have been identified that control the expression levels of protein components of this pathway (Jones and Lal, 2012; Mudhasani et al., 2008). Specifically, we will focus on the SA-MicroRNAs that regulate expression of p53, p21, MDM2, SIRT1, MYC, TERT, and MCD1 (Table 1).

2.1 p53

MiRNAs MiR-125b, MiR-504, MiR-25, and MiR-30d directly bind *p53* mRNA, suppress p53 protein levels, and inhibit senescence (Kumar et al.,

Table 1 SA-MicroRNAs in the p53/p21 Pathway

MiRNA	Target	References
MiR-125b, MiR-504, MiR-25, MiR-30d	p53	Kumar et al. (2011)
MiR-192, MiR-194, MiR-215, MiR-605	MDM2	Jones and Lal (2012)
MiR-34a, MiR-22, MiR-138, MiR-181a, MiR-217	SIRT1	Jazbutyte et al. (2013), Menghini et al. (2009), Ming et al. (2016), and Yamakuchi (2012)
MiR-34a	FoxM1 and MYC	Cui et al. (2016) and Xu et al. (2015)
MiR-29c-3p	CNOT6	Shang et al. (2016)
MiR-22	CDK6, SIRT1, SP1	Xu et al. (2011)
MiR-195	TERT, SIRT1	Okada et al. (2016)
MiR-22	AKT3	Zheng and Xu (2014)
MiR-200c	ZEB1, BMI1	Magenta et al. (2011)
MiR-181a, MiR-181b, MiR-138	SIRT1	Rivetti di Val Cervo et al. (2012) and Zhou et al. (2016)
MiR-130b, MiR-302a, MiR-302b, MiR302c, MiR-302d, MiR-515-3p	p21	Borgdorff et al. (2010)
MiR-106b	p21	Ivanovska et al. (2008) and Li et al. (2011)
MiR-20a	p21	Sokolova et al. (2015)
MiR-17	p21	Gibcus et al. (2011)
MiR-208	p21	Zhang et al. (2011)
MiR-663	p21	Yi et al. (2012)
MiR-519	DUT1, EXO1, RPA2, POLE4	Abdelmohsen et al. (2012) and Marasa et al. (2010)

2011). For example, lowering MiR-125b levels increased p53 levels and enabled senescence, while elevating MiR-125b levels suppressed p53 abundance and induced apoptosis in human lung fibroblasts responding to stress and developmental cues (Le et al., 2009) (Fig. 1).

2.2 p21

Several MiRNAs directly suppress p21 expression and thus influence cell senescence and proliferation. For instance, senescence-inhibitory MiRNAs including MiR-130b, MiR-302a, MiR-302b, MiR302c, MiR-302d, and MiR-515-3p attenuate RAS(G12V)-induced senescence of human mammary epithelial cells by preventing the RAS(G12V)-induced upregulation of p21 levels (Borgdorff et al., 2010). MiR-106b reduced p21-mediated cell cycle arrest in prostate cancer cells exhibiting radiation resistance, likely through direct targeting of MiR-106b to the *p21* mRNA (Ivanovska et al., 2008; Li et al., 2011). Similarly, MiR-20a was shown to target the 3′UTR of *p21* mRNA and blocked p21-mediated growth arrest in TGF-β-responsive colon carcinoma (Sokolova et al., 2015). MiR-208 promoted insulin-induced vascular smooth muscle cell proliferation by down-regulating p21 (Zhang et al., 2011), and MiR-663 suppressed p21 expression and promoted the proliferation and tumorigenesis of nasopharyngeal carcinoma cells (Yi et al., 2012).

Other MiRNAs indirectly enhance p21 expression leading to cell cycle arrest and senescence. For instance, MiR-519a, MiR-519b, and MiR-519c trigger p21-induced senescence by modulating multiple processes including DDR, calcium influx, and autophagy. The impact of MiR-519 on senescence was observed in HeLa cells and WI-38 human diploid fibroblasts (HDFs) (Abdelmohsen et al., 2012; Marasa et al., 2010). The cytoplasmic deadenylase CNOT6 is a component of the deadenylase complex, which contributes to the prevention of cell death and senescence (Mittal et al., 2011). MiR-29c-3p promoted senescence through the p53/p21 and p16/RB pathways by reducing CNOT6 expression levels in human mesenchymal stem cells (Shang et al., 2016).

2.3 MDM2

The murine double minute clone 2 (MDM2) is an oncogene that suppresses the expression of p53 by functioning as a ubiquitin ligase which promotes p53 ubiquitination and degradation by the 26S proteasome (Momand et al., 2000; Piette et al., 1997). Accordingly, the repression of MDM2

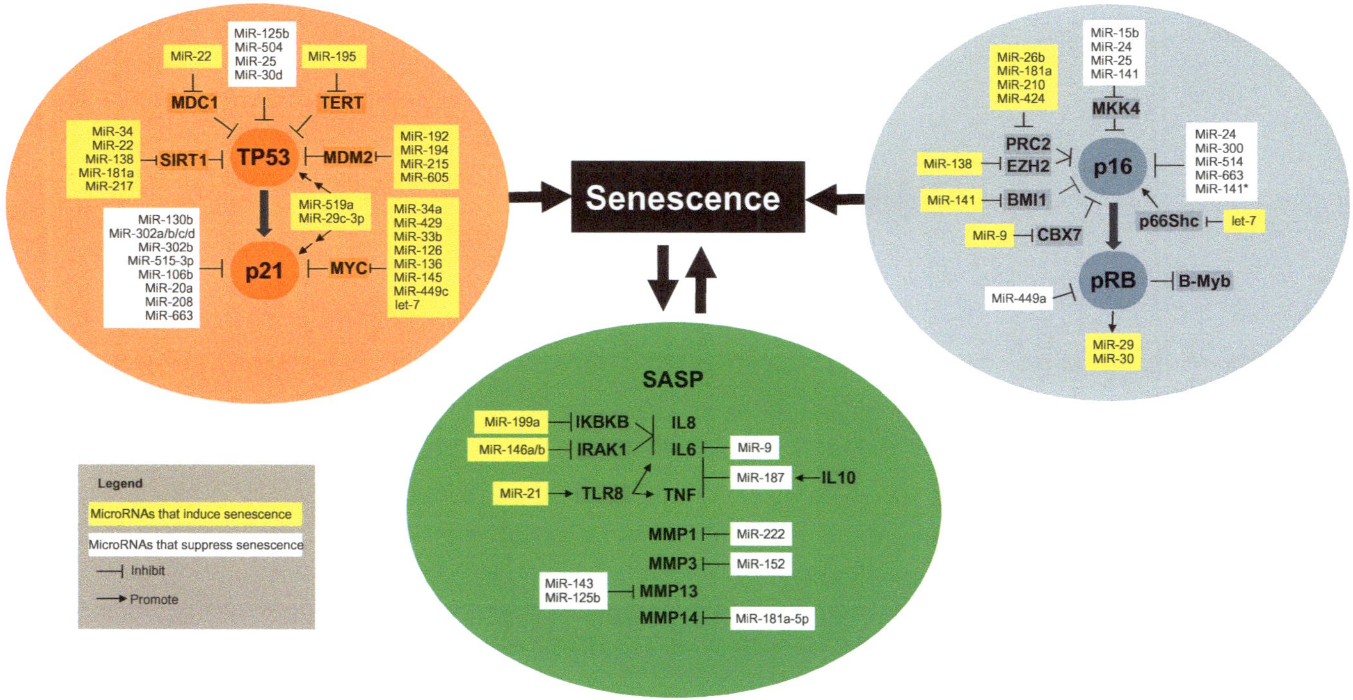

Fig. 1 SA-MicroRNAs implicated in cellular senescence. Schematic representation of the MicroRNAs that promote or inhibit senescence in three major domains of senescence: the p53/p21 pathway, the p16/RB pathway, and the trait SASP. MicroRNAs that promote senescence are in *yellow boxes*, MicroRNAs that suppress senescence are in *white boxes*.

expression by MiR-192, MiR-194, MiR-215, and MiR-605 enhanced p53 levels and senescence (Jones and Lal, 2012).

2.4 SIRT1

The silent mating type information regulation 2 homologue 1 (SIRT1) is a deacetylase that regulates metabolic activity in response to cellular stress. SIRT1-mediated deacetylation of p53 suppressed its transcriptional induction of p21 expression (Brooks and Gu, 2009). MiR-34a, MiR-22, MiR-138, MiR-181a, and MiR-217 decrease SIRT1 expression, thereby inducing p53 levels and cell senescence (Jazbutyte et al., 2013; Menghini et al., 2009; Ming et al., 2016; Xu et al., 2011; Yamakuchi, 2012). Interestingly, p53 amplified this effect by transcriptionally inducing the expression of MiR-34a, MiR-34b, and mir-34c, which triggered senescence in various cell systems (Bai et al., 2011; Cui et al., 2016; Kumamoto et al., 2008; Tazawa et al., 2007; Xu et al., 2015).

2.5 MYC

The oncoprotein MYC influences DNA replication, cell cycle progression, and apoptosis. MYC suppresses p21 expression levels, thereby modulating genotoxic and apoptotic responses (Seoane et al., 2002). Several MiRNAs, including MiR-34a, MiR-429, MiR-33b, MiR-126, MiR-136, MiR-145, MiR-449c, and let-7 were found to target the *MYC* mRNA, although it is not known if they all induce senescence (Deng and Sui, 2013). MiR-34a further impacts senescence via MYC-modulated telomerase activity in human hepatocellular carcinoma (Cui et al., 2016; Xu et al., 2015). Senescence of WI-38 and BJ-1 HDFs triggered by metformin treatment led to increased MiR-429 levels (Cufi et al., 2012), although it is not yet known if MiR-429 targets *MYC* mRNA in this paradigm.

2.6 TERT

The reverse transcriptase enzyme telomerase (TERT) is essential for maintaining telomere length and thus prevents senescence (Shawi and Autexier, 2008). MiR-195 induced senescence in old mesenchymal stem cells (OMSCs) by directly reducing the levels of TERT and SIRT1. Conversely, inhibiting MiR-195 enhanced the expression of TERT and SIRT1, reduced p53 levels, and thereby delayed senescence (Okada et al., 2016). MiR-1207-5p and MiR-1266 suppressed cell growth and gastric cancer cell invasion by targeting TERT, but it remains to be tested if they induce cell senescence (Chen et al., 2014).

2.7 MDC1

The mediator of DNA damage checkpoint 1 (MDC1) regulates p53 activity and in turn the DDR, DNA repair, cell survival, and cell death (Nakanishi et al., 2007). Accordingly, a rise in MiR-22, which directly represses MDC1 production, hindered the DDR machinery, and induced premature senescence (Coster and Goldberg, 2010; Lee et al., 2015).

3. SA-MicroRNAs ASSOCIATED WITH THE P16/RB PATHWAY

Several MiRNAs have been implicated in p16-mediated senescence. While some MiRNAs influence p16 expression directly through binding to the mRNA, other MiRNAs act upstream to influence p16 levels. Here, we will focus on SA-MiRNAs that affect key components of this pathway, including p16, PRC2, MKK4, and RB (Table 2).

3.1 p16

The SA-MiRNA MiR-24 binds the 3′UTR of *p16* mRNA and suppresses its translation in WI-38 HDFs and in HeLa cells (Lal et al., 2008). While MiR-24 was not found to inhibit senescence, its levels are inversely correlated with the levels of p16 in osteoarthritis-associated senescence (Philipot et al., 2014). In addition, MiR-24 promoted cell proliferation by targeting p16 and the CDK inhibitor p27 in various human cancer cell lines (Giglio et al., 2013). Other MiRNAs suppress p16 expression by targeting *p16* mRNA include MiR-300, MiR-514, MiR-663, and MiR-141, although their effect on senescence remains to be studied (Al-Khalaf et al., 2013; Kabir et al., 2015).

3.2 PRC2

p16 is encoded by the *INK4A-ARF* locus, which is transcriptionally suppressed by the Polycomb-repressive complex 2 (PRC2). Polycomb group (PcG) proteins were found to disassociate from *INK4A-ARF* locus in senescent cells, allowing p16 levels to increase via increased transcription (Bracken et al., 2007). MiRNAs such as MiR-26b, MiR-181a, MiR-210, and MiR-424 enhanced p16 expression by directly suppressing PRC2

Table 2 SA-MicroRNAs in the p16/RB Pathway

MicroRNA	Target	References
MiR-24s	p16	Lal et al. (2008) and Philipot et al. (2014)
MiR-26b	EED; EZH2	Overhoff et al. (2014)
MiR-181a	CBX7	Overhoff et al. (2014)
MiR-210	EED, EZH2, suz12	Overhoff et al. (2014)
MiR-424	EED, suz12	Overhoff et al. (2014)
MiR-138	EZH2	Liang et al. (2013)
MiR15b, MiR-24, MiR-25, MiR-141	MKK4	Marasa et al. (2009)
let-7	p66SHC	Xu et al. (2014)
MiR-141	BMI1	Dimri et al. (2013)
MiR-9	CBX7	O'Loghlen et al. (2015)
MiR-29, MiR-30	B-Myb	Martinez et al. (2011)
MiR-29c-3p	CNOT6	Shang et al. (2016)
MiR-203	E2F3A, E2F3B, ZBP89	Noguchi et al. (2012)
MiR-205	E2F1	Dar et al. (2011)
MiR-449a	CCND1, E2F1, E2F3	Noonan et al. (2010), Mao et al. (2016), and Ren et al. (2014)

factors like the PcG proteins CBX7, embryonic ectoderm development (EED), enhancer of zeste homologue 2 (EZH2), and suppressor of zeste 12 homologue (Suz12) (Overhoff et al., 2014). Additionally, in human renal cell carcinoma, MiR-138 induced p16 and cell senescence by targeting EZH2 (Liang et al., 2013).

3.3 MKK4

The mitogen-activated protein (MAP) kinase kinase 4 (MKK4) phosphorylates and thereby activates other MAP kinases in response to stress. High levels of MKK4 in senescent cells decreased cell proliferation and triggered senescence (Cunningham et al., 2010). *MKK4* mRNA is a target of four MiRNAs (MiR-15b, MiR-24, MiR-25, MiR-141) that jointly suppress MKK4 expression. Downregulation of MKK4 by these MiRNAs decreased p16 levels and delayed senescence, while their coordinated inhibition elevated MKK4 and p16 levels and induced senescence in WI-38 HDFs (Marasa et al., 2009).

3.4 RB

Activation of the RB pathway increased the levels of MiR-29 and MiR-30, which associated with the 3′UTR of the *MYBL2* (*B-Myb*) mRNA, repressed MYBL2 expression and triggered senescence (Martinez et al., 2011; Zhou et al., 2016). While DU-145 prostate cancer cells with mutant RB are resistant to MiR-449a-induced senescence, overexpression of wild-type RB induces senescence by restoring the function of MiR-449a (Noonan et al., 2010). Other studies point to additional MiR-449a targets in the RB pathway; for example, in prostate cancer cells, MiR-449a suppressed E2F1 and induced senescence following ionizing radiation (Mao et al., 2016), and in human lung cancer cells, MiR-449a lowered E2F3 abundance and thus induced senescence (Ren et al., 2014). Ectopic expression of MiR-203 in melanoma cells induced hypophosphorylated RB and elevated other senescence markers by suppressing the expression of E2F3a, E2F3b, and ZBP-89 (Noguchi et al., 2012).

3.5 Other SA-MiRNAs Regulating the 16/RB Pathway

In HDFs, let-7 delayed senescence by reducing the translation of the adaptor protein p66SHC, a key regulator of lifespan in mammals (Xu et al., 2014). In addition, MiR-141 induced p16 levels and consequently senescence by suppressing the production of BMI1 in HDFs (Dimri et al., 2013). In IMR-90 and WI-38 HDFs, MiR-9 was found to promote p16 expression by lowering the abundance of CBX7 (O'Loghlen et al., 2015). Finally, overexpression of MiR-205 in melanoma cells suppressed proliferation and induced markers of senescence including p16 (Dar et al., 2011).

4. SA-MicroRNAs IN SASP

The senescence-associated production and secretion of cytokines, growth factors, and proteases constitutes the SASP (Coppe et al., 2008). This

Table 3 SA-MicroRNAs in SASP

MicroRNA	Target	References
MiR-146a/b	IRAK1, TLR8	Bhaumik et al. (2009) and Gysler et al. (2016)
MiR-143	MMP13	Osaki et al. (2011)
MiR-335	PTEN	Kabir et al. (2016)
MiR-183	ITGB1	Li et al. (2010)
MiR-9	IL6	Zhang et al. (2016a)
MiR-222	MMP1	Liu et al. (2009)
MiR-125b	TNF, MMP13	Tili et al. (2007) and Xu et al. (2012)
MiR-152	MMP3	Zheng et al. (2013)
MiR-187	TNF, IL6	Rossato et al. (2012)

phenotype is governed by transcription factors, RBPs and ncRNAs, including both MiRNAs and lncRNAs. In this section, we focus on MiRNAs that regulate SASP, in particular interleukins (ILs), MMPs, and tumor necrosis factor alpha (TNF) (Table 3).

4.1 Interleukins

The production and secretion of ILs increases in senescent cells and creates a proinflammatory environment that promotes tumor progression (Coppe et al., 2010). In primary human fibroblasts, MiR-146a and MiR-146b directly suppressed the expression of interleukin 1 receptor-associated kinase 1 (IRAK1), a key regulator of the IL1A (IL-1α) receptor signaling, leading to inhibition of IL6 and IL8 production. Interestingly, high levels of secreted cytokines upregulated MiR-146a/b expression, which in turn prevented excessive SASP activity (Bhaumik et al., 2009). The tumor-suppressor MiR-9 inhibits IL6 expression in cervical adenocarcinoma (Zhang et al., 2016c), and MiR-187, induced by IL10, similarly suppressed the production of IL6 and TNF (Rossato et al., 2012). The circulating MiRNA MiR-21 is a ligand for TLR8 in immune cells, and via this alternative mechanism, it stimulates IL6 and TNF production (Fabbri et al., 2012). MiR-199a suppresses the production of IκB kinase-β (IKBKB), which is required for NF-κB activation, thereby lowering IL6 and IL8 production (Chen et al., 2008).

4.2 Matrix Metalloproteinases

The trait SASP includes the production of numerous MMPs, responsible for degrading the extracellular matrix and cleaving cell surface receptors and ligands. As senescent cells accumulate, the rising MMP activity promotes tissue damage through chronic inflammation and extracellular matrix remodeling caused by SASP (Verma and Hansch, 2007). MMP levels are regulated by different MiRNAs. MiR-222 directly suppresses MMP1 expression in oral tongue squamous cell carcinoma, and further suppresses the production of superoxide dismutase 2 (SOD2), which would otherwise induce MMP1. In addition, MiR-222 indirectly lowers MMP13 levels by downregulating the histone deacetylase HDAC4 (Liu et al., 2009; Song et al., 2015). MiR-152 targets the 3′UTR of *MMP3* mRNA, lowering MMP3 expression, and significantly reduces the invasiveness of glioma cells (Zheng et al., 2013). MiR-143 targets the *MMP13* 3′UTR and suppresses the expression of MMP13. In lung osteosarcoma, the robust downregulation of MiR-143 leads to a strong increase of MMP13 levels and cancer cell invasiveness (Osaki et al., 2011). The lower levels of MiR-125b in cutaneous squamous cell carcinoma (cSCC) compared with healthy skin allows higher production of MMP13 and accordingly promotes invasion of cSCC (Xu et al., 2012). Finally, MiR-181a-5p directly suppresses MMP14 expression in breast cancer cells (Roth and Cao, 2015).

4.3 TNF

The proinflammatory cytokine TNF is also regulated by MiRNAs. As mentioned earlier, the circulating MiR-21 was capable of binding and activating TLR8 in immune cells, thereby stimulating TNF and IL6 secretion (Fabbri et al., 2012). By contrast, in primary human monocytes, IL10 induced MiR-187, which in turn suppressed TNF and IL6 (Rossato et al., 2012), while in RAW 264.7 macrophages, LPS suppressed MiR-125b, which targets *TNF* mRNA, and thus enabled TNF production (Tili et al., 2007).

5. SA-MicroRNAs IN PATHOLOGY

As discussed earlier, there is a complex functional connection between senescence and several age-related pathologies. In this section, we discuss some of the major SA-MiRNAs implicated in disease processes.

5.1 Impact on Pathology by SA-MicroRNAs in the p53/p21 Pathway

Many MiRNAs having an impact on the p53/p21 pathway are implicated in diverse pathologies. For example, MiR-181a and MiR-30c synergistically regulate the p53/p21 pathway in diabetes-induced cardiac hypertrophy (Raut et al., 2016). The MiR-34a-mediated reduction of SIRT1, a suppressor of p53 activity, was implicated in tumorigenesis of the colon, lung, and other tissues (Kim et al., 2016; Lai et al., 2012; Misso et al., 2014; Rupaimoole and Slack, 2016; Wang et al., 2013). The Kruppel-like zinc finger protein ZNF224 transcriptionally increased MiR-663a, an MiRNA that lowers p53 and p21 production, and thus ZNF224 induced cell growth and resistance to apoptosis in human breast ductal carcinoma cells (Cho et al., 2016). MiR-621 suppressed production of the p53 inhibitor FBXO11, elevating p53 activity and sensitizing breast cancer cells to apoptosis (Xue et al., 2016), while MiR-24 increased metastasis and invasion of hepatocellular carcinoma cells by lowering p53 levels (Chen et al., 2016b). Collectively, these findings suggest that the network of MiRNAs affecting p53 function can influence pathologies, particularly cancer (Krell et al., 2013).

Recent studies suggest that the MiRNAs affecting p21 expression are particularly tightly involved in cancer. For instance, MiR-95-3p promoted tumorigenesis by targeting and reducing p21 levels in hepatocellular carcinoma (Ye et al., 2016), while inhibition of p21 by MiR-146b enhanced the proliferation of anaplastic thyroid cancer cells (Zhang et al., 2016a). The radiation-induced MiR-208a increased human lung cancer cell proliferation by lowering p21, and both MiR-106b and 208 enhanced radioresistance by lowering p21 expression (Tang et al., 2016; Zheng et al., 2015). In nasopharyngeal carcinoma, MiR-17-5p and MiR-663 promoted cell proliferation and tumorigenesis by lowering p21 abundance (Chen et al., 2016a; Yi et al., 2012), while in esophageal carcinoma, MiR-31 suppressed tumorigenesis in a p21-dependent manner (Ning et al., 2014). The MiRNA cluster MiR-17-92 regulated stem cell potential in MLL leukemia by reducing p21 expression (Wong et al., 2010), while upregulation of p21 by MiR-6734 induced cell cycle arrest and apoptosis in colon cancer cells (Kang et al., 2016). The impact of some of these MiRNAs on senescence remains to be investigated.

5.2 Impact on Pathology by SA-MicroRNAs in the p16/RB Pathway

MiR-24 inhibits p16 expression (Lal et al., 2008) and promoted cell growth, proliferation, and metastasis in gastric cancer cells, nonsmall cell lung cancer, and breast cancer, although a role for p16 in these processes await confirmation (Xu et al., 2013; Zhang et al., 2016b; Zhao et al., 2015). In another disease model, osteoarthritis, MiR-24 was identified as a negative regulator of p16 (Philipot et al., 2014). A rise in p16 levels following the loss of MiR-877-3p suppressed proliferation and tumorigenicity of bladder cancer by arresting cells in the G1 phase (Li et al., 2016b).

5.3 Impact on Pathology by SASP-Associated SA-MicroRNAs

Strong connections between SASP and human pathology are rapidly emerging. Among some prominent examples, lowering MiR-142-3p in macrophages contributed to increasing IL6 levels in aged mice and was proposed to enhance inflammation (Liu et al., 2016), and MiR-218-mediated suppression of the leucine-rich repeat-containing G protein-coupled receptor 4 (LGR4) was proposed to prevent the IL6-induced proliferation and invasion of prostate cancer cells (Li et al., 2016a). In addition, MiR-33a suppressed IL6-induced tumor metastasis of gallbladder cancer by inhibiting the expression of Twist, a transcription factor involved in epithelial-to-mesenchymal transition (Margetts, 2012; Zhang et al., 2016d). Interestingly, p16 suppressed tumorigenesis in breast stromal fibroblasts by repressing the expression and secretion of IL6. This effect was mediated by MiR-146b-5p, which bound to the 3'UTR of *IL6* mRNA and suppressed IL6 production (Al-Ansari and Aboussekhra, 2015). In patients with myasthenia gravis, low levels of MiR-181c were found to correlate with high serum levels of IL7, encoded by the MiR-181-target *IL7* mRNA (Zhang et al., 2016e). MiR-100-3p and MiR-877-3p suppressed production of IL8 and IL1B, respectively, in mesangial cells from IgA nephropathy patients (Liang et al., 2016). In a number of cultured neuroblastoma cells (SK-N-AS) and glioma cells (U251 and T98G), as well as in patients with glioma, MiR-93-5p suppressed the expression of IL8 and VEGF (Fabbri et al., 2015, 2016). The MiRNA that suppresses IRAK1 expression, MiR-146a, inhibited cell growth, lowered cell migration, and induced apoptosis in nonsmall cell lung cancer (Chen et al., 2013), although the role of IRAK1 in this phenotype was not examined. These reports highlight the extensive functional crosstalk among senescence pathways, inflammation, and disease by MiRNAs that modulate SASP.

The network of MiRNAs affecting senescence-associated MMPs has also been connected to cancer. MiR-143 and MiR-125b suppressed MMP13 expression and inhibited the growth of breast, ovarian, gastric, bladder, and other cancers (Han et al., 2013; Luo et al., 2015; Wang et al., 2016;

Wu et al., 2015; Zhai et al., 2016), although the contribution of MMP13 to these effects remains to be confirmed. In human osteosarcoma, MiR-143 lowered the expression of plasminogen activator inhibitor-1 (PAI-1, SERPINE1); the finding that PAI-1 promoted the production and secretion of MMP13, and thereby enhanced invasion and metastasis, is in keeping with the tumor-suppressor function of MiR-143 (Hirahata et al., 2016). Furthermore, as MiR-143 is a circulating MiRNA, it may serve as a biomarker for bladder cancer diagnosis (Motawi et al., 2016). Other MiRNAs like MiR-15b and MiR-152 reduced glioma cell invasion and angiogenesis by suppressing MMP3 and NRP2 (Zheng et al., 2013). MiR-181a-5p suppressed metastases by lowering MMP14 levels in breast cancer cells (Roth and Cao, 2015). It is worth noting that MiR-181a, by lowering SIRT1 and PRC2, was also implicated in activating p53 and inducing p16 levels, respectively, as explained earlier, suggesting that these pathways are coordinately regulated by MiRNAs.

6. CONCLUDING REMARKS AND PERSPECTIVES

The impact of senescent cells on tissue homeostasis is complex and evolves over time, as senescent cells accumulate in the human body with age. Senescent cells are beneficial in some instances, such as during wound healing, muscle regeneration, suppression of pancreatic and liver fibrosis, and prevention of tumorigenesis in young organisms. However, senescent cells can be detrimental in other instances, particularly in the elderly, as they contribute to the development of age-related declines and diseases such as cancer, diabetes, cardiovascular disorders, diminished immunity, and neurodegeneration. Thus, there is rising interest in understanding the molecular regulators of senescence so that we can develop therapeutic approaches that promote its beneficial actions and ameliorate its harmful effects. To help achieve this larger objective, we have reviewed the MiRNAs that promote or suppress senescence.

We discussed SA-MiRNAs and their impact on cellular senescence by focusing on MiRNAs that enhance or diminish the SASP as well as the main senescence pathways, p53/p21 and p16/RB (Tables 1–3). Some SA-MiRNAs function as direct regulators, for instance, MiR-25 and MiR-30d suppressed senescence by directly targeting the *p53* mRNA, while others function as indirect regulators, as shown for MiR-34a and MiR-22, which induced senescence by downregulating SIRT1 levels, which in turn enhanced p53 function, or for MiR-138, which targeted

EZH2 leading to transcriptional upregulation of p16 expression. Other SA-MiRNAs, such as MiR-146a and MiR-146b, can modulate SASP, although there is increasing evidence that SA-MiRNAs can coordinately influence multiple senescent traits (Fig. 1).

Importantly, the majority of SA-MiRNAs have been investigated in cell culture models, and so it is important to assess if they change with age and whether they have an impact upon aging and age-related diseases. The development of mouse models of senescence and MiRNA function is starting to enable the systematic analysis of SA-MiRNA function in vivo, although much work is still needed. Understanding the impact of SA-MiRNAs on aging- and age-related diseases will also illuminate their diagnostic and therapeutic potential. For example, SA-MiRNAs in bodily fluids such as blood, urine, and saliva could serve as diagnostic or prognostic markers, perhaps helping to identify disease processes influenced by senescence.

Targeting MiRNAs by gain- or loss-of-function for therapeutic purposes is also an attractive proposition. Despite progress in this area, important challenges remain with the delivery of therapeutic MiRNA-relevant reagents with high specificity and efficacy (Chen et al., 2015). The gain-of-function approach using MiRNA mimics is widely used; however, the outcome should be cautiously interpreted as they may lead to nonspecific changes in gene expression (Jin et al., 2015). The loss-of-function approach using antisense oligos has often been used to inhibit MiRNA activity, but they can also target other sequences with few mismatches resulting in nonspecific effects. To overcome the above limitations, gene knockout would be an ideal choice for MiRNA deletion for functional analysis. Recently, genome-editing technologies based on zinc finger nucleases (ZFNs), transcription activator-like effector nucleases (TALENs), and CRISPR-Cas9-RNA-guided endonucleases (RGENs) have been widely used to knockout genomic loci of interest (Cai and Yang, 2014). Expansion of these technologies to study MiRNAs is emerging (Chang et al., 2016; Kim et al., 2013). Thus, using such new techniques to target MiRNAs will provide an excellent opportunity to further elucidate the role of SA-MiRNAs in vitro and in vivo.

Understanding the full spectrum of transcriptional and posttranscriptional regulators of SA-MiRNAs will further help with accurate targeting of the senescent phenotype. The studies discussed here highlight the efforts underway to exploit senescence regulatory factors toward favorable clinical outcomes in cancer and other diseases influenced by senescent cells.

ACKNOWLEDGMENTS

This work was supported in full by the National Institute on Aging—Intramural Research Program of the National Institutes of Health.

REFERENCES

Abdelmohsen, K., Gorospe, M., 2015. Noncoding RNA control of cellular senescence. Wiley Interdiscip. Rev. RNA 6, 615–629.

Abdelmohsen, K., Srikantan, S., Tominaga, K., Kang, M.J., Yaniv, Y., Martindale, J.L., Yang, X.L., Park, S.S., Becker, K.G., Subramanian, M., Maudsley, S., Lal, A., Gorospe, M., 2012. Growth inhibition by miR-519 via multiple p21-inducing pathways. Mol. Cell. Biol. 32, 2530–2548.

Acosta, J.C., O'Loghlen, A., Banito, A., Guijarro, M.V., Augert, A., Raguz, S., Fumagalli, M., Da Costa, M., Brown, C., Popov, N., Takatsu, Y., Melamed, J., di Fagagna, F.D., Bernard, D., Hernando, E., Gil, J., 2008. Chemokine signaling via the CXCR2 receptor reinforces senescence. Cell 133, 1006–1018.

Al-Ansari, M.M., Aboussekhra, A., 2015. miR-146b-5p mediates p16-dependent repression of IL-6 and suppresses paracrine procarcinogenic effects of breast stromal fibroblasts. Oncotarget 6, 30006–30016.

Al-Khalaf, H.H., Mohideen, P., Nallar, S.C., Kalvakolanu, D.V., Aboussekhra, A., 2013. The cyclin-dependent kinase inhibitor p16INK4a physically interacts with transcription factor Sp1 and cyclin-dependent kinase 4 to transactivate microRNA-141 and microRNA-146b-5p spontaneously and in response to ultraviolet light-induced DNA damage. J. Biol. Chem. 288, 35511–35525.

Bai, X.Y., Ma, Y.X., Ding, R., Fu, B., Shi, S.Z., Chen, X.M., 2011. miR-335 and miR-34a promote renal senescence by suppressing mitochondrial antioxidative enzymes. J. Am. Soc. Nephrol. 22, 1252–1261.

Baker, D.J., Wijshake, T., Tchkonia, T., LeBrasseur, N.K., Childs, B.G., van de Sluis, B., Kirkland, J.L., van Deursen, J.M., 2011. Clearance of p16(Ink4a)-positive senescent cells delays ageing-associated disorders. Nature 479, 232–236.

Bartel, D.P., 2009. MicroRNAs: target recognition and regulatory functions. Cell 136, 215–233.

Ben-Porath, I., Weinberg, R.A., 2005. The signals and pathways activating cellular senescence. Int. J. Biochem. Cell Biol. 37, 961–976.

Bhaumik, D., Scott, G.K., Schokrpur, S., Patil, C.K., Orjalo, A.V., Rodier, F., Lithgow, G.J., Campisi, J., 2009. MicroRNAs miR-146a/b negatively modulate the senescence-associated inflammatory mediators IL-6 and IL-8. Aging 1, 402–411.

Borgdorff, V., Lleonart, M.E., Bishop, C.L., Fessart, D., Bergin, A.H., Overhoff, M.G., Beach, D.H., 2010. Multiple microRNAs rescue from Ras-induced senescence by inhibiting p21(Waf1/Cip1). Oncogene 29, 2262–2271.

Bracken, A.P., Kleine-Kohlbrecher, D., Dietrich, N., Pasini, D., Gargiulo, G., Beekman, C., Theilgaard-Monch, K., Minucci, S., Porse, B.T., Marine, J.C., Hansen, K.H., Helin, K., 2007. The polycomb group proteins bind throughout the INK4A-ARF locus and are disassociated in senescent cells. Genes Dev. 21, 525–530.

Braig, M., Lee, S., Loddenkemper, C., Rudolph, C., Peters, A.H.F.M., Schlegelberger, B., Stein, H., Dorken, B., Jenuwein, T., Schmitt, C.A., 2005. Oncogene-induced senescence as an initial barrier in lymphoma development. Nature 436, 660–665.

Brooks, C.L., Gu, W., 2009. How does SIRT1 affect metabolism, senescence and cancer? Nat. Rev. Cancer 9, 123–128.

Bushati, N., Cohen, S.M., 2007. microRNA functions. Annu. Rev. Cell Dev. Biol. 23, 175–205.

Cai, M., Yang, Y., 2014. Targeted genome editing tools for disease modeling and gene therapy. Curr. Gene Ther. 14, 2–9.
Campisi, J., 1997. Aging and cancer: the double-edged sword of replicative senescence. J. Am. Geriatr. Soc. 45, 482–488.
Campisi, J., 2005. Senescent cells, tumor suppression, and organismal aging: good citizens, bad neighbors. Cell 120, 513–522.
Catalanotto, C., Cogoni, C., Zardo, G., 2016. MicroRNA in control of gene expression: an overview of nuclear functions. Int. J. Mol. Sci. 17, pii: E1712.
Chang, H., Yi, B., Ma, R.X., Zhang, X.G., Zhao, H.Y., Xi, Y.G., 2016. CRISPR/cas9, a novel genomic tool to knock down microRNA in vitro and in vivo. Sci. Rep. 6, 22312.
Chen, R., Alvero, A.B., Silasi, D.A., Kelly, M.G., Fest, S., Visintin, I., Leiser, A., Schwartz, P.E., Rutherford, T., Mor, G., 2008. Regulation of IKKbeta by miR-199a affects NF-kappaB activity in ovarian cancer cells. Oncogene 27, 4712–4723.
Chen, G., Umelo, I.A., Lv, S.S., Teugels, E., Fostier, K., Kronenberger, P., Dewaele, A., Sadones, J., Geers, C., De Greve, J., 2013. miR-146a inhibits cell growth, cell migration and induces apoptosis in non-small cell lung cancer cells. Plos One 8, e60317.
Chen, L., Lu, M.H., Zhang, D., Hao, N.B., Fan, Y.H., Wu, Y.Y., Wang, S.M., Xie, R., Fang, D.C., Zhang, H., Hu, C.J., Yang, S.M., 2014. miR-1207-5p and miR-1266 suppress gastric cancer growth and invasion by targeting telomerase reverse transcriptase. Cell Death Dis. 5, e1034.
Chen, Y.C., Gao, D.Y., Huang, L., 2015. In vivo delivery of miRNAs for cancer therapy: challenges and strategies. Adv. Drug Deliver. Rev. 81, 128–141.
Chen, C., Lu, Z., Yang, J., Hao, W., Qin, Y., Wang, H., Xie, C., Xie, R., 2016a. MiR-17-5p promotes cancer cell proliferation and tumorigenesis in nasopharyngeal carcinoma by targeting p21. Cancer Med. 5, 3489–3499.
Chen, L., Luo, L., Chen, W., Xu, H.X., Chen, F., Chen, L.Z., Zeng, W.T., Chen, J.S., Huang, X.H., 2016b. MicroRNA-24 increases hepatocellular carcinoma cell metastasis and invasion by targeting p53: miR-24 targeted p53. Biomed. Pharmacother. 84, 1113–1118.
Childs, B.G., Durik, M., Baker, D.J., van Deursen, J.M., 2015. Cellular senescence in aging and age-related disease: from mechanisms to therapy. Nat. Med. 21, 1424–1435.
Chinta, S.J., Woods, G., Rane, A., Demaria, M., Campisi, J., Andersen, J.K., 2015. Cellular senescence and the aging brain. Exp. Gerontol. 68, 3–7.
Cho, J.G., Park, S., Lim, C.H., Kim, H.S., Song, S.Y., Roh, T.Y., Sung, J.H., Suh, W., Ham, S.J., Lim, K.H., Park, S.G., 2016. ZNF224, Kruppel like zinc finger protein, induces cell growth and apoptosis-resistance by down-regulation of p21 and p53 via miR-663a. Oncotarget 7, 31177–31190.
Choi, J., Shendrik, I., Peacocke, M., Peehl, D., Buttyan, R., Ikeguchi, E.F., Katz, A.E., Benson, M.C., 2000. Expression of senescence-associate beta-galactosidase in enlarged prostates from men with benign prostatic hyperplasia. Urology 56, 160–166.
Collado, M., Gil, J., Efeyan, A., Guerra, C., Schuhmacher, A.J., Barradas, M., Benguria, A., Zaballos, A., Flores, J.M., Barbacid, M., Beach, D., Serrano, M., 2005. Tumour biology—senescence in premalignant tumours. Nature 436, 642.
Coppe, J.P., Kauser, K., Campisi, J., Beausejour, C.M., 2006. Secretion of vascular endothelial growth factor by primary human fibroblasts at senescence. J. Biol. Chem. 281, 29568–29574.
Coppe, J.P., Patil, C.K., Rodier, F., Sun, Y., Munoz, D.P., Goldstein, J., Nelson, P.S., Desprez, P.Y., Campisi, J., 2008. Senescence-associated secretory phenotypes reveal cell-nonautonomous functions of oncogenic RAS and the p53 tumor suppressor. PLoS Biol. 6, 2853–2868.
Coppe, J.P., Desprez, P.Y., Krtolica, A., Campisi, J., 2010. The senescence-associated secretory phenotype: the dark side of tumor suppression. Annu. Rev. Pathol. Mech. Dis. 5, 99–118.

Coster, G., Goldberg, M., 2010. The cellular response to DNA damage: a focus on MDC1 and its interacting proteins. Nucleus 1, 166–178.

Cristofalo, V.J., Lorenzini, A., Allen, R.G., Torres, C., Tresini, M., 2004. Replicative senescence: a critical review. Mech. Ageing Dev. 125, 827–848.

Cufi, S., Vazquez-Martin, A., Oliveras-Ferraros, C., Quirantes, R., Segura-Carretero, A., Micol, V., Joven, J., Bosch-Barrera, J., Del Barco, S., Martin-Castillo, B., Vellon, L., Menendez, J.A., 2012. Metformin lowers the threshold for stress-induced senescence: a role for the microRNA-200 family and miR-205. Cell Cycle 11, 1235–1246.

Cui, H., Ge, J., Xie, N., Banerjee, S., Zhou, Y., Antony, V.B., Thannickal, V.J., Liu, G., 2016. miR-34a inhibits lung fibrosis by inducing lung fibroblast senescence. Am. J. Respir. Cell Mol. Biol. 56, 168–178.

Cunningham, S.C., Gallmeier, E., Kern, S.E., 2010. MKK4 as oncogene or tumor supressor: in cancer and senescence, the story's getting old. Aging 2, 752–753.

Dankort, D., Curley, D.P., Cartlidge, R.A., Nelson, B., Karnezis, A.N., Damsky, W.E., You, M.J., DePinho, R.A., McMahon, M., Bosenberg, M., 2009. Braf(V600E) cooperates with Pten loss to induce metastatic melanoma. Nat. Genet. 41, 544–552.

Dar, A.A., Majid, S., de Semir, D., Nosrati, M., Bezrookove, V., Kashani-Sabet, M., 2011. miRNA-205 suppresses melanoma cell proliferation and induces senescence via regulation of E2F1 protein. J. Biol. Chem. 286, 16606–16614.

Davalos, A.R., Kawahara, M., Malhotra, G.K., Schaum, N., Huang, J.H., Ved, U., Beausejour, C.M., Coppe, J.P., Rodier, F., Campisi, J., 2013. p53-dependent release of Alarmin HMGB1 is a central mediator of senescent phenotypes. J. Cell Biol. 201, 613–629.

Demaria, M., Ohtani, N., Youssef, S.A., Rodier, F., Toussaint, W., Mitchell, J.R., Laberge, R.M., Vijg, J., Van Steeg, H., Dolle, M.E.T., Hoeijmakers, J.H.J., de Bruin, A., Hara, E., Campisi, J., 2014. An essential role for senescent cells in optimal wound healing through secretion of PDGF-AA. Dev. Cell 31, 722–733.

Deng, G., Sui, G., 2013. Noncoding RNA in oncogenesis: a new era of identifying key players. Int. J. Mol. Sci. 14, 18319–18349.

Dimri, G.P., Lee, X.H., Basile, G., Acosta, M., Scott, C., Roskelley, C., Medrano, E.E., Linskens, M., Rubelj, I., Pereirasmith, O., Peacocke, M., Campisi, J., 1995. A biomarker that identifies senescent human-cells in culture and in aging skin in-vivo. Proc. Natl. Acad. Sci. U.S.A. 92, 9363–9367.

Dimri, M., Carroll, J.D., Cho, J.H., Dimri, G.P., 2013. microRNA-141 regulates BMI1 expression and induces senescence in human diploid fibroblasts. Cell Cycle 12, 3537–3546.

Fabbri, M., Paone, A., Calore, F., Galli, R., Gaudio, E., Santhanam, R., Lovat, F., Fadda, P., Mao, C., Nuovo, G.J., Zanesi, N., Crawford, M., Ozer, G.H., Wernicke, D., Alder, H., Caligiuri, M.A., Nana-Sinkam, P., Perrotti, D., Croce, C.M., 2012. MicroRNAs bind to Toll-like receptors to induce prometastatic inflammatory response. Proc. Natl. Acad. Sci. U.S.A. 109, E2110–E2116.

Fabbri, E., Brognara, E., Montagner, G., Ghimenton, C., Eccher, A., Cantu, C., Khalil, S., Bezzerri, V., Provezza, L., Bianchi, N., Finotti, A., Borgatti, M., Moretto, G., Chilosi, M., Cabrini, G., Gambari, R., 2015. Regulation of IL-8 gene expression in gliomas by microRNA miR-93. BMC Cancer 15, 661.

Fabbri, E., Montagner, G., Bianchi, N., Finotti, A., Borgatti, M., Lampronti, I., Cabrini, G., Gambari, R., 2016. MicroRNA miR-93-5p regulates expression of IL-8 and VEGF in neuroblastoma SK-N-AS cells. Oncol. Rep. 35, 2866–2872.

Fabian, M.R., Sonenberg, N., Filipowicz, W., 2010. Regulation of mRNA translation and stability by microRNAs. Annu. Rev. Biochem. 79, 351–379.

Freund, A., Orjalo, A.V., Desprez, P.Y., Campisi, J., 2010. Inflammatory networks during cellular senescence: causes and consequences. Trends Mol. Med. 16, 238–246.

Freund, A., Laberge, R.M., Demaria, M., Campisi, J., 2012. Lamin B1 loss is a senescence-associated biomarker. Mol. Biol. Cell 23, 2066–2075.

Gey, C., Seeger, K., 2013. Metabolic changes during cellular senescence investigated by proton NMR-spectroscopy. Mech. Ageing Dev. 134, 130–138.

Giglio, S., Cirombella, R., Amodeo, R., Portaro, L., Lavra, L., Vecchione, A., 2013. MicroRNA miR-24 promotes cell proliferation by targeting the CDKs inhibitors p27Kip1 and p16INK4a. J. Cell. Physiol. 228, 2015–2023.

Goldstein, S., 1990. Replicative senescence: the human fibroblast comes of age. Science 249, 1129–1133.

Gorospe, M., Abdelmohsen, K., 2011. MicroRegulators come of age in senescence. Trends Genet. 27, 233–241.

Han, Y., Liu, Y., Zhang, H., Wang, T., Diao, R., Jiang, Z., Gui, Y., Cai, Z., 2013. Hsa-miR-125b suppresses bladder cancer development by down-regulating oncogene SIRT7 and oncogenic long non-coding RNA MALAT1. FEBS Lett. 587, 3875–3882.

Hayflick, L., 1965. The limited in vitro lifetime of human diploid cell strains. Exp. Cell Res. 37, 614–636.

Hayflick, L., 1992. Aging, longevity, and immortality in vitro. Exp. Gerontol. 27, 363–368.

Hayflick, L., Moorhead, P.S., 1961. The serial cultivation of human diploid cell strains. Exp. Cell Res. 25, 585–621.

Herbig, U., Jobling, W.A., Chen, B.P., Chen, D.J., Sedivy, J.M., 2004. Telomere shortening triggers senescence of human cells through a pathway involving ATM, p53, and p21 (CIP1), but not p16(INK4a). Mol. Cell 14, 501–513.

Herbig, U., Ferreira, M., Condel, L., Carey, D., Sedivy, J.M., 2006. Cellular senescence in aging primates. Science 311, 1257.

Hirahata, M., Osaki, M., Kanda, Y., Sugimoto, Y., Yoshioka, Y., Kosaka, N., Takeshita, F., Fujiwara, T., Kawai, A., Ito, H., Ochiya, T., Okada, F., 2016. PAI-1, a target gene of miR-143, regulates invasion and metastasis by upregulating MMP-13 expression of human osteosarcoma. Cancer Med. 5, 892–902.

Ivanovska, I., Ball, A.S., Diaz, R.L., Magnus, J.F., Kibukawa, M., Schelter, J.M., Kobayashi, S.V., Lim, L., Burchard, J., Jackson, A.L., Linsley, P.S., Cleary, M.A., 2008. MicroRNAs in the miR-106b family regulate p21/CDKN1A and promote cell cycle progression. Mol. Cell. Biol. 28, 2167–2174.

Jazbutyte, V., Fiedler, J., Kneitz, S., Galuppo, P., Just, A., Holzmann, A., Bauersachs, J., Thum, T., 2013. MicroRNA-22 increases senescence and activates cardiac fibroblasts in the aging heart. Age 35, 747–762.

Jeyapalan, J.C., Ferreira, M., Sedivy, J.A., Herbig, U., 2007. Accumulation of senescent cells in mitotic tissue of aging primates. Mech. Ageing Dev. 128, 36–44.

Jin, H.Y., Gonzalez-Martin, A., Miletic, A.V., Lai, M.Y., Knight, S., Sabouri-Ghomi, M., Head, S.R., Macauley, M.S., Rickert, R.C., Xiao, C.C., 2015. Transfection of microRNA mimics should be used with caution. Front. Genet. 6, 340.

Jones, M., Lal, A., 2012. MicroRNAs, wild-type and mutant p53: more questions than answers. RNA Biol. 9, 781–791.

Kabir, F.M.L., DeInnocentes, P., Bird, R.C., 2015. Altered microRNA expression profiles and regulation of INK4A/CDKN2A tumor suppressor genes in canine breast cancer models. J. Cell. Biochem. 116, 2956–2969.

Kang, M.R., Park, K.H., Yang, J.O., Lee, C.W., Oh, S.J., Yun, J., Lee, M.Y., Han, S.B., Kang, J.S., 2016. miR-6734 up-regulates p21 gene expression and induces cell cycle arrest and apoptosis in colon cancer cells. PloS One 11, e0160961.

Kim, V.N., Han, J., Siomi, M.C., 2009. Biogenesis of small RNAs in animals. Nat. Rev. Mol. Cell. Biol. 10, 126–139.

Kim, Y.K., Wee, G., Park, J., Kim, J., Baek, D., Kim, J.S., Kim, V.N., 2013. TALEN-based knockout library for human microRNAs. Nat. Struct. Mol. Biol. 20, 1458–1464.

Kim, Y.H., Lee, W.K., Lee, E.B., Son, J.W., Kim, D.S., Park, J.Y., 2016. Combined effect of metastasis-related microRNA, miR-34 and miR-124 family, methylation on prognosis of non-small-cell lung cancer. Clin. Lung Cancer 18, e13–e20.

Kloosterman, W.P., Plasterk, R.H.A., 2006. The diverse functions of MicroRNAs in animal development and disease. Dev. Cell 11, 441–450.

Krell, J., Frampton, A.E., Colombo, T., Gall, T.M., De Giorgio, A., Harding, V., Stebbing, J., Castellano, L., 2013. The p53 miRNA interactome and its potential role in the cancer clinic. Epigenomics 5, 417–428.

Krishnamurthty, J., Torrice, C., Ramsey, M.R., Kovalev, G.I., Al-Regaiey, K., Su, L.S., Sharpless, N.E., 2004. Ink4a/Arf expression is a biomarker of aging. J. Clin. Invest. 114, 1299–1307.

Krishnamurthy, J., Ramsey, M.R., Ligon, K.L., Torrice, C., Koh, A., Bonner-Weir, S., Sharpless, N.E., 2006. p16INK4a induces an age-dependent decline in islet regenerative potential. Nature 443, 453–457.

Krizhanovsky, V., Yon, M., Dickins, R.A., Hearn, S., Simon, J., Miething, C., Yee, H., Zender, L., Lowe, S.W., 2008. Senescence of activated stellate cells limits liver fibrosis. Cell 134, 657–667.

Kuilman, T., Peeper, D.S., 2009. Senescence-messaging secretome: SMS-ing cellular stress. Nat. Rev. Cancer 9, 81–94.

Kuilman, T., Michaloglou, C., Vredeveld, L.C.W., Douma, S., van Doom, R., Desmet, C.J., Aarden, L.A., Mooi, W.J., Peeper, D.S., 2008. Oncogene-induced senescence relayed by an interleukin-dependent inflammatory network. Cell 133, 1019–1031.

Kuilman, T., Michaloglou, C., Mooi, W.J., Peeper, D.S., 2010. The essence of senescence. Genes Dev. 24, 2463–2479.

Kumamoto, K., Spillare, E.A., Fujita, K., Horikawa, I., Yamashita, T., Appella, E., Nagashima, M., Takenoshita, S., Yokota, J., Harris, C.C., 2008. Nutlin-3a activates p53 to both down-regulate inhibitor of growth 2 and up-regulate mir-34a, mir-34b, and mir-34c expression, and induce senescence. Cancer Res. 68, 3193–3203.

Kumar, M., Lu, Z., Takwi, A.A., Chen, W., Callander, N.S., Ramos, K.S., Young, K.H., Li, Y., 2011. Negative regulation of the tumor suppressor p53 gene by microRNAs. Oncogene 30, 843–853.

Lai, X., Wolkenhauer, O., Vera, J., 2012. Modeling miRNA regulation in cancer signaling systems: miR-34a regulation of the p53/Sirt1 signaling module. Methods Mol. Biol. 880, 87–108.

Lal, A., Kim, H.H., Abdelmohsen, K., Kuwano, Y., Pullmann, R., Srikantan, S., Subrahmanyam, R., Martindale, J.L., Yang, X.L., Ahmed, F., Navarro, F., Dykxhoorn, D., Lieberman, J., Gorospe, M., 2008. p16(INK4a) translation suppressed by miR-24. Plos One 3, e1864.

Le, M.T.N., Teh, C., Shyh-Chang, N., Xie, H.M., Zhou, B.Y., Korzh, V., Lodish, H.F., Lim, B., 2009. MicroRNA-125b is a novel negative regulator of p53. Genes Dev. 23, 862–876.

Lee, J.H., Park, S.J., Jeong, S.Y., Kim, M.J., Jun, S., Lee, H.S., Chang, I.Y., Lim, S.C., Yoon, S.P., Yong, J., You, H.J., 2015. MicroRNA-22 suppresses DNA repair and promotes genomic instability through targeting of MDC1. Cancer Res. 75, 1298–1310.

Li, B., Shi, X.B., Nori, D., Chao, C.K., Chen, A.M., Valicenti, R., White Rde, V., 2011. Down-regulation of microRNA 106b is involved in p21-mediated cell cycle arrest in response to radiation in prostate cancer cells. Prostate 71, 567–574.

Li, F., Gu, C., Tian, F., Jia, Z., Meng, Z., Ding, Y., Yang, J., 2016a. MiR-218 impedes IL-6-induced prostate cancer cell proliferation and invasion via suppression of LGR4 expression. Oncol. Rep. 35, 2859–2865.

Li, S., Zhu, Y., Liang, Z., Wang, X., Meng, S., Xu, X., Xu, X., Wu, J., Ji, A., Hu, Z., Lin, Y., Chen, H., Mao, Y., Wang, W., Zheng, X., Liu, B., Xie, L., 2016b.

Up-regulation of p16 by miR-877-3p inhibits proliferation of bladder cancer. Oncotarget 7, 51773–51783.
Liang, J., Zhang, Y., Jiang, G., Liu, Z., Xiang, W., Chen, X., Chen, Z., Zhao, J., 2013. MiR-138 induces renal carcinoma cell senescence by targeting EZH2 and is down-regulated in human clear cell renal cell carcinoma. Oncol. Res. 21, 83–91.
Liang, Y., Zhao, G., Tang, L., Zhang, J., Li, T., Liu, Z., 2016. MiR-100-3p and miR-877-3p regulate overproduction of IL-8 and IL-1beta in mesangial cells activated by secretory IgA from IgA nephropathy patients. Exp. Cell Res. 347, 312–321.
Lingor, P., 2010. Regulation of Cell Death and Survival by RNA Interference—The Roles of miRNA and siRNA. In: Apoptosome: An up-and-Coming Therapeutical Tool, Springer, New York, pp. 95–117.
Liu, X., Yu, J., Jiang, L., Wang, A., Shi, F., Ye, H., Zhou, X., 2009. MicroRNA-222 regulates cell invasion by targeting matrix metalloproteinase 1 (MMP1) and manganese superoxide dismutase 2 (SOD2) in tongue squamous cell carcinoma cell lines. Cancer Genomics Proteomics 6, 131–139.
Liu, Y., Song, X., Meng, S., Jiang, M., 2016. Downregulated expression of miR-142-3p in macrophages contributes to increased IL-6 levels in aged mice. Mol. Immunol. 80, 11–16.
Loaiza, N., Demaria, M., 2016. Cellular senescence and tumor promotion: is aging the key? Biochim. Biophys. Acta 1865, 155–167.
Luo, X.G., Ding, J.Q., Chen, S.D., 2010. Microglia in the aging brain: relevance to neurodegeneration. Mol. Neurodegener. 5, 12.
Luo, S., Wang, J., Ma, Y., Yao, Z., Pan, H., 2015. PPARgamma inhibits ovarian cancer cells proliferation through upregulation of miR-125b. Biochem. Biophys. Res. Commun. 462, 85–90.
Mao, A., Liu, Y., Wang, Y., Zhao, Q., Zhou, X., Sun, C., Di, C., Si, J., Gan, L., Zhang, H., 2016. miR-449a enhances radiosensitivity through modulating pRb/E2F1 in prostate cancer cells. Tumour Biol. 37, 4831–4840.
Marasa, B.S., Srikantan, S., Masuda, K., Abdelmohsen, K., Kuwano, Y., Yang, X., Martindale, J.L., Rinker-Schaeffer, C.W., Gorospe, M., 2009. Increased MKK4 abundance with replicative senescence is linked to the joint reduction of multiple microRNAs. Sci. Signal. 2, ra69.
Marasa, B.S., Srikantan, S., Martindale, J.L., Kim, M.M., Lee, E.K., Gorospe, M., Abdelmohsen, K., 2010. MicroRNA profiling in human diploid fibroblasts uncovers miR-519 role in replicative senescence. Aging 2, 333–343.
Margetts, P.J., 2012. Twist: a new player in the epithelial-mesenchymal transition of the peritoneal mesothelial cells. Nephrol. Dial. Transplant. 27, 3978–3981.
Martinez, I., Cazalla, D., Almstead, L.L., Steitz, J.A., DiMaio, D., 2011. miR-29 and miR-30 regulate B-Myb expression during cellular senescence. Proc. Natl. Acad. Sci. U.S.A. 108, 522–527.
Massague, J., 2008. TGF beta in cancer. Cell 134, 215–230.
Mehta, I.S., Figgitt, M., Clements, C.S., Kill, I.R., Bridger, J.M., 2007. Alterations to nuclear architecture and genome behavior in senescent cells. Ann. N. Y. Acad. Sci. 1100, 250–263.
Menghini, R., Casagrande, V., Cardellini, M., Martelli, E., Terrinoni, A., Amati, F., Vasa-Nicotera, M., Ippoliti, A., Novelli, G., Melino, G., Lauro, R., Federici, M., 2009. MicroRNA 217 modulates endothelial cell senescence via silent information regulator 1. Circulation 120, 1524–1532.
Michaloglou, C., Vredeveld, L.C.W., Soengas, M.S., Denoyelle, C., Kuilman, T., van der Horst, C.M.A.M., Majoor, D.M., Shay, J.W., Mooi, W.J., Peeper, D.S., 2005. BRAF (E600)-associated senescence-like cell cycle arrest of human naevi. Nature 436, 720–724.

Ming, G.F., Wu, K., Hu, K., Chen, Y., Xiao, J., 2016. NAMPT regulates senescence, proliferation, and migration of endothelial progenitor cells through the SIRT1 AS lncRNA/miR-22/SIRT1 pathway. Biochem. Biophys. Res. Commun. 478, 1382–1388.

Misso, G., Di Martino, M.T., De Rosa, G., Farooqi, A.A., Lombardi, A., Campani, V., Zarone, M.R., Gulla, A., Tagliaferri, P., Tassone, P., Caraglia, M., 2014. Mir-34: a new weapon against cancer? Mol. Ther. Nucleic Acids 3, e194.

Mittal, S., Aslam, A., Doidge, R., Medica, R., Winkler, G.S., 2011. The Ccr4a (CNOT6) and Ccr4b (CNOT6L) deadenylase subunits of the human Ccr4-Not complex contribute to the prevention of cell death and senescence. Mol. Biol. Cell 22, 748–758.

Molofsky, A.V., Slutsky, S.G., Joseph, N.M., He, S., Pardal, R., Krishnamurthy, J., Sharpless, N.E., Morrison, S.J., 2006. Increasing p16INK4a expression decreases forebrain progenitors and neurogenesis during ageing. Nature 443, 448–452.

Momand, J., Wu, H.H., Dasgupta, G., 2000. MDM2—master regulator of the p53 tumor suppressor protein. Gene 242, 15–29.

Motawi, T.K., Rizk, S.M., Ibrahim, T.M., Ibrahim, I.A., 2016. Circulating microRNAs, miR-92a, miR-100 and miR-143, as non-invasive biomarkers for bladder cancer diagnosis. Cell Biochem. Funct. 34, 142–148.

Mudhasani, R., Zhu, Z., Hutvagner, G., Eischen, C.M., Lyle, S., Hall, L.L., Lawrence, J.B., Imbalzano, A.N., Jones, S.N., 2008. Loss of miRNA biogenesis induces p19Arf-p53 signaling and senescence in primary cells. J. Cell Biol. 181, 1055–1063.

Nakanishi, M., Ozaki, T., Yamamoto, H., Hanamoto, T., Kikuchi, H., Furuya, K., Asaka, M., Delia, D., Nakagawara, A., 2007. NFBD1/MDC1 associates with p53 and regulates its function at the crossroad between cell survival and death in response to DNA damage. J. Biol. Chem. 282, 22993–23004.

Narita, M., Nunez, S., Heard, E., Narita, M., Lin, A.W., Hearn, S.A., Spector, D.L., Hannon, G.J., Lowe, S.W., 2003. Rb-mediated heterochromatin formation and silencing of E2F target genes during cellular senescence. Cell 113, 703–716.

Navarro, A., Lopez-Cepero, J.M., del Pino, M.J.S., 2001. Skeletal muscle and aging. Front. Biosci. 6, D26–D44.

Newman, M.A., Hammond, S.M., 2010. Emerging paradigms of regulated microRNA processing. Gene Dev. 24, 1086–1092.

Ning, Z., Zhu, H., Li, F., Liu, Q., Liu, G., Tan, T., Zhang, B., Chen, S., Li, G., Huang, D., Meltzer, S.J., Zhang, H., 2014. Tumor suppression by miR-31 in esophageal carcinoma is p21-dependent. Genes Cancer 5, 436–444.

Noguchi, S., Mori, T., Otsuka, Y., Yamada, N., Yasui, Y., Iwasaki, J., Kumazaki, M., Maruo, K., Akao, Y., 2012. Anti-oncogenic microRNA-203 induces senescence by targeting E2F3 protein in human melanoma cells. J. Biol. Chem. 287, 11769–11777.

Noonan, E.J., Place, R.F., Basak, S., Pookot, D., Li, L.C., 2010. miR-449a causes Rb-dependent cell cycle arrest and senescence in prostate cancer cells. Oncotarget 1, 349–358.

Ohtani, N., Takahashi, A., Mann, D.J., Hara, E., 2012. Cellular senescence: a double-edged sword in the fight against cancer. Exp. Dermatol. 21 (Suppl. 1), 1–4.

Okada, M., Kim, H.W., Matsu-ura, K., Wang, Y.G., Xu, M., Ashraf, M., 2016. Abrogation of age-induced microRNA-195 rejuvenates the senescent mesenchymal stem cells by reactivating telomerase. Stem Cells 34, 148–159.

O'Loghlen, A., Brookes, S., Martin, N., Rapisarda, V., Peters, G., Gil, J., 2015. CBX7 and miR-9 are part of an autoregulatory loop controlling p16(INK) (4a). Aging Cell 14, 1113–1121.

Osaki, M., Takeshita, F., Sugimoto, Y., Kosaka, N., Yamamoto, Y., Yoshioka, Y., Kobayashi, E., Yamada, T., Kawai, A., Inoue, T., Ito, H., Oshimura, M., Ochiya, T., 2011. MicroRNA-143 regulates human osteosarcoma metastasis by regulating matrix metalloprotease-13 expression. Mol. Ther. 19, 1123–1130.

Overhoff, M.G., Garbe, J.C., Koh, J., Stampfer, M.R., Beach, D.H., Bishop, C.L., 2014. Cellular senescence mediated by p16INK4A-coupled miRNA pathways. Nucleic Acids Res. 42, 1606–1618.

Philipot, D., Guerit, D., Platano, D., Chuchana, P., Olivotto, E., Espinoza, F., Dorandeu, A., Pers, Y.M., Piette, J., Borzi, R.M., Jorgensen, C., Noel, D., Brondello, J.M., 2014. p16(INK4a) and its regulator miR-24 link senescence and chondrocyte terminal differentiation-associated matrix remodeling in osteoarthritis. Arthritis Res. Ther. 16, R58.

Piette, J., Neel, H., Marechal, V., 1997. Mdm2: keeping p53 under control. Oncogene 15, 1001–1010.

Raut, S.K., Singh, G.B., Rastogi, B., Saikia, U.N., Mittal, A., Dogra, N., Singh, S., Prasad, R., Khullar, M., 2016. miR-30c and miR-181a synergistically modulate p53-p21 pathway in diabetes induced cardiac hypertrophy. Mol. Cell. Biochem. 417, 191–203.

Ren, X.S., Yin, M.H., Zhang, X., Wang, Z., Feng, S.P., Wang, G.X., Luo, Y.J., Liang, P.Z., Yang, X.Q., He, J.X., Zhang, B.L., 2014. Tumor-suppressive microRNA-449a induces growth arrest and senescence by targeting E2F3 in human lung cancer cells. Cancer Lett. 344, 195–203.

Rossato, M., Curtale, G., Tamassia, N., Castellucci, M., Mori, L., Gasperini, S., Mariotti, B., De Luca, M., Mirolo, M., Cassatella, M.A., Locati, M., Bazzoni, F., 2012. IL-10-induced microRNA-187 negatively regulates TNF-alpha, IL-6, and IL-12p40 production in TLR4-stimulated monocytes. Proc. Natl. Acad. Sci. U.S.A. 109, E3101–E3110.

Rosso, A., Balsamo, A., Gambino, R., Dentelli, P., Falcioni, R., Cassader, M., Pegoraro, L., Pagano, G., Brizzi, M.F., 2006. p53 mediates the accelerated onset of senescence of endothelial progenitor cells in diabetes. J. Biol. Chem. 281, 4339–4347.

Roth, E., Cao, J., 2015. miR-181 suppresses metastasis via MMP-14. Aging 7, 740–741.

Rupaimoole, R., Slack, F.J., 2016. A role for miR-34 in colon cancer stem cell homeostasis. Stem Cell Investig. 3, 42.

Saeidimehr, S., Ebrahimi, A., Saki, N., Goodarzi, P., Rahim, F., 2016. MicroRNA-based linkage between aging and cancer: from epigenetics view point. Cell J. 18, 117–126.

Seoane, J., Le, H.V., Massague, J., 2002. Myc suppression of the p21(Cip1) Cdk inhibitor influences the outcome of the p53 response to DNA damage. Nature 419, 729–734.

Shah, M.Y., Ferrajoli, A., Sood, A.K., Lopez-Berestein, G., Calin, G.A., 2016. microRNA therapeutics in cancer—an emerging concept. EBioMedicine 12, 34–42.

Shang, J., Yao, Y., Fan, X., Shangguan, L., Li, J., Liu, H., Zhou, Y., 2016. miR-29c-3p promotes senescence of human mesenchymal stem cells by targeting CNOT6 through p53-p21 and p16-pRB pathways. Biochim. Biophys. Acta 1863, 520–532.

Shawi, M., Autexier, C., 2008. Telomerase, senescence and ageing. Mech. Ageing Dev. 129, 3–10.

Sokolova, V., Fiorino, A., Zoni, E., Crippa, E., Reid, J.F., Gariboldi, M., Pierotti, M.A., 2015. The effects of miR-20a on p21: two mechanisms blocking growth arrest in TGF-beta-responsive colon carcinoma. J. Cell. Physiol. 230, 3105–3114.

Song, J., Jin, E.H., Kim, D., Kim, K.Y., Chun, C.H., Jin, E.J., 2015. MicroRNA-222 regulates MMP-13 via targeting HDAC-4 during osteoarthritis pathogenesis. BBA Clin. 3, 79–89.

Sousa-Victor, P., Gutarra, S., Garcia-Prat, L., Rodriguez-Ubreva, J., Ortet, L., Ruiz-Bonilla, V., Jardi, M., Ballestar, E., Gonzalez, S., Serrano, A.L., Perdiguero, E., Munoz-Canoves, P., 2014. Geriatric muscle stem cells switch reversible quiescence into senescence. Nature 506, 316–321.

Tang, Y., Cui, Y., Li, Z., Jiao, Z., Zhang, Y., He, Y., Chen, G., Zhou, Q., Wang, W., Zhou, X., Luo, J., Zhang, S., 2016. Radiation-induced miR-208a increases the proliferation and radioresistance by targeting p21 in human lung cancer cells. J. Exp. Clin. Cancer Res. 35, 7.

Tazawa, H., Tsuchiya, N., Izumiya, M., Nakagama, H., 2007. Tumor-suppressive miR-34a induces senescence-like growth arrest through modulation of the E2F pathway in human colon cancer cells. Proc. Natl. Acad. Sci. U.S.A. 104, 15472–15477.

Tili, E., Michaille, J.J., Cimino, A., Costinean, S., Dumitru, C.D., Adair, B., Fabbri, M., Alder, H., Liu, C.G., Calin, G.A., Croce, C.M., 2007. Modulation of miR-155 and miR-125b levels following lipopolysaccharide/TNF-alpha stimulation and their possible roles in regulating the response to endotoxin shock. J. Immunol. 179, 5082–5089.

Verma, R.P., Hansch, C., 2007. Matrix metalloproteinases (MMPs): chemical-biological functions and (Q)SARs. Bioorg. Med. Chem. 15, 2223–2268.

Wang, R., Ma, J., Wu, Q., Xia, J., Miele, L., Sarkar, F.H., Wang, Z., 2013. Functional role of miR-34 family in human cancer. Curr. Drug Targets 14, 1185–1191.

Wang, L., He, J., Xu, H., Xu, L., Li, N., 2016. MiR-143 targets CTGF and exerts tumor-suppressing functions in epithelial ovarian cancer. Am. J. Transl. Res. 8, 2716–2726.

Wei, S., Wei, W.Y., Sedivy, J.M., 1999. Expression of catalytically active telomerase does not prevent premature senescence caused by overexpression of oncogenic Ha-Ras in normal human fibroblasts. Cancer Res. 59, 1539–1543.

Wong, P., Iwasaki, M., Somervaille, T.C., Ficara, F., Carico, C., Arnold, C., Chen, C.Z., Cleary, M.L., 2010. The miR-17-92 microRNA polycistron regulates MLL leukemia stem cell potential by modulating p21 expression. Cancer Res. 70, 3833–3842.

Wu, S., Liu, F., Xie, L., Peng, Y., Lv, X., Zhu, Y., Zhang, Z., He, X., 2015. miR-125b suppresses proliferation and invasion by targeting MCL1 in gastric cancer. BioMed Res. Int. 2015, 365273.

Xu, D., Takeshita, F., Hino, Y., Fukunaga, S., Kudo, Y., Tamaki, A., Matsunaga, J., Takahashi, R.U., Takata, T., Shimamoto, A., Ochiya, T., Tahara, H., 2011. miR-22 represses cancer progression by inducing cellular senescence. J. Cell Biol. 193, 409–424.

Xu, N., Zhang, L., Meisgen, F., Harada, M., Heilborn, J., Homey, B., Grander, D., Stahle, M., Sonkoly, E., Pivarcsi, A., 2012. MicroRNA-125b down-regulates matrix metallopeptidase 13 and inhibits cutaneous squamous cell carcinoma cell proliferation, migration, and invasion. J. Biol. Chem. 287, 29899–29908.

Xu, Z., Wang, X., Fang, L., 2013. MicroRNA miR-24 enhances tumor angiogenesis and metastasis in breast cancer. Cancer Res. 73, P4-07-19.

Xu, F., Pang, L., Cai, X., Liu, X., Yuan, S., Fan, X., Jiang, B., Zhang, X., Dou, Y., Gorospe, M., Wang, W., 2014. let-7-repressesed Shc translation delays replicative senescence. Aging Cell 13, 185–192.

Xu, X., Chen, W., Miao, R., Zhou, Y., Wang, Z., Zhang, L., Wan, Y., Dong, Y., Qu, K., Liu, C., 2015. miR-34a induces cellular senescence via modulation of telomerase activity in human hepatocellular carcinoma by targeting FoxM1/c-Myc pathway. Oncotarget 6, 3988–4004.

Xue, J., Chi, Y., Chen, Y., Huang, S., Ye, X., Niu, J., Wang, W., Pfeffer, L.M., Shao, Z.M., Wu, Z.H., Wu, J., 2016. MiRNA-621 sensitizes breast cancer to chemotherapy by suppressing FBXO11 and enhancing p53 activity. Oncogene 35, 448–458.

Yamakuchi, M., 2012. MicroRNA regulation of SIRT1. Front. Physiol. 3, 68.

Yao, S., 2016. MicroRNA biogenesis and their functions in regulating stem cell potency and differentiation. Biol. Proced. Online 18, 8.

Ye, J., Yao, Y., Song, Q., Li, S., Hu, Z., Yu, Y., Hu, C., Da, X., Li, H., Chen, Q., Wang, Q.K., 2016. Up-regulation of miR-95-3p in hepatocellular carcinoma promotes tumorigenesis by targeting p21 expression. Sci. Rep. 6, 34034.

Yi, C., Wang, Q., Wang, L., Huang, Y., Li, L., Liu, L., Zhou, X., Xie, G., Kang, T., Wang, H., Zeng, M., Ma, J., Zeng, Y., Yun, J.P., 2012. MiR-663, a microRNA targeting p21(WAF1/CIP1), promotes the proliferation and tumorigenesis of nasopharyngeal carcinoma. Oncogene 31, 4421–4433.

Zhai, L., Ma, C., Li, W., Yang, S., Liu, Z., 2016. miR-143 suppresses epithelial-mesenchymal transition and inhibits tumor growth of breast cancer through down-regulation of ERK5. Mol. Carcinog. 55, 1990–2000.
Zhang, R.G., Poustovoitov, M.V., Ye, X.F., Santos, H.A., Chen, W., Daganzo, S.M., Erzberger, J.P., Serebriiskii, I.G., Canutescu, A.A., Dunbrack, R.L., Pehrson, J.R., Berger, J.M., Kaufman, P.D., Adams, P.D., 2005. Formation of MacroH2A-containing senescence-associated heterochromatin foci and senescence driven by ASF1a and HIRA. Dev. Cell 8, 19–30.
Zhang, R.G., Chen, W., Adams, P.D., 2007. Molecular dissection of formation of senescence-associated heterochromatin foci. Mol. Cell. Biol. 27, 2343–2358.
Zhang, Y., Wang, Y., Wang, X., Zhang, Y., Eisner, G.M., Asico, L.D., Jose, P.A., Zeng, C., 2011. Insulin promotes vascular smooth muscle cell proliferation via microRNA-208-mediated downregulation of p21. J. Hypertens. 29, 1560–1568.
Zhang, D., Song, Y., Wang, Y., Liu, X., Liu, X., Ma, X., 2016a. Insight of in vitro small-interfering RNA release from chitosan nanoparticles under enzymolysis with forster resonance energy transfer analysis. J. Pharm. Sci. 105, 301–307.
Zhang, H.Y., Duan, J.J., Qu, Y.J., Deng, T., Liu, R., Zhang, L., Bai, M., Li, J.L., Ning, T., Ge, S.H., Wang, X., Wang, Z.Z., Fan, Q., Li, H.L., Ying, G.G., Huang, D.Z., Ba, Y., 2016b. Onco-miR-24 regulates cell growth and apoptosis by targeting BCL2L11 in gastric cancer. Protein Cell 7, 141–151.
Zhang, J., Jia, J., Zhao, L., Li, X., Xie, Q., Chen, X., Wang, J., Lu, F., 2016c. Down-regulation of microRNA-9 leads to activation of IL-6/Jak/STAT3 pathway through directly targeting IL-6 in HeLa cell. Mol. Carcinog. 55, 732–742.
Zhang, M., Gong, W., Zuo, B., Chu, B., Tang, Z., Zhang, Y., Yang, Y., Zhou, D., Weng, M., Qin, Y., Ma, M., Jiang, A., Ma, F., Quan, Z., 2016d. The microRNA miR-33a suppresses IL-6-induced tumor progression by binding twist in gallbladder cancer. Oncotarget 7, 78640–78652.
Zhang, Y., Guo, M., Xin, N., Shao, Z., Zhang, X., Zhang, Y., Chen, J., Zheng, S., Fu, L., Wang, Y., Zhou, D., Chen, H., Huang, Y., Dong, R., Xiao, C., Liu, Y., Geng, D., 2016e. Decreased microRNA miR-181c expression in peripheral blood mononuclear cells correlates with elevated serum levels of IL-7 and IL-17 in patients with myasthenia gravis. Clin. Exp. Med. 16, 413–421.
Zhao, G.B., Liu, L.J., Zhao, T.S., Jin, S.D., Jiang, S.B., Cao, S.Q., Han, J.Q., Xin, Y.Z., Dong, Q., Liu, X., Cui, J., 2015. Upregulation of miR-24 promotes cell proliferation by targeting NAIF1 in non-small cell lung cancer. Tumor Biol. 36, 3693–3701.
Zheng, X., Chopp, M., Lu, Y., Buller, B., Jiang, F., 2013. MiR-15b and miR-152 reduce glioma cell invasion and angiogenesis via NRP-2 and MMP-3. Cancer Lett. 329, 146–154.
Zheng, L., Zhang, Y., Liu, Y., Zhou, M., Lu, Y., Yuan, L., Zhang, C., Hong, M., Wang, S., Li, X., 2015. MiR-106b induces cell radioresistance via the PTEN/PI3K/AKT pathways and p21 in colorectal cancer. J. Transl. Med. 13, 252.
Zhou, Z., Yin, Y., Chang, Q., Sun, G., Lin, J., Dai, Y., 2016. Downregulation of B-myb promotes senescence via the ROS-mediated p53/p21 pathway, in vascular endothelial cells. Cell Prolif. 50, e12319.

CHAPTER FIVE

MiRNA Deregulation in Cardiac Aging and Associated Disorders

Robin Verjans, Marc van Bilsen, Blanche Schroen[1]
Cardiovascular Research Institute Maastricht (CARIM), Maastricht University, Maastricht, The Netherlands
[1]Corresponding author: e-mail address: b.schroen@maastrichtuniversity.nl

Contents

Abstract

The prevalence of age-related diseases is increasing dramatically, among which cardiac disease represents the leading cause of death. Aging of the heart is characterized by various molecular and cellular hallmarks impairing both cardiomyocytes and non-cardiomyocytes, and resulting in functional deteriorations of the cardiac system. The aging process includes desensitization of β-adrenergic receptor (βAR)-signaling and

International Review of Cell and Molecular Biology, Volume 334
ISSN 1937-6448
http://dx.doi.org/10.1016/bs.ircmb.2017.03.004

decreased calcium handling, altered growth signaling and cardiac hypertrophy, mitochondrial dysfunction and impaired autophagy, increased programmed cell death, low-grade inflammation of noncanonical inflammatory cells, and increased ECM deposition.

MiRNAs play a fundamental role in regulating the processes underlying these detrimental changes in the cardiac system, indicating that MiRNAs are crucially involved in aging. Among others, MiR-34, MiR-146a, and members of the MiR-17-92 cluster, are deregulated during senescence and drive cardiac aging processes. It is therefore suggested that MiRNAs form possible therapeutic targets to stabilize the aged failing myocardium.

1. INTRODUCTION

1.1 Heart Failure Is an Ageing Disease

The global population ages dramatically as the average life expectancy increased the past century. In the United States, currently ±12% of the population is over 65 years and is estimated to further increase to ±20% by 2030 (Yazdanyar and Newman, 2009). This increase in global age is accompanied by a decrease in quality of life in the elderly population and a dramatically growing prevalence of age-related diseases. Cardiac disease represents the leading cause of death among these age-related diseases, accounting for ±40% of all deaths in the elderly population (Go et al., 2014; Yazdanyar and Newman, 2009). Heart failure (HF), the inability of the heart to pump sufficient blood into the body, is associated with frequent hospitalizations (Jencks et al., 2009), resulting in high consummation of medical resources and a poor 5-year survival rate (Stewart et al., 2002). HF in particular is generally accepted as a typical age-associated cardiac disease and is reaching epidemic proportions, with ±88% of HF deaths and >75% of all HF hospitalizations in the United States occurring in adults aged over 65 years (Fang et al., 2008; Lakatta and Levy, 2003). HF is estimated to form an even greater problem in the future since, besides the fact that the increasing aged population has a high HF prevalence, effective treatment is still lacking. Besides the increasing age-associated prevalence of several cardiac risk factors including hypertension, obesitas, and insulin resistance, aging itself is a major independent risk factor for cardiac disease (Bursi et al., 2006a; Lakatta, 1993), stressing the importance of understanding the processes underlying intrinsic myocardial aging.

MicroRNAs (MiRNAs) are a class of small ±22 nucleotides long non-coding RNA molecules that negatively regulate target gene expression (Ha and Kim, 2014; Kunz et al., 2015). MiRNAs are implicated in cardiac disease and increasing evidence links MiRNAs to cardiac aging. Regulatory functions of age-associated MiRNAs may potentially explain observed age-related deleterious phenotypes.

1.2 Cardiac Aging

Due to the lack of proliferative capacity of cardiomyocytes, it is suggested that the heart is one of the organs most affected by detrimental age-associated changes. The senescent heart is characterized by structural remodeling at both organs, cellular and molecular level. Remodeling of the aging cardiac system includes an increase in left ventricular wall thickness, stiffening of the heart, increased deposition of collagen, and collagen cross-linking (Vasan et al., 2003b). On a cellular level, aging is associated with prolongation of cardiomyocyte action potential (Liu et al., 2000), decreased cardiomyocyte survival, as well as a decline in the number of mitochondria and cardiac stem cells (Dai and Rabinovitch, 2009; Lakatta, 2003; Shih et al., 2011). Furthermore, endothelial dysfunction, migration/proliferation of vascular smooth muscle cells, increased fibroblast proliferation, and proinflammatory signaling of inflammatory cells are additional cellular processes underlying cardiac aging (Wang et al., 2006, 2014b). These structural changes lead to an altered function of the aged myocardium, including an increase in the prevalence of atrial fibrillation (Lakatta and Levy, 2003), a lowered cardiac reserve (Lakatta, 2003), as well as declined diastolic function due to increased left ventricular diastolic stiffness, decreased filling (Dai and Rabinovitch, 2009; Dai et al., 2009), and decreased contraction and relaxation velocity (Bursi et al., 2006b; Chen et al., 1998; Liu et al., 2000; Redfield et al., 2005).

The associated molecular processes include (1) desensitization of β-adrenergic receptor (βAR)-signaling and decreased calcium handling, (2) altered growth signaling, (3) mitochondrial dysfunction and impaired autophagy, (4) increased programmed cell death, (5) low-grade inflammation of noncanonical inflammatory cells, and (6) increased extracellular matrix (ECM) deposition (Dai and Rabinovitch, 2009; Lakatta, 2003; Lakatta and Levy, 2003; Loffredo et al., 2014). Thus, the intrinsic mechanisms associated with age-induced impairment of the myocardium are identified and mimic those implicated in HF. Unfortunately, precise knowledge on what initiates these cellular changes and the role of MiRNAs therein is still lacking. This review provides an overview of the cardiac molecular and cellular processes driving myocardial aging listed earlier, and describes the regulating role of senescence-associated MiRNAs in these processes. Given the similarity in molecular mechanisms of the aging and failing heart, we will as well discuss MiRNAs implicated in HF of other causes than aging, focusing on MiRNAs that show differential expression in the aged heart.

2. βAR DESENSITIZATION AND CALCIUM HANDLING

The sympathetic nervous system undergoes a series of rearrangements with aging. With normal aging, circulating norepinephrine levels increase by 10%–15% per decade (Esler et al., 1981) and cardiac norepinephrine levels increase due to diminished reuptake and increased tissue spill over (Esler et al., 1995). Chronically increased βAR occupancy by norepinephrine leads to βAR desensitization (Janczewski and Lakatta, 2010; White et al., 1994; Xiao et al., 1998) and impairs cardiomyocyte calcium handling, eventually leading to deficits in cellular contraction and relaxation (Davies et al., 1996; Xiao et al., 1994). Prolonged myocardial relaxation time forms a hallmark of age-related diastolic dysfunction and is caused, among others, by impaired active relaxation of cardiomyocytes due to depressed Ca^{2+}-uptake and Ca^{2+}-release activities in the aging heart (Kirchhefer et al., 2004; Zhu et al., 2005; Zile and Brutsaert, 2002). In the normal situation, βAR activation regulates Ca^{2+} homeostasis and contraction through accumulation of the second messenger cAMP, which activates protein kinase A (PKA). In its turn, PKA phosphorylates phospholamban (PLB) and relieves its inhibitory effect on calcium-transporter SERCA2a, promoting sarcoplasmic reticulum cycling (Ishikawa and Homcy, 1997; Rockman et al., 2002) (Fig. 1). Decreased expression (Dai et al., 2009) and activity (Janczewski and Lakatta, 2010) of SERCA2a are thought to be the primary mechanisms for the impaired cardiac relaxation in the aged myocardium. In addition, age-associated alterations in proteins able to regulate SERCA2a activity, including PLB (Cain et al., 1998; Meyer and Dillmann, 1998; Schmidt et al., 2000; Tschope et al., 2004), PKA (Jiang et al., 1993), and CAMKII (Xu and Narayanan, 1998), have also been documented in the aged heart. Thus, βAR signaling plays a crucial role in maintaining a balanced calcium homeostasis and regulating cardiomyocyte contraction and relaxation, and dysregulated βAR signaling forms one of the major cellular hallmarks of the senescent myocardium.

Evidence shows that altered βAR activity shapes cardiac MiRNA profiles while specific MiRNAs, including βAR-responsive ones, are able to influence βAR signaling. This implies an important role for MiRNAs in age-associated desensitization of βAR signaling. Hou et al. (2012) profiled

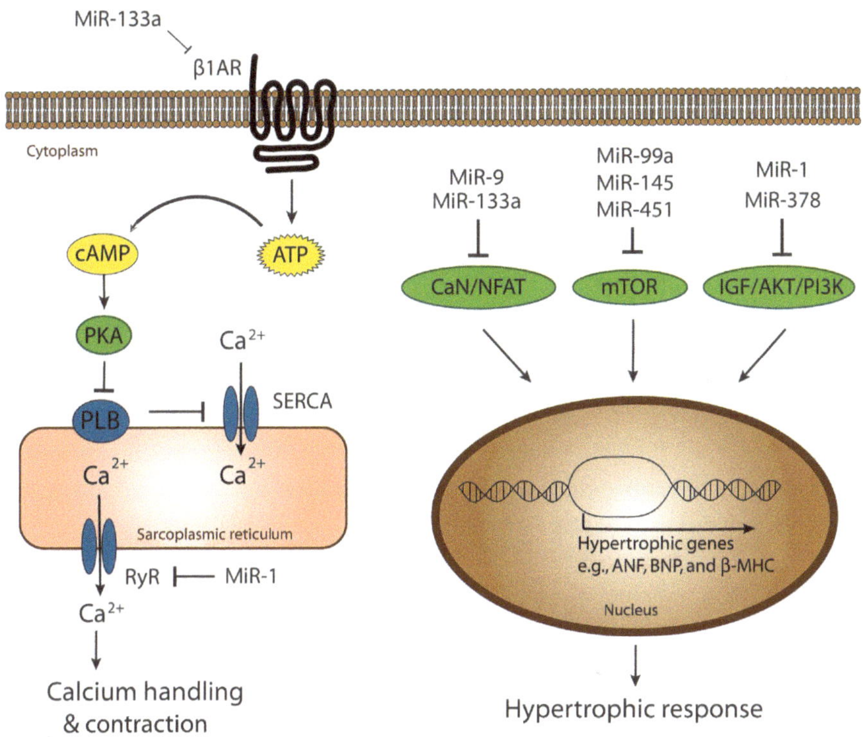

Fig. 1 Age-associated MiRNAs in control of the cardiomyocyte β-adrenergic signaling and the hypertrophic response. Activation of β1-adrenergic receptor (β1-AR) controls cardiomyocyte calcium homeostasis via accumulation of the second messenger cAMP, which activates the phospholamban (PLB) inhibitor, protein kinase A (PKA). Inhibition of PLB derepresses the calcium-transporter SERCA2a and promotes calcium (Ca^{2+}) entry into the sarcoplasmic reticulum. Sarcoplasmic Ca^{2+} is in its turn exported to the cytoplasm by ryanodine receptor 2 (RyR2) to regulate cardiomyocyte contraction. Stimulation of major growth signaling pathways, including the calcineurin/NFAT, mTOR, and IGF/AKT/PI3K pathway, induces a hypertrophic response characterized by transcription of hypertrophic genes: atrial natriuretic factor (ANF), brain natriuretic factor (BNP), and β-myosin heavy chain (β-MHC). At every level, MiRNAs are implicated in regulating these processes.

rat cardiac MiRNA expression under conditions of βAR activation and inhibition, identifying MiRNAs controlling βAR pathway activity. Activation of βAR signaling upregulated 43 while downregulating 9 MiRNAs. On the other hand, βAR inhibition upregulated and downregulated 5 and 28 MiRNAs, respectively (Hou et al., 2012). Similarly, Carrillo et al. (2011) found that the expression of 33 MiRNAs was significantly changed in cardiac lysates from rats stimulated with the βAR agonist isoproterenol (ISO). The expression levels of MiR-21, MiR-132, MiR-214, and MiR-222 were more than sevenfold increased. MiR-21, MiR-132, and MiR-222 are also directly involved in controlling calcium levels via targeting, beta-2 subunit of the voltage-dependent calcium channel (CACNB2) (Carrillo et al., 2011). Moreover, Ling et al. (2013) have shown that MiR-26b, MiR-21, and MiR-320 levels are affected by changes in βAR activity.

βAR signaling is shown to control one of the most abundantly cardiac expressed MiRNA clusters, the MiR-1/133 cluster via a cAMP response element in the promoter region of muscle specific MiRNA cluster MiR-1/MiR-133a (Care et al., 2007; Lu et al., 2009; Matkovich et al., 2010; Trajkovski et al., 2012). Intriguingly, MiR-1 potentiates βAR signaling, causing hyperphosphorylation of ryanodine receptor 2 (RyR2) channels thereby increasing excitation contraction coupling (Terentyev et al., 2009). On the contrary, MiR-133 inhibits βAR-signaling at multiple levels. Transgenic cardiac-specific MiR-133 expression preserves cardiac performance and attenuates cardiac remodeling upon pressure overload (Castaldi et al., 2014). This protective MiR-133 effect is associated with decreased cAMP/PKA-dependent phosphorylation of PLB, and resulting from MiR-133 directed targeting of the β1AR (ADBR1) and several other components of the β1AR signaling cascade (Castaldi et al., 2014). Thus, while both MiRNAs are under the regulation of βAR-signaling, these MiRNAs have opposite roles. Exactly how their balance is regulated requires further investigation.

The cardioprotective effect of biased β-blocker carvedilol (Carv) involves enhanced β-arrestin-mediated signaling, thereby affecting pri-miRNA processing of specific MiRNAs by Drosha. As a consequence, the maturation of MiR-125a-5p, MiR-125b-5p, MiR-150, MiR-199a-3p, and MiR-214 is upregulated in the mouse heart (Kim et al., 2014; Zhang et al., 2015b). These Carv-responsive MiRNAs may potentially explain the cardioprotective effect of biased βAR signaling (Noma et al., 2007). In addition, their expression is altered in the aged heart (Dimitrakopoulou et al., 2015; Zhang et al., 2015b), suggesting their potential involvement in the cardiac aging process.

Collectively, the aging myocardium is characterized by βAR desensitization and impaired calcium handling. Particularly MiR-1/133a seems to function in opposite βAR feedback loops, and the exact biological reasoning remains to be clarified.

3. ALTERED GROWTH SIGNALING AND HYPERTROPHY

Cardiac hypertrophy is a hallmark of cardiac aging. Although this adaptive response minimizes myocardial wall stress and maintains cardiac function on the short term, persistent hypertrophic signaling on the long term is detrimental and underlies cardiac pathology (Anversa et al., 1986; Olivetti et al., 1991). Cellular changes of cardiac hypertrophy are characterized by increased cardiomyocyte volume, expression of the fetal gene program and increased protein synthesis (Berk et al., 2007; Chen et al., 2002; Heineke and Molkentin, 2006).

Numerous MiRNA profiling studies have been performed in human hypertrophic patients and rodent models of cardiac hypertrophy, identifying various MiRNAs which actively participate in cardiomyocyte hypertrophic signaling, including MiR-1/MiR-133 (Townley-Tilson et al., 2010) MiR-208 (Callis et al., 2009; van Rooij et al., 2007, 2009), MiR-18 (Tatsuguchi et al., 2007), MiR-98/let-7 (Yang et al., 2011), MiR-9 (Wang et al., 2010a), MiR-214 (Tang et al., 2016; Yang et al., 2013), MiR-199a (Haghikia et al., 2011; Rane et al., 2010), MiR-199b (da Costa Martins et al., 2010), MiR-23a (Lin et al., 2009), and more recently the MiR-212/132 family (Strum et al., 2009; Ucar et al., 2012; Zhang et al., 2014a).

Deregulation of major growth signaling pathways, including the mTOR/P70/S6K (Johnson et al., 2013), IGF/AKT/PI3K (Antos et al., 2002; Naga Prasad et al., 2000), and calcineurin/NFAT pathway (Molkentin et al., 1998), have been implicated in hypertrophic growth of aging cardiomyocytes (Fig. 1).

3.1 MiRNAs Affecting mTOR/P70/S6K Signaling

The mammalian target of rapamycin (mTOR) pathway plays a critical role in cardiomyocyte hypertrophy, function, and structure via integration of nutrient and hormonal signals to regulate growth and is a major modulator of aging and age-related disease (Johnson et al., 2013). Inhibition of the prohypertrophic mTOR/P70/S6K signaling pathway has cardioprotective effects when the heart ages or suffers from ischemia or overload (Kato et al., 2012; McMullen et al., 2004; Sen et al., 2013; Shioi et al., 2003). These findings suggest that MiRNAs targeting mTOR signaling in the failing or senescent heart could protect the heart from detrimental remodeling. Indeed, several different MiRNAs are shown to suppress mTOR, directly or indirectly, thereby exerting cardioprotective functions in vitro and in vivo.

For example, MiR-99a was shown to negatively regulate mTOR expression and P70/S6K activation, preventing cardiomyocyte hypertrophy under stress conditions in vitro and in vivo (Li et al., 2016c). MiR-145 also exerts its antihypertrophic effect by decreasing the activity of the mTOR/P70/S6K signaling cascade. MiR-145 expression is proven to be significantly upregulated in end-stage HF patients (Naga Prasad et al., 2009). Li et al. (2013) have shown that MiR-145 attenuates hypertrophy of cardiomyocytes induced by the βAR agonist ISO by suppressing the expression and nuclear translocation of GATA6. In its turn, prohypertrophic transcription factor GATA6 has been found to be involved in the mTOR-induced pathological hypertrophy (Liang et al., 2001; van Berlo et al., 2010) activating transcription of the fetal genes ANF, BNP, and β-MHC (Charron et al., 1999) (Fig. 1). These results indicate that MiR-145 attenuates the cardiac hypertrophic response by decreasing mTOR activity indirectly. MiR-451 levels were significantly increased in type 2 diabetes mellitus murine hearts and in neonatal rat cardiac myocytes stimulated with palmitic acid. Loss of MiR-451 function reduced palmitate-induced lipotoxicity in nRCMs. HFD-fed, cardiomyocyte-specific, MiR-451 knockout mice were protected from hypertrophy and displayed lower levels of phosphorylated mTOR, implying that MiR-451 is involved in diabetic cardiomyopathy, at least partly, through suppression of mTOR activity (Kuwabara et al., 2015).

Thus, different studies suggest that MiRNA-exerted regulation of mTOR plays an important role in cardiac aging and its inhibition protects the heart from detrimental hypertrophic remodeling under pathological conditions.

3.2 MiRNAs Affecting IGF/AKT/PI3K Signaling

Insulin/IGF signaling plays a critical role in aging. Human serum IGF-1 levels decline over time, correlating with increased risk of HF (Vasan et al., 2003a). Both Akt and the class IA isoform of phophoinositide 3-kinase (PI3K) are important factors downstream of insulin and IGF receptors, and regulate cell growth, survival, and metabolism (Cantley, 2002).

Muscle-specific MiR-1 is one of the most cardiac abundant MiRNAs, and different studies have shown an inverse correlation between MiR-1 expression and cardiac hypertrophy (Care et al., 2007; Sayed et al., 2007). Several MiR-1 cardiac targets involved in the control of hypertrophy have been identified, including HDAC4 (Chen et al., 2006), Hand2

(Zhao et al., 2005), ras GTPase-activating protein (Ras-GAP), and ras homolog enriched in brain (Rheb) (Sayed et al., 2007). Besides, MiR-1 additionally modulates the cardiac hypertrophic response via direct targeting of IGF-1 (Elia et al., 2009; Shan et al., 2009) and IGFR-1 (Elia et al., 2009), influencing insulin signaling. The prohypertrophic IGF-1 acts as a negative regulator of cardiac enriched MiRNA-378. In turn, this MiRNA regulates cardiac remodeling by both direct targeting of IGF-1R and indirect inhibition of IGF/PI3K/AKT signaling through targeting of Grb-2, a known cardiac hypertrophy modulator and an essential component of the Ras signaling pathway that is downstream of IGF-1 and other growth factors (Ganesan et al., 2013; Nagalingam et al., 2013). In the end, MiR-378 blocks growth activity, leading to suppression of fetal gene expression and inhibition of hypertrophy of cardiomyocytes (Ganesan et al., 2013; Nagalingam et al., 2013).

3.3 MiRNAs Affecting Calcineurin/NFAT Signaling

Calcineurin-dependent NFAT (nuclear factor of activated T cells) activation is part of a central prohypertrophic signaling cascade. NFAT functions as a calcineurin effector, trafficking between the cytosol and the nucleus to regulate the cardiac hypertrophic response (Molkentin et al., 1998; Wilkins and Molkentin, 2004). Increased cardiac activation of calcineurin/NFAT signaling is observed in aging mice (Dai et al., 2009). Two MiRNAs, MiR-133a and MiR-9, have so far been shown to be able to decrease calcineurin/NFAT signaling and ameliorate hypertrophic cell growth. First, myocardial expression of MiR-133a is decreased in human and animal hypertrophy (Care et al., 2007; Sayed et al., 2007; van Rooij et al., 2006), and is inversely related to cardiac hypertrophy (Care et al., 2007). MiR-133a directly targets calcineurin (Dong et al., 2010) and NFAT (Li et al., 2010b), preventing phenylephrine-induced hypertrophic cell growth and repression of the fetal gene program (Liu et al., 2008). In line, loss of MiR-133a increased angiotensinogen expression and hypertrophy (Myers et al., 2015). The second calcineurin/NFAT-associated MiRNA, MiR-9 indirectly suppresses NFAT through translational suppression of myocardin, which is necessary for nuclear activation of NFAT (Wang et al., 2010a).

Although the majority of identified hypertrophy-regulating MiRNAs has not been studied in the process of aging, these reports indicate that many MiRNAs play a crucial role in controlling the hypertrophic response in cardiomyocytes. MiRNAs exert their hypertrophy-regulatory function via impacting major growth signaling pathways, which are also associated with age-related diseases. Cellular hypertrophic signaling as a result of senescence could potentially be caused by dysregulation of these hypertrophy-regulating MiRNAs. Further research should be conducted to determine if the age-associated hypertrophic response is directly caused by altered expression of these MiRNAs, or a secondary result.

4. MAINTAINING METABOLIC AND MITOCHONDRIAL FUNCTION WITH AGE

Metabolic dysfunction is associated with aging and many age-related diseases (Hu and Liu, 2014). Whereas fatty acid oxidation is the primary source of energy in the adult healthy myocardium, both the senescent and failing heart display a striking impairment of mitochondrial fatty acid oxidation and oxidative phosphorylation and an increased reliance on glucose metabolism (Stanley et al., 2005; Taegtmeyer et al., 2004). Despite intense research in this area, there is little consensus on the causes of this "metabolic substrate shift" or whether the shift in substrate utilization in failing and aging myocardium is detrimental nore compensatory (van Bilsen et al., 2009). Also the role of MiRNAs in the metabolic shift needs further clarification.

By feeding mice different diets, it was demonstrated that high fat feeding induced distinct cardiac MiRNA expression patterns as assessed by RNA deep sequencing (Guedes et al., 2016). This included down-regulation of senescence-associated MiRNAs, MiR-29b, and MiR-15. Moreover, high fat feeding upregulated expression levels of aging MiRNAs MiR-21a, MiR-29c, MiR-22, and MiR-223 (Guedes et al., 2016). Not only fatty acid levels affect MiRNA profiles, recent studies have shown that high glucose significantly reduced expression levels of MiR-126 and MiR-375 (El Ouaamari et al., 2008; Keller et al., 2012; Zampetaki et al., 2010). These findings suggest that age-associated alterations in the metabolic environment may cause a distinct MiRNA profile. More importantly, a growing body of evidence suggests that numerous MiRNAs are implicated in the regulation of metabolic homeostasis by targeting genes with a prominent role in (1) fatty acid or (2) glucose metabolism, and (3) energy production via controlling mitochondrial-biogenesis, -function, and -autophagy.

4.1 MiRNAs Regulating Fatty Acid Metabolism

The healthy adult myocardium relies primarily on mitochondrial fatty acid oxidation for its energy production. Numerous mitochondrial enzymes involved in fatty acid β-oxidation have been shown to be under the control of MiRNAs (Fig. 2). The MiR-199a/214 cluster has been shown to play a regulatory role in the metabolic switch from mitochondrial fatty acid oxidation in the healthy myocardium toward glucose utilization at the onset of HF, by actively repressing PPARβ/δ (el Azzouzi et al., 2013). In addition, MiR-199a suppresses PPARα indirectly, resulting in repression of different PPARα-target genes involved within the β-oxidation (Li et al., 2014a). Both MiR-199a and MiR-214 are increased in failing human and rodent hearts (van Rooij et al., 2006), and both decrease reliance on fatty acid

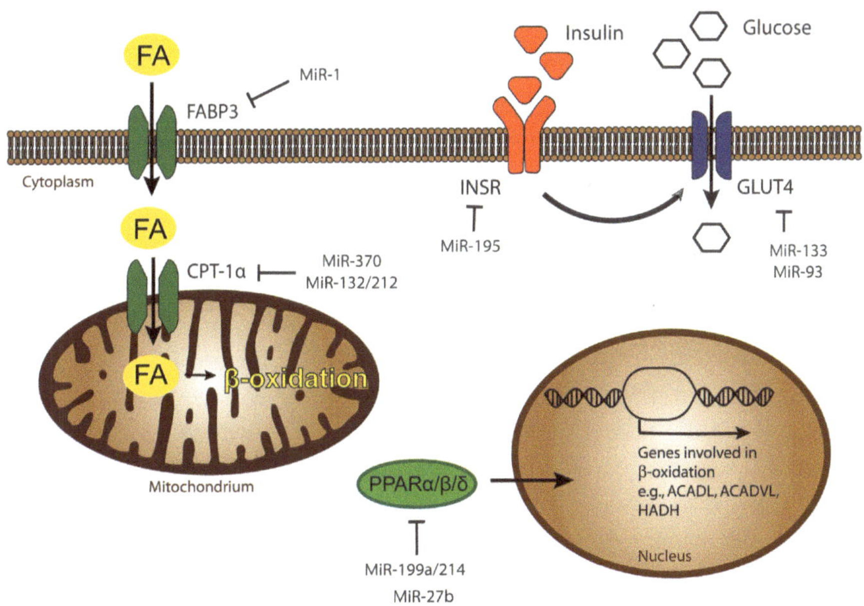

Fig. 2 MiRNAs regulating fatty acid and glucose metabolism in the aging myocardium. Fatty acid (FA) metabolism is facilitated by fatty acid binding protein 3 (FABP3), which transports FAs into the cell. Here, FAs undergo β-oxidation after CPT1α-dependent transport into the mitochondrion. Nuclear translocation of different peroxisome proliferator-activated receptors (PPARα/β/δ) induces transcription of various genes involved in β-oxidation, such as long chain- and very long chain-acyl-CoA dehydrogenase (ACADL and ACADVL), and hydroxyacyl-CoA dehydrogenase (HADH). Glucose metabolism is highly dependent on insulin-mediated activation of insulin receptor (INSR) which stimulates glucose transport by GLUT4. Central factors in these metabolic pathways are regulated by MiRNAs.

oxidation through inhibition of PPARδ (el Azzouzi et al., 2013). Overexpression results in mitochondrial dysfunction and decreased mitochondrial fatty acid oxidation. Conversely, silencing of MiR-199a/214 in mice subjected to pressure overload derepressed cardiac PPARδ levels, normalized mitochondrial fatty acid oxidation, and ultimately lead to improved cardiac function (el Azzouzi et al., 2013).

MiR-370 regulates the rate of β-oxidation by suppressing expression of carnitine palmitoyltransferase 1a (CPT1a), an enzyme involved in the import of fatty acids into the mitochondria. In addition, other genes important for fatty acid utilization, such as diacylglycerol *O*-acyltransferase 2 (DGAT2), and fatty acid synthase are also targeted by MiR-370 strengthening its functional participation in fatty acid metabolism (Iliopoulos et al., 2010). Also, MiR-132 and MiR-212 decreased the transporting rate of long-chain acylcarnitines into the mitochondria for β-oxidation via translational regulation of CPT1a expression (Soni et al., 2014; Spillmann et al., 2015). MiR-33a and MiR-33b inhibit the expression of various metabolic enzymes, marking them as crucial players in controlling fatty acid β-oxidation (Rottiers et al., 2011). MiR-1, which we already described as central in hypertrophy and contraction, negatively regulates the expression of heart-type fatty acid binding protein 3 (FABP3), leading to decreased transport of fatty acids within the cardiomyocyte and subsequent impaired β-oxidation (Varrone et al., 2013). MiR-27b targets PPARγ thereby regulating the cardiac metabolic shift through suppression of fatty acid fuelling of mitochondria (Wang et al., 2012c). MiR-27b transgenic mice displayed lower cardiac levels of PPARγ while silencing of endogenous MiR-27b in mice showed a higher expression of cardiac PPARγ, which resulted in increased fatty acid oxidation and attenuation of cardiac dysfunction in a pressure overload model (Wang et al., 2012c).

These results reveal the intricate regulation of fatty acid β-oxidation by MiRNAs, promoting the metabolic switch toward increased reliance on glycolysis. Whether similar processes are involved in the metabolic substrate shift driving myocardial aging, remains to be elucidated.

4.2 MiRNAs Regulating Glucose Metabolism

In addition to fatty acid oxidation (Fig. 2), various MiRNAs able to impact cardiac glucose metabolism have been identified. Larger reliance on glucose utilization for energy production is a consistent feature of the failing and aging heart (Barger and Kelly, 2000; Razeghi et al., 2001) which has a major impact on the available energy levels due to the aerobic nature of the heart (Di Lisa et al., 2007). Interestingly, the age-associated MiRNA, MiR-34a, directly targets several glycolytic enzymes, including hexokinase 1, hexokinase 2, glucose-6-phosphate isomerase, and pyruvate dehydrogenase kinase 1, thereby repressing cellular glycolysis, potentially attempting to attenuate the age-associated increased cardiac reliance on glucose metabolism (Kim et al., 2013).

Additional MiRNAs shown to inhibit glucose metabolism include let-7, MiR-208, MiR-195, MiR-143, MiR-133, MiR-93, and MiR-103/107, while MiR-223 stimulates glycolysis. The lin28/let 7 pathway functions as a central regulator of mammalian glucose metabolism. Expression of members of the let-7 family was enhanced in the diabetic heart and overexpression induces glucose intolerance due to diminished glucose-induced insulin secretion as a result of translational inhibition of IGF-1R, insulin receptor (INSR) and GLUT4 (Frost and Olson, 2011; Li et al., 2016b; Zhu et al., 2011a). Moreover, RNA-binding proteins Lin28a and Lin28b increase insulin sensitivity and resistance to diabetes through blocking of let-7 maturation and subsequent de-repression of multiple let-7 targets (Rybak et al., 2008; Viswanathan et al., 2008; Zhang et al., 2015a; Zhu et al., 2011a). Similarly, heart specific MiR-208 negatively regulates med13, a regulator of energy expenditure and numerous other genes involved in energy balance in the heart. Indeed, inhibition of MiR-208 in mice conferred resistance to obesity, improved systemic insulin sensitivity, and glucose tolerance (Grueter et al., 2012). MiR-195 suppresses the expression of INSR impairing insulin signaling cascade and glycogen synthesis (Yang et al., 2014b). Silencing of MiR-103/107 modulates expression of INSR regulator caveolin 1 (CAV1) and improves insulin sensitivity and glucose homeostasis, whereas forced expression of MiR-103/107 impaired glucose homeostasis (Trajkovski et al., 2011). Moreover, MiR-143 (Jordan et al., 2011), MiR-133 (Horie et al., 2009), and MiR-93 (Chen et al., 2013) have also been shown to control glucose homeostasis, in which the latter two are shown to inhibit cardiomyocyte glucose uptake via direct targeting of glucose-transporter GLUT4.

On the other hand, MiR-223 overexpression increased GLUT4 expression, resulting in an increase in glucose uptake and an adaptive response to restore normal glucose uptake in the diabetic heart. Conversely, systemic inhibitor-mediated silencing of MiR-223 resulted in a significant decrease in cardiac GLUT4 expression (Lu et al., 2010). Collectively, these studies highlight the central role of MiRNAs in glucose metabolism. MiRNAs exerting a glucose-stimulating function may potentially be responsible for the observed age-induced increased reliance on glucose metabolism.

4.3 Mitochondrial Biogenesis and Function Under the Control of MiRNAs

Not only the "metabolic substrate shift" plays a prominent role in the aging heart, mitochondrial dysfunction, and dysregulation are also associated with aging and many age-related diseases (Hu and Liu, 2011). Due to the constant demand for high ATP levels for maintaining contractile activity in the heart, the loss of mitochondrial number and function during cellular aging has adverse impact on cardiac muscle (Boengler et al., 2010; Paradies et al., 2004). The main causes of age-associated mitochondrial impairment are reduced mitochondrial biogenesis and increased number of dysfunctional mitochondria, the latter being partly due to decreased autophagy (Hoshino et al., 2013). Mitochondrial biogenesis is regulated by several pathways, most of which rely on peroxisome proliferator-activated receptor y coactivator 1α (PCG-1α) (Handschin and Spiegelman, 2006). PCG-1α enhances mitochondrial function but cardiac expression is repressed in the failing heart (Wenz, 2011). PGC-1α is regulated by MiR-696 is skeletal muscle in response to physical activity, and mitochondrial biogenesis and fatty acid oxidation were decreased following MiR-696 overexpression (Aoi et al., 2010). Further research needs to be conducted using cardiomyocytes to interrogate the possible detrimental impact of MiR-696 on mitochondrial biogenesis during myocardial senescence.

In addition to decreased formation of mitochondria, impaired mitochondrial functioning forms a driving factor of cellular aging in the heart. Different MiRNAs have been identified to influence ATP generation by inhibiting mitochondrial function (Fig. 3). Members of the aging-miRNA family, MiR-15a and MiR-15b, reduced ATP generation and mitochondrial integrity in rat cardiomyocytes (Nishi et al., 2010; Sun et al., 2011). Also, MiR-181 family members affect ATP generation. MiR-181c causes electron transport chain complex IV remodeling through regulation of the mitochondrial cytochrome *c* oxidase subunit 1 (mt COX1) at the translational level (Das et al., 2014), while MiR-181a modulates the citric acid cycle (Chu et al., 2015). Moreover, both MiR-338 and MiR-210 modulate ATP production by depressing the expression of the main components in the electron transfer chain, cytochrome *c* oxidase IV (COXIV) and cytochrome *c* oxidase assembly protein 10 (COX10) (Aschrafi et al., 2008). Finally, MiR-484 negatively regulates mitochondrial fission in the heart

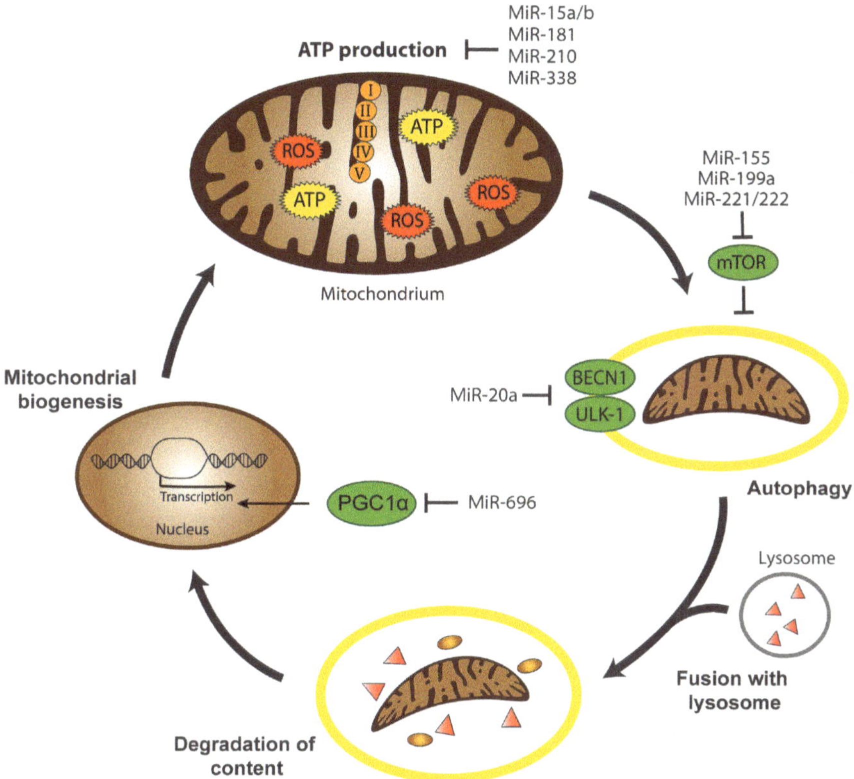

Fig. 3 Aging MiRNAs negatively regulate mitochondrial function via controlling mitochondrial ATP production, biogenesis, and autophagy. Mitochondria generate energy in the form of ATP, which is accompanied by the production of reactive oxygen species (ROS). Beclin 1- (BECN1) and unc-51-like autophagy-activating kinase 1 (ULK1) initiate the formation of autophagosomes to protect the cell against dysfunctional, ROS-producing mitochondria. Autophagosomal content undergoes lysation after fusion with a lysosome. Degraded components can be recycled during PPARG Coactivator 1 Alpha (PGC1-α)-mediated mitochondrial biogenesis.

via targeting mitochondrial fission protein (Fis1), thereby disrupting maintenance of normal mitochondrial function (Wang et al., 2012d).

Impaired mitochondrial function does not only limit ATP production, dysfunctional mitochondria also form the most important source of ROS (Fig. 3). Although generation of ROS is an inevitable by-product of aerobic metabolism and low levels of ROS are necessary for intracellular signaling (Hekimi et al., 2011), the aged and failing myocardium are characterized with an excess of ROS produced by various types of cardiac cells, that seems to be under the control of MiRNAs as well

(Haendeler et al., 2004; Magenta et al., 2014; Tsutsui et al., 2009). MiR-210, which expression is induced in senescent fibroblasts (Faraonio et al., 2012), induces the generation of ROS through impairment of mitochondrial function via suppressing activity of OXPHOS complex I and II (Chan et al., 2009; Puissegur et al., 2011). Similarly, MiR-494 induction in senescent fibroblasts increases ROS production (Faraonio et al., 2012). ROS can also be generated by various enzymes, including the nicotinamide adenine dinucleotide phosphate (NADPH)-oxidases (Kyrychenko et al., 2015). MiR-448 is shown to be in control of ROS production through repression of the NOX2 regulatory subunit Ncf1. Downregulation of cardiac MiR-448 led to increased Ncf1 expression and abnormally elevated ROS production and ultimately dilated cardiomyopathy and arrhythmias (Kyrychenko et al., 2015).

The cellular defense mechanism to counter the overproduction of ROS is the scavenging of free radicals by an array of antioxidants, such as superoxide dismutase (SOD) (Nojiri et al., 2006; Tsutsui et al., 2011). Both MiR-34a and MiR-335 are upregulated in kidneys of old rodents and target ROS-scavenging enzymes SOD2 and Thrioredoxin Reductase 2 (Txnrd2), thereby causing oxidative stress (Bai et al., 2011). Similarly, MiR-21 increases oxidative stress via targeting of SOD2 in angiogenic progenitor cells (Fleissner et al., 2010). MiR-146a downregulates the expression of NOX4, the major catalytic subunit of NADPH oxidase (Vasa-Nicotera et al., 2011). Aged cells display increased expression of NOX4 and decreased MiR-146a levels and overexpression of MiR-146a decreases the rate of HUVEC senescence (Vasa-Nicotera et al., 2011). Moreover, MiR-1 modulates oxidative stress in cardiomyocytes via targeting multiple antioxidant enzymes, including SOD1, the catalytic subunit of the glutamate cysteine ligase (Gclc), and glucose-6-phosphate dehydrogenase (G6PD) (Wang et al., 2015b). Gclc is a positive regulator of glutathione (GSH), promoting scavenging of ROS. MiR-1 transgenic mice hearts display repressed expression of SOD1, Gclc, and G6PD, contributing to the increased ROS level and susceptibility to oxidative stress within these hearts (Wang et al., 2015b). In addition, MiR-144 levels are repressed in cultured cardiomyocytes exposed to high-glucose and in hearts of streptozotocin (STZ)-induced diabetic mice. MiR-144 promotes ROS-induced oxidative damage through directly targeting of the central regulator of the response to oxidative stress, nuclear factor-erythroid 2-related factor 2 (Nrf2) (Yu et al., 2015).

Remarkably, MiRNAs themselves can become oxidatively modified by ROS, which may in turn alter their stability, integrity, binding affinity, and functionality, thereby contributing to altered redox signaling and disease mechanisms (Poulsen et al., 2012; Wang et al., 2015a). More specifically, MiR-184 is shown to interact with the 3′ UTRs of Bcl-xL and Bcl-w only after being subjected to oxidative modification, initiating apoptosis in the heart cell line H9c2 (Wang et al., 2015a). These results suggest that MiRNAs have a pivotal role in maintaining cardiac mitochondrial function and energy metabolism in the presence and absence of oxidative stress, preserving myocardial homeostasis.

4.4 MiRNAs Orchestrating Autophagy

As described earlier, the senescent myocardium is associated with defective mitochondria (Dai and Rabinovitch, 2009; Khrapko et al., 1999; Linnane et al., 1989) due to the infrequent replacement of cardiomyocytes and their high susceptibility to mitochondrial damage (Rezzani et al., 2012; Terzioglu and Larsson, 2007). Senescent cardiomyocytes counteract this detrimental process via autophagy, which is characterized by the delivery of intracellular components to the lysosome, maintaining cardiomyocyte homeostasis and mitochondrial activity through the renewal/recycling of cytoplasmic materials and organelles, removal of aggregated proteins, and the provision of energy and biomolecules to cells (Dutta et al., 2012; Rubinsztein et al., 2011; Wang and Klionsky, 2011) (Fig. 3). Although autophagy plays a relatively minor role in proliferating noncardiomyocyte cells (such as endothelial cells (ECs) and fibroblasts), it has been shown that a progressive decline in autophagy efficiency in senescent cardiomyocytes plays an important role in age-related cardiac diseases (Inuzuka et al., 2009). Due to the loss of autophagy, dysfunctional mitochondria accumulate as the cells ages, and promote the formation of ROS, leading to further accumulation of damaged mitochondria and ROS creating a positive-feedback loop that results in the "aging" phenotype with a decline in autophagy at its core (Rezzani et al., 2012; Terman et al., 2010).

Strikingly, Gibbings and coworkers have shown that autophagic activity is crucial in regulating MiRNA processing through engaging posttranscriptional regulation of DICER and establishes a checkpoint required for continued loading of MiRNAs into AGO2 (Gibbings et al., 2012). Autophagy-deficient cells exhibit decreased ago loading with MiRNAs, diminished MiRNA levels, and reduced silencing activity of exogenous siRNA duplexes (Gibbings et al., 2012). Thus, autophagy is required for continued loading of MiRNA duplexes into AGO and for their durable activity. This is supported by the fact that DICER protein accumulates in cells depleted of different autophagy-related genes (Sakurai et al., 2011). These results suggest that autophagy influences the expression of genes crucial in MiRNA processing and thereby maturation of MiRNAs. An age-related decline in autophagy may therefore affect MiRNA processing and thus mature MiRNA profiles in the heart.

The role of MiRNAs in autophagy is gaining more interest and one examples of an age-associated MiRNA that promotes autophagy is MiR-155. MiR-155 negatively regulates the autophagy-inhibitor mTOR through targeting of multiple genes involved within this pathway, including RHEB, RPTOR Independent Companion Of MTOR Complex 2 (RICTOR) and Ribosomal Protein S6 Kinase B2 (RPS6KB2) (Wan et al., 2014). These results suggest that MiR-155 possesses the potential to exert a beneficial role in the aging myocardium via prevention of an age-induced deleterious decrease in autophagy.

On the other hand, several studies report that different forms of stress can provoke an increase in cardiomyocyte autophagic activity and link this prolonged/excessive upregulation of autophagy with the development of cardiac hypertrophy (Cao et al., 2011; Nakai et al., 2007). In these situations, MiRNAs functioning as negative regulators of autophagy may benefit cardiac remodeling and function. Furthermore, MiR-30a decreases autophagic activity through suppression of Beclin 1 (Becn1) expression, which is able to regulate the expression of the main autophagy gene LC3B (Zhu et al., 2009; Zou et al., 2012). Strikingly, MiR-30 family members are downregulated in human and mouse hypertrophic hearts (Duisters et al., 2009; Pan et al., 2013) and AngII treatment is shown to result in autophagy-induced myocardial hypertrophy, at least partly, via decreasing the expression of MiR-30 (Zhu et al., 2009). Initiation of autophagy by the unc-51-like autophagy-activating kinase 1 (ULK1) complex is suppressed by MiR-20a and MiR-106b (Wu et al., 2012). Other negative regulators of autophagy include MiR-101 (Frankel et al., 2011; Wu et al., 2015), MiR-199a (Li et al., 2015d), MiR-221 (Su et al., 2015; Wang et al., 2012a), MiR-222 (Su et al., 2016), and MiR-374a (Huang et al., 2012). Collectively, MiRNAs able to increase autophagic activity potentially have beneficial a role during myocardial aging, whereas autophagy suppressing-miRNAs may aggravate age-related myocardial phenotypes.

5. PROGRAMMED CELL DEATH

Cardiomyocyte renewal occurs at a low rate (±1%/year) in youth and seems to slow down even further with advancing age (Bergmann et al., 2009). This nonmitotic nature of cardiomyocytes marks the importance of a balanced homeostasis to prevent cell death. Aging in the mammalian heart is characterized by a gradual loss of myocytes and hypertrophy of the remaining myocytes (Olivetti et al., 1991). Cardiomyocyte loss is primarily due to necrosis and apoptosis (Fig. 4). Deregulation of both cell death pathways has been observed in heart disease (Whelan et al., 2010) as well as during normal myocardial senescence (Kajstura et al., 1996; Kung et al., 2011).

5.1 MiRNAs Regulating Necrotic Cell Death

Necrosis is morphologically characterized by an increase in cell volume, dilation of organelles, rupture of the plasma membrane, and subsequent loss

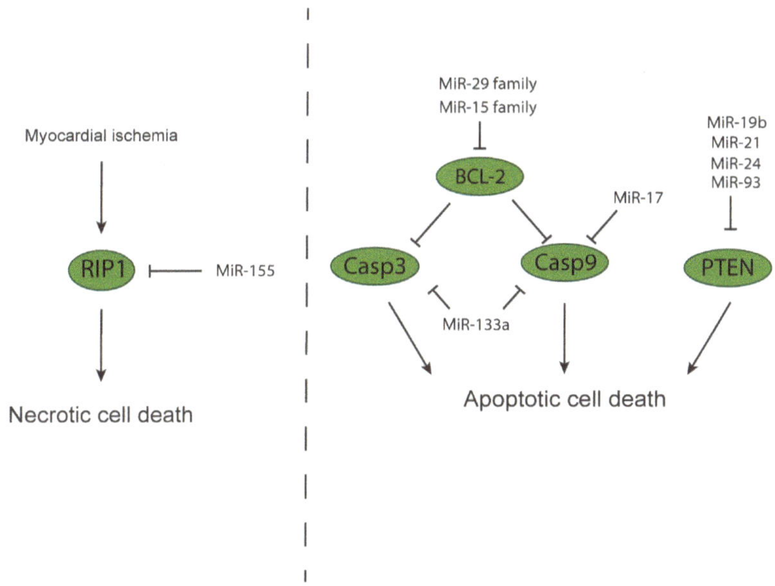

Fig. 4 Both cell death signaling pathways, apoptosis and necrosis, are affected by aging MiRNAs. Necrotic cell death can be caused by myocardial ischemia via activation of receptor interacting protein 1 (RIP1). Apoptotic cell death is regulated by B-cell lymphoma 2 (BCL-2) and executed by various caspases, including Casp3 and Casp9, and proapoptotic gene phosphatase and tensin homolog (PTEN).

of intracellular contents (Chiong et al., 2011; Kroemer et al., 2009). Cardiomyocytes can undergo necrosis when being exposed to excessive stress due to, for example, the lack of oxygen and nutrients, as during myocardial ischemia (Guerra et al., 1999). So far, only limited evidence has been gathered on the role of MiRNAs in regulation of cardiac necrosis. Age-associated MiR-155 was reported to attenuate hydrogen peroxide (H_2O_2)-induced necrotic cell death in cardiomyocyte progenitor cells (CMPCs) (Liu et al., 2011). Strikingly, transplantation of CMPCs overexpressing MiR-155 in the infarcted myocardium induced an antinecrotic response through repression of receptor-interacting protein 1 (RIP1) (Liu et al., 2011). Although certain MiRNAs have been shown to influence cardiac necrotic cell death in vitro as well as in vivo, the role of MiRNAs in regulation of necrotic cell death pathways is under investigated, especially in the context of age-associated necrosis.

5.2 MiRNAs Promoting Apoptotic Cell Death

Unlike necrosis, myocardial apoptotic cell death and the role of MiRNAs therein have been well documented. Apoptosis is a highly regulated process in which signaling is mediated by the mitochondria (Crow et al., 2004), regulated by a family of both proapoptotic and antiapoptotic member proteins called BCL-2 (Gustafsson and Gottlieb, 2008), and executed by a family of cysteine proteases, called caspases (Hengartner, 2000).

A large body of work describes the impact of MiRNAs on cardiac apoptosis, including MiRNAs implicated in myocardial aging, such as MiR-34a. MiR-34a is upregulated in cardiomyocyte apoptosis induced by doxorubicin or excessive ROS accumulation (Matsumoto et al., 2013), and its inhibition has cardioprotective effects, partially diminishing apoptosis both in vitro and in vivo (Boon et al., 2013; Desai et al., 2014; Matsumoto et al., 2013; Yang et al., 2015). Interestingly, MiR-34a has been identified as a predictive plasma marker of future HF in patients after acute MI (Matsumoto et al., 2013). Moreover, MiR-92a, a member of the aging MiRNA cluster MiR-17-92, has been proven to regulate cardiomyocyte apoptosis. Circulating expression of MiR-92a is elevated in acute MI patients as well as in post-MI cardiac and blood samples of rodents (Bonauer et al., 2009). Knockdown of MiR-92a has cardioprotective effects through reduction of apoptosis while overexpression has the opposite effect (Bonauer et al., 2009; Jiang et al., 2014; Liu et al., 2016; Zhang et al., 2014b).

Two different age-associated MiRNA families, the MiR-15- and MiR-29 family, are proven to regulate apoptotic signaling through direct inhibition of antiapoptotic gene BCL-2. Various MiR-15 family members were found to be upregulated in the overloaded heart (Tijsen et al., 2014) while being downregulated in both border and infarct zone of mice subjected to MI (Porrello et al., 2013). Several of these members are proven to promote apoptosis, including MiR-15a, MiR-15b, MiR-16, MiR-195, and MiR-497, and worsen myocardial ischemia-induced cardiac injury (Hullinger et al., 2012; Li et al., 2015c; Thum et al., 2007; van Rooij et al., 2006; Zhu et al., 2011b). Furthermore, MiR-29 family members, MiR-29a/c protect cultured cardiomyocytes from I/R injury, evidenced by increased cell survival and decreased Casp3 activity after inhibition (Ye et al., 2010).

Various other MiRNAs have been identified to modify cell death via targeting of additional negative regulators of the apoptotic pathway. For example, apoptosis-induced MiR-140 promotes apoptosis via targeting of antiapoptotic mitofusin 1 (Li et al., 2014b). Silencing of MiR-140 in vitro and in vivo attenuates apoptosis and reduces infarct size in a mouse model of I/R injury (Li et al., 2014b).

MiR-155 is, in addition to prevention of necrosis, suggested to stimulate apoptosis. Inhibition of MiR-155 in LPS-treated mice attenuated cardiac apoptosis and preserved cardiac function, while forced expression of MiR-155 aggravated cardiac dysfunction (Wang et al., 2016). These results are indicative for a proapoptotic function of MiR-155 in the heart during sepsis. The discrepancy between the observed antinecrotic and proapoptotic effect can possibly be explained by the fact that the role of MiR-155 was investigated in different cell types. Next to that, Wang et al. studied the effect of the well-known proinflammatory MiR-155 on cardiomyocyte apoptosis in a LPS-induced sepsis model. It is possible that the decreased apoptosis after inhibition of MiR-155 is rather due to a diminished inflammatory response than a direct effect on apoptotic signaling.

Additional MiRNAs enhancing apoptosis via targeting negative regulators/factors of the intrinsic apoptotic pathway include MiR-1 (Zhu et al., 2016), MiR-100 (Chen et al., 2015), and MiR-132 (Hong et al., 2015). These results suggest that various MiRNAs are able to provoke apoptosis in cardiomyocytes, in most, but not all, cases via direct or indirect suppression of activity of antiapoptotic gene BCL-2, thereby aggravating myocardial function.

5.3 MiRNAs Inhibiting Apoptotic Cell Death

Other MiRNA have been shown to inhibit apoptosis, including the senescence-associated MiRNA cluster MiR-17-92. Hypoxia decreases expression of various MiR-17-92 cluster members, and overexpression of this cluster markedly attenuates apoptosis (Yan et al., 2009). For example, MiR-17 exert its antiapoptotic function through suppression of Casp9 and apoptotic protease-activating factor 1 (Apaf-1) (Song et al., 2015) while MiR-19b prevents apoptosis via phosphatase and tensin homolog (PTEN) suppression (Xu et al., 2016).

One of the best-studied antiapoptotic MiRNAs is MiR-21. MiR-21 displays increased expression levels in border areas in infarcted hearts while being downregulated in infarcted areas (Cheng and Zhang, 2010; Dong et al., 2009). Silencing of MiR-21 aggravates H_2O_2-induced apoptotic whereas overexpression of MiR-21 protects against apoptosis and subsequently decreases MI size and myocardial remodeling (Dong et al., 2009; Yang et al., 2016a). The antiapoptotic effect of MiR-21 is established via targeting of different proapoptotic genes, such as PTEN (Meng et al., 2007; Yang et al., 2014a), PDCD4 (Asangani et al., 2008; Cheng et al., 2009), and TPM1 (Zhu et al., 2007). Interestingly, Huang and coworkers showed that MiR-21/MiR-146a synergistically reduced ischemia/hypoxia-induced cardiomyocyte apoptosis in vitro and in a murine MI-model, the combination of the MiRNAs being superior to each MiRNA alone (Huang et al., 2016).

Similarly, MiR-24 and MiR-93 are reported to target, among others, PTEN and attenuate cardiomyocyte apoptosis and myocardial injury in vivo (Ke et al., 2016; Qian et al., 2011; Yang et al., 2016c). Although MiR-24 prevents apoptosis in cardiomyocytes, forced expression of MiR-24 in ECs promotes apoptosis through direct targeting of GATA2 and PAK4, which in their turn, impinge on BCL-2-associated agonist of cell death (Fiedler et al., 2011). Silencing of endothelial MiR-24 in mice with MI attenuated cardiac EC apoptosis, reduced infarct size and preserved cardiac function (Fiedler et al., 2011).

Caspases are crucial in apoptotic signaling and MiR-133 is in control of multiple of these key players, exerting an antiapoptotic function. Hearts from cardiac-specific MiR-133 transgenic mice displayed reduced Casp3 activation and apoptosis during pressure overload (Castaldi et al., 2014). In addition, MiR-133 protects cardiomyocytes from cell death through suppression of Casp9 explaining, at least partly, the beneficial effect of ischemic postconditioning (He et al., 2011; Xu et al., 2014).

Several members of the MiR-30 family, MiR-30e in particular, display reduced expression in response to doxorubicin-induced cardiotoxicity and were found to mediate apoptosis through inhibition of apoptotic signaling by mitochondria (Li et al., 2010a; Roca-Alonso et al., 2015). In addition, MiR-30e inhibits the expression of BCL-2/adenovirus E1B-interacting protein 3-like (BNIP3L or NIX) (Roca-Alonso et al., 2015).

Many additional MiRNAs have been shown to attenuate apoptotic cell death in cardiomyocytes exposed to H_2O_2 or doxorubicin, including MiR-199a (Rane et al., 2009), MiR-22 (Li et al., 2016a; Yang et al., 2016b), MiR-23a (Mao et al., 2014; Ruan et al., 2012), MiR-125b (Ma et al., 2016a), MiR-132 (Hong et al., 2015), MiR-150 (Tang et al., 2015), and MiR-206 (Song et al., 2009). These reports demonstrate that many different MiRNAs are able to inhibit apoptotic cell death in the myocardium, including various age-associated MiRNAs. Decreased expression of these antiapoptotic MiRNAs in the aging heart may explain the observed increase in apoptotic cell death.

6. INFLAMMATORY ENDOTHELIAL ACTIVATION AND MACROPHAGE RECRUITMENT

Aging is associated with a systemic proinflammatory state, sometimes referred to as "inflammaging," which can activate signaling cascades leading to structural and functional remodeling of the heart (Collier et al., 2011; Lam et al., 2011; Paulus and Tschope, 2013). A growing body of evidence indicates that proinflammatory signaling is a key factor driving remodeling of the heart with advancing age (reviewed in Lakatta, 2015). Allostatic-associated proinflammatory signaling molecules are upregulated whereas antiinflammatory molecules are downregulated in the myocardium. Importantly, the aging-associated proinflammatory response is not performed by "professional immune cells"; aged cardiac cells, including ECs, smooth muscle cells, and fibroblasts, acquire a senescence-associated secretory proinflammatory phenotype (Lakatta et al., 2009; Wang et al., 2014b). In particular, senescence of ECs and the accompanying dysfunction promotes the pathogenesis of age-related coronary disease, including atherosclerosis (Brandes et al., 2005; Campisi et al., 2011; Freund et al., 2011). Disturbed flow enhances cell senescence and primes ECs for inflammatory gene expression, and might promote accumulation of monocyte-derived macrophages at damaged areas in the arterial intima (Hajra et al., 2000;

Jongstra-Bilen et al., 2006; Warboys et al., 2014). Thus, MiRNAs controlling endothelial activation/senescence and the potential following recruitment of macrophages can influence cellular aging and the progression of age-related cardiac diseases (Fig. 5).

6.1 MiRNAs Regulating Endothelial Activation

Interestingly, among the most abundant MiRNAs in ECs are age-deregulated MiRNAs such as MiR-146a, MiR-21, MiR-126a, and members of the MiR-17-92 cluster (Harris et al., 2008). Moreover, shear stress forms the main cause of accelerated cellular senescence and shapes the MiRNA signature of ECs (Kuehbacher et al., 2007; Menghini et al., 2009; Suarez et al., 2007), altering the expression of senescence MiRNAs (Hergenreider et al., 2012; Qin et al., 2010). Among these MiRNAs, MiR-146a represses NFkB signaling via direct inhibition of IRAK 1 and TRAF6 and, thus, cellular enrichment of MiR-146a through systemic delivery of mimetics suppresses NFkB activation and reduces macrophage recruitment (Li et al., 2015a). In addition, MiR-146a$^{-/-}$ mice are resistant to endothelial activation following treatment with IL-1β because of reduced induction of proinflammatory adhesion molecules (Cheng et al., 2013). Conversely, overexpression of MiR-146a suppresses both chemokine production and monocyte adhesion to the endothelium, supporting the hypothesis that MiR-146a ameliorates endothelial inflammation and the progression of atherosclerosis (Ma et al., 2016b). Similarly, EC-enriched MiRNA MiR-126a is shown to be involved in endothelial activation via targeting vascular cell adhesion molecule-1 (VCAM-1) expression, preventing leukocyte adhesion to ECs (Harris et al., 2008; Hu et al., 2015; Zernecke et al., 2009).

Shear stress-inducible MiR-19a and MiR-21 promote endothelial inflammation and disease progression. Circulating MiR-21 may be an inflammatory biomarker linking the aging process to cardiac disease (Olivieri et al., 2012), displaying an age-related increase in the plasma MiR-21 levels (Schetter et al., 2009). Here, MiR-21 expression and IL-6 and IL-8 levels correlated, suggesting a potential role of MiR-21 in inflammatory gene expression. On the contrary, disturbed blood flow-induced upregulation of MiR-21 in ECs suppresses translation of PPARα, inducing the expression of vascular VCAM-1 and CCL2, and promoting endothelial inflammation (Zhou et al., 2011). Moreover, MiR-19a is upregulated in atherosclerotic ECs and increases NFkB activity

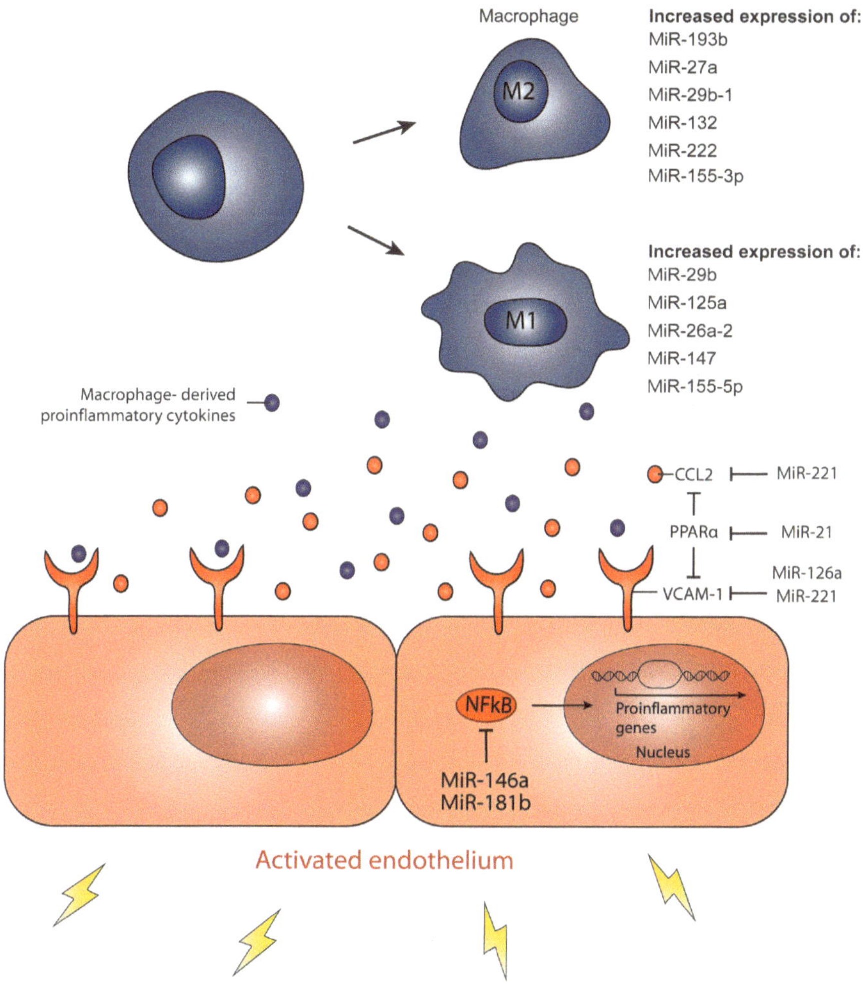

Fig. 5 Macrophage polarization shapes their MiRNA profile, subsequently affecting endothelium activation and their inflammatory response. Macrophages can polarize into either proinflammatory M1 macrophages or antiinflammatory M2 macrophages, which display distinct MiRNA profiles. NFκB signaling plays a central role in endothelium activation, increasing the expression of proinflammatory cytokines and chemokines, including vascular cell adhesion molecule-1 (VCAM-1), and C—C motif chemokine ligand 2 (CCL2). Expression of CCL2 leads to the recruitment of M1 macrophages, which in their turn secrete proinflammatory cytokines to promote further endothelial activation.

by targeting several negative regulators of NFkB, functioning as a proatherogenic factor (Akhtar et al., 2015).

Additional senescent-associated MiRNAs induced by endothelial shear stress include MiR-181b, MiR-23b, and MiR-150, whereas shear stress downregulates expression of MiR-24, MiR-103, MiR-210, and MiR-221 (Hergenreider et al., 2012; Son et al., 2013; Wang et al., 2010b, 2014a). MiR-181b dampens the inflammatory response in the endothelium by limiting the translocation of NFkB from the cytoplasm to the nucleus, decreasing NFkB activity and contributing to the antiinflammatory phenotype of ECs at athero-protected regions of (coronary) arteries (Ma et al., 2016b; Sun et al., 2012, 2014). Furthermore, members of the let-7 MiRNA family are upregulated in ECs at athero-protected sites, whereas MiR-103 and MiR-221 are upregulated at athero-prone sites (Loyer et al., 2014). More specifically, overexpression of let-7g and let-7i has been shown to suppress endothelial and cardiac inflammatory activation, whereas a proinflammatory phenotype was acquired in the absence of let-7g or let-7i (Liao et al., 2014; Wang et al., 2015c). Similarly, increased MiR-103 expression enhances endothelial inflammation, while blocking of MiR-103 reduces macrophage accumulation and chemokine expression, suggesting a proinflammatory role for MiR-103 (Hartmann et al., 2016a, b). Finally, MiR-221 is able to decrease the expression of NFkB target genes such as VCAM-1 and CCL2 via targeting transcription factor ETS-1 (Kanters et al., 2003).

6.2 MiRNAs Influence Macrophage Polarization and Function

Acquisition of a senescence-associated secretory proinflammatory phenotype during endothelial aging is followed by recruitment of inflammatory cells. Following their adhesion to the endothelium, inflammatory cells can migrate into the intima, differentiate into macrophages, and enhance the local inflammatory response. Aging is associated with changes in cardiac tissue macrophage composition, gene expression, and function, altering the inflammatory response and increasing myocardial susceptibility (Pinto et al., 2014). More specifically, senescent mice show a suppressed postinfarct inflammatory response, evidenced by a decreased and delayed neutrophil and macrophage infiltration, markedly reduced cytokine and chemokine expression and impaired phagocytosis of death cardiomyocytes (Bujak et al., 2008). It is now recognized that MiRNAs can profoundly influence the biology of inflammatory cells, in particular the differentiation of monocytes into polarized macrophages. A large number of age-associated MiRNAs are differentially regulated during the differentiation of monocytes into macrophages, for example, MiR-24, MiR-30b, MiR-142-3p, and MiR-199a-5p are downregulated in this process (Fordham et al., 2015; Lin et al., 2014). In particular, MiR-24, MiR-30b, and MiR-142-3p impair macrophage differentiation and activation, as evidenced by reduced macrophage phagocytosis and reduced expression of inflammatory cytokines upon overexpression (Bettencourt et al., 2013; Fordham et al., 2015; Naqvi et al., 2015).

Once differentiated, recruited macrophages can take on two extreme states (Das et al., 2015; Sica et al., 2015). The initiation phase of an inflammatory response is characterized by proinflammatory macrophages, also known as classically activated or M1 macrophages, which secrete inflammatory cytokines, such as IL-12, IL-1β, and tumor necrosis factor (Lichtnekert et al., 2013; Sica et al., 2015). Antiinflammatory macrophages, also known as alternatively activated or M2 macrophages, play important roles in inflammation resolution and tissue repair by secreting antiinflammatory cytokines such as IL-10 (Lichtnekert et al., 2013). Proinflammatory M1 macrophages have increased expression of age-associated MiRNAs MiR-29b, MiR-125a, MiR-26a-2, MiR-147, and MiR-155-5p, whereas antiinflammatory M2 macrophages have increased expression of MiR-193b or MiR-27a, MiR-29b-1, MiR-132, MiR-222, and MiR-155-3p, respectively (Graff et al., 2012; Nazari-Jahantigh et al., 2012; O'Connell et al., 2007) (Fig. 5). Moreover, forced expression of MiRNAs affects macrophage polarization status. For example, in vitro overexpression of MiR-155-5p and MiR-29b induced differentiation toward a M1 macrophage expression pattern (Graff et al., 2012; Nazari-Jahantigh et al., 2012).

One of the most well-investigated MiRNAs in cardiac inflammatory disease, MiR-155, is upregulated in the peripheral blood of aged women (Sredni et al., 2011), in human B cells from elderly individuals (Frasca et al., 2015) and is suggested to form a senescence-associated secretory marker (Frasca et al., 2016). Moreover, Corsten et al. (2012) have shown that elevated MiR-155 levels were found in cardiac tissue of both human and mice with viral-induced myocarditis (Corsten et al., 2012). MiR-155 functions, among others, as a positive regulator of TLR-signaling and its silencing during viral-induced myocarditis mice models resulted in attenuation of cardiac inflammation (Heymans et al., 2013). MiR-155 is highly expressed in both mouse and human aortic atherosclerotic lesions but is reported to have a controversial role in atherosclerotic inflammation, reported to function both anti- and proatherogenic. Transplantation of MiR-155$^{-/-}$ bone marrow cells into LDLr$^{-/-}$ mice was reported to amplify macrophage inflammation, increase atherosclerotic lesion size, and decrease plaque stability (Donners et al., 2012), while both whole body MiR-155$^{-/-}$, ApoE$^{-/-}$ mice and ApoE$^{-/-}$ mice receiving bone marrow transplant of MiR-155$^{-/-}$ cells, showed markedly reduced macrophage inflammation and decreased atherosclerotic lesions (Du et al., 2014; Nazari-Jahantigh et al., 2012). Importantly, the differences in the ability of macrophage MiR-155 to reduce lesion progression may greatly depend on the stage of atherosclerotic development (i.e., early vs advanced) when the expression of key target genes is stage specific (Wei et al., 2015).

Aging-related MiRNAs MiR-21 and MiR-146b were found to be upregulated in myocarditis patients. These MiRNAs also have a central role in immune activation and inflammation (Corsten et al., 2012). Silencing of these MiRNAs by specific inhibitors showed a strong attenuation of viral-induced myocarditis mice (Liu et al., 2013). The exact role of these MiRNAs in the pathology of myocarditis is not elucidated; however, it is shown that the expression of MiR-21 and MiR-146b is correlated with IL-17 expression, implicating an immune modulating function for both age-associated MiRNAs (Liu et al., 2013).

Additionally, antiinflammatory MiRNAs, MiR-106a and MiR-93, were downregulated in the acute phase of human myocarditis (Corsten et al., 2012), while being upregulated in the senescent heart (Zhang et al., 2012). Both are involved in antiinflammatory responses by inhibiting the production of inflammatory cytokines (Schroen and Heymans, 2012; Zhu et al., 2013a).

Strikingly, MiR-181b is also increased in EC-extracellular vesicles-treated monocytic cells, while this MiRNAs is shown to suppress monocyte activation upon overexpression (Njock et al., 2015). These results suggest that crosstalk between the endothelium and monocytes/macrophages may be mediated via secreted extracellular vesicles containing antiinflammatory MiRNAs. This finding marks the important function of MiRNAs in controlling endothelial activation and the inflammatory response. We can conclude from these studies that MiRNA are key regulators of endothelial activation, monocyte differentiation, and macrophage activation and polarization. Dysregulation of these MiRNAs affects the aging process of the myocardium, potentially aggravating the processes driving cellular senescence.

7. ECM REMODELING

The ECM is a complex collection of proteins, including fibronectin and different types of collagens, and functions as the structural backbone of the heart in addition to being a critical regulator of cellular processes like proliferation, migration, and differentiation, through the control of signaling via growth factors and bioactive molecules (DeQuach et al., 2010). Importantly, excessive ECM hallmarks the senescent myocardium, increasing cardiac stiffness and mediating diastolic dysfunction (Barasch et al., 2011; Borlaug and Paulus, 2011; Gupta et al., 2014). In the heart, fibroblasts are the most abundant cell type (Gaspard et al., 2014; Lu et al., 2013) and they play a crucial role in myocardial homeostasis by preserving the balance of ECM synthesis and degradation (Tomoda et al., 2001). In response to various stresses, such as myocardial injury and exposure to cytokines or growth factors, fibroblasts undergo transformation to myofibroblasts (Cieslik et al., 2011; Ghosh et al., 2010a; Herum et al., 2013; Van Linthout et al., 2014; Westermann et al., 2011), which control ECM composition by regulating the secretion and activity of proteolytic enzymes, including members of the ECM degrading superfamily of matrix metalloproteinase and their tissue inhibitors (Ghosh et al., 2010b; Vasquez et al., 2011). The main profibrogenic growth factor identified as a central player of myofibroblasts transformation and ECM remodeling is transforming growth factorβ (TGFβ) (Chen et al., 2000; Vasquez et al., 2011), which signals through MAPK or SMAD proteins to drive fibrotic gene expression (Fig. 6).

TGFβ signaling does not only induce fibroblast activation and differentiation, it also shapes the fibroblast MiRNA profile. SMAD proteins have a dual function: (1) they bind to the DNA to regulate the transcription of MiRNAs and (2) they interact with the Drosha/DGCR8 complex to influence MiRNA maturation (Davis et al., 2008). For example, TGFβ1 induces expression of MiR-21 (Kumarswamy et al., 2012), while reducing expression of the MiR-30 family (Volkmann et al., 2013). Additionally, MiR-212/132 levels were increased after TGFβ treatment in ECs (Kumarswamy et al., 2014). These results suggest a possible link between the fibrotic response and MiRNAs. Different MiRNAs have been shown to posses the ability to regulate pathological fibrosis, including MiRNAs displaying an age-associated dysregulation.

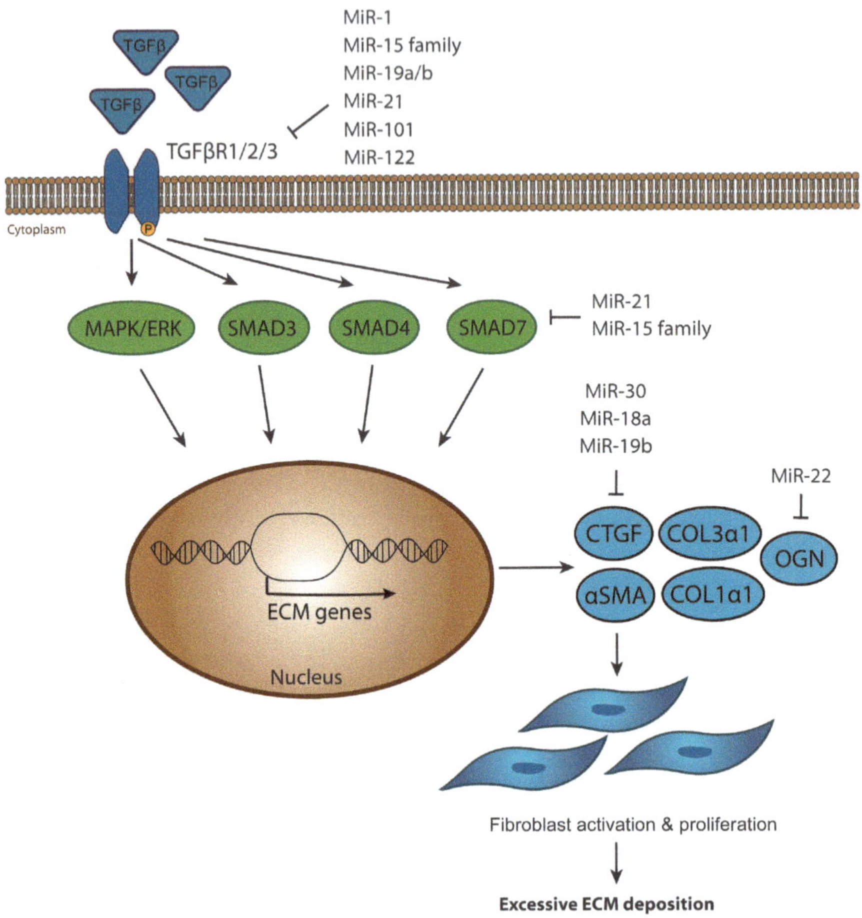

Fig. 6 Senescence-associated MiRNAs alter cardiac fibroblast activation and proliferation via targeting of the MAPK and TGFβ pathway, resulting in disbalance of the extracellular matrix (ECM). Activation of transforming growth factor beta receptors (TGFβR1/2/3) promotes MAPK/ERK and/or SMAD signaling. Both pathways stimulate the fibrotic response, which is characterized by increased transcription of connective tissue growth factor (CTGF), alpha smooth muscle actin (αSMA), osteoglycin (OGN), and different collagens (COL1α1 and COL3α1). Increased expression of these profibrotic genes leads to excessive ECM deposition through activation and proliferation of cardiac fibroblasts.

7.1 MiRNAs Suppressing ECM Production

Members of the age-associated MiR-17-92 MiRNA cluster, including MiR-18a and MiR-19b, have been implicated as potential regulators of ECM production during aging cardiomyopathy, through their targets

CTGF and TSP-1 (van Almen et al., 2011). Moreover, MiR-19a/b are shown to be a negative regulator of TGFβ signaling via direct targeting of TGFβR2, resulting in inhibition of interstitial fibrosis and attenuation of cardiac remodeling (Zou et al., 2016). MiR-22, another age-associated MiRNA, is one of the most abundantly expressed MiRNAs in the heart (Hu et al., 2012) whose expression increases during aging (Jazbutyte et al., 2013). This MiRNA promotes senescence and activation of cultured cardiac fibroblasts via repression of its target, the small leucine-rich proteoglycan osteoglycin (OGN) (Jazbutyte et al., 2013). Although MiR-22 knockout mice do not display any cardiac abnormalities under basal conditions, upon cardiac stress, they quickly develop dilated cardiomyopathy accompanied with excessive cardiac fibrosis (Gurha et al., 2012; Huang et al., 2013). MiR-22 inhibition might therefore be a potential therapeutic target to block fibrosis development in the elderly heart patient.

The well-documented MiR-34a has been implicated in cardiac aging and fibrosis formation. It is part of a positive-feedback loop, induced by TGFβ and itself promoting TGFβ-induced activation of cardiac fibroblasts by targeting co-smad and SMAD4 (Huang et al., 2014). MiR-34a is upregulated in senescent murine hearts as well as in aged human myocardial biopsies (Boon et al., 2013). Silencing of MiR-34a protected the heart against excessive cardiac fibrosis and remodeling upon pressure overload or MI (Bernardo et al., 2012; Boon et al., 2013; Emanueli and Thum, 2013; Loffredo et al., 2013).

The most well-studied MiRNA for its role in fibrosis is age-associated MiR-21. This cardiac fibroblast-enriched MiRNA is increased in human and murine HF (Thum et al., 2008; van Rooij et al., 2006; Villar et al., 2013), and both its myocardial and plasma levels correlated with intracardiac fibrosis in arotic stenosis patients (Villar et al., 2013). MiR-21 has multiple targets implicated in matrix deposition, including sprout homologye 1 (Thum et al., 2008), PTEN (Roy et al., 2009), SMAD7 (Lorenzen et al., 2015), and TGFβR3 (Liang et al., 2012), and their inhibition synergizes to provoke a pronounced cardiac fibrosis phenotype.

Furthermore, the aging-associated MiR-29 family is downregulated in response to cardiac stress (Kriegel et al., 2012; van Rooij et al., 2006). Notably, circulating levels of MiR-29a correlated inversely with cardiac fibrosis (Dawson et al., 2013; Roncarati et al., 2014). This family protects against cardiac fibrosis via direct targeting of a number of ECM genes and subsequent inhibition of MAPK- and TGFβ/SMAD3-signaling, preventing fibroblast activation and proliferation (Abonnenc et al., 2013; Tao et al.,

2016; Zhang et al., 2014c; Zhu et al., 2013b). Strikingly, rescue of MiR-29b expression 14 days after AngII treatment reduced cardiac fibrosis and restored cardiac function, indicating the therapeutic potential in established hypertensive heart disease (Zhang et al., 2014c).

The MiR-15 family (MiR-15a, MiR-15b, MiR-16, MiR-195, MiR-497, and MiR-322) is induced in response to cardiac stress (Hullinger et al., 2012; van Rooij et al., 2006) and targets TGFβR1, SMAD3, and SMAD7 among others, presumably to counteract TGFβ activity and to inhibit ECM remodeling (Tijsen et al., 2014). Indeed, inhibition of MiR-15b specifically aggravated fibrosis in pressure overloaded mice (Tijsen et al., 2014). On the other hand, upon experimental ischemia–reperfusion, inhibition of the complete MiR-15 family yielded smaller infarct sizes and less fibrosis (Hullinger et al., 2012). The basis for the different outcomes of MiR-15 inhibition remains unclear, but may be owed to the different nature of the insults to the heart and inhibition of the complete MiR-15 family vs only one of the family members.

Cardiac fibrosis during experimental pressure overload is associated with the downregulation of MiR-133a (Castoldi et al., 2012) and concomitant derepression of its targets fibronectin, Col1a1, and Col4a1 (Castoldi et al., 2012; Chen et al., 2014; Matkovich et al., 2010). Also, the senescence-associated MiR-30 is a fibroblast-enriched MiRNA, downregulated in the failing human and rodent heart (van Rooij et al., 2006), exerting its antifibrotic function via targeting of CTGF (Duisters et al., 2009).

Many other MiRNAs associated with aging have been identified as negative regulators of the cardiac fibrotic response by inhibiting crucial profibrotic signaling cascades, such as the TGFβ and MAPK/ERK pathways, including MiR-24, MiR-122, MiR-101, MiR-19a/b, and MiR-150. MiR-24 is downregulated shortly after MI and negatively regulates cardiac fibrosis by controlling the expression of Furin, which is necessary for TGFβ excretion (Wang et al., 2012b). Overexpression of MiR-24 into the mouse myocardium before inducing MI led to a reduced infarct size and improved cardiac function (Wang et al., 2012b). The same protective effect of MiR-122 on myocardial fibrosis in human AOS was revealed and was dependent on the targeting of TGFβ1 (Beaumont et al., 2014). Adenovirus-mediated overexpression of MiR-101 can attenuate interstitial fibrosis formation and inhibits MI-induced cardiac dysfunction in rats via silencing of TGFβR1 and the transcription factor c-Fos, suppressing proliferation of cardiac fibroblasts (Pan et al., 2012; Zhao et al., 2015).

Next to TGFβ, MAPK/ERK signaling plays a critical role in activation of cardiac fibroblasts and collagen formation (Fan et al., 2014). Interestingly, MiR-7a/b inhibits both the TGFβ and MAPK pathways and subsequently reduces collagen I expression and inhibits cardiac fibroblast proliferation and migration (Li et al., 2015b). MiR-150 is decreased in murine hearts suffering from pressure overload and inhibits cardiac fibroblast activation upon overexpression in vitro, while MiR-150 KO mice display a more severe deterioration of cardiac function due to excessive cardiac fibrosis in comparison with wild-type mice (Deng et al., 2016).

7.2 MiRNAs Promoting ECM Production

Age-associated and TGFβ-induced MiR-125b promotes fibrosis through inhibition of the antifibrotic secreted frizzled-related protein 5 (SFRP5), resulting in both proliferation and activation of cardiac fibroblasts and derepression of αSMA and collagen I and III (Nagpal et al., 2016).

Overall, age-associated changes in the molecular environment, such as increased TGFβ signaling, are accompanied with altered MiRNA expression profiles. In turn, these MiRNAs are able to influence fibroblast activation, differentiation, and collagen production, marking them as potential therapeutic targets to fine-tune ECM deposition.

8. CONCLUSION

Various cellular processes are identified to drive cardiac senescence resulting in age-associated functional deterioration. Therewith, described molecular hallmarks of cardiac aging also drive the progression of HF, implying that HF is an accelerated state of myocardial aging, involving similar processes. Still, "healthy aging" suggests the presence of a heart that is functional enough to meet the body's demand and such heart will by definition differ from a failing heart. Therefore, using HF-MiRNA studies to understand the cardiac aging process better as we did here is acceptable. Notably, recent studies suggest that age-associated functional and structural changes in the heart are, at least partly, caused by exhaustion of the stem cell pool in the myocardium, leading to impaired repair and regeneration in the aged heart (reviewed in Anversa et al., 2005). However, the role of stem cells in cardiac aging was not discussed within this review since the abundance and biological relevance of stem cells in the heart is still under debate. Future research should be conducted to study the role of MiRNAs more extensively in the context of the healthy aging heart.

As highlighted throughout this review, cardiac and circulating MiRNA expression profiles change during aging. The dysregulation of these responsive MiRNAs can be the cause of age-induced alterations in the microenvironment, including desensitization of β-adrenergic signaling, local proinflammatory stimulation, ECM-deriving growth factors, and induction of apoptosis. In their turn, MiRNAs are able to orchestrate the processes underlying senescence and potentially aggravate or counterbalance the cardiac aging phenotypes. This bidirectional connection between the cellular hallmarks of aging and MiRNA function stresses the importance to gain knowledge about (1) the effect of microenvironmental aging on MiRNA expression and (2) the exact regulatory function of MiRNAs in cellular processes driving cardiac aging.

REFERENCES

Abonnenc, M., Nabeebaccus, A.A., Mayr, U., Barallobre-Barreiro, J., Dong, X., Cuello, F., Sur, S., Drozdov, I., Langley, S.R., Lu, R., Stathopoulou, K., Didangelos, A., Yin, X., Zimmermann, W.H., Shah, A.M., Zampetaki, A., Mayr, M., 2013. Extracellular matrix secretion by cardiac fibroblasts: role of microRNA-29b and microRNA-30c. Circ. Res. 113, 1138–1147.

Akhtar, S., Hartmann, P., Karshovska, E., Rinderknecht, F.A., Subramanian, P., Gremse, F., Grommes, J., Jacobs, M., Kiessling, F., Weber, C., Steffens, S., Schober, A., 2015. Endothelial hypoxia-inducible factor-1alpha promotes atherosclerosis and monocyte recruitment by upregulating microRNA-19a. Hypertension 66, 1220–1226.

Antos, C.L., McKinsey, T.A., Frey, N., Kutschke, W., Mcanally, J., Shelton, J.M., Richardson, J.A., Hill, J.A., Olson, E.N., 2002. Activated glycogen synthase-3 beta suppresses cardiac hypertrophy in vivo. Proc. Natl. Acad. Sci. U.S.A. 99, 907–912.

Anversa, P., Hiler, B., Ricci, R., Guideri, G., Olivetti, G., 1986. Myocyte cell loss and myocyte hypertrophy in the aging rat heart. J. Am. Coll. Cardiol. 8, 1441–1448.

Anversa, P., Rota, M., Urbanek, K., Hosoda, T., Sonnenblick, E.H., Leri, A., Kajstura, J., Bolli, R., 2005. Myocardial aging—a stem cell problem. Basic Res. Cardiol. 100, 482–493.

Aoi, W., Naito, Y., Mizushima, K., Takanami, Y., Kawai, Y., Ichikawa, H., Yoshikawa, T., 2010. The microRNA miR-696 regulates PGC-1{alpha} in mouse skeletal muscle in response to physical activity. Am. J. Physiol. Endocrinol. Metab. 298, E799–E806.

Asangani, I.A., Rasheed, S.A., Nikolova, D.A., Leupold, J.H., Colburn, N.H., Post, S., Allgayer, H., 2008. MicroRNA-21 (miR-21) post-transcriptionally downregulates tumor suppressor Pdcd4 and stimulates invasion, intravasation and metastasis in colorectal cancer. Oncogene 27, 2128–2136.

Aschrafi, A., Schwechter, A.D., Mameza, M.G., Natera-Naranjo, O., Gioio, A.E., Kaplan, B.B., 2008. MicroRNA-338 regulates local cytochrome c oxidase IV mRNA levels and oxidative phosphorylation in the axons of sympathetic neurons. J. Neurosci. 28, 12581–12590.

Bai, X.Y., Ma, Y., Ding, R., Fu, B., Shi, S., Chen, X.M., 2011. miR-335 and miR-34a promote renal senescence by suppressing mitochondrial antioxidative enzymes. J. Am. Soc. Nephrol. 22, 1252–1261.

Barasch, E., Gottdiener, J.S., Aurigemma, G., Kitzman, D.W., Han, J., Kop, W.J., Tracy, R.P., 2011. The relationship between serum markers of collagen turnover and cardiovascular outcome in the elderly: the Cardiovascular Health Study. Circ. Heart Fail. 4, 733–739.

Barger, P.M., Kelly, D.P., 2000. PPAR signaling in the control of cardiac energy metabolism. Trends Cardiovasc. Med. 10, 238–245.

Beaumont, J., Lopez, B., Hermida, N., Schroen, B., San Jose, G., Heymans, S., Valencia, F., Gomez-Doblas, J.J., de Teresa, E., Diez, J., Gonzalez, A., 2014. microRNA-122 down-regulation may play a role in severe myocardial fibrosis in human aortic stenosis through TGF-beta1 up-regulation. Clin. Sci. (Lond.) 126, 497–506.

Bergmann, O., Bhardwaj, R.D., Bernard, S., Zdunek, S., Barnabe-Heider, F., Walsh, S., Zupicich, J., Alkass, K., Buchholz, B.A., Druid, H., Jovinge, S., Frisen, J., 2009. Evidence for cardiomyocyte renewal in humans. Science 324, 98–102.

Berk, B.C., Fujiwara, K., Lehoux, S., 2007. ECM remodeling in hypertensive heart disease. J. Clin. Invest. 117, 568–575.

Bernardo, B.C., Gao, X.M., Winbanks, C.E., Boey, E.J., Tham, Y.K., Kiriazis, H., Gregorevic, P., Obad, S., Kauppinen, S., Du, X.J., Lin, R.C., Mcmullen, J.R., 2012. Therapeutic inhibition of the miR-34 family attenuates pathological cardiac remodeling and improves heart function. Proc. Natl. Acad. Sci. U. S. A. 109, 17615–17620.

Bettencourt, P., Marion, S., Pires, D., Santos, L.F., Lastrucci, C., Carmo, N., Blake, J., Benes, V., Griffiths, G., Neyrolles, O., Lugo-Villarino, G., Anes, E., 2013. Actin-binding protein regulation by microRNAs as a novel microbial strategy to modulate phagocytosis by host cells: the case of N-Wasp and miR-142-3p. Front. Cell. Infect. Microbiol. 3, 19.

Boengler, K., Hilfiker-Kleiner, D., Heusch, G., Schulz, R., 2010. Inhibition of permeability transition pore opening by mitochondrial STAT3 and its role in myocardial ischemia/reperfusion. Basic Res. Cardiol. 105, 771–785.

Bonauer, A., Carmona, G., Iwasaki, M., Mione, M., Koyanagi, M., Fischer, A., Burchfield, J., Fox, H., Doebele, C., Ohtani, K., Chavakis, E., Potente, M., Tjwa, M., Urbich, C., Zeiher, A.M., Dimmeler, S., 2009. MicroRNA-92a controls angiogenesis and functional recovery of ischemic tissues in mice. Science 324, 1710–1713.

Boon, R.A., Iekushi, K., Lechner, S., Seeger, T., Fischer, A., Heydt, S., Kaluza, D., Treguer, K., Carmona, G., Bonauer, A., Horrevoets, A.J., Didier, N., Girmatsion, Z., Biliczki, P., Ehrlich, J.R., Katus, H.A., Muller, O.J., Potente, M., Zeiher, A.M., Hermeking, H., Dimmeler, S., 2013. MicroRNA-34a regulates cardiac ageing and function. Nature 495, 107–110.

Borlaug, B.A., Paulus, W.J., 2011. Heart failure with preserved ejection fraction: pathophysiology, diagnosis, and treatment. Eur. Heart J. 32, 670–679.

Brandes, R.P., Fleming, I., Busse, R., 2005. Endothelial aging. Cardiovasc. Res. 66, 286–294.

Bujak, M., Kweon, H.J., Chatila, K., Li, N., Taffet, G., Frangogiannis, N.G., 2008. Aging-related defects are associated with adverse cardiac remodeling in a mouse model of reperfused myocardial infarction. J. Am. Coll. Cardiol. 51, 1384–1392.

Bursi, F., Rocca, W.A., Killian, J.M., Weston, S.A., Knopman, D.S., Jacobsen, S.J., Roger, V.L., 2006a. Heart disease and dementia: a population-based study. Am. J. Epidemiol. 163, 135–141.

Bursi, F., Weston, S.A., Redfield, M.M., Jacobsen, S.J., Pakhomov, S., Nkomo, V.T., Meverden, R.A., Roger, V.L., 2006b. Systolic and diastolic heart failure in the community. JAMA 296, 2209–2216.

Cain, B.S., Meldrum, D.R., Joo, K.S., Wang, J.F., Meng, X., Cleveland Jr., J.C., Banerjee, A., Harken, A.H., 1998. Human SERCA2a levels correlate inversely with age in senescent human myocardium. J. Am. Coll. Cardiol. 32, 458–467.

Càllis, T.E., Pandya, K., Seok, H.Y., Tang, R.H., Tatsuguchi, M., Huang, Z.P., Chen, J.F., Deng, Z., Gunn, B., Shumate, J., Willis, M.S., Selzman, C.H., Wang, D.Z., 2009. MicroRNA-208a is a regulator of cardiac hypertrophy and conduction in mice. J. Clin. Invest. 119, 2772–2786.

Campisi, J., Andersen, J.K., Kapahi, P., Melov, S., 2011. Cellular senescence: a link between cancer and age-related degenerative disease? Semin. Cancer Biol. 21, 354–359.

Cantley, L.C., 2002. The phosphoinositide 3-kinase pathway. Science 296, 1655–1657.

Cao, D.J., Wang, Z.V., Battiprolu, P.K., Jiang, N., Morales, C.R., Kong, Y., Rothermel, B.A., Gillette, T.G., Hill, J.A., 2011. Histone deacetylase (HDAC) inhibitors attenuate cardiac hypertrophy by suppressing autophagy. Proc. Natl. Acad. Sci. U.S.A. 108, 4123–4128.

Care, A., Catalucci, D., Felicetti, F., Bonci, D., Addario, A., Gallo, P., Bang, M.L., Segnalini, P., Gu, Y., Dalton, N.D., Elia, L., Latronico, M.V., Hoydal, M., Autore, C., Russo, M.A., Dorn 2nd, G.W., Ellingsen, O., Ruiz-Lozano, P., Peterson, K.L., Croce, C.M., Peschle, C., Condorelli, G., 2007. MicroRNA-133 controls cardiac hypertrophy. Nat. Med. 13, 613–618.

Carrillo, E.D., Escobar, Y., Gonzalez, G., Hernandez, A., Galindo, J.M., Garcia, M.C., Sanchez, J.A., 2011. Posttranscriptional regulation of the beta2-subunit of cardiac L-type Ca2+ channels by MicroRNAs during long-term exposure to isoproterenol in rats. J. Cardiovasc. Pharmacol. 58, 470–478.

Castaldi, A., Zaglia, T., Di Mauro, V., Carullo, P., Viggiani, G., Borile, G., Di Stefano, B., Schiattarella, G.G., Gualazzi, M.G., Elia, L., Stirparo, G.G., Colorito, M.L., Pironti, G., Kunderfranco, P., Esposito, G., Bang, M.L., Mongillo, M., Condorelli, G., Catalucci, D., 2014. MicroRNA-133 modulates the beta1-adrenergic receptor transduction cascade. Circ. Res. 115, 273–283.

Castoldi, G., Di Gioia, C.R., Bombardi, C., Catalucci, D., Corradi, B., Gualazzi, M.G., Leopizzi, M., Mancini, M., Zerbini, G., Condorelli, G., Stella, A., 2012. MiR-133a regulates collagen 1A1: potential role of miR-133a in myocardial fibrosis in angiotensin II-dependent hypertension. J. Cell. Physiol. 227, 850–856.

Chan, S.Y., Zhang, Y.Y., Hemann, C., Mahoney, C.E., Zweier, J.L., Loscalzo, J., 2009. MicroRNA-210 controls mitochondrial metabolism during hypoxia by repressing the iron-sulfur cluster assembly proteins ISCU1/2. Cell Metab. 10, 273–284.

Charron, F., Paradis, P., Bronchain, O., Nemer, G., Nemer, M., 1999. Cooperative interaction between GATA-4 and GATA-6 regulates myocardial gene expression. Mol. Cell. Biol. 19, 4355–4365.

Chen, C.H., Nakayama, M., Nevo, E., Fetics, B.J., Maughan, W.L., Kass, D.A., 1998. Coupled systolic-ventricular and vascular stiffening with age: implications for pressure regulation and cardiac reserve in the elderly. J. Am. Coll. Cardiol. 32, 1221–1227.

Chen, M.M., Lam, A., Abraham, J.A., Schreiner, G.F., Joly, A.H., 2000. CTGF expression is induced by TGF- beta in cardiac fibroblasts and cardiac myocytes: a potential role in heart fibrosis. J. Mol. Cell. Cardiol. 32, 1805–1819.

Chen, F., Kook, H., Milewski, R., Gitler, A.D., LU, M.M., LI, J., Nazarian, R., Schnepp, R., Jen, K., Biben, C., Runke, G., Mackay, J.P., Novotny, J., Schwartz, R.J., Harvey, R.P., Mullins, M.C., Epstein, J.A., 2002. Hop is an unusual homeobox gene that modulates cardiac development. Cell 110, 713–723.

Chen, J.F., Mandel, E.M., Thomson, J.M., Wu, Q., Callis, T.E., Hammond, S.M., Conlon, F.L., Wang, D.Z., 2006. The role of microRNA-1 and microRNA-133 in skeletal muscle proliferation and differentiation. Nat. Genet. 38, 228–233.

Chen, Y.H., Heneidi, S., Lee, J.M., Layman, L.C., Stepp, D.W., Gamboa, G.M., Chen, B.S., Chazenbalk, G., Azziz, R., 2013. miRNA-93 inhibits GLUT4 and is overexpressed in adipose tissue of polycystic ovary syndrome patients and women with insulin resistance. Diabetes 62, 2278–2286.

Chen, S., Puthanveetil, P., Feng, B., Matkovich, S.J., Dorn 2nd, G.W., Chakrabarti, S., 2014. Cardiac miR-133a overexpression prevents early cardiac fibrosis in diabetes. J. Cell. Mol. Med. 18, 415–421.

Chen, A., Li, G., Chen, L., Guo, J., Liu, Y., 2015. Downregulation of microRNA-100 protects H_2O_2-induced apoptosis in neonatal cardiomyocytes. Int. J. Clin. Exp. Pathol. 8, 5491–5496.

Cheng, Y., Zhang, C., 2010. MicroRNA-21 in cardiovascular disease. J. Cardiovasc. Transl. Res. 3, 251–255.

Cheng, Y., Liu, X., Zhang, S., Lin, Y., Yang, J., Zhang, C., 2009. MicroRNA-21 protects against the H(2)O(2)-induced injury on cardiac myocytes via its target gene PDCD4. J. Mol. Cell. Cardiol. 47, 5–14.

Cheng, H.S., Sivachandran, N., Lau, A., Boudreau, E., Zhao, J.L., Baltimore, D., Delgado-Olguin, P., Cybulsky, M.I., Fish, J.E., 2013. MicroRNA-146 represses endothelial activation by inhibiting pro-inflammatory pathways. EMBO Mol. Med. 5, 1017–1034.

Chiong, M., Wang, Z.V., Pedrozo, Z., Cao, D.J., Troncoso, R., Ibacache, M., Criollo, A., Nemchenko, A., Hill, J.A., Lavandero, S., 2011. Cardiomyocyte death: mechanisms and translational implications. Cell Death Dis. 2 e244.

Chu, B., Wu, T., Miao, L., MEI, Y., Wu, M., 2015. MiR-181a regulates lipid metabolism via IDH1. Sci. Rep. 5, 8801.

Cieslik, K.A., Taffet, G.E., Carlson, S., Hermosillo, J., Trial, J., Entman, M.L., 2011. Immune-inflammatory dysregulation modulates the incidence of progressive fibrosis and diastolic stiffness in the aging heart. J. Mol. Cell. Cardiol. 50, 248–256.

Collier, P., Watson, C.J., Voon, V., Phelan, D., Jan, A., Mak, G., Martos, R., Baugh, J.A., Ledwidge, M.T., McDonald, K.M., 2011. Can emerging biomarkers of myocardial remodelling identify asymptomatic hypertensive patients at risk for diastolic dysfunction and diastolic heart failure? Eur. J. Heart Fail. 13, 1087–1095.

Corsten, M.F., Papageorgiou, A., Verhesen, W., Carai, P., Lindow, M., Obad, S., Summer, G., Coort, S.L., Hazebroek, M., Van Leeuwen, R., Gijbels, M.J., Wijnands, E., Biessen, E.A., De Winther, M.P., Stassen, F.R., Carmeliet, P., Kauppinen, S., Schroen, B., Heymans, S., 2012. MicroRNA profiling identifies microRNA-155 as an adverse mediator of cardiac injury and dysfunction during acute viral myocarditis. Circ. Res. 111, 415–425.

Crow, M.T., Mani, K., Nam, Y.J., Kitsis, R.N., 2004. The mitochondrial death pathway and cardiac myocyte apoptosis. Circ. Res. 95, 957–970.

da Costa Martins, P.A., Salic, K., Gladka, M.M., Armand, A.S., Leptidis, S., el Azzouzi, H., Hansen, A., Coenen-de Roo, C.J., Bierhuizen, M.F., van der Nagel, R., van Kuik, J., de Weger, R., de Bruin, A., Condorelli, G., Arbones, M.L., Eschenhagen, T., de Windt, L.J., 2010. MicroRNA-199b targets the nuclear kinase Dyrk1a in an auto-amplification loop promoting calcineurin/NFAT signalling. Nat. Cell Biol. 12, 1220–1227.

Dai, D.F., Rabinovitch, P.S., 2009. Cardiac aging in mice and humans: the role of mitochondrial oxidative stress. Trends Cardiovasc. Med. 19, 213–220.

Dai, D.F., Santana, L.F., Vermulst, M., Tomazela, D.M., Emond, M.J., Maccoss, M.J., Gollahon, K., Martin, G.M., Loeb, L.A., Ladiges, W.C., Rabinovitch, P.S., 2009. Overexpression of catalase targeted to mitochondria attenuates murine cardiac aging. Circulation 119, 2789–2797.

Das, S., Bedja, D., Campbell, N., Dunkerly, B., Chenna, V., Maitra, A., Steenbergen, C., 2014. miR-181c regulates the mitochondrial genome, bioenergetics, and propensity for heart failure in vivo. PLoS One 9, e96820.

Das, A., Sinha, M., Datta, S., Abas, M., Chaffee, S., Sen, C.K., Roy, S., 2015. Monocyte and macrophage plasticity in tissue repair and regeneration. Am. J. Pathol. 185, 2596–2606.

Davies, C.H., Ferrara, N., Harding, S.E., 1996. Beta-adrenoceptor function changes with age of subject in myocytes from non-failing human ventricle. Cardiovasc. Res. 31, 152–156.
Davis, B.N., Hilyard, A.C., Lagna, G., Hata, A., 2008. SMAD proteins control DROSHA-mediated microRNA maturation. Nature 454, 56–61.
Dawson, K., Wakili, R., Ordog, B., Clauss, S., Chen, Y., Iwasaki, Y., Voigt, N., Qi, X.Y., Sinner, M.F., Dobrev, D., Kaab, S., Nattel, S., 2013. MicroRNA29: a mechanistic contributor and potential biomarker in atrial fibrillation. Circulation 127, 1466–1475. 1475e1–28.
Deng, P., Chen, L., Liu, Z., Ye, P., WANG, S., WU, J., YAO, Y., Sun, Y., Huang, X., Ren, L., Zhang, A., Wang, K., Wu, C., Yue, Z., Xu, X., Chen, M., 2016. MicroRNA-150 inhibits the activation of cardiac fibroblasts by regulating c-Myb. Cell. Physiol. Biochem. 38, 2103–2122.
Dequach, J.A., Mezzano, V., Miglani, A., Lange, S., Keller, G.M., Sheikh, F., Christman, K.L., 2010. Simple and high yielding method for preparing tissue specific extracellular matrix coatings for cell culture. PLoS One 5, e13039.
Desai, V.G., Kwekel, C.J., Vijay, V., Moland, C.L., Herman, E.H., Lee, T., Han, T., Lewis, S.M., Davis, K.J., Muskhelishvili, L., Kerr, S., Fuscoe, J.C., 2014. Early biomarkers of doxorubicin-induced heart injury in a mouse model. Toxicol. Appl. Pharmacol. 281, 221–229.
Di Lisa, F., Canton, M., Menabo, R., Kaludercic, N., Bernardi, P., 2007. Mitochondria and cardioprotection. Heart Fail. Rev. 12, 249–260.
Dimitrakopoulou, K., Vrahatis, A.G., Bezerianos, A., 2015. Integromics network meta-analysis on cardiac aging offers robust multi-layer modular signatures and reveals micronome synergism. BMC Genomics 16, 147.
Dong, S., Cheng, Y., Yang, J., Li, J., Liu, X., Wang, X., Wang, D., Krall, T.J., Delphin, E.S., Zhang, C., 2009. MicroRNA expression signature and the role of microRNA-21 in the early phase of acute myocardial infarction. J. Biol. Chem. 284, 29514–29525.
Dong, D.L., Chen, C., Huo, R., Wang, N., Li, Z., Tu, Y.J., Hu, J.T., Chu, X., Huang, W., Yang, B.F., 2010. Reciprocal repression between microRNA-133 and calcineurin regulates cardiac hypertrophy: a novel mechanism for progressive cardiac hypertrophy. Hypertension 55, 946–952.
Donners, M.M., Wolfs, I.M., Stoger, L.J., Van Der Vorst, E.P., Pottgens, C.C., Heymans, S., Schroen, B., Gijbels, M.J., De Winther, M.P., 2012. Hematopoietic miR155 deficiency enhances atherosclerosis and decreases plaque stability in hyperlipidemic mice. PLoS One 7, e35877.
Du, F., Yu, F., Wang, Y., Hui, Y., Carnevale, K., Fu, M., Lu, H., Fan, D., 2014. MicroRNA-155 deficiency results in decreased macrophage inflammation and attenuated atherogenesis in apolipoprotein E-deficient mice. Arterioscler. Thromb. Vasc. Biol. 34, 759–767.
Duisters, R.F., Tijsen, A.J., Schroen, B., Leenders, J.J., Lentink, V., Van Der Made, I., Herias, V., Van Leeuwen, R.E., Schellings, M.W., Barenbrug, P., Maessen, J.G., Heymans, S., Pinto, Y.M., Creemers, E.E., 2009. miR-133 and miR-30 regulate connective tissue growth factor: implications for a role of microRNAs in myocardial matrix remodeling. Circ. Res. 104, 170–178. 6 p following 178.
Dutta, D., Calvani, R., Bernabei, R., Leeuwenburgh, C., Marzetti, E., 2012. Contribution of impaired mitochondrial autophagy to cardiac aging: mechanisms and therapeutic opportunities. Circ. Res. 110, 1125–1138.
El Azzouzi, H., Leptidis, S., Dirkx, E., Hoeks, J., Van Bree, B., Brand, K., McClellan, E.A., Poels, E., Sluimer, J.C., Van Den Hoogenhof, M.M., Armand, A.S., Yin, X., Langley, S., Bourajjaj, M., Olieslagers, S., Krishnan, J., Vooijs, M., Kurihara, H.,

Stubbs, A., Pinto, Y.M., Krek, W., Mayr, M., Da Costa Martins, P.A., Schrauwen, P., De Windt, L.J., 2013. The hypoxia-inducible microRNA cluster miR-199a approximately 214 targets myocardial PPARdelta and impairs mitochondrial fatty acid oxidation. Cell Metab. 18, 341–354.

El Ouaamari, A., Baroukh, N., Martens, G.A., Lebrun, P., Pipeleers, D., Van Obberghen, E., 2008. miR-375 targets 3′-phosphoinositide-dependent protein kinase-1 and regulates glucose-induced biological responses in pancreatic beta-cells. Diabetes 57, 2708–2717.

Elia, L., Contu, R., Quintavalle, M., Varrone, F., Chimenti, C., Russo, M.A., Cimino, V., De Marinis, L., Frustaci, A., Catalucci, D., Condorelli, G., 2009. Reciprocal regulation of microRNA-1 and insulin-like growth factor-1 signal transduction cascade in cardiac and skeletal muscle in physiological and pathological conditions. Circulation 120, 2377–2385.

Emanueli, C., Thum, T., 2013. miRNAGE-34 induces cardiac damAGE. Cell Res. 23, 866–867.

Esler, M., Jackman, G., Leonard, P., Skews, H., Bobik, A., Jennings, G., 1981. Effect of propranolol on noradrenaline kinetics in patients with essential hypertension. Br. J. Clin. Pharmacol. 12, 375–380.

Esler, M., Kaye, D., Thompson, J., Jennings, G., Cox, H., Turner, A., Lambert, G., Seals, D., 1995. Effects of aging on epinephrine secretion and regional release of epinephrine from the human heart. J. Clin. Endocrinol. Metab. 80, 435–442.

Fan, H., Ma, L., Fan, B., Wu, J., Yang, Z., Wang, L., 2014. Role of PDGFR-beta/PI3K/AKT signaling pathway in PDGF-BB induced myocardial fibrosis in rats. Am. J. Transl. Res. 6, 714–723.

Fang, J., Mensah, G.A., Croft, J.B., Keenan, N.L., 2008. Heart failure-related hospitalization in the U.S., 1979 to 2004. J. Am. Coll. Cardiol. 52, 428–434.

Faraonio, R., Salerno, P., Passaro, F., Sedia, C., Iaccio, A., Bellelli, R., Nappi, T.C., Comegna, M., Romano, S., Salvatore, G., Santoro, M., Cimino, F., 2012. A set of miRNAs participates in the cellular senescence program in human diploid fibroblasts. Cell Death Differ. 19, 713–721.

Fiedler, J., Jazbutyte, V., Kirchmaier, B.C., Gupta, S.K., Lorenzen, J., Hartmann, D., Galuppo, P., Kneitz, S., Pena, J.T., Sohn-Lee, C., Loyer, X., Soutschek, J., Brand, T., Tuschl, T., Heineke, J., Martin, U., Schulte-Merker, S., Ertl, G., Engelhardt, S., Bauersachs, J., Thum, T., 2011. MicroRNA-24 regulates vascularity after myocardial infarction. Circulation 124, 720–730.

Fleissner, F., Jazbutyte, V., Fiedler, J., Gupta, S.K., Yin, X., XU, Q., Galuppo, P., Kneitz, S., Mayr, M., Ertl, G., Bauersachs, J., Thum, T., 2010. Short communication: asymmetric dimethylarginine impairs angiogenic progenitor cell function in patients with coronary artery disease through a microRNA-21-dependent mechanism. Circ. Res. 107, 138–143.

Fordham, J.B., Naqvi, A.R., Nares, S., 2015. Regulation of miR-24, miR-30b, and miR-142-3p during macrophage and dendritic cell differentiation potentiates innate immunity. J. Leukoc. Biol. 98, 195–207.

Frankel, L.B., Wen, J., Lees, M., Hoyer-Hansen, M., Farkas, T., Krogh, A., Jaattela, M., Lund, A.H., 2011. microRNA-101 is a potent inhibitor of autophagy. EMBO J. 30, 4628–4641.

Frasca, D., Diaz, A., Romero, M., Ferracci, F., Blomberg, B.B., 2015. MicroRNAs miR-155 and miR-16 decrease AID and E47 in B cells from elderly individuals. J. Immunol. 195, 2134–2140.

Frasca, D., Diaz, A., Romero, M., Blomberg, B.B., 2016. Human peripheral late/exhausted memory B cells express a senescent-associated secretory phenotype and preferentially utilize metabolic signaling pathways. Exp. Gerontol. 87, 113–120.

Freund, A., Patil, C.K., Campisi, J., 2011. p38MAPK is a novel DNA damage response-independent regulator of the senescence-associated secretory phenotype. EMBO J. 30, 1536–1548.

Frost, R.J., Olson, E.N., 2011. Control of glucose homeostasis and insulin sensitivity by the Let-7 family of microRNAs. Proc. Natl. Acad. Sci. U.S.A. 108, 21075–21080.

Ganesan, J., Ramanujam, D., Sassi, Y., Ahles, A., Jentzsch, C., Werfel, S., Leierseder, S., Loyer, X., Giacca, M., Zentilin, L., Thum, T., Laggerbauer, B., Engelhardt, S., 2013. MiR-378 controls cardiac hypertrophy by combined repression of mitogen-activated protein kinase pathway factors. Circulation 127, 2097–2106.

Gaspard, G.J., Maclean, J., Rioux, D., Pasumarthi, K.B., 2014. A novel beta-adrenergic response element regulates both basal and agonist-induced expression of cyclin-dependent kinase 1 gene in cardiac fibroblasts. Am. J. Physiol. Cell Physiol. 306, C540–C550.

Ghosh, A.K., Bradham, W.S., Gleaves, L.A., De Taeye, B., Murphy, S.B., Covington, J.W., Vaughan, D.E., 2010a. Genetic deficiency of plasminogen activator inhibitor-1 promotes cardiac fibrosis in aged mice: involvement of constitutive transforming growth factor-beta signaling and endothelial-to-mesenchymal transition. Circulation 122, 1200–1209.

Ghosh, S.S., Salloum, F.N., Abbate, A., Krieg, R., Sica, D.A., Gehr, T.W., Kukreja, R.C., 2010b. Curcumin prevents cardiac remodeling secondary to chronic renal failure through deactivation of hypertrophic signaling in rats. Am. J. Physiol. Heart Circ. Physiol. 299, H975–H984.

Gibbings, D., Mostowy, S., Jay, F., Schwab, Y., Cossart, P., Voinnet, O., 2012. Selective autophagy degrades DICER and AGO2 and regulates miRNA activity. Nat. Cell Biol. 14, 1314–1321.

Go, A.S., Mozaffarian, D., Roger, V.L., Benjamin, E.J., Berry, J.D., Blaha, M.J., Dai, S., Ford, E.S., Fox, C.S., Franco, S., Fullerton, H.J., Gillespie, C., Hailpern, S.M., Heit, J.A., Howard, V.J., Huffman, M.D., Judd, S.E., Kissela, B.M., Kittner, S.J., Lackland, D.T., Lichtman, J.H., Lisabeth, L.D., Mackey, R.H., Magid, D.J., Marcus, G.M., Marelli, A., Matchar, D.B., Mcguire, D.K., Mohler 3rd, E.R., Moy, C.S., Mussolino, M.E., Neumar, R.W., Nichol, G., Pandey, D.K., Paynter, N.P., Reeves, M.J., Sorlie, P.D., Stein, J., Towfighi, A., Turan, T.N., Virani, S.S., Wong, N.D., Woo, D., Turner, M.B., American Heart Association Statistics Committee and Stroke Statistics Subcommittee, 2014. Heart disease and stroke statistics—2014 update: a report from the American Heart Association. Circulation 129, e28–e292.

Graff, J.W., Dickson, A.M., Clay, G., McCaffrey, A.P., Wilson, M.E., 2012. Identifying functional microRNAs in macrophages with polarized phenotypes. J. Biol. Chem. 287, 21816–21825.

Grueter, C.E., Van Rooij, E., Johnson, B.A., Deleon, S.M., Sutherland, L.B., QI, X., Gautron, L., Elmquist, J.K., Bassel-Duby, R., Olson, E.N., 2012. A cardiac microRNA governs systemic energy homeostasis by regulation of MED13. Cell 149, 671–683.

Guedes, E.C., Franca, G.S., Lino, C.A., Koyama, F.C., Moreira Ldo, N., Alexandre, J.G., Barreto-Chaves, M.L., Galante, P.A., Diniz, G.P., 2016. MicroRNA expression signature Is altered in the cardiac remodeling induced by high fat diets. J. Cell. Physiol. 231, 1771–1783.

Guerra, S., Leri, A., Wang, X., Finato, N., Di Loreto, C., Beltrami, C.A., Kajstura, J., Anversa, P., 1999. Myocyte death in the failing human heart is gender dependent. Circ. Res. 85, 856–866.

Gupta, A., Kaur, C.D., Jangdey, M., Saraf, S., 2014. Matrix metalloproteinase enzymes and their naturally derived inhibitors: novel targets in photocarcinoma therapy. Ageing Res. Rev. 13, 65–74.

Gurha, P., Abreu-Goodger, C., Wang, T., Ramirez, M.O., Drumond, A.L., Van Dongen, S., Chen, Y., Bartonicek, N., Enright, A.J., Lee, B., Kelm Jr., R.J., Reddy, A.K., Taffet, G.E., Bradley, A., Wehrens, X.H., Entman, M.L., Rodriguez, A., 2012. Targeted deletion of microRNA-22 promotes stress-induced cardiac dilation and contractile dysfunction. Circulation 125, 2751–2761.

Gustafsson, A.B., Gottlieb, R.A., 2008. Heart mitochondria: gates of life and death. Cardiovasc. Res. 77, 334–343.

Ha, M., Kim, V.N., 2014. Regulation of microRNA biogenesis. Nat. Rev. Mol. Cell Biol. 15, 509–524.

Haendeler, J., Hoffmann, J., Diehl, J.F., Vasa, M., Spyridopoulos, I., Zeiher, A.M., Dimmeler, S., 2004. Antioxidants inhibit nuclear export of telomerase reverse transcriptase and delay replicative senescence of endothelial cells. Circ. Res. 94, 768–775.

Haghikia, A., Missol-Kolka, E., Tsikas, D., Venturini, L., Brundiers, S., Castoldi, M., Muckenthaler, M.U., Eder, M., Stapel, B., Thum, T., Haghikia, A., Petrasch-Parwez, E., Drexler, H., Hilfiker-Kleiner, D., Scherr, M., 2011. Signal transducer and activator of transcription 3-mediated regulation of miR-199a-5p links cardiomyocyte and endothelial cell function in the heart: a key role for ubiquitin-conjugating enzymes. Eur. Heart J. 32, 1287–1297.

Hajra, L., Evans, A.I., Chen, M., Hyduk, S.J., Collins, T., Cybulsky, M.I., 2000. The NF-kappa B signal transduction pathway in aortic endothelial cells is primed for activation in regions predisposed to atherosclerotic lesion formation. Proc. Natl. Acad. Sci. U.S.A. 97, 9052–9057.

Handschin, C., Spiegelman, B.M., 2006. Peroxisome proliferator-activated receptor gamma coactivator 1 coactivators, energy homeostasis, and metabolism. Endocr. Rev. 27, 728–735.

Harris, T.A., Yamakuchi, M., Ferlito, M., Mendell, J.T., Lowenstein, C.J., 2008. MicroRNA-126 regulates endothelial expression of vascular cell adhesion molecule 1. Proc. Natl. Acad. Sci. U.S.A. 105, 1516–1521.

Hartmann, P., Zhou, Z., Natarelli, L., Wei, Y., Nazari-Jahantigh, M., Zhu, M., Grommes, J., Steffens, S., Weber, C., Schober, A., 2016a. Corrigendum: endothelial Dicer promotes atherosclerosis and vascular inflammation by miRNA-103-mediated suppression of KLF4. Nat. Commun. 7, 11907.

Hartmann, P., Zhou, Z., Natarelli, L., Wei, Y., Nazari-Jahantigh, M., Zhu, M., Grommes, J., Steffens, S., Weber, C., Schober, A., 2016b. Endothelial Dicer promotes atherosclerosis and vascular inflammation by miRNA-103-mediated suppression of KLF4. Nat. Commun. 7, 10521.

He, B., Xiao, J., Ren, A.J., Zhang, Y.F., Zhang, H., Chen, M., Xie, B., Gao, X.G., Wang, Y.W., 2011. Role of miR-1 and miR-133a in myocardial ischemic postconditioning. J. Biomed. Sci. 18, 22.

Heineke, J., Molkentin, J.D., 2006. Regulation of cardiac hypertrophy by intracellular signalling pathways. Nat. Rev. Mol. Cell Biol. 7, 589–600.

Hekimi, S., Lapointe, J., Wen, Y., 2011. Taking a "good" look at free radicals in the aging process. Trends Cell Biol. 21, 569–576.

Hengartner, M.O., 2000. The biochemistry of apoptosis. Nature 407, 770–776.

Hergenreider, E., Heydt, S., Treguer, K., Boettger, T., Horrevoets, A.J., Zeiher, A.M., Scheffer, M.P., Frangakis, A.S., Yin, X., Mayr, M., Braun, T., Urbich, C., Boon, R.A., Dimmeler, S., 2012. Atheroprotective communication between endothelial cells and smooth muscle cells through miRNAs. Nat. Cell Biol. 14, 249–256.

Herum, K.M., Lunde, I.G., Skrbic, B., Florholmen, G., Behmen, D., Sjaastad, I., Carlson, C.R., Gomez, M.F., Christensen, G., 2013. Syndecan-4 signaling via NFAT regulates extracellular matrix production and cardiac myofibroblast differentiation in response to mechanical stress. J. Mol. Cell. Cardiol. 54, 73–81.

Heymans, S., Corsten, M.F., Verhesen, W., Carai, P., Van Leeuwen, R.E., Custers, K., Peters, T., Hazebroek, M., Stoger, L., Wijnands, E., Janssen, B.J., Creemers, E.E., Pinto, Y.M., Grimm, D., Schurmann, N., Vigorito, E., Thum, T., Stassen, F., Yin, X., Mayr, M., De Windt, L.J., Lutgens, E., Wouters, K., De Winther, M.P., Zacchigna, S., Giacca, M., Van Bilsen, M., Papageorgiou, A.P., Schroen, B., 2013. Macrophage microRNA-155 promotes cardiac hypertrophy and failure. Circulation 128, 1420–1432.

Hong, S., Lee, J., Seo, H.H., Lee, C.Y., Yoo, K.J., Kim, S.M., Lee, S., Hwang, K.C., Choi, E., 2015. Na(+)-Ca(2+) exchanger targeting miR-132 prevents apoptosis of cardiomyocytes under hypoxic condition by suppressing Ca(2+) overload. Biochem. Biophys. Res. Commun. 460, 931–937.

Horie, T., Ono, K., Nishi, H., Iwanaga, Y., Nagao, K., Kinoshita, M., Kuwabara, Y., Takanabe, R., Hasegawa, K., Kita, T., Kimura, T., 2009. MicroRNA-133 regulates the expression of GLUT4 by targeting KLF15 and is involved in metabolic control in cardiac myocytes. Biochem. Biophys. Res. Commun. 389, 315–320.

Hoshino, A., Mita, Y., Okawa, Y., Ariyoshi, M., Iwai-Kanai, E., Ueyama, T., Ikeda, K., Ogata, T., Matoba, S., 2013. Cytosolic p53 inhibits Parkin-mediated mitophagy and promotes mitochondrial dysfunction in the mouse heart. Nat. Commun. 4, 2308.

Hou, Y., Sun, Y., Shan, H., Li, X., Zhang, M., Zhou, X., Xing, S., Sun, H., Chu, W., Qiao, G., Lu, Y., 2012. beta-adrenoceptor regulates miRNA expression in rat heart. Med. Sci. Monit. 18, BR309–BR314.

Hu, F., Liu, F., 2011. Mitochondrial stress: a bridge between mitochondrial dysfunction and metabolic diseases? Cell. Signal 23, 1528–1533.

Hu, F., Liu, F., 2014. Targeting tissue-specific metabolic signaling pathways in aging: the promise and limitations. Protein Cell 5, 21–35.

Hu, Y., Matkovich, S.J., Hecker, P.A., Zhang, Y., Edwards, J.R., Dorn 2nd, G.W., 2012. Epitranscriptional orchestration of genetic reprogramming is an emergent property of stress-regulated cardiac microRNAs. Proc. Natl. Acad. Sci. U.S.A. 109, 19864–19869.

Hu, J., Zeng, L., Huang, J., Wang, G., Lu, H., 2015. miR-126 promotes angiogenesis and attenuates inflammation after contusion spinal cord injury in rats. Brain Res. 1608, 191–202.

Huang, Y., Guerrero-Preston, R., Ratovitski, E.A., 2012. Phospho-DeltaNp63alpha-dependent regulation of autophagic signaling through transcription and micro-RNA modulation. Cell Cycle 11, 1247–1259.

Huang, Z.P., Chen, J., Seok, H.Y., Zhang, Z., Kataoka, M., Hu, X., Wang, D.Z., 2013. MicroRNA-22 regulates cardiac hypertrophy and remodeling in response to stress. Circ. Res. 112, 1234–1243.

Huang, Y., Qi, Y., Du, J.Q., Zhang, D.F., 2014. MicroRNA-34a regulates cardiac fibrosis after myocardial infarction by targeting Smad4. Expert Opin. Ther. Targets 18, 1355–1365.

Huang, W., Tian, S.S., Hang, P.Z., Sun, C., Guo, J., Du, Z.M., 2016. Combination of microRNA-21 and microRNA-146a attenuates cardiac dysfunction and apoptosis during acute myocardial infarction in mice. Mol. Ther. Nucleic Acids 5, e296.

Hullinger, T.G., Montgomery, R.L., Seto, A.G., Dickinson, B.A., Semus, H.M., Lynch, J.M., Dalby, C.M., Robinson, K., Stack, C., Latimer, P.A., Hare, J.M., Olson, E.N., Van Rooij, E., 2012. Inhibition of miR-15 protects against cardiac ischemic injury. Circ. Res. 110, 71–81.

Iliopoulos, D., Drosatos, K., Hiyama, Y., Goldberg, I.J., Zannis, V.I., 2010. MicroRNA-370 controls the expression of microRNA-122 and Cpt1alpha and affects lipid metabolism. J. Lipid Res. 51, 1513–1523.

Inuzuka, Y., Okuda, J., Kawashima, T., Kato, T., Niizuma, S., Tamaki, Y., Iwanaga, Y., Yoshida, Y., Kosugi, R., Watanabe-Maeda, K., Machida, Y., Tsuji, S.,

Aburatani, H., Izumi, T., Kita, T., Shioi, T., 2009. Suppression of phosphoinositide 3-kinase prevents cardiac aging in mice. Circulation 120, 1695–1703.

Ishikawa, Y., Homcy, C.J., 1997. The adenylyl cyclases as integrators of transmembrane signal transduction. Circ. Res. 80, 297–304.

Janczewski, A.M., Lakatta, E.G., 2010. Modulation of sarcoplasmic reticulum Ca(2+) cycling in systolic and diastolic heart failure associated with aging. Heart Fail. Rev. 15, 431–445.

Jazbutyte, V., Fiedler, J., Kneitz, S., Galuppo, P., Just, A., Holzmann, A., Bauersachs, J., Thum, T., 2013. MicroRNA-22 increases senescence and activates cardiac fibroblasts in the aging heart. Age (Dordr.) 35, 747–762.

Jencks, S.F., Williams, M.V., Coleman, E.A., 2009. Rehospitalizations among patients in the Medicare fee-for-service program. N. Engl. J. Med. 360, 1418–1428.

Jiang, M.T., Moffat, M.P., Narayanan, N., 1993. Age-related alterations in the phosphorylation of sarcoplasmic reticulum and myofibrillar proteins and diminished contractile response to isoproterenol in intact rat ventricle. Circ. Res. 72, 102–111.

Jiang, C., Ji, N., Luo, G., Ni, S., Zong, J., Chen, Z., Bao, D., Gong, X., Fu, T., 2014. The effects and mechanism of miR-92a and miR-126 on myocardial apoptosis in mouse ischemia-reperfusion model. Cell Biochem. Biophys. 70, 1901–1906.

Johnson, S.C., Rabinovitch, P.S., Kaeberlein, M., 2013. mTOR is a key modulator of ageing and age-related disease. Nature 493, 338–345.

Jongstra-Bilen, J., Haidari, M., Zhu, S.N., Chen, M., Guha, D., Cybulsky, M.I., 2006. Low-grade chronic inflammation in regions of the normal mouse arterial intima predisposed to atherosclerosis. J. Exp. Med. 203, 2073–2083.

Jordan, S.D., Kruger, M., Willmes, D.M., Redemann, N., Wunderlich, F.T., Bronneke, H.S., Merkwirth, C., Kashkar, H., Olkkonen, V.M., Bottger, T., Braun, T., Seibler, J., Bruning, J.C., 2011. Obesity-induced overexpression of miRNA-143 inhibits insulin-stimulated AKT activation and impairs glucose metabolism. Nat. Cell Biol. 13, 434–446.

Kajstura, J., Cheng, W., Sarangarajan, R., Li, P., Li, B., Nitahara, J.A., Chapnick, S., Reiss, K., Olivetti, G., Anversa, P., 1996. Necrotic and apoptotic myocyte cell death in the aging heart of Fischer 344 rats. Am. J. Physiol. 271, H1215–H1228.

Kanters, E., Pasparakis, M., Gijbels, M.J., Vergouwe, M.N., Partouns-Hendriks, I., Fijneman, R.J., Clausen, B.E., Forster, I., Kockx, M.M., Rajewsky, K., Kraal, G., Hofker, M.H., De Winther, M.P., 2003. Inhibition of NF-kappaB activation in macrophages increases atherosclerosis in LDL receptor-deficient mice. J. Clin. Invest. 112, 1176–1185.

Kato, H., Nakajima, S., Saito, Y., Takahashi, S., Katoh, R., Kitamura, M., 2012. mTORC1 serves ER stress-triggered apoptosis via selective activation of the IRE1-JNK pathway. Cell Death Differ. 19, 310–320.

Ke, Z.P., Xu, P., Shi, Y., Gao, A.M., 2016. MicroRNA-93 inhibits ischemia-reperfusion induced cardiomyocyte apoptosis by targeting PTEN. Oncotarget 7, 28796–28805.

Keller, D.M., Clark, E.A., Goodman, R.H., 2012. Regulation of microRNA-375 by cAMP in pancreatic beta-cells. Mol. Endocrinol. 26, 989–999.

Khrapko, K., Bodyak, N., Thilly, W.G., Van Orsouw, N.J., Zhang, X., Coller, H.A., Perls, T.T., Upton, M., Vijg, J., Wei, J.Y., 1999. Cell-by-cell scanning of whole mitochondrial genomes in aged human heart reveals a significant fraction of myocytes with clonally expanded deletions. Nucleic Acids Res. 27, 2434–2441.

Kim, H.R., Roe, J.S., Lee, J.E., Cho, E.J., Youn, H.D., 2013. p53 regulates glucose metabolism by miR-34a. Biochem. Biophys. Res. Commun. 437, 225–231.

Kim, I.M., Wang, Y., Park, K.M., Tang, Y., Teoh, J.P., Vinson, J., Traynham, C.J., Pironti, G., Mao, L., 2014. Beta-arrestin1-biased beta1-adrenergic receptor signaling regulates microRNA processing. Circ. Res. 114, 833–844.

Kirchhefer, U., Baba, H.A., Hanske, G., Jones, L.R., Kirchhof, P., Schmitz, W., Neumann, J., 2004. Age-dependent biochemical and contractile properties in atrium of transgenic mice overexpressing junctin. Am. J. Physiol. Heart Circ. Physiol. 287, H2216–H2225.

Kriegel, A.J., Liu, Y., Fang, Y., Ding, X., Liang, M., 2012. The miR-29 family: genomics, cell biology, and relevance to renal and cardiovascular injury. Physiol. Genomics 44, 237–244.

Kroemer, G., Galluzzi, L., Vandenabeele, P., Abrams, J., Alnemri, E.S., Baehrecke, E.H., Blagosklonny, M.V., El-Deiry, W.S., Golstein, P., Green, D.R., Hengartner, M., Knight, R.A., Kumar, S., Lipton, S.A., Malorni, W., Nunez, G., Peter, M.E., Tschopp, J., Yuan, J., Piacentini, M., Zhivotovsky, B., Melino, G., Nomenclature Committee On Cell Death 2009, 2009. Classification of cell death: recommendations of the Nomenclature Committee on Cell Death 2009. Cell Death Differ. 16, 3–11.

Kuehbacher, A., Urbich, C., Zeiher, A.M., Dimmeler, S., 2007. Role of Dicer and Drosha for endothelial microRNA expression and angiogenesis. Circ. Res. 101, 59–68.

Kumarswamy, R., Volkmann, I., Jazbutyte, V., Dangwal, S., Park, D.H., Thum, T., 2012. Transforming growth factor-beta-induced endothelial-to-mesenchymal transition is partly mediated by microRNA-21. Arterioscler. Thromb. Vasc. Biol. 32, 361–369.

Kumarswamy, R., Volkmann, I., Beermann, J., Napp, L.C., Jabs, O., Bhayadia, R., Melk, A., Ucar, A., Chowdhury, K., Lorenzen, J.M., Gupta, S.K., Batkai, S., Thum, T., 2014. Vascular importance of the miR-212/132 cluster. Eur. Heart J. 35, 3224–3231.

Kung, G., Konstantinidis, K., Kitsis, R.N., 2011. Programmed necrosis, not apoptosis, in the heart. Circ. Res. 108, 1017–1036.

Kunz, M., Xiao, K., Liang, C., Viereck, J., Pachel, C., Frantz, S., Thum, T., Dandekar, T., 2015. Bioinformatics of cardiovascular miRNA biology. J. Mol. Cell. Cardiol. 89, 3–10.

Kuwabara, Y., Horie, T., Baba, O., Watanabe, S., Nishiga, M., Usami, S., Izuhara, M., Nakao, T., Nishino, T., Otsu, K., Kita, T., Kimura, T., Ono, K., 2015. MicroRNA-451 exacerbates lipotoxicity in cardiac myocytes and high-fat diet-induced cardiac hypertrophy in mice through suppression of the LKB1/AMPK pathway. Circ. Res. 116, 279–288.

Kyrychenko, S., Kyrychenko, V., Badr, M.A., Ikeda, Y., Sadoshima, J., Shirokova, N., 2015. Pivotal role of miR-448 in the development of ROS-induced cardiomyopathy. Cardiovasc. Res. 108, 324–334.

Lakatta, E.G., 1993. Cardiovascular regulatory mechanisms in advanced age. Physiol. Rev. 73, 413–467.

Lakatta, E.G., 2003. Arterial and cardiac aging: major shareholders in cardiovascular disease enterprises. Part III: cellular and molecular clues to heart and arterial aging. Circulation 107, 490–497.

Lakatta, E.G., 2015. So! What's aging? Is cardiovascular aging a disease? J. Mol. Cell. Cardiol. 83, 1–13.

Lakatta, E.G., Levy, D., 2003. Arterial and cardiac aging: major shareholders in cardiovascular disease enterprises. Part II: the aging heart in health: links to heart disease. Circulation 107, 346–354.

Lakatta, E.G., Wang, M., Najjar, S.S., 2009. Arterial aging and subclinical arterial disease are fundamentally intertwined at macroscopic and molecular levels. Med. Clin. North Am. 93, 583–604. Table of Contents.

Lam, C.S., Lyass, A., Kraigher-Krainer, E., Massaro, J.M., Lee, D.S., Ho, J.E., Levy, D., Redfield, M.M., Pieske, B.M., Benjamin, E.J., Vasan, R.S., 2011. Cardiac dysfunction and noncardiac dysfunction as precursors of heart failure with reduced and preserved ejection fraction in the community. Circulation 124, 24–30.

Li, J., Donath, S., Li, Y., Qin, D., Prabhakar, B.S., Li, P., 2010a. miR-30 regulates mitochondrial fission through targeting p53 and the dynamin-related protein-1 pathway. PLoS Genet. 6, e1000795.

Li, Q., Lin, X., Yang, X., Chang, J., 2010b. NFATc4 is negatively regulated in miR-133a-mediated cardiomyocyte hypertrophic repression. Am. J. Physiol. Heart Circ. Physiol. 298, H1340–H1347.

Li, R., Yan, G., Zhang, Q., Jiang, Y., Sun, H., Hu, Y., Sun, J., Xu, B., 2013. miR-145 inhibits isoproterenol-induced cardiomyocyte hypertrophy by targeting the expression and localization of GATA6. FEBS Lett. 587, 1754–1761.

Li, B., Zhang, Z., Zhang, H., Quan, K., Lu, Y., Cai, D., Ning, G., 2014a. Aberrant miR199a-5p/caveolin1/PPARalpha axis in hepatic steatosis. J. Mol. Endocrinol. 53, 393–403.

Li, J., Li, Y., Jiao, J., Wang, J., Li, Y., Qin, D., Li, P., 2014b. Mitofusin 1 is negatively regulated by microRNA 140 in cardiomyocyte apoptosis. Mol. Cell. Biol. 34, 1788–1799.

Li, K., Ching, D., Luk, F.S., Raffai, R.L., 2015a. Apolipoprotein E enhances microRNA-146a in monocytes and macrophages to suppress nuclear factor-kappaB-driven inflammation and atherosclerosis. Circ. Res. 117, e1–e11.

Li, R., Xiao, J., Qing, X., Xing, J., Xia, Y., Qi, J., Liu, X., Zhang, S., Sheng, X., Zhang, X., Ji, X., 2015b. Sp1 mediates a therapeutic role of MiR-7a/b in Angiotensin II-induced cardiac fibrosis via mechanism involving the TGF-beta and MAPKs pathways in cardiac fibroblasts. PLoS One 10, e0125513.

Li, X., Zeng, Z., Li, Q., Xu, Q., Xie, J., Hao, H., Luo, G., Liao, W., Bin, J., Huang, X., Liao, Y., 2015c. Inhibition of microRNA-497 ameliorates anoxia/reoxygenation injury in cardiomyocytes by suppressing cell apoptosis and enhancing autophagy. Oncotarget 6, 18829–18844.

Li, Z., Song, Y., Liu, L., Hou, N., An, X., Zhan, D., Li, Y., Zhou, L., Li, P., Yu, L., Xia, J., Zhang, Y., Wang, J., Yang, X., 2015d. miR-199a impairs autophagy and induces cardiac hypertrophy through mTOR activation. Cell Death Differ. [Epub ahead of print].

Li, G., Wang, G., Ma, L., Guo, J., Song, J., Ma, L., Zhao, X., 2016a. miR-22 regulates starvation-induced autophagy and apoptosis in cardiomyocytes by targeting p38alpha. Biochem. Biophys. Res. Commun. 478, 1165–1172.

Li, J., Ren, Y., Shi, E., Tan, Z., Xiong, J., Yan, L., Jiang, X., 2016b. Inhibition of the let-7 family MicroRNAs induces cardioprotection against Ischemia-reperfusion injury in diabetic rats. Ann. Thorac. Surg. 102, 829–835.

Li, Q., Xie, J., Wang, B., Li, R., Bai, J., Ding, L., Gu, R., Wang, L., Xu, B., 2016c. Overexpression of microRNA-99a attenuates cardiac hypertrophy. PLoS One 11, e0148480.

Liang, Q., De Windt, L.J., Witt, S.A., Kimball, T.R., Markham, B.E., Molkentin, J.D., 2001. The transcription factors GATA4 and GATA6 regulate cardiomyocyte hypertrophy in vitro and in vivo. J. Biol. Chem. 276, 30245–30253.

Liang, H., Zhang, C., Ban, T., Liu, Y., Mei, L., Piao, X., Zhao, D., Lu, Y., Chu, W., Yang, B., 2012. A novel reciprocal loop between microRNA-21 and TGFbetaRIII is involved in cardiac fibrosis. Int. J. Biochem. Cell Biol. 44, 2152–2160.

Liao, Y.C., Wang, Y.S., Guo, Y.C., Lin, W.L., Chang, M.H., Juo, S.H., 2014. Let-7g improves multiple endothelial functions through targeting transforming growth factor-beta and SIRT-1 signaling. J. Am. Coll. Cardiol. 63, 1685–1694.

Lichtnekert, J., Kawakami, T., Parks, W.C., Duffield, J.S., 2013. Changes in macrophage phenotype as the immune response evolves. Curr. Opin. Pharmacol. 13, 555–564.

Lin, Z., Murtaza, I., Wang, K., Jiao, J., Gao, J., Li, P.F., 2009. miR-23a functions downstream of NFATc3 to regulate cardiac hypertrophy. Proc. Natl. Acad. Sci. U.S.A. 106, 12103–12108.

Lin, H.S., Gong, J.N., Su, R., Chen, M.T., Song, L., Shen, C., Wang, F., Ma, Y.N., Zhao, H.L., Yu, J., Li, W.W., Huang, L.X., Xu, X.H., Zhang, J.W., 2014. miR-199a-5p inhibits monocyte/macrophage differentiation by targeting the activin A type 1B receptor gene and finally reducing C/EBPalpha expression. J. Leukoc. Biol. 96, 1023–1035.

Ling, S., Nanhwan, M., Qian, J., Kodakandla, M., Castillo, A.C., Thomas, B., Liu, H., Ye, Y., 2013. Modulation of microRNAs in hypertension-induced arterial remodeling through the beta1 and beta3-adrenoreceptor pathways. J. Mol. Cell. Cardiol. 65, 127–136.

Linnane, A.W., Marzuki, S., Ozawa, T., Tanaka, M., 1989. Mitochondrial DNA mutations as an important contributor to ageing and degenerative diseases. Lancet 1, 642–645.

Liu, S.J., Wyeth, R.P., Melchert, R.B., Kennedy, R.H., 2000. Aging-associated changes in whole cell K(+) and L-type Ca(2+) currents in rat ventricular myocytes. Am. J. Physiol. Heart Circ. Physiol. 279, H889–H900.

Liu, N., Bezprozvannaya, S., Williams, A.H., Qi, X., Richardson, J.A., Bassel-Duby, R., Olson, E.N., 2008. microRNA-133a regulates cardiomyocyte proliferation and suppresses smooth muscle gene expression in the heart. Genes Dev. 22, 3242–3254.

Liu, J., Van Mil, A., Vrijsen, K., Zhao, J., Gao, L., Metz, C.H., Goumans, M.J., Doevendans, P.A., Sluijter, J.P., 2011. MicroRNA-155 prevents necrotic cell death in human cardiomyocyte progenitor cells via targeting RIP1. J. Cell. Mol. Med. 15, 1474–1482.

Liu, Y.L., Wu, W., Xue, Y., Gao, M., Yan, Y., Kong, Q., Pang, Y., Yang, F., 2013. MicroRNA-21 and -146b are involved in the pathogenesis of murine viral myocarditis by regulating TH-17 differentiation. Arch. Virol. 158, 1953–1963.

Liu, H., Li, G., Zhao, W., Hu, Y., 2016. Inhibition of MiR-92a may protect endothelial cells after acute myocardial infarction in rats: role of KLF2/4. Med. Sci. Monit. 22, 2451–2462.

Loffredo, F.S., Pancoast, J.R., Lee, R.T., 2013. Keep PNUTS in your heart. Circ. Res. 113, 97–99.

Loffredo, F.S., Nikolova, A.P., Pancoast, J.R., Lee, R.T., 2014. Heart failure with preserved ejection fraction: molecular pathways of the aging myocardium. Circ. Res. 115, 97–107.

Lorenzen, J.M., Schauerte, C., Hubner, A., Kolling, M., Martino, F., Scherf, K., Batkai, S., Zimmer, K., Foinquinos, A., Kaucsar, T., Fiedler, J., Kumarswamy, R., Bang, C., Hartmann, D., Gupta, S.K., Kielstein, J., Jungmann, A., Katus, H.A., Weidemann, F., Muller, O.J., Haller, H., Thum, T., 2015. Osteopontin is indispensible for AP1-mediated angiotensin II-related miR-21 transcription during cardiac fibrosis. Eur. Heart J. 36, 2184–2196.

Loyer, X., Potteaux, S., Vion, A.C., Guerin, C.L., Boulkroun, S., Rautou, P.E., Ramkhelawon, B., Esposito, B., Dalloz, M., Paul, J.L., Julia, P., Maccario, J., Boulanger, C.M., Mallat, Z., Tedgui, A., 2014. Inhibition of microRNA-92a prevents endothelial dysfunction and atherosclerosis in mice. Circ. Res. 114, 434–443.

Lu, Y., Zhang, Y., Shan, H., Pan, Z., Li, X., Li, B., Xu, C., Zhang, B., Zhang, F., Dong, D., Song, W., Qiao, G., Yang, B., 2009. MicroRNA-1 downregulation by propranolol in a rat model of myocardial infarction: a new mechanism for ischaemic cardioprotection. Cardiovasc. Res. 84, 434–441.

Lu, H., Buchan, R.J., Cook, S.A., 2010. MicroRNA-223 regulates Glut4 expression and cardiomyocyte glucose metabolism. Cardiovasc. Res. 86, 410–420.

Lu, D., Aroonsakool, N., Yokoyama, U., Patel, H.H., Insel, P.A., 2013. Increase in cellular cyclic AMP concentrations reverses the profibrogenic phenotype of cardiac myofibroblasts: a novel therapeutic approach for cardiac fibrosis. Mol. Pharmacol. 84, 787–793.

Ma, H., Wang, X., Ha, T., Gao, M., Liu, L., Wang, R., Yu, K., Kalbfleisch, J.H., Kao, R.L., Williams, D.L., Li, C., 2016a. MicroRNA-125b prevents cardiac dysfunction in polymicrobial sepsis by targeting TRAF6-mediated nuclear factor κB activation and p53-mediated apoptotic signaling. J. Infect. Dis. 214, 1773–1783.

Ma, S., Tian, X.Y., Zhang, Y., Mu, C., Shen, H., Bismuth, J., Pownall, H.J., Huang, Y., Wong, W.T., 2016b. E-selectin-targeting delivery of microRNAs by microparticles ameliorates endothelial inflammation and atherosclerosis. Sci. Rep. 6, 22910.

Magenta, A., Greco, S., Capogrossi, M.C., Gaetano, C., Martelli, F., 2014. Nitric oxide, oxidative stress, and p66Shc interplay in diabetic endothelial dysfunction. Biomed. Res. Int. 2014, 193095.

Mao, J., Lv, Z., Zhuang, Y., 2014. MicroRNA-23a is involved in tumor necrosis factor-alpha induced apoptosis in mesenchymal stem cells and myocardial infarction. Exp. Mol. Pathol. 97, 23–30.

Matkovich, S.J., Wang, W., Tu, Y., Eschenbacher, W.H., Dorn, L.E., Condorelli, G., Diwan, A., Nerbonne, J.M., Dorn 2nd, G.W., 2010. MicroRNA-133a protects against myocardial fibrosis and modulates electrical repolarization without affecting hypertrophy in pressure-overloaded adult hearts. Circ. Res. 106, 166–175.

Matsumoto, S., Sakata, Y., Suna, S., Nakatani, D., Usami, M., Hara, M., Kitamura, T., Hamasaki, T., Nanto, S., Kawahara, Y., Komuro, I., 2013. Circulating p53-responsive microRNAs are predictive indicators of heart failure after acute myocardial infarction. Circ. Res. 113, 322–326.

McMullen, J.R., Sherwood, M.C., Tarnavski, O., Zhang, L., Dorfman, A.L., Shioi, T., Izumo, S., 2004. Inhibition of mTOR signaling with rapamycin regresses established cardiac hypertrophy induced by pressure overload. Circulation 109, 3050–3055.

Meng, F., Henson, R., Wehbe-Janek, H., Ghoshal, K., Jacob, S.T., Patel, T., 2007. MicroRNA-21 regulates expression of the PTEN tumor suppressor gene in human hepatocellular cancer. Gastroenterology 133, 647–658.

Menghini, R., Casagrande, V., Cardellini, M., Martelli, E., Terrinoni, A., Amati, F., Vasa-Nicotera, M., Ippoliti, A., Novelli, G., Melino, G., Lauro, R., Federici, M., 2009. MicroRNA 217 modulates endothelial cell senescence via silent information regulator 1. Circulation 120, 1524–1532.

Meyer, M., Dillmann, W.H., 1998. Sarcoplasmic reticulum Ca(2+)-ATPase overexpression by adenovirus mediated gene transfer and in transgenic mice. Cardiovasc. Res. 37, 360–366.

Molkentin, J.D., Lu, J.R., Antos, C.L., Markham, B., Richardson, J., Robbins, J., Grant, S.R., Olson, E.N., 1998. A calcineurin-dependent transcriptional pathway for cardiac hypertrophy. Cell 93, 215–228.

Myers, R., Timofeyev, V., Li, N., Kim, C., Ledford, H.A., Sirish, P., Lau, V., Zhang, Y., Fayyaz, K., Singapuri, A., Lopez, J.E., Knowlton, A.A., Zhang, X.D., Chiamvimonvat, N., 2015. Feedback mechanisms for cardiac-specific microRNAs and cAMP signaling in electrical remodeling. Circ. Arrhythm. Electrophysiol. 8, 942–950.

Naga Prasad, S.V., Esposito, G., Mao, L., Koch, W.J., Rockman, H.A., 2000. Gbetagamma-dependent phosphoinositide 3-kinase activation in hearts with in vivo pressure overload hypertrophy. J. Biol. Chem. 275, 4693–4698.

Naga Prasad, S.V., Duan, Z.H., Gupta, M.K., Surampudi, V.S., Volinia, S., Calin, G.A., Liu, C.G., Kotwal, A., Moravec, C.S., Starling, R.C., Perez, D.M., Sen, S., Wu, Q., Plow, E.F., Croce, C.M., Karnik, S., 2009. Unique microRNA profile in end-stage heart failure indicates alterations in specific cardiovascular signaling networks. J. Biol. Chem. 284, 27487–27499.

Nagalingam, R.S., Sundaresan, N.R., Gupta, M.P., Geenen, D.L., Solaro, R.J., Gupta, M., 2013. A cardiac-enriched microRNA, miR-378, blocks cardiac hypertrophy by targeting Ras signaling. J. Biol. Chem. 288, 11216–11232.

Nagpal, V., Rai, R., Place, A.T., Murphy, S.B., Verma, S.K., Ghosh, A.K., Vaughan, D.E., 2016. MiR-125b is critical for fibroblast-to-myofibroblast transition and cardiac fibrosis. Circulation 133, 291–301.

Nakai, A., Yamaguchi, O., Takeda, T., Higuchi, Y., Hikoso, S., Taniike, M., Omiya, S., Mizote, I., Matsumura, Y., Asahi, M., Nishida, K., Hori, M., Mizushima, N., Otsu, K., 2007. The role of autophagy in cardiomyocytes in the basal state and in response to hemodynamic stress. Nat. Med. 13, 619–624.

Naqvi, A.R., Fordham, J.B., Nares, S., 2015. miR-24, miR-30b, and miR-142-3p regulate phagocytosis in myeloid inflammatory cells. J. Immunol. 194, 1916–1927.

Nazari-Jahantigh, M., Wei, Y., Noels, H., Akhtar, S., Zhou, Z., Koenen, R.R., Heyll, K., Gremse, F., Kiessling, F., Grommes, J., Weber, C., Schober, A., 2012. MicroRNA-155 promotes atherosclerosis by repressing Bcl6 in macrophages. J. Clin. Invest. 122, 4190–4202.

Nishi, H., Ono, K., Iwanaga, Y., Horie, T., Nagao, K., Takemura, G., Kinoshita, M., Kuwabara, Y., Mori, R.T., Hasegawa, K., Kita, T., Kimura, T., 2010. MicroRNA-15b modulates cellular ATP levels and degenerates mitochondria via Arl2 in neonatal rat cardiac myocytes. J. Biol. Chem. 285, 4920–4930.

Njock, M.S., Cheng, H.S., Dang, L.T., Nazari-Jahantigh, M., Lau, A.C., Boudreau, E., Roufaiel, M., Cybulsky, M.I., Schober, A., Fish, J.E., 2015. Endothelial cells suppress monocyte activation through secretion of extracellular vesicles containing antiinflammatory microRNAs. Blood 125, 3202–3212.

Nojiri, H., Shimizu, T., Funakoshi, M., Yamaguchi, O., Zhou, H., Kawakami, S., Ohta, Y., Sami, M., Tachibana, T., Ishikawa, H., Kurosawa, H., Kahn, R.C., Otsu, K., Shirasawa, T., 2006. Oxidative stress causes heart failure with impaired mitochondrial respiration. J. Biol. Chem. 281, 33789–33801.

Noma, T., Lemaire, A., Naga Prasad, S.V., Barki-Harrington, L., Tilley, D.G., Chen, J., Le Corvoisier, P., Violin, J.D., Wei, H., Lefkowitz, R.J., Rockman, H.A., 2007. Beta-arrestin-mediated beta1-adrenergic receptor transactivation of the EGFR confers cardioprotection. J. Clin. Invest. 117, 2445–2458.

O'connell, R.M., Taganov, K.D., Boldin, M.P., Cheng, G., Baltimore, D., 2007. MicroRNA-155 is induced during the macrophage inflammatory response. Proc. Natl. Acad. Sci. U.S.A. 104, 1604–1609.

Olivetti, G., Melissari, M., Capasso, J.M., Anversa, P., 1991. Cardiomyopathy of the aging human heart. Myocyte loss and reactive cellular hypertrophy. Circ. Res. 68, 1560–1568.

Olivieri, F., Spazzafumo, L., Santini, G., Lazzarini, R., Albertini, M.C., Rippo, M.R., Galeazzi, R., Abbatecola, A.M., Marcheselli, F., Monti, D., Ostan, R., Cevenini, E., Antonicelli, R., Franceschi, C., Procopio, A.D., 2012. Age-related differences in the expression of circulating microRNAs: miR-21 as a new circulating marker of inflammaging. Mech. Ageing Dev. 133, 675–685.

Pan, Z., Sun, X., Shan, H., Wang, N., Wang, J., Ren, J., Feng, S., Xie, L., Lu, C., Yuan, Y., Zhang, Y., Wang, Y., Lu, Y., Yang, B., 2012. MicroRNA-101 inhibited postinfarct cardiac fibrosis and improved left ventricular compliance via the FBJ osteosarcoma oncogene/transforming growth factor-beta1 pathway. Circulation 126, 840–850.

Pan, W., Zhong, Y., Cheng, C., Liu, B., Wang, L., Li, A., Xiong, L., Liu, S., 2013. MiR-30-regulated autophagy mediates angiotensin II-induced myocardial hypertrophy. PLoS One 8, e53950.

Paradies, G., Petrosillo, G., Pistolese, M., Di Venosa, N., Federici, A., Ruggiero, F.M., 2004. Decrease in mitochondrial complex I activity in ischemic/reperfused rat heart: involvement of reactive oxygen species and cardiolipin. Circ. Res. 94, 53–59.

Paulus, W.J., Tschope, C., 2013. A novel paradigm for heart failure with preserved ejection fraction: comorbidities drive myocardial dysfunction and remodeling through coronary microvascular endothelial inflammation. J. Am. Coll. Cardiol. 62, 263–271.

Pinto, A.R., Godwin, J.W., Chandran, A., Hersey, L., Ilinykh, A., Debuque, R., Wang, L., Rosenthal, N.A., 2014. Age-related changes in tissue macrophages precede cardiac functional impairment. Aging (Albany NY) 6, 399–413.

Porrello, E.R., Mahmoud, A.I., Simpson, E., Johnson, B.A., Grinsfelder, D., Canseco, D., Mammen, P.P., Rothermel, B.A., Olson, E.N., Sadek, H.A., 2013. Regulation of neonatal and adult mammalian heart regeneration by the miR-15 family. Proc. Natl. Acad. Sci. U.S.A. 110, 187–192.

Poulsen, H.E., Specht, E., Broedbaek, K., Henriksen, T., Ellervik, C., Mandrup-Poulsen, T., Tonnesen, M., Nielsen, P.E., Andersen, H.U., Weimann, A., 2012. RNA modifications by oxidation: a novel disease mechanism? Free Radic. Biol. Med. 52, 1353–1361.

Puissegur, M.P., Mazure, N.M., Bertero, T., Pradelli, L., Grosso, S., Robbe-Sermesant, K., Maurin, T., Lebrigand, K., Cardinaud, B., Hofman, V., Fourre, S., Magnone, V., Ricci, J.E., Pouyssegur, J., Gounon, P., Hofman, P., Barbry, P., Mari, B., 2011. miR-210 is overexpressed in late stages of lung cancer and mediates mitochondrial alterations associated with modulation of HIF-1 activity. Cell Death Differ. 18, 465–478.

Qian, L., Van Laake, L.W., Huang, Y., Liu, S., Wendland, M.F., Srivastava, D., 2011. miR-24 inhibits apoptosis and represses Bim in mouse cardiomyocytes. J. Exp. Med. 208, 549–560.

Qin, X., Wang, X., Wang, Y., Tang, Z., Cui, Q., Xi, J., Li, Y.S., Chien, S., Wang, N., 2010. MicroRNA-19a mediates the suppressive effect of laminar flow on cyclin D1 expression in human umbilical vein endothelial cells. Proc. Natl. Acad. Sci. U.S.A. 107, 3240–3244.

Rane, S., He, M., Sayed, D., Vashistha, H., Malhotra, A., Sadoshima, J., Vatner, D.E., Vatner, S.F., Abdellatif, M., 2009. Downregulation of miR-199a derepresses hypoxia-inducible factor-1alpha and Sirtuin 1 and recapitulates hypoxia preconditioning in cardiac myocytes. Circ. Res. 104, 879–886.

Rane, S., He, M., Sayed, D., Yan, L., Vatner, D., Abdellatif, M., 2010. An antagonism between the AKT and beta-adrenergic signaling pathways mediated through their reciprocal effects on miR-199a-5p. Cell. Signal. 22, 1054–1062.

Razeghi, P., Young, M.E., Alcorn, J.L., Moravec, C.S., Frazier, O.H., Taegtmeyer, H., 2001. Metabolic gene expression in fetal and failing human heart. Circulation 104, 2923–2931.

Redfield, M.M., Jacobsen, S.J., Borlaug, B.A., Rodeheffer, R.J., Kass, D.A., 2005. Age- and gender-related ventricular-vascular stiffening: a community-based study. Circulation 112, 2254–2262.

Rezzani, R., Stacchiotti, A., Rodella, L.F., 2012. Morphological and biochemical studies on aging and autophagy. Ageing Res. Rev. 11, 10–31.

Roca-Alonso, L., Castellano, L., Mills, A., Dabrowska, A.F., Sikkel, M.B., Pellegrino, L., Jacob, J., Frampton, A.E., Krell, J., Coombes, R.C., Harding, S.E., Lyon, A.R., Stebbing, J., 2015. Myocardial MiR-30 downregulation triggered by doxorubicin drives alterations in beta-adrenergic signaling and enhances apoptosis. Cell Death Dis. 6, e1754.

Rockman, H.A., Koch, W.J., Lefkowitz, R.J., 2002. Seven-transmembrane-spanning receptors and heart function. Nature 415, 206–212.

Roncarati, R., Viviani Anselmi, C., Losi, M.A., Papa, L., Cavarretta, E., Da Costa Martins, P., Contaldi, C., Saccani Jotti, G., Franzone, A., Galastri, L., Latronico, M.V., Imbriaco, M., Esposito, G., De Windt, L., Betocchi, S., Condorelli, G., 2014. Circulating miR-29a, among other up-regulated microRNAs, is the only biomarker for both hypertrophy and fibrosis in patients with hypertrophic cardiomyopathy. J. Am. Coll. Cardiol. 63, 920–927.

Rottiers, V., Najafi-Shoushtari, S.H., Kristo, F., Gurumurthy, S., Zhong, L., Li, Y., Cohen, D.E., Gerszten, R.E., Bardeesy, N., Mostoslavsky, R., Naar, A.M., 2011. MicroRNAs in metabolism and metabolic diseases. Cold Spring Harb. Symp. Quant. Biol. 76, 225–233.
Roy, S., Khanna, S., Hussain, S.R., Biswas, S., Azad, A., Rink, C., Gnyawali, S., Shilo, S., Nuovo, G.J., Sen, C.K., 2009. MicroRNA expression in response to murine myocardial infarction: miR-21 regulates fibroblast metalloprotease-2 via phosphatase and tensin homologue. Cardiovasc. Res. 82, 21–29.
Ruan, W., Xu, J.M., Li, S.B., Yuan, L.Q., Dai, R.P., 2012. Effects of down-regulation of microRNA-23a on TNF-alpha-induced endothelial cell apoptosis through caspase-dependent pathways. Cardiovasc. Res. 93, 623–632.
Rubinsztein, D.C., Marino, G., Kroemer, G., 2011. Autophagy and aging. Cell 146, 682–695.
Rybak, A., Fuchs, H., Smirnova, L., Brandt, C., Pohl, E.E., Nitsch, R., Wulczyn, F.G., 2008. A feedback loop comprising lin-28 and let-7 controls pre-let-7 maturation during neural stem-cell commitment. Nat. Cell Biol. 10, 987–993.
Sakurai, K., Amarzguioui, M., Kim, D.H., Alluin, J., Heale, B., Song, M.S., Gatignol, A., Behlke, M.A., Rossi, J.J., 2011. A role for human Dicer in pre-RISC loading of siRNAs. Nucleic Acids Res. 39, 1510–1525.
Sayed, D., Hong, C., Chen, I.Y., Lypowy, J., Abdellatif, M., 2007. MicroRNAs play an essential role in the development of cardiac hypertrophy. Circ. Res. 100, 416–424.
Schetter, A.J., Nguyen, G.H., Bowman, E.D., Mathe, E.A., Yuen, S.T., Hawkes, J.E., Croce, C.M., Leung, S.Y., Harris, C.C., 2009. Association of inflammation-related and microRNA gene expression with cancer-specific mortality of colon adenocarcinoma. Clin. Cancer Res. 15, 5878–5887.
Schmidt, U., Del Monte, F., Miyamoto, M.I., Matsui, T., Gwathmey, J.K., Rosenzweig, A., Hajjar, R.J., 2000. Restoration of diastolic function in senescent rat hearts through adenoviral gene transfer of sarcoplasmic reticulum Ca(2+)-ATPase. Circulation 101, 790–796.
Schroen, B., Heymans, S., 2012. Small but smart—microRNAs in the centre of inflammatory processes during cardiovascular diseases, the metabolic syndrome, and ageing. Cardiovasc. Res. 93, 605–613.
Sen, S., Kundu, B.K., Wu, H.C., Hashmi, S.S., Guthrie, P., Locke, L.W., Roy, R.J., Matherne, G.P., Berr, S.S., Terwelp, M., Scott, B., Carranza, S., Frazier, O.H., Glover, D.K., Dillmann, W.H., Gambello, M.J., Entman, M.L., Taegtmeyer, H., 2013. Glucose regulation of load-induced mTOR signaling and ER stress in mammalian heart. J. Am. Heart Assoc. 2, e004796.
Shan, Z.X., Lin, Q.X., Fu, Y.H., Deng, C.Y., Zhou, Z.L., Zhu, J.N., Liu, X.Y., Zhang, Y.Y., Li, Y., Lin, S.G., Yu, X.Y., 2009. Upregulated expression of miR-1/miR-206 in a rat model of myocardial infarction. Biochem. Biophys. Res. Commun. 381, 597–601.
Shih, H., Lee, B., Lee, R.J., Boyle, A.J., 2011. The aging heart and post-infarction left ventricular remodeling. J. Am. Coll. Cardiol. 57, 9–17.
Shioi, T., McMullen, J.R., Tarnavski, O., Converso, K., Sherwood, M.C., Manning, W.J., Izumo, S., 2003. Rapamycin attenuates load-induced cardiac hypertrophy in mice. Circulation 107, 1664–1670.
Sica, A., Erreni, M., Allavena, P., Porta, C., 2015. Macrophage polarization in pathology. Cell. Mol. Life Sci. 72, 4111–4126.
Son, D.J., Kumar, S., Takabe, W., Kim, C.W., Ni, C.W., Alberts-Grill, N., Jang, I.H., Kim, S., Kim, W., Won Kang, S., Baker, A.H., Woong Seo, J., Ferrara, K.W., Jo, H., 2013. The atypical mechanosensitive microRNA-712 derived from pre-

ribosomal RNA induces endothelial inflammation and atherosclerosis. Nat. Commun. 4, 3000.
Song, G., Zhang, Y., Wang, L., 2009. MicroRNA-206 targets notch3, activates apoptosis, and inhibits tumor cell migration and focus formation. J. Biol. Chem. 284, 31921–31927.
Song, S., Seo, H.H., Lee, S.Y., Lee, C.Y., Lee, J., Yoo, K.J., Yoon, C., Choi, E., Hwang, K.C., Lee, S., 2015. MicroRNA-17-mediated down-regulation of apoptotic protease activating factor 1 attenuates apoptosome formation and subsequent apoptosis of cardiomyocytes. Biochem. Biophys. Res. Commun. 465, 299–304.
Soni, M.S., Rabaglia, M.E., Bhatnagar, S., Shang, J., Ilkayeva, O., Mynatt, R., Zhou, Y.P., Schadt, E.E., Thornberry, N.A., Muoio, D.M., Keller, M.P., Attie, A.D., 2014. Downregulation of carnitine acyl-carnitine translocase by miRNAs 132 and 212 amplifies glucose-stimulated insulin secretion. Diabetes 63, 3805–3814.
Spillmann, F., Trimpert, C., Peng, J., Eckerle, L.G., Staudt, A., Warstat, K., Felix, S.B., Pieske, B., Tschope, C., Van Linthout, S., 2015. High-density lipoproteins reduce palmitate-induced cardiomyocyte apoptosis in an AMPK-dependent manner. Biochem. Biophys. Res. Commun. 466, 272–277.
Sredni, S.T., Gadd, S., Jafari, N., Huang, C.C., 2011. A parallel study of mRNA and microRNA profiling of peripheral blood in young adult women. Front. Genet. 2, 49.
Stanley, W.C., Recchia, F.A., Okere, I.C., 2005. Metabolic therapies for heart disease: fish for prevention and treatment of cardiac failure? Cardiovasc. Res. 68, 175–177.
Stewart, S., Jenkins, A., Buchan, S., McGuire, A., Capewell, S., McMurray, J.J., 2002. The current cost of heart failure to the National Health Service in the UK. Eur. J. Heart Fail. 4, 361–371.
Strum, J.C., Johnson, J.H., Ward, J., Xie, H., Feild, J., Hester, A., Alford, A., Waters, K.M., 2009. MicroRNA 132 regulates nutritional stress-induced chemokine production through repression of SirT1. Mol. Endocrinol. 23, 1876–1884.
Su, M., Wang, J., Wang, C., Wang, X., Dong, W., Qiu, W., Wang, Y., Zhao, X., Zou, Y., Song, L., Zhang, L., Hui, R., 2015. MicroRNA-221 inhibits autophagy and promotes heart failure by modulating the p27/CDK2/mTOR axis. Cell Death Differ. 22, 986–999.
Su, M., Chen, Z., Wang, C., Song, L., Zou, Y., Zhang, L., Hui, R., Wang, J., 2016. Cardiac-specific overexpression of miR-222 induces heart failure and inhibits autophagy in mice. Cell. Physiol. Biochem. 39, 1503–1511.
Suarez, Y., Fernandez-Hernando, C., Pober, J.S., Sessa, W.C., 2007. Dicer dependent microRNAs regulate gene expression and functions in human endothelial cells. Circ. Res. 100, 1164–1173.
Sun, L.L., Jiang, B.G., Li, W.T., Zou, J.J., Shi, Y.Q., Liu, Z.M., 2011. MicroRNA-15a positively regulates insulin synthesis by inhibiting uncoupling protein-2 expression. Diabetes Res. Clin. Pract. 91, 94–100.
Sun, X., Icli, B., Wara, A.K., Belkin, N., He, S., Kobzik, L., Hunninghake, G.M., Vera, M.P., Registry, M., Blackwell, T.S., Baron, R.M., Feinberg, M.W., 2012. MicroRNA-181b regulates NF-kappaB-mediated vascular inflammation. J. Clin. Invest. 122, 1973–1990.
Sun, X., He, S., Wara, A.K., Icli, B., Shvartz, E., Tesmenitsky, Y., Belkin, N., Li, D., Blackwell, T.S., Sukhova, G.K., Croce, K., Feinberg, M.W., 2014. Systemic delivery of microRNA-181b inhibits nuclear factor-kappaB activation, vascular inflammation, and atherosclerosis in apolipoprotein E-deficient mice. Circ. Res. 114, 32–40.
Taegtmeyer, H., Golfman, L., Sharma, S., Razeghi, P., Van Arsdall, M., 2004. Linking gene expression to function: metabolic flexibility in the normal and diseased heart. Ann. N. Y. Acad. Sci. 1015, 202–213.

Tang, Y., Wang, Y., Park, K.M., Hu, Q., Teoh, J.P., Broskova, Z., Ranganathan, P., Jayakumar, C., Li, J., Su, H., Tang, Y., Ramesh, G., Kim, I.M., 2015. MicroRNA-150 protects the mouse heart from ischaemic injury by regulating cell death. Cardiovasc. Res. 106, 387–397.

Tang, C.M., Liu, F.Z., Zhu, J.N., Fu, Y.H., Lin, Q.X., Deng, C.Y., Hu, Z.Q., Yang, H., Zheng, X.L., Cheng, J.D., Wu, S.L., Shan, Z.X., 2016. Myocyte-specific enhancer factor 2C: a novel target gene of miR-214-3p in suppressing angiotensin II-induced cardiomyocyte hypertrophy. Sci. Rep. 6, 36146.

Tao, H., Chen, Z.W., Yang, J.J., Shi, K.H., 2016. MicroRNA-29a suppresses cardiac fibroblasts proliferation via targeting VEGF-A/MAPK signal pathway. Int. J. Biol. Macromol. 88, 414–423.

Tatsuguchi, M., Seok, H.Y., Callis, T.E., Thomson, J.M., Chen, J.F., Newman, M., Rojas, M., Hammond, S.M., Wang, D.Z., 2007. Expression of microRNAs is dynamically regulated during cardiomyocyte hypertrophy. J. Mol. Cell. Cardiol. 42, 1137–1141.

Terentyev, D., Belevych, A.E., Terentyeva, R., Martin, M.M., Malana, G.E., Kuhn, D.E., Abdellatif, M., Feldman, D.S., Elton, T.S., Gyorke, S., 2009. miR-1 overexpression enhances Ca(2+) release and promotes cardiac arrhythmogenesis by targeting PP2A regulatory subunit B56alpha and causing CaMKII-dependent hyperphosphorylation of RyR2. Circ. Res. 104, 514–521.

Terman, A., Kurz, T., Navratil, M., Arriaga, E.A., Brunk, U.T., 2010. Mitochondrial turnover and aging of long-lived postmitotic cells: the mitochondrial-lysosomal axis theory of aging. Antioxid. Redox Signal. 12, 503–535.

Terzioglu, M., Larsson, N.G., 2007. Mitochondrial dysfunction in mammalian ageing. Novartis Found. Symp. 287, 197–208. discussion 208-13.

Thum, T., Galuppo, P., Wolf, C., Fiedler, J., Kneitz, S., Van Laake, L.W., Doevendans, P.A., Mummery, C.L., Borlak, J., Haverich, A., Gross, C., Engelhardt, S., Ertl, G., Bauersachs, J., 2007. MicroRNAs in the human heart: a clue to fetal gene reprogramming in heart failure. Circulation 116, 258–267.

Thum, T., Gross, C., Fiedler, J., Fischer, T., Kissler, S., Bussen, M., Galuppo, P., Just, S., Rottbauer, W., Frantz, S., Castoldi, M., Soutschek, J., Koteliansky, V., Rosenwald, A., Basson, M.A., Licht, J.D., Pena, J.T., Rouhanifard, S.H., Muckenthaler, M.U., Tuschl, T., Martin, G.R., Bauersachs, J., Engelhardt, S., 2008. MicroRNA-21 contributes to myocardial disease by stimulating MAP kinase signalling in fibroblasts. Nature 456, 980–984.

Tijsen, A.J., Van Der Made, I., Van Den Hoogenhof, M.M., Wijnen, W.J., Van Deel, E.D., De Groot, N.E., Alekseev, S., Fluiter, K., Schroen, B., Goumans, M.J., Van Der Velden, J., Duncker, D.J., Pinto, Y.M., Creemers, E.E., 2014. The microRNA-15 family inhibits the TGFbeta-pathway in the heart. Cardiovasc. Res. 104, 61–71.

Tomoda, Y., Kikumoto, K., Isumi, Y., Katafuchi, T., Tanaka, A., Kangawa, K., Dohi, K., Minamino, N., 2001. Cardiac fibroblasts are major production and target cells of adrenomedullin in the heart in vitro. Cardiovasc. Res. 49, 721–730.

Townley-Tilson, W.H., Callis, T.E., Wang, D., 2010. MicroRNAs 1, 133, and 206: critical factors of skeletal and cardiac muscle development, function, and disease. Int. J. Biochem. Cell Biol. 42, 1252–1255.

Trajkovski, M., Hausser, J., Soutschek, J., Bhat, B., Akin, A., Zavolan, M., Heim, M.H., Stoffel, M., 2011. MicroRNAs 103 and 107 regulate insulin sensitivity. Nature 474, 649–653.

Trajkovski, M., Ahmed, K., Esau, C.C., Stoffel, M., 2012. MyomiR-133 regulates brown fat differentiation through Prdm16. Nat. Cell Biol. 14, 1330–1335.

Tschope, C., Spillmann, F., Rehfeld, U., Koch, M., Westermann, D., Altmann, C., Dendorfer, A., Walther, T., Bader, M., Paul, M., Schultheiss, H.P., Vetter, R., 2004.

Improvement of defective sarcoplasmic reticulum Ca2+ transport in diabetic heart of transgenic rats expressing the human kallikrein-1 gene. FASEB J. 18, 1967–1969.
Tsutsui, M., Shimokawa, H., Otsuji, Y., Ueta, Y., Sasaguri, Y., Yanagihara, N., 2009. Nitric oxide synthases and cardiovascular diseases: insights from genetically modified mice. Circ. J. 73, 986–993.
Tsutsui, H., Kinugawa, S., Matsushima, S., 2011. Oxidative stress and heart failure. Am. J. Physiol. Heart Circ. Physiol. 301, H2181–H2190.
Ucar, A., Gupta, S.K., Fiedler, J., Erikci, E., Kardasinski, M., Batkai, S., Dangwal, S., Kumarswamy, R., Bang, C., Holzmann, A., Remke, J., Caprio, M., Jentzsch, C., Engelhardt, S., Geisendorf, S., Glas, C., Hofmann, T.G., Nessling, M., Richter, K., Schiffer, M., Carrier, L., Napp, L.C., Bauersachs, J., Chowdhury, K., Thum, T., 2012. The miRNA-212/132 family regulates both cardiac hypertrophy and cardiomyocyte autophagy. Nat. Commun. 3, 1078.
Van Almen, G.C., Verhesen, W., Van Leeuwen, R.E., Van De Vrie, M., Eurlings, C., Schellings, M.W., Swinnen, M., Cleutjens, J.P., Van Zandvoort, M.A., Heymans, S., Schroen, B., 2011. MicroRNA-18 and microRNA-19 regulate CTGF and TSP-1 expression in age-related heart failure. Aging Cell 10, 769–779.
Van Berlo, J.H., Elrod, J.W., Van Den Hoogenhof, M.M., York, A.J., Aronow, B.J., Duncan, S.A., Molkentin, J.D., 2010. The transcription factor GATA-6 regulates pathological cardiac hypertrophy. Circ. Res. 107, 1032–1040.
Van Bilsen, M., Van Nieuwenhoven, F.A., Van Der Vusse, G.J., 2009. Metabolic remodelling of the failing heart: beneficial or detrimental? Cardiovasc. Res. 81, 420–428.
Van Linthout, S., Miteva, K., Tschope, C., 2014. Crosstalk between fibroblasts and inflammatory cells. Cardiovasc. Res. 102, 258–269.
Van Rooij, E., Sutherland, L.B., Liu, N., Williams, A.H., Mcanally, J., Gerard, R.D., Richardson, J.A., Olson, E.N., 2006. A signature pattern of stress-responsive microRNAs that can evoke cardiac hypertrophy and heart failure. Proc. Natl. Acad. Sci. U.S.A. 103, 18255–18260.
Van Rooij, E., Sutherland, L.B., Qi, X., Richardson, J.A., Hill, J., Olson, E.N., 2007. Control of stress-dependent cardiac growth and gene expression by a microRNA. Science 316, 575–579.
Van Rooij, E., Quiat, D., Johnson, B.A., Sutherland, L.B., Qi, X., Richardson, J.A., Kelm Jr., R.J., Olson, E.N., 2009. A family of microRNAs encoded by myosin genes governs myosin expression and muscle performance. Dev. Cell 17, 662–673.
Varrone, F., Gargano, B., Carullo, P., Di Silvestre, D., De Palma, A., Grasso, L., Di Somma, C., Mauri, P., Benazzi, L., Franzone, A., Jotti, G.S., Bang, M.L., Esposito, G., Colao, A., Condorelli, G., Catalucci, D., 2013. The circulating level of FABP3 is an indirect biomarker of microRNA-1. J. Am. Coll. Cardiol. 61, 88–95.
Vasan, R.S., Sullivan, L.M., D'agostino, R.B., Roubenoff, R., Harris, T., Sawyer, D.B., Levy, D., Wilson, P.W., 2003a. Serum insulin-like growth factor I and risk for heart failure in elderly individuals without a previous myocardial infarction: the Framingham Heart Study. Ann. Intern. Med. 139, 642–648.
Vasan, R.S., Sullivan, L.M., Roubenoff, R., Dinarello, C.A., Harris, T., Benjamin, E.J., Sawyer, D.B., Levy, D., Wilson, P.W., D'agostino, R.B., Framingham Heart, S., 2003b. Inflammatory markers and risk of heart failure in elderly subjects without prior myocardial infarction: the Framingham Heart Study. Circulation 107, 1486–1491.
Vasa-Nicotera, M., Chen, H., Tucci, P., Yang, A.L., Saintigny, G., Menghini, R., Mahe, C., Agostini, M., Knight, R.A., Melino, G., Federici, M., 2011. miR-146a is modulated in human endothelial cell with aging. Atherosclerosis 217, 326–330.
Vasquez, C., Benamer, N., Morley, G.E., 2011. The cardiac fibroblast: functional and electrophysiological considerations in healthy and diseased hearts. J. Cardiovasc. Pharmacol. 57, 380–388.

Villar, A.V., Garcia, R., Merino, D., Llano, M., Cobo, M., Montalvo, C., Martin-Duran, R., Hurle, M.A., Nistal, J.F., 2013. Myocardial and circulating levels of microRNA-21 reflect left ventricular fibrosis in aortic stenosis patients. Int. J. Cardiol. 167, 2875–2881.
Viswanathan, S.R., Daley, G.Q., Gregory, R.I., 2008. Selective blockade of microRNA processing by Lin28. Science 320, 97–100.
Volkmann, I., Kumarswamy, R., Pfaff, N., Fiedler, J., Dangwal, S., Holzmann, A., Batkai, S., Geffers, R., Lother, A., Hein, L., Thum, T., 2013. MicroRNA-mediated epigenetic silencing of sirtuin1 contributes to impaired angiogenic responses. Circ. Res. 113, 997–1003.
Wan, G., Xie, W., Liu, Z., Xu, W., Lao, Y., Huang, N., Cui, K., Liao, M., He, J., Jiang, Y., Yang, B.B., Xu, H., Xu, N., Zhang, Y., 2014. Hypoxia-induced MIR155 is a potent autophagy inducer by targeting multiple players in the MTOR pathway. Autophagy 10, 70–79.
Wang, K., Klionsky, D.J., 2011. Mitochondria removal by autophagy. Autophagy 7, 297–300.
Wang, M., Zhao, D., Spinetti, G., Zhang, J., Jiang, L.Q., Pintus, G., Monticone, R., Lakatta, E.G., 2006. Matrix metalloproteinase 2 activation of transforming growth factor-beta1 (TGF-beta1) and TGF-beta1-type II receptor signaling within the aged arterial wall. Arterioscler. Thromb. Vasc. Biol. 26, 1503–1509.
Wang, K., Long, B., Zhou, J., Li, P.F., 2010a. miR-9 and NFATc3 regulate myocardin in cardiac hypertrophy. J. Biol. Chem. 285, 11903–11912.
Wang, K.C., Garmire, L.X., Young, A., Nguyen, P., Trinh, A., Subramaniam, S., Wang, N., Shyy, J.Y., Li, Y.S., Chien, S., 2010b. Role of microRNA-23b in flow-regulation of Rb phosphorylation and endothelial cell growth. Proc. Natl. Acad. Sci. U.S.A. 107, 3234–3239.
Wang, C., Wang, S., Zhao, P., Wang, X., Wang, J., Wang, Y., Song, L., Zou, Y., Hui, R., 2012a. MiR-221 promotes cardiac hypertrophy in vitro through the modulation of p27 expression. J. Cell. Biochem. 113, 2040–2046.
Wang, J., Huang, W., Xu, R., Nie, Y., Cao, X., Meng, J., Xu, X., Hu, S., Zheng, Z., 2012b. MicroRNA-24 regulates cardiac fibrosis after myocardial infarction. J. Cell. Mol. Med. 16, 2150–2160.
Wang, J., Song, Y., Zhang, Y., Xiao, H., Sun, Q., Hou, N., Guo, S., Wang, Y., Fan, K., Zhan, D., Zha, L., Cao, Y., Li, Z., Cheng, X., Zhang, Y., Yang, X., 2012c. Cardiomyocyte overexpression of miR-27b induces cardiac hypertrophy and dysfunction in mice. Cell Res. 22, 516–527.
Wang, K., Long, B., Jiao, J.Q., Wang, J.X., Liu, J.P., Li, Q., Li, P.F., 2012d. miR-484 regulates mitochondrial network through targeting Fis1. Nat. Commun. 3, 781.
Wang, K.C., Nguyen, P., Weiss, A., Yeh, Y.T., Chien, H.S., Lee, A., Teng, D., Subramaniam, S., Li, Y.S., Chien, S., 2014a. MicroRNA-23b regulates cyclin-dependent kinase-activating kinase complex through cyclin H repression to modulate endothelial transcription and growth under flow. Arterioscler. Thromb. Vasc. Biol. 34, 1437–1445.
Wang, M., Jiang, L., Monticone, R.E., Lakatta, E.G., 2014b. Proinflammation: the key to arterial aging. Trends Endocrinol. Metab. 25, 72–79.
Wang, J.X., Gao, J., Ding, S.L., Wang, K., Jiao, J.Q., Wang, Y., Sun, T., Zhou, L.Y., Long, B., Zhang, X.J., Li, Q., Liu, J.P., Feng, C., Liu, J., Gong, Y., Zhou, Z., Li, P.F., 2015a. Oxidative modification of miR-184 enables It to target Bcl-xL and Bcl-w. Mol. Cell 59, 50–61.
Wang, L., Yuan, Y., Li, J., Ren, H., Cai, Q., Chen, X., Liang, H., Shan, H., Fu, Z.D., Gao, X., Lv, Y., Yang, B., Zhang, Y., 2015b. MicroRNA-1 aggravates cardiac oxidative stress by post-transcriptional modification of the antioxidant network. Cell Stress Chaperones 20, 411–420.
Wang, X., Wang, H.X., Li, Y.L., Zhang, C.C., Zhou, C.Y., Wang, L., Xia, Y.L., Du, J., Li, H.H., 2015c. MicroRNA Let-7i negatively regulates cardiac inflammation and fibrosis. Hypertension 66, 776–785.

Wang, H., Bei, Y., Huang, P., Zhou, Q., Shi, J., Sun, Q., Zhong, J., Li, X., Kong, X., Xiao, J., 2016. Inhibition of miR-155 protects against LPS-induced cardiac dysfunction and apoptosis in mice. Mol. Ther. Nucleic Acids 5, e374.

Warboys, C.M., De Luca, A., Amini, N., Luong, L., Duckles, H., Hsiao, S., White, A., Biswas, S., Khamis, R., Chong, C.K., Cheung, W.M., Sherwin, S.J., Bennett, M.R., Gil, J., Mason, J.C., Haskard, D.O., Evans, P.C., 2014. Disturbed flow promotes endothelial senescence via a p53-dependent pathway. Arterioscler. Thromb. Vasc. Biol. 34, 985–995.

Wei, Y., Zhu, M., Corbalan-Campos, J., Heyll, K., Weber, C., Schober, A., 2015. Regulation of Csf1r and Bcl6 in macrophages mediates the stage-specific effects of microRNA-155 on atherosclerosis. Arterioscler. Thromb. Vasc. Biol. 35, 796–803.

Wenz, T., 2011. Mitochondria and PGC-1alpha in aging and age-associated diseases. J. Aging Res. 2011, 810619.

Westermann, D., Lindner, D., Kasner, M., Zietsch, C., Savvatis, K., Escher, F., Von Schlippenbach, J., Skurk, C., Steendijk, P., Riad, A., Poller, W., Schultheiss, H.P., Tschope, C., 2011. Cardiac inflammation contributes to changes in the extracellular matrix in patients with heart failure and normal ejection fraction. Circ. Heart Fail. 4, 44–52.

Whelan, R.S., Kaplinskiy, V., Kitsis, R.N., 2010. Cell death in the pathogenesis of heart disease: mechanisms and significance. Annu. Rev. Physiol. 72, 19–44.

White, M., Roden, R., Minobe, W., Khan, M.F., Larrabee, P., Wollmering, M., Port, J.D., Anderson, F., Campbell, D., Feldman, A.M., et al., 1994. Age-related changes in beta-adrenergic neuroeffector systems in the human heart. Circulation 90, 1225–1238.

Wilkins, B.J., Molkentin, J.D., 2004. Calcium-calcineurin signaling in the regulation of cardiac hypertrophy. Biochem. Biophys. Res. Commun. 322, 1178–1191.

Wu, H., Wang, F., Hu, S., Yin, C., Li, X., Zhao, S., Wang, J., Yan, X., 2012. MiR-20a and miR-106b negatively regulate autophagy induced by leucine deprivation via suppression of ULK1 expression in C2C12 myoblasts. Cell. Signal. 24, 2179–2186.

Wu, D., Jiang, H., Chen, S., Zhang, H., 2015. Inhibition of microRNA-101 attenuates hypoxia/reoxygenationinduced apoptosis through induction of autophagy in H9c2 cardiomyocytes. Mol. Med. Rep. 11, 3988–3994.

Xiao, R.P., Spurgeon, H.A., O'connor, F., Lakatta, E.G., 1994. Age-associated changes in beta-adrenergic modulation on rat cardiac excitation-contraction coupling. J. Clin. Invest. 94, 2051–2059.

Xiao, R.P., Tomhave, E.D., Wang, D.J., Ji, X., Boluyt, M.O., Cheng, H., Lakatta, E.G., Koch, W.J., 1998. Age-associated reductions in cardiac beta1- and beta2-adrenergic responses without changes in inhibitory G proteins or receptor kinases. J. Clin. Invest. 101, 1273–1282.

Xu, A., Narayanan, N., 1998. Effects of aging on sarcoplasmic reticulum Ca2+−cycling proteins and their phosphorylation in rat myocardium. Am. J. Physiol. 275, H2087–H2094.

Xu, C., Hu, Y., Hou, L., Ju, J., Li, X., Du, N., Guan, X., Liu, Z., Zhang, T., Qin, W., Shen, N., Bilal, M.U., Lu, Y., Zhang, Y., Shan, H., 2014. beta-Blocker carvedilol protects cardiomyocytes against oxidative stress-induced apoptosis by up-regulating miR-133 expression. J. Mol. Cell. Cardiol. 75, 111–121.

Xu, J., Tang, Y., Bei, Y., Ding, S., Che, L., Yao, J., Wang, H., Lv, D., Xiao, J., 2016. miR-19b attenuates H2O2-induced apoptosis in rat H9C2 cardiomyocytes via targeting PTEN. Oncotarget 7, 10870–10878.

Yan, H.L., Xue, G., Mei, Q., Wang, Y.Z., Ding, F.X., Liu, M.F., Lu, M.H., Tang, Y., Yu, H.Y., Sun, S.H., 2009. Repression of the miR-17-92 cluster by p53 has an important function in hypoxia-induced apoptosis. EMBO J. 28, 2719–2732.

Yang, Y., Ago, T., Zhai, P., Abdellatif, M., Sadoshima, J., 2011. Thioredoxin 1 negatively regulates angiotensin II-induced cardiac hypertrophy through upregulation of miR-98/let-7. Circ. Res. 108, 305–313.

Yang, T., Zhang, G.F., Chen, X.F., Gu, H.H., Fu, S.Z., Xu, H.F., Feng, Q., Ni, Y.M., 2013. MicroRNA-214 provokes cardiac hypertrophy via repression of EZH2. Biochem. Biophys. Res. Commun. 436, 578–584.
Yang, W.M., Jeong, H.J., Park, S.Y., Lee, W., 2014a. Saturated fatty acid-induced miR-195 impairs insulin signaling and glycogen metabolism in HepG2 cells. FEBS Lett. 588, 3939–3946.
Yang, Q., Yang, K., Li, A., 2014b. microRNA-21 protects against ischemia-reperfusion and hypoxia-reperfusion-induced cardiocyte apoptosis via the phosphatase and tensin homolog/Akt-dependent mechanism. Mol. Med. Rep. 9, 2213–2220.
Yang, Y., Cheng, H.W., Qiu, Y., Dupee, D., Noonan, M., Lin, Y.D., Fisch, S., Unno, K., Sereti, K.I., Liao, R., 2015. MicroRNA-34a plays a key role in cardiac repair and regeneration following myocardial infarction. Circ. Res. 117, 450–459.
Yang, F., Liu, W., Yan, X., Zhou, H., Zhang, H., Liu, J., Yu, M., Zhu, X., Ma, K., 2016a. Effects of mir-21 on cardiac microvascular endothelial cells after acute myocardial infarction in rats: role of Phosphatase and Tensin Homolog (PTEN)/Vascular Endothelial Growth Factor (VEGF) signal pathway. Med. Sci. Monit. 22, 3562–3575.
Yang, J., Chen, L., Ding, J., Zhang, J., Fan, Z., Yang, C., Yu, Q., Yang, J., 2016b. Cardioprotective effect of miRNA-22 on hypoxia/reoxygenation induced cardiomyocyte injury in neonatal rats. Gene 579, 17–22.
Yang, X., Qin, Y., Shao, S., Yu, Y., Zhang, C., Dong, H., Lv, G., Dong, S., 2016c. MicroRNA-214 inhibits left ventricular remodeling in an acute myocardial infarction rat model by suppressing cellular apoptosis via the Phosphatase and Tensin Homolog (PTEN). Int. Heart J. 57, 247–250.
Yazdanyar, A., Newman, A.B., 2009. The burden of cardiovascular disease in the elderly: morbidity, mortality, and costs. Clin. Geriatr. Med. 25, 563–577. vii.
Ye, Y., Hu, Z., Lin, Y., Zhang, C., Perez-Polo, J.R., 2010. Downregulation of microRNA-29 by antisense inhibitors and a PPAR-gamma agonist protects against myocardial ischaemia-reperfusion injury. Cardiovasc. Res. 87, 535–544.
Yu, M., Liu, Y., Zhang, B., Shi, Y., Cui, L., Zhao, X., 2015. Inhibiting microRNA-144 abates oxidative stress and reduces apoptosis in hearts of streptozotocin-induced diabetic mice. Cardiovasc. Pathol. 24, 375–381.
Zampetaki, A., Kiechl, S., Drozdov, I., Willeit, P., Mayr, U., Prokopi, M., Mayr, A., Weger, S., Oberhollenzer, F., Bonora, E., Shah, A., Willeit, J., Mayr, M., 2010. Plasma microRNA profiling reveals loss of endothelial miR-126 and other microRNAs in type 2 diabetes. Circ. Res. 107, 810–817.
Zernecke, A., Bidzhekov, K., Noels, H., Shagdarsuren, E., Gan, L., Denecke, B., Hristov, M., Koppel, T., Jahantigh, M.N., Lutgens, E., Wang, S., Olson, E.N., Schober, A., Weber, C., 2009. Delivery of microRNA-126 by apoptotic bodies induces CXCL12-dependent vascular protection. Sci. Signal 2, ra81.
Zhang, X., Azhar, G., Wei, J.Y., 2012. The expression of microRNA and microRNA clusters in the aging heart. PLoS One 7, e34688.
Zhang, L., Huang, D., Wang, Q., Shen, D., Wang, Y., Chen, B., Zhang, J., Gai, L., 2014a. MiR-132 inhibits expression of SIRT1 and induces pro-inflammatory processes of vascular endothelial inflammation through blockade of the SREBP-1c metabolic pathway. Cardiovasc. Drugs Ther. 28, 303–311.
Zhang, L., Zhou, M., Qin, G., Weintraub, N.L., Tang, Y., 2014b. MiR-92a regulates viability and angiogenesis of endothelial cells under oxidative stress. Biochem. Biophys. Res. Commun. 446, 952–958.
Zhang, Y., Huang, X.R., Wei, L.H., Chung, A.C., Yu, C.M., Lan, H.Y., 2014c. miR-29b as a therapeutic agent for angiotensin II-induced cardiac fibrosis by targeting TGF-beta/Smad3 signaling. Mol. Ther. 22, 974–985.
Zhang, X., Azhar, G., Williams, E.D., Rogers, S.C., Wei, J.Y., 2015a. MicroRNA clusters in the adult mouse heart: age-associated changes. Biomed. Res. Int. 2015, 732397.

Zhang, M., Sun, D., Li, S., Pan, X., Zhang, X., Zhu, D., Li, C., Zhang, R., Gao, E., Wang, H., 2015b. Lin28a protects against cardiac ischaemia/reperfusion injury in diabetic mice through the insulin-PI3K-mTOR pathway. J. Cell. Mol. Med. 19, 1174–1182.

Zhao, Y., Samal, E., Srivastava, D., 2005. Serum response factor regulates a muscle-specific microRNA that targets Hand2 during cardiogenesis. Nature 436, 214–220.

Zhao, X., Wang, K., Liao, Y., Zeng, Q., Li, Y., Hu, F., Liu, Y., Meng, K., Qian, C., Zhang, Q., Guan, H., Feng, K., Zhou, Y., Du, Y., Chen, Z., 2015. MicroRNA-101a inhibits cardiac fibrosis induced by hypoxia via targeting TGFbetaRI on cardiac fibroblasts. Cell. Physiol. Biochem. 35, 213–226.

Zhou, J., Wang, K.C., Wu, W., Subramaniam, S., Shyy, J.Y., Chiu, J.J., Li, J.Y., Chien, S., 2011. MicroRNA-21 targets peroxisome proliferators-activated receptor-alpha in an autoregulatory loop to modulate flow-induced endothelial inflammation. Proc. Natl. Acad. Sci. U.S.A. 108, 10355–10360.

Zhu, X., Altschafl, B.A., Hajjar, R.J., Valdivia, H.H., Schmidt, U., 2005. Altered Ca2+ sparks and gating properties of ryanodine receptors in aging cardiomyocytes. Cell Calcium 37, 583–591.

Zhu, S., Si, M.L., Wu, H., Mo, Y.Y., 2007. MicroRNA-21 targets the tumor suppressor gene tropomyosin 1 (TPM1). J. Biol. Chem. 282, 14328–14336.

Zhu, H., Wu, H., Liu, X., Li, B., Chen, Y., Ren, X., Liu, C.G., Yang, J.M., 2009. Regulation of autophagy by a beclin 1-targeted microRNA, miR-30a, in cancer cells. Autophagy 5, 816–823.

Zhu, H., Shyh-Chang, N., Segre, A.V., Shinoda, G., Shah, S.P., Einhorn, W.S., Takeuchi, A., Engreitz, J.M., Hagan, J.P., Kharas, M.G., Urbach, A., Thornton, J.E., Triboulet, R., Gregory, R.I., Consortium, D., Investigators, M., Altshuler, D., Daley, G.Q., 2011a. The Lin28/let-7 axis regulates glucose metabolism. Cell 147, 81–94.

Zhu, H., Yang, Y., Wang, Y., Li, J., Schiller, P.W., Peng, T., 2011b. MicroRNA-195 promotes palmitate-induced apoptosis in cardiomyocytes by down-regulating Sirt1. Cardiovasc. Res. 92, 75–84.

Zhu, D., Pan, C., Li, L., Bian, Z., Lv, Z., Shi, L., Zhang, J., Li, D., Gu, H., Zhang, C.Y., Liu, Y., Zen, K., 2013a. MicroRNA-17/20a/106a modulate macrophage inflammatory responses through targeting signal-regulatory protein alpha. J. Allergy Clin. Immunol. 132. 426–36.e8.

Zhu, J.N., Chen, R., Fu, Y.H., Lin, Q.X., Huang, S., Guo, L.L., Zhang, M.Z., Deng, C.Y., Zou, X., Zhong, S.L., Yang, M., Zhuang, J., Yu, X.Y., Shan, Z.X., 2013b. Smad3 inactivation and MiR-29b upregulation mediate the effect of carvedilol on attenuating the acute myocardium infarction-induced myocardial fibrosis in rat. PLoS One 8, e75557.

Zhu, W.S., Guo, W., Zhu, J.N., Tang, C.M., Fu, Y.H., Lin, Q.X., Tan, N., Shan, Z.X., 2016. Hsp90aa1: a novel target gene of miR-1 in cardiac ischemia/reperfusion injury. Sci. Rep. 6, 24498.

Zile, M.R., Brutsaert, D.L., 2002. New concepts in diastolic dysfunction and diastolic heart failure. Part II: causal mechanisms and treatment. Circulation 105, 1503–1508.

Zou, Z., Wu, L., Ding, H., Wang, Y., Zhang, Y., Chen, X., Chen, X., Zhang, C.Y., Zhang, Q., Zen, K., 2012. MicroRNA-30a sensitizes tumor cells to cis-platinum via suppressing beclin 1-mediated autophagy. J. Biol. Chem. 287, 4148–4156.

Zou, M., Wang, F., Gao, R., Wu, J., Ou, Y., Chen, X., Wang, T., Zhou, X., Zhu, W., Li, P., Qi, L.W., Jiang, T., Wang, W., Li, C., Chen, J., He, Q., Chen, Y., 2016. Autophagy inhibition of hsa-miR-19a-3p/19b-3p by targeting TGF-beta R II during TGF-beta1-induced fibrogenesis in human cardiac fibroblasts. Sci. Rep. 6, 24747.

CHAPTER SIX

MicroRNA Dysregulation in Aging and Pathologies of the Skeletal Muscle

Rachel McCormick[1], Katarzyna Goljanek-Whysall[1]
Musculoskeletal Biology II, Centre for Integrated Research into Musculoskeletal Aging, Institute of Aging and Chronic Disease, University of Liverpool, Liverpool, United Kingdom
[1]Corresponding authors: e-mail address: r.mccormick@hotmail.co.uk; k.whysall@liverpool.ac.uk

Contents

Abstract

Skeletal muscle is one of the biggest organs of the body with important mechanistic and metabolic functions. Muscle homeostasis is controlled by environmental, genetic, and epigenetic factors. Indeed, MiRNAs, small noncoding RNAs robust regulators of gene expression, have and have been shown to regulate muscle homeostasis on several levels: through controlling myogenesis, muscle growth (hypertrophy) and atrophy, as well as interactions of muscle with other tissues. Given the large number of MiRNA target genes and the important role of MiRNAs in most physiological processes and various diseases, MiRNAs may have an enormous potential as therapeutic targets against

International Review of Cell and Molecular Biology, Volume 334
ISSN 1937-6448
http://dx.doi.org/10.1016/bs.ircmb.2017.03.005

numerous disorders, including pathologies of muscle. The purpose of this review is to present the current knowledge of the role of MiRNAs in skeletal muscle homeostasis and pathologies and the potential of MiRNAs as therapeutics for skeletal muscle wasting, with particular focus on the age- and disease-related loss of muscle mass and function.

ABBREVIATIONS

ALS amyotrophic lateral sclerosis
BAF barrier to autointergration factor
Col4a1 collagen type IV alpha 1 chain
COPD chronic obstructive pulmonary disease
COXIV cytochrome *c* oxidase subunit IV
DMD Duchenne muscular dystrophy
ERRα estrogen-related receptor alpha
Ezh2 enhancer of zeste homolog 2
Fbn1 fibrillin 1
FGF fibroblasts growth factor
Fzd7 frizzled class receptor 7
HDAC4 histone deacetylase 4
HOXA-11 homeobox protein A-11
Hspg2 heparan sulfate proteoglycan 2
IFGBP5 insulin-like growth factor-binding protein 5
IGF-1 insulin growth factor 1
IGF1R insulin growth factor receptor 1
IL interleukin
Lamc1 laminin subunit
MAP4K3 mitogen-activated protein kinase
Meox2 mesenchyme homeobox 2
MHC myosin heavy chain
MicroRNAs MiRNAs, MiRs
MRFs myogenic regulatory factors
Myf4 myogenic factor 4
Myf5 myogenic regulatory factor 5
MyoD myogenic differentiation 1
MyoR myogenic repressor
NF-κβ nuclear factor kappa beta
Nid2 NFAT5-nuclear factor of activated T-cells 5
NMD neuromuscular degeneration
NMJ neuromuscular junction
NRF1 nuclear respiratory factor 1
PARP2 poly(ADP-ribose) polymerase 2
Pax3/7 paired box 3/7
PGC-1α peroxisome proliferator-activated receptor gamma coactivator 1-alpha
RARB retinoic acid receptor
RISC RNA-induced silencing complex

ROS reactive oxygen species
Sirt1 Sirtuin 1
SOCS1 suppressor of cytokine signalling 1
Sparc secreted protein acidic and cysteine rich
SRF serum response factor
TGF-β transforming growth factor β
TNF-α tumor necrosis factor alpha
TWEAK tumor necrosis factor-like weak inducer of apoptosis
US United States
YY1 Yin-yang 1

1. INTRODUCTION

Skeletal muscle accounts for approximately 40% of total body weight and is vital to the correct function of the body with important roles in locomotion, stabilizing body positions, and producing heat and energy. Muscle homeostasis depends on a balance between muscle hypertrophy, regeneration, and atrophy (Moylan and Reid, 2007). Skeletal muscle is terminally differentiated; nevertheless, it shows great plasticity during exercise or mechanical loading, and this plasticity declines with age and disease. Given the importance of skeletal muscle for overall function, the progressive loss of muscle mass and strength as a result of aging and degenerative disorders, not only muscle-specific but also disorders such as cancer, presents a huge socioeconomic burden on society. This is due to irreversible muscle loss often leading to hospitalization, loss of physical independence, and an increase in other morbidities.

Sarcopenia is the loss of muscle mass and function during aging (Rosenberg, 1989) and has been shown to occur in both human and animal models (Lushaj et al., 2008; Sousa-Victor et al., 2014). In humans, sarcopenia affects individuals from approximately the 4–5th decade of life (Lexell et al., 1988), with a decrease of 30%–50% in skeletal muscle mass and function by the time individuals reach approximately 80 years of age (Akima et al., 2001). With the ever growing aging population, demographic analysis predicts that the world population of over 60s will reach 2 billion by 2050, nearly tripling from the beginning of the millennium, resulting in the over 80s being the fastest growing subpopulation in the developed world (WHO, 2014). Although there have been improvements in lifespan, the same advances have not been made in health span, meaning that the extra years of living are under poor health. Sarcopenia is a major contributor to frailty in the older population resulting in further immobility with a loss of independence, as well as increasing the risk of other chronic diseases and morbidity (Coin et al., 2013). Although there is no available data for the cost to the UK or other European countries, in 2000 the US spent 1.5% of their national budget ($18.5 billion) on sarcopenia (Janssen et al., 2004). This means just a 10% reduction in sarcopenia prevalence would result in a saving of $1.1 billion per year in US healthcare costs, signifying the importance of finding a treatment or preventative therapy for sarcopenia.

Several aspects of age-related loss of muscle mass and function are common with other muscle-wasting diseases. Amyotrophic lateral sclerosis (ALS, motor neuron disease) is a rare degenerative disease that is associated with motor neuron degeneration and loss of muscle, with an incidence rate of 2–4/100,000 per year in Europe (Gladman and Zinman, 2015; Lopez-Bastida et al., 2009; van der Steen et al., 2009). Although the direct cost of ALS to the NHS is unknown, studies have shown that in the US, ALS costs an individual approximately $64,000 a year (Larkindale et al., 2014), costing $1 billion a year nationally. This cost was estimated as €36,380 annually per patient in Europe (Schepelmann et al., 2010). Furthermore, a study on Canadian patients has shown direct health care costs of $32,337; however, the indirect healthcare costs to ALS patients and their families can also reach over $50,000 annually (Gladman et al., 2014).

Cachexia is the loss of muscle mass and function due to a chronic disease and is a powerful prognostic factor for morbidity and mortality. One of the most common forms of cachexia is cancer cachexia, with approxiamtely 50%–80% of advanced cancer patients suffering from muscle loss and estimates state that 1% of the entire population are affected by cachexia. Cancer cachexia is associated with increased mortality rates up to 80% and estimated total costs more than €100 billion per year in Europe (von Haehling and Anker, 2014).

Other diseases resulting in loss of muscle mass and function, such as muscular dystrophy and myasthenia gravis, have also been shown to cost tens of thousands of dollars per patient annually in the US (Schepelmann et al., 2010).

Despite the clear problem of muscle loss and disease, the only intervention currently available for muscle loss is exercise; however, exercise as a treatment for muscle loss is not possible for all affected by muscle wasting. Exercise is not a suitable practice for all patients suffering from severe muscle loss. Lifelong exercise was unable to prevent the loss in muscle mass and function in mice (Kayani et al., 2008) and lifelong master athletes were shown to still undergo a loss in strength, power, and endurance with age (Grassi et al., 1991). Therefore, there is currently no cure for skeletal muscle loss in mass or function; however, due to increase in the aging population and the significant costs associated with caring for patients with muscle wasting, there is an urgent need to understand the mechanisms of muscle wasting and design novel and effective therapies.

1.1 Mechanisms of Muscle Wasting

Despite the numerous classified diseases which result in muscle loss, a lot of the underlying mechanisms are similar between the different diseases, whereby muscle mass and function, as well as the ability of skeletal muscle to adapt to stimuli are lost. Several proposed processes include: decrease in protein synthesis (Welle et al., 1993), infiltration of fat and connective tissue into muscle (Addison et al., 2014; Brack et al., 2007), dysregulation of proteasomal degradation pathways (Chondrogiannia et al., 2000; Cuervo and Dice, 2000), mitochondrial dysfunction (Sakellariou et al., 2013; Short et al., 2005), reduced number and/or function and increased senescence of satellite cells (Conboy et al., 2005; Parker, 2015; Sousa-Victor et al., 2014), disrupted ROS homeostasis (Broome et al., 2006; McDonagh, 2016; McDonagh et al., 2014, 2015, 2016; Palomero et al., 2013), and increased inflammation (Fagiolo et al., 1993). These dysfunctions are proposed to contribute to a decrease in fiber number and size, neuromuscular junction (NMJ) deterioration, decrease in force production and poor regeneration seen in models of muscle loss (Lexell et al., 1988; Lightfoot et al., 2014; Sakellariou et al., 2014; Vasilaki et al., 2016).

2. MiRNAs IN MUSCLE

MicroRNAs (MiRNAs, MiRs) are small noncoding RNAs that regulate gene expression in a posttranscriptional manner. MiRNAs work through binding to target sites within, usually, the 3′ UTR of the target transcript. This leads to usually downregulation of gene expression through inhibition of translation and/or mRNA degradation. Each MiRNA is predicted to have multiple target genes and approximately two-thirds of the human genome is predicted to be targeted by MiRNAs. MiRNAs have been demonstrated to be robust regulators of development and adult organism homeostasis. Moreover, MiRNA expression has been shown to be altered in many disorders; therefore, it is likely that MiRNAs may be key regulators of various diseases, including diseases of the muscle. Their small size and efficient delivery into muscle also makes them potential attractive therapeutic candidates against muscle wasting.

2.1 MiRNA Discovery

MiRNAs were first discovered in 1993 when Lee et al. revealed that the gene encoding lin-4 regulated the translation of LIN-14 protein post-transcriptionally through an antisense mechanism in *Caenorhabditis elegans* (Lee et al., 1993; Wightman et al., 1993). It was not until several years later that the next MiRNA, Let-7, was discovered, and this MiRNA again was shown to play a role in larval development (Reinhart et al., 2000). Since, thousands of MiRNAs and over 50,000 MiRNA:mRNA interactions have been identified, with pivotal roles in many different physiological processes and numerous disease states.

2.2 MiRNA Biogenesis and Mechanism of Action

Depending on the origin of their sequence, MiRNAs are either intergenic or introgenic. Intergenic MiRNAs are transcribed by their own promoter as independent transcripts, whereas introgenic MiRNAs are encoded within the introns of protein coding genes, and these are acquired through transcription of their host genes (Reinhart et al., 2000). MiRNA biogenesis is a multistep process which begins with the transcription of a long precursor transcript known as pri-miRNA (primary MiRNA) by RNA polymerase II or polymerase III (Bartel, 2004; Borchert et al., 2006). Pri-miRNAs have a characteristic local hairpin that is cleaved by an RNAse III enzyme Drosha, in conjunction with its partner protein, DGCR8 (DiGeorge syndrome chromosomal region 8) into pre-miRNA (precursor MiRNA).

Pre-miRNA is actively transported from the nucleus into the cytoplasm by Exportin-5 in a Ran guanosine triphosphate-dependent manner (Lund et al., 2004). Following arrival into the nucleus, pre-miRNA is further cleaved by another type III RNAse, Dicer, which gives rise to the MiRNA-3p and -5p strands (Fig. 1).

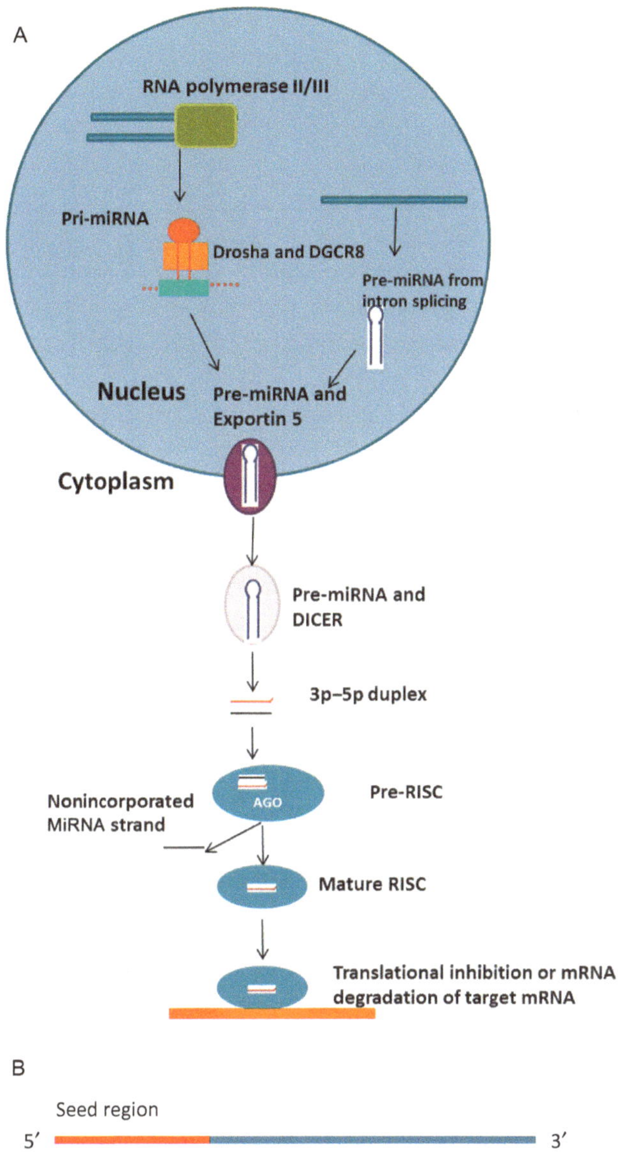

Fig. 1 (A). MiRNA biogenesis. pri-miRNA is produced by RNA polymerase II/III and is cleaved by Drosha in association with DGCR8 into pre-miRNA. Pre-miRNA is actively transferred into the cytoplasm by Exportin-5, where it is cleaved by the enzyme Dicer into the MiRNA-3p and -5p strands. The mature MiRNA is incorporated into the RISC containing an AGO protein. MiRNA guides the RISC to the target sequences within the target mRNA(s), and this results in inhibition of protein translation and/or mRNA degradation. The nonincorporated MiRNA strand is usually degraded. (B) General structure of mature MiRNA. The seed region, shown in *red*, is usually 2–7 base pairs from the 5′ end and is essential for MiRNA–mRNA binding. The seed region has a completely complementary match to the target mRNA, although the rest of the MiRNA may not be complementary to the target mRNA.

An alternative pathway for MiRNA biogenesis has been described, in which some introns, which have been debranched mimic the structural features of pre-miRNAs to enter the MiRNA-processing pathway bypassing Drosha-mediated cleavage. These MiRNAs are known as miRtrons (Ruby et al., 2007).

The function of MiRNAs is to posttranscriptionally affect gene expression, in most cases through gene silencing. For gene silencing process to occur, mainly based on the thermostability of the 5′ end of the cleaved MiRNA, either the -3p or -5p strand, or in some cases both strands, forms mature MiRNA that is incorporated into the RISC (RNA-induced silencing complex) and guides the RISC, which includes the endonuclease Argonaute 2 protein, to the complementary sites, usually within the 3′ UTR within the target transcript(s). MiRNA binding to its target mRNA(s) occurs via sequence complementarity. The first 6–8 nucleotides of the MiRNA (seed sequence) bind to the complementary target sites within the mRNA(s). Additional pairing usually occurs between the remaining MiRNA sequence and the mRNA, and this additional pairing has been suggested to influence the strength of the MiRNA:target interactions and the mechanism of MiRNA inhibition of gene expression (Bartel, 2009). The binding of MiRNA to its target mRNA results in the suppression of translation or degradation of the mRNA transcript, thus resulting in the protein product being absent in the cells (Bartel, 2009) (Fig. 1). Both of these mechanisms have been shown to be involved in MiRNA function (Krol et al., 2010).

Following biogenesis of MiRNAs, MiRNAs are considered to be highly stable compared to most of the RNA family with a study showing the average half-life of MiRNAs in fibroblasts is 119 h (Gantier et al., 2011). However, the stability of individual MiRNAs varies; for example, MiR-125b has been shown to have a half-life of 225 h in fibroblasts (Gantier et al., 2011), whereas MiR-29b has been shown to have a half-life of 4 h in human and mouse cells (Hwang et al., 2007).

Despite the fact that MiRNAs has been shown to be stable molecules, a microarray study has demonstrated a negative correlation between MiRNA stability and MiRNA expression level (Li et al., 2013), suggesting that the levels detected and the activity of MiRNA may vary. The susceptibility of MiRNAs to degradation can be affected by other factors, for example, MiR-122 stability is increased with the binding of a DNA/RNA-binding protein, Translin in vivo (Yu and Hecht, 2008). AGO proteins are also known to protect MiRNAs from degradation and other factors, such as target mRNAs, nucleotide content and environmental factors have also been shown to affect the stability of MiRNAs (Li et al., 2013).

Furthermore, MiRNAs are easily detectable in both blood and plasma, and have been shown to be stable following incubations at room temperature, through numerous cycles of freeze-thawing and through long periods of freezing (Balzano et al., 2015; Mitchell et al., 2008). Due to these features, there are many studies which show evidence that MiRNAs in the blood can be used as biomarkers for diseases (Cacchiarelli et al., 2011; Chen et al., 2008; Sørensen et al., 2016).

Whether MiRNAs can be absorbed from food is debatable. Baier et al. demonstrated that MiR-29b and MiR-200c were absorbed the blood cells of humans following the ingestion of milk (Baier et al., 2014). However, further studies on the same set of samples failed to detect any changes in MiRNA levels in the plasma or red blood cells of the patients, nor were any bovine-specific MiRNAs detected (Auerbach et al., 2016). Furthermore, no changes in the levels of MiR-30b in the blood and numerous tissues of mice fed by transgenic females were detected (Laubier et al., 2015). MiR-375 and MiR-200c knockout pups also did not uptake these MiRNAs into the blood or tissues through feeding by wild-type dams (Title et al., 2015). Moreover, Snow et al. found that following ingestion of plant MiRNAs by healthy atheletes, plant MiRNAs were not detected in the plasma of the athletes (Snow et al., 2013).

The study by Title et al. is the only study so far to distinguish between endogenous and exogenous MiRNAs, suggesting that dietary MiRNAs may be absorbed, with MiRNAs: MiR-29b and MiR-200 (Bagge et al., 2012; Hennessy et al., 2010) being linked to glucose sensing. However, the changes in these MiRNAs reported by Baier et al. were suggested to be a response to dietary intake and not exogenous MiRNAs (Witwer, 2014).

2.3 MiRNAs in Embryonic and Adulthood Myogenesis

Skeletal muscle is one of the biggest organs of the body and is characterized by multiperipheral nuclei and striations. Skeletal muscle is comprised primarily of contractile materials but is also associated with nerves, blood vessels, and connective tissue—all of which contribute to the function of skeletal muscle. Skeletal muscle plays mechanistic and metabolic functions in the body, both of which have been shown to be regulated by MiRNAs through controlling myogenesis, muscle growth (hypertrophy), and atrophy, as well as interactions of muscle with other tissues (Goljanek-Whysall et al., 2011; Hitachi and Tsuchida, 2013). Myogenesis, a process of muscle formation, is a complex process that occurs during embryonic development and in adulthood during muscle regeneration. The cells involved in myogenesis include mainly, however, not only muscle stem/progenitor cells from within the myotome and satellite cells but also adult skeletal muscle stem cells (Mauro, 1961; Munsterberg et al., 1995).

During the formation of skeletal muscle, progenitor cells proliferate and differentiate to form mature muscle. There is also a population of progenitor cells that becomes a reserve pool of quiescent progenitor cells required for muscle regeneration, and these are known as satellite cells (Buckingham et al., 2003; Mauro, 1961). The process of myogenesis is controlled by Pax (paired box genes) proteins: Pax3 and Pax7 and myogenic regulatory factors (MRFs), such as myogenic differentiation 1 (MyoD), myogenic regulatory factor 5 (Myf5), myogenic factor 4 (Myf4), and myogenin (Cheng et al., 1993; Munsterberg et al., 1995). MRFs have been shown to regulate MiRNA expression during myogenesis (Gagan et al., 2011; Goljanek-Whysall et al., 2011; Koutsoulidou et al., 2011; Sweetman et al., 2008), and MiRNAs have been shown to control myogenesis through regulation of proliferation, differentiation, and maintenance of the quiescent state of satellite cells (Cheung et al., 2012; Goljanek-Whysall et al., 2011, 2012b; Huang et al., 2012; Townley-Tilson et al., 2010). Many studies have also shown the differential expression of MiRNAs during the different stages of myogenesis (Goljanek-Whysall et al., 2011; Guo et al., 2016; Iwasaki et al., 2015; Koning et al., 2012; Wong and Tellam, 2008).

A selection of MiRNAs which are highly enriched in muscle, known as myomiRs, has been identified. These include MiR-1, -133a, -133b, -206, -208b, -486, and -499 (Kim et al., 2006; Sempere et al., 2004; Small et al., 2010; van Rooij et al., 2007, 2009). These account for approximately 25% of all MiRNA expression in the muscle (Nielsen et al., 2010). Regulation of myomiRs is controlled by MRFs, and myomiRs can also regulate MRFs expression (Gagan et al., 2011; Sweetman et al., 2008).

myomiRs have been shown to control all aspects of muscle homeostasis, with MiR-1, MiR-133, and MiR-206 controlling myogenesis during embryonic development and in adulthood, by targeting Pax3 and Pax7 gene expression (Crist et al., 2009; Goljanek-Whysall et al., 2011),

MiR-208b and MiR-499 controlling myofibre type (van Rooij et al., 2009), and MiR-1, MiR-128a, MiR-133, MiR-206, and MiR-208 affecting muscle hypertrophy and atrophy (Hitachi and Tsuchida, 2013). Moreover, myomiRs have been shown to be differentially expressed in muscle in response to exercise (Nielsen et al., 2010) and disease (Di Filippo et al., 2016; Eisenberg et al., 2007), identifying an important role of MiRNAs, in particular of muscle-specific or muscle-enriched MiRNAs in regulating the plasticity, adaptation, and dysfunction of skeletal muscle.

2.4 MiRNAs in Embryonic Myogenesis

MiRNAs are some of the key regulators of myogenesis. Deletion of the enzyme Dicer in myogenic progenitor cells early in development, resulting in loss of expression of several MiRNAs, has been associated with delayed myogenesis during mouse embryogenesis (Goljanek-Whysall et al., 2011). Moreover, myomiRs: MiR-1, MiR-206, MiR-133, and MiR-27b have been shown to regulate early muscle development through controlling Pax3 expression in myogenic progenitors in the somites and through regulation of chromatin regulatory mechanism, the SWI/SNF components, during embryonic myogenesis (Crist et al., 2009; Goljanek-Whysall et al., 2011, 2014). Other myogenesis mechanisms have also been shown to be regulated by myomiRs: MiRs: MiR-1, MiR-206, and MiR-133, including splicing during embryonic myogenesis (Boutz et al., 2007) and mechanisms associated with proliferation and differentiation of myogenic progenitors during embryonic development (Chen et al., 2006). Interestingly, individual genetic deletion of myomiRs: MiR-206 and MiR-133 in mice shows no robust developmental phenotypes (Liu et al., 2008, 2012), suggesting that these MiRNAs may have a very specific role during embryonic myogenesis or potential compensatory mechanisms exist in the absence of these MiRNAs (Boettger et al., 2014).

2.5 MiRNAs in Myoblast Proliferation and Differentiation

Myoblast proliferation and differentiation are mutually exclusive processes, also key aspects of myogenesis. The involvement of MiRNAs in myoblast proliferation through control of key signaling pathways and transcription factors has been researched however is not fully understood. For example, MiRNA-133a has been shown to enhance myoblast proliferation through decreasing the expression of serum response factor (SRF) (Chen et al., 2006) and insulin-like growth factor 1 (IGF-1) (Huang et al., 2011).

Furthermore, MiRNA-27a has been demonstrated to increase myoblast proliferation through the inhibition of myostatin expression (Huang et al., 2012; Yang et al., 2014). In contrast, MiRs: MiR-195 and MiR-497 have been shown to inhibit myoblast proliferation through targeting the insulin growth factor receptor (IGFR). These MiRNAs, along with MiR-1 and MiR-206, have been shown to control the expression of important cell cyclins, therefore regulate myoblast proliferation (Goljanek-Whysall et al., 2011; Wei et al., 2016a).

In contrast to the role of MiRNAs in myoblast proliferation, the involvement of MiRNAs in muscle differentiation is well studied, and numerous MiRNAs, in particular myomiRs, have been implicated in the process.

Pax7 and Pax3 are critical factors for myogenesis that are expressed in myogenic progenitors during embryonic development and in satellite cells during adulthood. However for myogenesis to occur, MRFs must be activated, and this requires the downregulation of both Pax3 and Pax7 expression (Epstein et al., 1995; Kumar et al., 2009). Numerous MiRNAs have been shown to regulate Pax3 expression, such as MiR-1, MiR-206 (Goljanek-Whysall et al., 2011), and MiR-27b (Crist et al., 2009). Interestingly, inhibition of MiR-1 and MiR-206 during embryogenesis in vivo (Goljanek-Whysall et al., 2011) and of MiR-27b in cultured satellite cells (Crist et al., 2009) led to delayed myogenesis, suggested to be associated with delayed downregulation of Pax3 expression. MiRNAs: MiR-1, MiR-206, and MiR-486 have also been shown to target Pax7 expression during myogenic differentiation (Chen et al., 2010). Furthermore, the overexpression of MiR-206 and MiR-1 resulted in premature myogenic differentiation, while the inhibition of these MiRs led to delayed myogenesis (Dey et al., 2011). The importance of the role of MiR-206 in myogenesis has been further demonstrated through the overexpression of validated MiR-206 targets in C2C12 cells resulting in increased expression of genes related to alternative cell fates, in particular osteogenesis and chondrogenesis (Goljanek-Whysall et al., 2012b). These data show that MiRNAs are not only vital for myogenesis to occur but also for preventing the progression of muscle progenitors into alternative cell fates.

Another MiRNA shown to negatively regulate myogenesis through inhibiting homeobox protein Hox-A11 expression is MiR-181 (Naguibneva et al., 2006). However, in a different study, MiRNA-181 has been shown to increase myogenic differentiation through upregulation of ER stress proteins: ATF6, BCL2, CASP12, CHOP, and EIF-2α and leading to an increase in MyoD and myogenin (Wei et al., 2016b).

HDAC4 (histone deacetylase 4), a negative regulator of myogenesis that inhibits MRFs through the inhibition of the myocytes enhancer factor-2 family (MEF2) (Lu et al., 2000; Miska et al., 1999), has been validated as a MiR-1 target. MiR-1-mediated downregulation of HDAC4 expression resulted in an increase in MyoD and Myogenin content in myoblasts and enhanced myogenic differentiation (Chen et al., 2006). Another regulator of the cell cycle, DNA polymerase α is regulated by MiR-206 resulting in cell cycle arrest and inhibition of myoblast proliferation. Moreover, MyoR and Id1–3, both negative regulators of myogenesis through repression of MRFs, are indirectly regulated by MiR-206 (Kim et al., 2006).

The expression of another myomiR, MiR-133, has also been shown to be upregulated during myogenesis (Chen et al., 2006; Koutsoulidou et al., 2011). However, the role of this MiRNA is more complex with MiR-133 having been shown to have the effects opposite to those of MiR-1 and MiR-206. Specifically, overexpression of MiR-133 led to the decrease in myogenin and myosin heavy chain (MHC) expression (Chen et al., 2006). This is in agreement with data showing positive effects of MiR-133 on myoblast proliferation rather than differentiation. However, others have demonstrated that the MiR-133-mediated inhibition of myoblast proliferation and initiation of differentiation is through the ERK1/2 pathway (Feng et al., 2013). Moreover, MiR-133 overexpression has been shown to prevent the de-differentiation of C2C12 myoblasts through regulating the levels of reversine (Kim et al., 2014b). The role of MiR-133 in regulation of proliferation and differentiation remains to be established, potentially using in vivo models.

Other MiRNAs have also been shown to enhance differentiation of myoblasts, for example, MiR-26a through targeting the histone methyltransferase, enhancer of zeste homolog 2 (Ezh2), a repressor of myogenesis (Wong and Tellam, 2008).

To summarize, the role of MiRNAs in myoblast proliferation and differentiation, as well as control of cell fate, has been thoroughly demonstrated by various groups (summarized in Table 1 and Fig. 2). It remains to be established, whether this regulation is a fine-tuning mechanism or a driving force of the regulated processed, with both having been suggested.

2.6 MiRNAs in Satellite Cell Biology and Muscle Regeneration

One of the most impressive characteristics of skeletal muscle is the ability to regenerate following an injury. Muscle regeneration is a similar process as

Table 1 The Role of MiRNAs in Maintaining Processes Associated With Muscle Homeostasis

MiRNA	Process	Target Genes (If Evidenced in Study)	References
MiR-133	Increases myoblast proliferation in vitro and in vivo	SRF	Chen et al. (2006)
		IGF-1	Huang et al. (2011)
	Increases myogenesis in vitro	ERK 1/2 pathway, SRF	Koutsoulidou et al. (2011), Feng et al. (2013), and Kim et al. (2014b)
	Inhibits myogenesis in vivo	SRF	Chen et al. (2006)
	Regulates myogenesis in vivo	BAF chromatin remodeling complexes	Goljanek-Whysall et al. (2014)
	Rhabdomyosarcoma in vitro	Microarray with lots of different changes, see paper for details	Rao et al. (2010)
MiR-195	Inhibits myoblast proliferation in vitro	IGF1R, insulin receptor, and cyclin genes	Wei et al. (2016a)
MiR-1	Increases regeneration in vitro and in vivo	MyoD, myogenin and Pax7, GDF-8	Nakasa et al. (2010) and Kim et al. (2016)
	Promotes myogenic differentiation in vivo	Pax3, BAF chromatin remodeling complexes	Goljanek-Whysall et al. (2011, 2014)
	Control cell fate of undifferentiated myoblasts in vitro	CLCN3, Smarcd2, Smarcb1, MAP4K3, Fzd7, Meox2, RARB, NFAT5	Goljanek-Whysall et al. (2012b)
MiR-206	Promotes myogenesis in vivo	Pax3, BAF chromatin remodeling complexes	Goljanek-Whysall et al. (2011, 2014)
	Muscle cell fate in vitro	CLCN3, Smarcd2, Smarcb1, MAP4K3, Fzd7, Meox2, RARB, NFAT5	Goljanek-Whysall et al. (2012b)
	Increases regeneration in vitro and in vivo	MyoD, myogenin and Pax7, GDF-8	Nakasa et al. (2010), Liu et al. (2012), Maciotta et al. (2012), and Kim et al. (2016)
	Regulation of neuromuscular junction in vivo	HDAC4, FGF pathway	McCarthy et al. (2007) and Williams et al. (2009c)

Table 1 The Role of MiRNAs in Maintaining Processes Associated With Muscle Homeostasis—cont'd

MiRNA	Process	Target Genes (If Evidenced in Study)	References
MiR-27a	Increases proliferation in vitro	Myostatin	Huang et al. (2012) and Yang et al. (2014)
MiR-27b	Regulates Pax7 expression in vivo	Pax3/7	Crist et al. (2009)
	Enhances myogenic differentiation in vivo	Pax3	Crist et al. (2009)
MiR-486	Increases myoblast differentiation in vitro	Pax7	Dey et al. (2011)
MiR-26a	Increases muscle regeneration in vivo	SMAD4	Dey et al. (2012)
	Increase myogenesis in vitro	Ezh2	Wong and Tellam (2008)
MiR-181	Enhances myogenesis in vivo	HOX-A11	Naguibneva et al. (2006)
	Regulates myofibres size in vitro	Sirt-1	Soriano-Arroquia et al. (2016b)
MiR-489	Controls satellite cell quiescence in vivo	DEK	Cheung et al. (2012)
MiR-31	Controls satellite cell quiescence ex vivo and in vivo	Myf5	Crist et al. (2012)
MiR-155	Increase regeneration in vivo	SOCS1	Nie et al. (2016a)
MiR-29a	Increase regeneration in vitro and in vivo	Fgf2, Fbn1, Lamc1, Nid2, Col4a1, Hspg2, and Sparc	Galimov et al. (2016)
MiR-675	Increase regeneration and myogenesis in vitro	BMP pathway, Cdc6	Dey et al. (2014)
MiR-431	Increase regeneration in vivo	Pax7	Wu et al. (2015)
MiR-208	Myofibre typing in vivo	Myosin genes	van Rooij et al., (2009)
MiR-378	Enhances myogenesis in vitro	MyoR	Gagan et al. (2011)
MiR-143	Regulation of satellite cell senescence in vitro	IFGBP5	Soriano-Arroquia et al. (2016a)

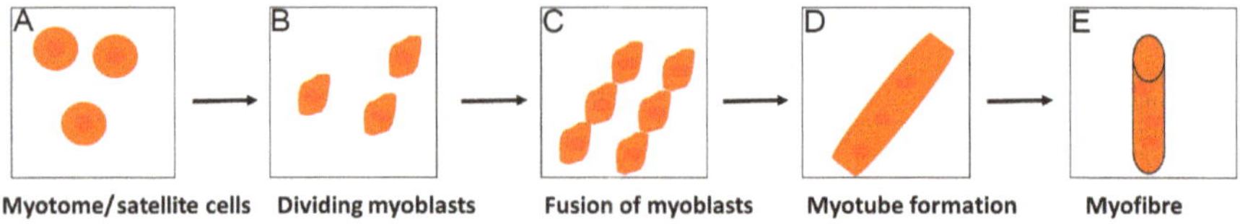

Fig. 2 Process of myogenesis. Myotome/satellite cells are precursors of myoblasts and are activated by intra- and extracellular stimuli to become committed myoblasts, undifferentiated and single-nucleated muscle cells. Myoblasts divide and fuse to become myotubes which eventually mature into myofibres.

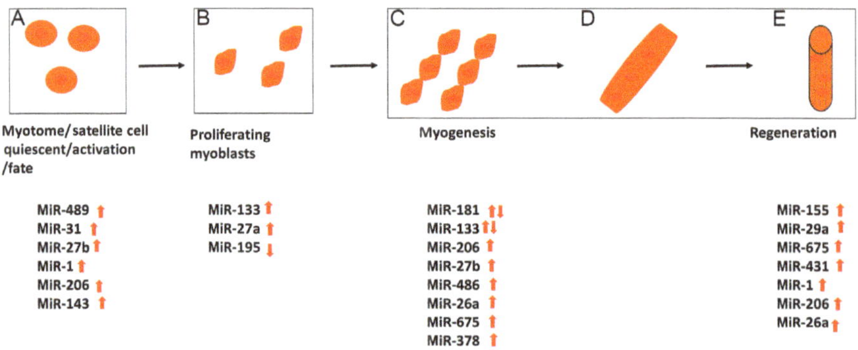

Fig. 3 The role of specifc MiRNAs. The involvement of specific MiRNAs in controlling myogenic progenitor/satellite cell activation into proliferating myoblasts and in myogenic differentiation and muscle regeneration is annotated below the relevant process and whether they have been shown to enhance (*upward facing arrows*) or inhibit (*downward facing arrows*) is indicated.

myogenesis, whereby quiescent satellite cell, adult muscle stem cells, become activated, proliferate, and differentiate thus contribute to the repair of the myofibres (Fig. 3). Following the formation of muscle, the population of reserve satellite cells remain quiescent until they are required and become activated. Satellite cells quiescence is vital for retaining the pool of stem cells and ability of muscle to regenerate following injury. The expression of many MiRNAs has been shown to be downregulated in quiescent satellite cells (Koning et al., 2012); nevertheless, there is evidence that MiRNAs play a key role in retaining the dormant state of satellite cells. Satellite cells of mice, which do not express the enzyme Dicer responsible for production of the majority mature MiRNAs, have been shown to exit the quiescent state and enter the cell cycle. Importantly, these activated satellite cells became apoptotic and following injury, the muscle of these mice did not regenerate efficiently (Cheung et al., 2012). The role of MiRNAs in muscle regeneration has been further validated in a study by Chen et al., which demonstrated the inability of satellite cells with Dicer knockout to reach terminal differentiation (Chen et al., 2010) and a decreased skeletal muscle mass, abnormal myofibre morphology, and perinatal death of muscle fibers in a muscle-specific knockout of Dicer (O'Rourke et al., 2007).

Another myomiR highly expressed in satellite cells, MiR-489, has been shown to regulate the maintenance satellite cell quiescence through targeting the oncogene DEK (Cheung et al., 2012), which regulates cell proliferation (Khodadoust et al., 2009). Moreover, MiRNA-31 has been shown to prevent the activation of myogenesis through the sequestration of Myf5 in messenger ribonucleoprotein particles (Crist et al., 2012). As discussed previously, HDAC4, a target of numerous MiRNAs expressed in muscle, has been shown to initiate the activation of Pax7 and the subsequent activation of satellite cells (Choi et al., 2014). This data gives ample evidence that MiRNAs are involved in the control of both the quiescence, activation and differentiation of satellite cells, and therefore regulation of muscle regeneration.

Indeed, the expression of several MiRNAs has been shown to change during muscle regeneration with similar patterns of MiRNA expression having been observed during embryonic myogenesis and muscle regeneration, e.g., MiR-1, MiR-499, and MiR-682 (however, it is noteworthy that MiR-682 is only expressed in mice and not humans) (Chen et al., 2011).

Muscle regeneration is reliant on complicated interactions of satellite cells and other cell types. MiR-155 has been shown to facilitate muscle regeneration through appropriate activation of macrophages in response to injury ensuring the correct balance between pro- and antiinflammatory cytokines. MiR-155 functions via inhibiting the repression of SOCS1 on the JAK/Stat pathway. Inhibition of this MiRNA led to delayed muscle regeneration (Nie et al., 2016a). Pharmacological or genetic deletion of another MiRNA, MiR-29a also resulted in decreased muscle regeneration mediated by a decrease in myoblast proliferation due to the inhibitory effect of MiR-29a on basement membrane protein expression, such as Col4a1 and Lamc1 (Galimov et al., 2016). These data demonstrate a role of MiR-29 in dismantling the basement membrane in the satellite cell niche, which is required for the further proliferation of satellite cells during muscle regeneration.

MyomiRs have also been shown to regulate muscle regeneration. Specifically, MiR-206 expression is upregulated in the muscle of mice following muscle injury and has been shown to be important in muscle regeneration. MiR-206 depletion results in delayed regeneration due to a decrease in fusion and differentiation through inhibiting Igfbp5 and Notch 3 (Liu et al., 2012). Moreover, MiR-206 targets high mobility group box 3 (Hmgb3) during muscle regeneration (Maciotta et al., 2012). Hmgb3 has been shown to prevent activation of Wnt signaling, which is important for the switching of proliferation of cells to differentiation in muscle (Tanaka et al., 2011).

Furthermore, the treatment of mice with a combination of three myomiRs: mir-1, MiR-133, and MiR-206 led to enhanced muscle regeneration and prevented fibrosis of regenerating muscle through increasing the expression of the MRFs: MyoD and myogenin, following muscle injury (Nakasa et al., 2010). Moreover, treatment of barium chloride-induced injured muscle of rats with MiR-206 and MiR-1 in combination with a myostatin-targeted siRNA increased regeneration of muscle with an observed increase in MRF expression (Kim et al., 2016). Treatment of muscle with other MiRNAs: MiR-431, MiR-675, and MiR-26a also increased muscle regeneration, with MiR-431 targeting Pax7 and increasing MRFs expression (Wu et al., 2015), MiR-675 inhibiting the TFG-β pathway (Dey et al., 2014), and MiR-26a and MiR-431 targeting the SMAD4, which affects the regenerative capacity of satellite cells (Dey et al., 2012; Lee et al., 2015). Collectively these results show that MiRNAs are important in muscle regeneration and myogenesis.

3. DYSREGULATION OF MiRNAs IN MUSCLE DISORDERS

Data from the previous sections have given vast amounts of evidence that MiRNAs are important in maintaining skeletal muscle homeostasis during embryonic and adult life. MiRNAs have also been shown to regulate the ability of skeletal muscle to adapt to contractions (Nielsen et. al., 2010). Therefore, recent work in the field has moved to identify changes in MiRNAs expression that occur during muscle disease, the effect that this has on muscle homeostasis and the therapeutic potential of MiRNAs in muscle-wasting diseases.

The general importance of MiRNAs in muscular skeletal disorders has been demonstrated by Eisenberg et al. through identifying differential expression patterns of MiRNAs in primary muscular skeletal disorders. The authors characterized five MiRNAs: MiR-146b, MiR-221, MiR-155, MiR-214, and MiR-222 that were consistently dysregulated in all of the muscle samples from all of muscle disorders studied, suggesting a common pathway of MiRNA dysregulation throughout muscular disease (Eisenberg et al., 2007). Interestingly, however, Soares et al. have shown that the function of MiRNAs similarly dysregulated in different muscle disorders may be context-dependent (Soares et al., 2014). As very few functional studies have been performed investigating the role of specific MiRNAs in muscle wasting during aging and disease, this question remains to be answered.

To identify the changes in MiRNA expression in sarcopenia, age-related loss of muscle mass and function, numerous microarray, and sequencing studies have been performed using muscle from young and older humans and model organisms. Studies in the muscle of humans identified 18 differentially expressed MiRNAs in the muscle of older individuals when compared to their younger counter parts. MiRNAs: let-7a, -b, -e, and -f, as well as MiR-25, MiR-98, MiR-195, and MiR-1268 were upregulated in the muscle of older people, and MiRs: MiR-22, MiR-24, MiR-27a, MiR-27b, MiR-30d, MiR-133a, MiR-133b, MiR-223, MiR-378, MiR-378* were downregulated. The upregulation of the let MiRNAs led to a decrease in the expression of cell cycle controllers in the muscle of older individuals, consistent with the let MiRNAs known to inhibit cell proliferation (Drummond et al., 2011).

Numerous studies have shown the changes of MiRNA expression in the muscle of old and young mice. Hamrick et al. have shown the differential expression of 57 MiRNAs between muscle of old and young mice, including the upregulation of MiR-7, MiR-542, MiR-486, and MiR-686 and the downregulation of MiR-181, MiR-434, MiR-455, and MiR-221 (Hamrick et al., 2010). In another study by Kim et al., the differential expression of MiRNAs in muscle of young and old mice has been demonstrated. Eight dysregulated have also been shown to be significantly overexpressed in muscle affected by eight muscular diseases, such as Duchenne muscular dystrophy (DMD), facioscapulohumeral muscular dystrophy, limb-girdle muscular dystrophies types 2A. This analysis may therefore represent a meaningful relationship between the age-regulated MiRNAs identified in mouse skeletal muscle and human muscular diseases and their potential pathogenic roles in muscle (Kim et al., 2014a). However, a microarray investigating MiRNA expression in four different models of muscle wasting, denervation, fasting, cancer cachexia, and diabetes, showed there was no common MiRNA expression pattern changes (Soares et al., 2014), suggesting that the different diseases may be characterized by their own unique MiRNA signatures. It remains to be established, whether the differences in the sets of dysregulated MiRNAs in muscle during aging and disease in different studies are the result of different platforms used or the limitations of the model organisms used, or yet other factors.

Other studies have found that MiR-143 (Soriano-Arroquia et al., 2016a) and MiR-1 is downregulated in satellite cells during aging (Di Filippo et al., 2016), whereas the expression of some MiRNAs is increased during aging: with MiR-26 MiR-29, MiR-30C, MiR-34b, MiR-465, and MiR-499 elevated in the muscle of old rodents (Hu et al., 2014b; Soriano-Arroquia et al., 2016b) and MiR-133 in older humans (Di Filippo et al., 2016).

Similarly, as during aging, the expression of MiRNAs: MiR-23a, MiR-29b, MiR-206, and MiR-455 has been shown to be increased in skeletal muscle of ALS patients, also leading to the disruption of the mitochondrial-related gene expression in the muscle (Russell et al., 2013). Furthermore, Williams et al. have shown the changes in MiR-206 expression in an ALS mouse model accelerated the disease progression (Williams et al., 2009c). Despite some knowledge of MiRNA expression in ALS patients, studies identifying changes in MiRNAs in the skeletal muscle of ALS patients are sparse.

MiRNAs are also differentially expressed in another disease associated with muscle wasting, chronic obstructive pulmonary disease (COPD) (Barreiro, 2016). The expression of myomiR MiR-1 has been shown to be downregulated in the muscle of COPD patients but decreased in the plasma of COPD patients. In the contrary, the expression of the myomiRs: MiR-1, MiR-133, MiR-206, and MiR-499 was elevated in the plasma of COPD patients (Donaldson et al., 2013). Furthermore, MiR-542 and MiR-422 levels have been correlated with the changes in the rectus femoris cross-sectional area in the muscle of COPD patients undergoing high-risk cardiothoracic surgery (Paul et al., 2015), and plasma levels of MiR-181 have been associated with acute muscle wasting (Bloch et al., 2015).

Muscular dystrophy is a devastating disease caused by the loss of the functional dystrophin protein, which gradually contributes to muscle wasting. There are numerous types of muscular dystrophy and MiRNAs have been implicated in the majority of these. The most common form of muscular dystrophy is DMD. Greco et al. have demonstrated the disruption of expression of 10 MiRNAs in patients and in mouse models of this disease (Greco et al., 2009). MiRNAs differentially expressed were divided into three categories: MiRNAs involved in muscle regeneration, degeneration, and those involved in inflammation. Most notably, MiR-206 was upregulated in the serum of muscular dystrophy patients, similarly as in other types of muscular dystrophy (Cacchiarelli et al., 2011; Koutsoulidou et al., 2015). Furthermore, the expression of MiRs: MiR-206, MiR-1, and MiR-133 has been shown to upregulated in the serum of patients affected by a different type of muscular dystrophy, myotonic muscular dystrophy

(Cacchiarelli et al., 2011). Moreover, another study has shown that MiRNA-1, MiR-133a, MiR-133b, and MiR-206 serum levels were elevated in muscular dystrophy patients with progressive muscle wasting compared to diseased patients with less severe muscle wasting (Koutsoulidou et al., 2015). High levels of these MiRNAs corresponded to low muscle strength function and in turn a low quality of life (Hu et al., 2014a).

The levels of expression of other MiRNAs: MiR-21, MiR-29, MiR-31, MiR-208a, MiR-208b, and MiR-499 have also been shown to change in the serum and muscle of models of muscular dystrophy (Cacchiarelli et al., 2011; Li et al., 2014a; Zanotti et al., 2015). Consequently, MiRNA levels in the serum and/or plasma have been proposed as biomarkers for the muscle diseases in numerous studies (Cacchiarelli et al., 2011; Giordani et al., 2014). However, care must be taken in translation of these results as Hildyard et al. have shown that despite MiR-31 having promising positive effects in an in vitro model of DMD, inhibiting MiR-31 did not successfully translate to in vivo modulation of DMD (Hildyard and Wells, 2016).

3.1 MiRNAs in Fibrosis of Muscle

The inability of muscle to completely regenerate results in the infiltration of fibrotic tissue into the muscle and the formation of scar tissue, and this can contribute to the loss of muscle mass and function (Ramaswamy et al., 2011). Infiltration of other tissues into muscle is associated with numerous muscle-wasting models, most notably sarcopenia (Evans et al., 1995; Song et al., 2004) and muscular dystrophy (Klingler et al., 2012). The majority of work investigating the effects of MiRNAs on fibrosis in muscle has been carried out in muscular dystrophy rodent models. As fibrosis is usually a secondary disease in muscle pathology, the role of MiRNAs directly in fibrosis in skeletal muscle has been proposed to be limited. Nevertheless, Ardite et al. identified a transforming growth factor β (TGF-β)-induced increase in MiR-21 expression in fibrotic tissue in the muscle of muscular dystrophic mice, as well as mice which have undergone a laceration, a known inducer of fibrosis. Furthermore, overexpression of MiR-21 caused an increase in the expression of fibrosis-associated genes in fibroblasts isolated from muscle, whereas inhibition of MiR-21 reverted the fibrotic response to TGF-β in vivo (Ardite et al., 2012). Another MiRNA, MiR-199 also has a well-established role in fibrosis through targeting the Wnt signaling pathway and this is further evidenced in pathologies such as muscular dystrophy, in which the levels of MiR-199 are dysregulated (Alexander et al., 2013).

Other MiRNAs have been linked to fibrosis in other tissues, such as MiR-29 in cardiac fibrosis (Williams et al., 2009a), and MiR-21, MiR-455, MiR-151-3p, MiR-486-5p, and MiR-3107 have been shown to be dysregulated in pulmonary fibrosis (Ji et al., 2015); however, the relevance of these MiRNAs to fibrosis in skeletal muscle is unknown.

3.2 MiRNAs in Denervation of Muscle Fibers

The NMJ is the chemical synapse between the nerve and the muscle. An intact NMJ is vital for the contraction of skeletal muscle. The degeneration of the NMJ, the resulting denervation of the muscle fiber, and the inability to reinnervate the muscle fibers are characteristic to numerous skeletal muscle disorders, most notably aging and in ALS (Ansved and Larsson, 1990; Blijham et al., 2007; Tomlinson and Irving, 1977; Vaughan, 1992; Williams et al., 2009c). The role of MiRNAs in denervation of muscle fibers is still unclear; nevertheless, the myomiR MiR-206, expression of which is upregulated following muscle denervation, has been shown to have numerous roles in the NMJ maintenance and development (Anderson et al., 2006). MiR-206 promotes NMJ maintenance and repair and inhibition of MiR-206 following the denervation by the damage of the sciatic nerve results in an increase in the time of reinnervation of muscle fibers (McCarthy et al., 2007; Williams et al., 2009c). Deletion of MiR-206 in a mouse ALS model has been shown to decrease the lifespan of the mice and enhance disease progression, whereas overexpression of MiR-206 has been demonstrated to slow down the progression of the disease.

MiR-133b has also been shown to be enriched at the NMJ and elevated following denervation (Williams et al., 2009c). MiR-133b has been proposed to have a role in the NMJ maintenance; however, a MiR-133b null mice showed no difference in phenotype of the NMJ and in the reinnervation of the muscle fibers (Valdez et al., 2014).

The expression of HDAC4, a histone deacetylase negatively correlated with NMJ function and to be involved in denervation, has been shown to increase following the denervation of muscle fibers. In the presence of neuromuscular defects, the inactivation of HDAC4 results in a suppression of muscle atrophy in a denervation model (Williams et al., 2009c; Winbanks et al., 2013). Expression levels of HDAC4 are negatively correlated with skeletal muscle reinnervation through a decrease in the levels of gene related to atrophy (atrogenes). Levels of HDAC4 are upregulated in NMD mice (a spinal muscular atrophy (SMARD1) model), ALS patients, and an ALS mouse model (Bruneteau et al., 2013; Cohen et al., 2007). Furthermore overexpression of HDAC4 results in denervation of the muscle fiber (Cohen et al., 2007); thus, HDAC4 has been hypothesized as a potential therapeutic target for neurodegenerative diseases (Kazantsev and Thompson, 2008; Mielcarek et al., 2013). Interestingly, HDAC4 has been shown to be a target of numerous MiRNAs, in particular, MiR-206 (Williams et al., 2009c), suggesting a potential mechanism through how MiRNAs may carry out a role in neurodegeneration and reinnervation. However, MiR-206 expression has not been correlated with the beneficial functional outcomes of HDAC4 which it has been shown to negatively affect (Bruneteau et al., 2013).

3.3 MiRNAs in Anabolic Resistance

Anabolic resistance is the inability to effectively initiate protein synthesis following the ingestion of protein or exercise. Anabolic resistance has been well characterized during sarcopenia (Cuthbertson et al., 2005); however, it has also been shown to occur in critically ill patients which undertake long periods of immobilization (Rennie, 2009). The expression of MiRNAs has been shown to be upregulated following exercise and following protein ingestion (Camera et al., 2016; Drummond et al., 2008; Nielsen et al., 2010); however, the role of MiRNAs in anabolic resistance is not well understood.

Drummond et al. have shown differential expression of MiRNAs following exercise and ingestion of essential amino acids between old and younger people. MiR-1 expression was decreased in the muscle of younger, but not older, individuals following exercise. The primary MiR-133a and MiR-206 transcripts were also differentially regulated in the muscle of young and older people; however, these changes were not reflected in the levels of the mature MiRNA. The levels of mRNA of Drosha and Exportin-5 were also altered with higher expression in the muscle of older people (Drummond et al., 2008). These data suggest that the lack of decrease in expression of MiR-1 in the muscle of older people following exercise and protein ingestion may contribute to anabolic resistance. Despite this, the same study showed no difference in the levels of the predicted targets of MiR-1 involved in muscle regulation and growth; however, muscle biopsies were taken just 3 and 6 h following ingestion; therefore, it is possible that this was too early to observe the downstream changes in the mRNA of target genes (Drummond et al., 2008).

3.4 MiRNAs in Mitochondrial Dysfunction

Mitochondria are essential for providing the ATP required for muscle contraction and are central to maintaining redox homeostasis and quality control of the cell, therefore essential for the viability of muscle. Skeletal muscle requires a large amount of energy, and the amount and function of mitochondria within skeletal muscle are related to their individual metabolizing capacity and thus vital to the adaption of muscle to exercise. Given the essential role of mitochondria in skeletal muscle maintenance and survival, alterations in mitochondria are considered a contributor to muscle wasting in numerous diseases (Bua et al., 2006; Dupuis et al., 2004; Percival et al., 2013). MiRNAs have been shown to regulate the mitochondria homeostasis (Li et al., 2012) and to target the mitochondrial genome in skeletal muscle (Shinde and Bhadra, 2015).

Despite the role of MiRNAs in mitochondria homeostasis in muscle being fairly well understood (Duarte et al., 2014; Shinde and Bhadra, 2015), the studies identifying the role of MiRNAs in skeletal muscle mitochondrial dysfunction have focused on peroxisome proliferator-activated receptor gamma coactivator (PGC-1α). PGC-1α is a master regulator of the mitochondria homeostasis, involved in energy metabolism and mitochondrial biogenesis through activation of mitochondrial proteins (Liang and Ward, 2006). Overexpression of PGC-1α protected mice from sarcopenia onset demonstrating an important role of PGC-1α in skeletal muscle aging (Wenz et al., 2009). Soares et al. identified a global decrease in MiR-21 and MiR-206 in numerous models of muscle atrophy and identified Yin-Yang 1 (YY1) as a target (Soares et al., 2014). YY1 is involved in mitochondrial biogenesis in skeletal muscle through PGC-1 activation (Cunningham et al., 2007).

MiRNA-696 also targets PGC-1α and its downstream effectors: pyruvate dehydrogenase kinase-4 (PDK4) and cytochrome *c* oxidase subunit IV (COXIV) (Aoi et al., 2010). MiR-696 also decreased fatty acid oxidation and mitochondrial DNA levels. An increase in MiR-696 expression occurs during immobilization model of muscle atrophy (Aoi et al., 2010). This suggests that MiR-696 dysregulation may be detrimental to skeletal muscle homeostasis. In a different study, inhibition of MiR-696 also increased muscle differentiation and increased the production of fast and slow MHC mRNA in engineered skeletal muscle tissues (Cheng et al., 2016). This study also investigated the effects of increased expression of PGC-1α following combined inhibition of MiR-133a and MiR-696. Interestingly, the inhibition of these two MiRNAs led to an increase in the levels of the oxidative mitochondrial dense type I MHC, suggesting that these two MiRNAs can affect the fiber type content of the muscle fibers. However, despite evidence for a negative role of MiR-133a on the mitochondria homeostasis in the muscle, MiR-133a-deficient mice show lower levels of PGC-1α and NRF1 resulting in a decreased mitochondrial mass and exercise tolerance (Nie et al., 2016b), phenotypically similar to sarcopenia, and thus demonstrates a vital role of MiR-133a to the mitochondria homeostasis in skeletal muscle.

Russell et al. elegantly showed a disrupted mitochondrial network with a decrease in the levels and activity of PGC-1α and the further downstream effectors: NRF1, COXIV, ERRα in the mouse model of ALS. MiR-23 has been shown to be upregulated in different mouse models of ALS (Russell et al., 2013; Williams et al., 2009c), and the overexpression of MiR-23 in wild-type mice has been shown to result in mitochondrial dysfunction phenotype similar to the animal models of ALS (Russell et al., 2013). Despite these data demonstrating the potential causative effect of MiR-23 in ALS, MiR-23a has been shown to decrease the expression of Atrogin-1 and Murf1, two ubiquitin ligases which are upregulated in numerous models of atrophy, and overexpression of MiR-23 protected against glucocorticoid-induced atrophy (Wada et al., 2011). Furthermore, MiR-23 expression is decreased in a diabetic model of atrophy in mice and a dexamethasone-induced atrophy of C2C12 myotubes (Hudson et al., 2014). This suggests that the mechanisms responsible for the different models of atrophy may differ; however, it may also suggest the difference in the functioning of the mitochondria in the studies, as the mitochondria of the mice with lower levels of MiR-23 were not studied.

Another MiRNA, MiR-149, has been shown to indirectly regulate PGC-1α expression through targeting and inhibiting PARP2 expression resulting in an increase in mitochondria biogenesis (Mohamed et al., 2014). Dysregulation of MiR-149 has been demonstrated in the muscles of different forms of muscular dystrophies (Israeli et al., 2016; Vignier et al., 2013) and in the blood of ALS patients (Maciotta et al., 2013).

Furthermore, MiR-494 has been shown to inhibit mitochondrial biogenesis and function through targeting mitochondrial proteins, such as mitochondrial transcription factor A and Forkhead box j3, resulting in the loss of mitochondria oxidative capacity and a decrease in the fast type II muscles fibers, which are proposed to be the most affected by sarcopenia (Greco et al., 2009). Furthermore, upregulation of MiR-494 is associated with a decrease in basal oxygen consumption, inhibition of myotube formation, and a decrease in highly oxidative type IIa expression in human muscle cells (Iwasaki et al., 2015). Tumor necrosis factor alpha (TNF-α)-induced increase in MiR-494 expression in C2C12 cells has been shown to result in downregulation of the insulin pathway, which is vital for muscle protein synthesis (Lee et al., 2013). This suggests a detrimental role of MiR-494 for the mitochondria homeostasis and myogenesis, and the discussed overexpression of MiR-494 is phenotypically similar to some aspects of sarcopenia. Furthermore, MiR-494 is the most upregulated MiRNA in the brains from ALS patients (Wakabayashi et al., 2014) and is upregulated in muscle of muscular dystrophy patients (Greco et al., 2009), suggesting the potential role in driving the pathogenic changes associated with muscle wasting.

The dysregulation of MiRNAs involved in mitochondrial biogenesis in numerous muscular diseases suggests a role for MiRNAs in mitochondrial dysfunction, and in particular in regulating changes in fiber type composition. It is also worthy to note that a lot of MiRNAs upregulated in muscle diseases are downregulated during exercise. During sarcopenia, muscle has been shown to be unable to evoke the transcription events need for the muscle to adapt to exercise, perhaps also MiRNA transcription. This has been shown in a study which characterized the different changes in expression levels of several MiRNAs pre- and postexercise in the muscle of young and older people, suggesting another potential detrimental effect of dysregulation of MiRNAs (Zacharewicz et al., 2014).

3.5 MiRNAs in Inflammation

Inflammation is the body's response to injury or the presence of foreign bodies and is vital for the protection of cells and tissues. However, chronic low grade inflammation is a well-characterized phenomenon in many muscular diseases. In sarcopenia, serum levels of proinflammatory cytokines: TNF-α, interleukin (IL)6, and C-reactive protein are increased with age, and this is correlated with a decrease in muscle mass (Aleman et al., 2011; Pedersen et al., 2003), performance (Thalacker-Mercer et al., 2010), function (Bautmans et al., 2011), strength (Norman et al., 2014; Tiainen et al., 2010), and fitness (Levinger et al., 2010). There is also an age-related increase in NF-κβ, a major mediator of inflammatory markers, within the muscle (Vasilaki et al., 2006) and an increased inflammatory state is also seen in muscular dystrophy (Nonaka, 1998), ALS and cachexia patients (de Matos-Neto et al., 2015). Another class of inflammatory muscular diseases is myositis, associated with the infiltration of inflammatory cells in the muscle and accompanied by muscle weakness (Bohan and Peter, 1975; Georgantas et al., 2014).

MiRNAs have been shown to be critical regulators of the systemic immune response (Baltimore et al., 2008) and involved in inflammatory diseases of muscle. The majority of studies investigating the effects of MiRNAs in muscle inflammation has used different models of myositis. In muscle from myositis patients that showed increased levels of numerous inflammation markers, the levels of MiR-1, MiR-133a, MiR-133b, MiR-206, and MiR-378 were decreased. In vitro work has shown that the overexpression of MiR-1, MiR-133a, MiR-133b, and MiR-206 in C2C12 cells, in which differentiation was inhibited by TNF-α, was able to reverse this block on differentiation (Georgantas et al., 2014). Furthermore, a microarray performed on the serum from myositis patients revealed the differential abundance of 87 MiRNAs with 38 of these MiRNAs being related to muscular activity (Remáková et al., 2013).

Numerous MiRNAs have been shown to regulate the inflammatory response, for example, MiR-155 (Li et al., 2016; Nie et al., 2016a) and MiR-181a (Xie et al., 2013). MiR-155 is involved in the antiinflammatory response, whereas MiR-181a expression levels are similar to the levels of proinflammatory markers IL-1α, IL-1β, IL-6, and TNF-α following an inflammatory stimuli (Xie et al., 2013). The expression of these MiRNAs has been shown to be increased in muscle during aging (Kim et al., 2014a; Soriano-Arroquia et al., 2016b); for example, IL6 levels are increased during aging in human myoblasts and are associated with decreased levels of MiR-143 and increased levels of its target gene Igfbp5, potentially leading to satellite cell senescence (Soriano-Arroquia et al., 2016a), suggesting these MiRNAs may play a role in sarcopenic inflammation.

Another inflammatory marker, tumor necrosis factor-like weak inducer of apoptosis (TWEAK), which has been related to skeletal muscle atrophy and inhibition of muscle cell differentiation in vitro (Dogra et al., 2006), has been shown to be upregulated in numerous diseases such as cachexia, ALS, and sarcopenia (Bowerman et al., 2015; Johnston et al., 2015; Tajrishi et al., 2014) and to regulate the expression numerous MiRNAs. Specifically, treatment of C2C12 myotubes with TWEAK led to downregulation of MiR-1, MiR-23, MiR-27a, MiR-27b, MiR-93, MiR-107, MiR-133a and MiR-133b, MiR-192, MiR-199 expression and upregulation of the expression of MiR-98, MiR-146a, MiR-322, MiR-455, MiR-470, and MiR-715.

TWEAK treatment in mice also led to downregulation of MiR-1, MiR-133a, and MiR-133b expression and upregulation of MiR-146a expression in the gastrocnemius muscle (Panguluri et al., 2010).

Let-7 and MiR-146a have both been shown to inhibit the secretion of proinflammatory cytokines in muscle (Bhaumik et al., 2009; Jiang et al., 2013) and are both upregulated in skeletal muscle from older individuals (Drummond et al., 2011).

Despite the evidence for MiRNA involvement in muscle inflammation, many of the MiRNAs, expression of which is increased during inflammation, show antiinflammatory properties, suggesting that these MiRNAs may be acting as a compensatory mechanism against the raised level of inflammation. Whether this compensatory mechanism is beneficial or not is debateable. For example, Soriano et al. described the age-related decrease in MiR-143 as a potential mechanism to improve the efficiency of myogenesis; however, this led to an increase in cell senescence (Soriano-Arroquia et al., 2016a). This is consistent with similar compensatory mechanisms previously described in muscle aging and ALS (Williams et al., 2009b) giving further evidence that disruption of the levels of MiRNA can be detrimental for muscle function.

3.6 MiRNAs as Potential Therapeutic for Muscle Wasting

Since one MiRNA can target numerous target genes, MiRNAs have critical roles in many physiological processes. As described in the previous sections, the dysregulation of MiRNAs likely contributes to muscle disease, and this disruption of the MiRNA signature has also been shown to contribute to many other diseases. MiRNA-based therapeutics are already in clinical trials for several diseases: Miravirsen (SPC3649), which inhibits the liver-specific MiRNA, MiR-122, is currently in phase 2a trials for use as a therapeutic against hepatitis C (Janssen et al., 2013).

The previous sections have shown great evidence for the dysregulation of MiRNA in numerous muscle-wasting diseases thus giving the manipulation of the levels of MiRNAs great potential as a therapeutic intervention for muscle wasting. There is ample data showing changes in mRNA and protein signatures during the loss of muscle mass and function and aberrant expression of myomiRs can lead to dysfunctional signaling pathways and eventually, pathologies (Goljanek-Whysall et al., 2012a, 2016). These data give great evidence that MiRNAs have a role in the loss of muscle mass and therefore MiRNAs may have potential therapeutic benefits for the loss of muscle mass and function.

Despite the evidence for a beneficial role of the manipulation of MiRNAs in skeletal muscle as discussed in previous sections, MiRNA therapy is still to move into clinical application. Furthermore, Mercken et al. described how calorie restriction reversed age-related changes in MiRNA expression of skeletal muscle (Mercken et al., 2013), and although it was not shown in this study, calorie restriction has been shown to protect against sarcopenia (Colman et al., 2008). Calorie-restricted mice also showed improved regeneration in a model of DMD (Cerletti et al., 2012), and we and others have shown that sarcopenia is associated with significant changes in MiRNA expression (Hamrick et al., 2010; Soriano-Arroquia et al., 2016b).

The therapeutic application of MiRNAs can be achieved in two ways. One strategy results in a gain of function of MiRNA targets and aims to inhibit MiRNAs by using inhibitors of MiRNA. These include anti-miRs, locked nucleic acids, or antagomiRs (Hönig et al., 2016; Krutzfeldt et al., 2005, 2006). These inhibitors are oligonucleotides with sequences that are complementary to the endogenous MiRNAs, which are chemically modified to enhance the affinity for the target MiRNA. The endogenous MiRNA is then unable to be processed by RISC or results in degradation of the endogenous MiRNA. The second strategy is known as MiRNA replacement and involves the reintroduction of a MiRNA mimic in the hope of restoring a loss of MiRNA target function. This proof of concept has been shown in a great number of animal studies. The majority of earlier work targeted MiRNAs dysregulated in cancer. However, this review has shown many examples of how MiRNA mimics and antagomiRs have been used in in vivo models of muscle-wasting diseases.

Despite the evidence of the potential of MiRNAs as powerful therapeutics, there are several obstacles required to be overcome before MiRNAs are used in clinical application. The use of chemical modifications is essential to maximize RNA-binding efficiency; however, these modifications have been linked with some toxicity such as the activation of immune cells (Bennett and Swayze, 2010). In another study, antagomiRs have shown liver toxicity (Swayze et al., 2007). There is also the threat of inhibiting MiRNAs that target oncogenes, thus activating cancerous pathways. Moreover, there is the hazard of antagomiRs binding nonspecifically, in particular as antagomiRs rarely distinguish between different family members, particularly if they have the same seed region (Li and Rana, 2014; Li et al., 2011). This, as well as the requirement for high doses of oligonucleotides being required to achieve inhibition of the MiRNAs, is likely to result in some unwanted off-target effects. Furthermore, there is still work to be done

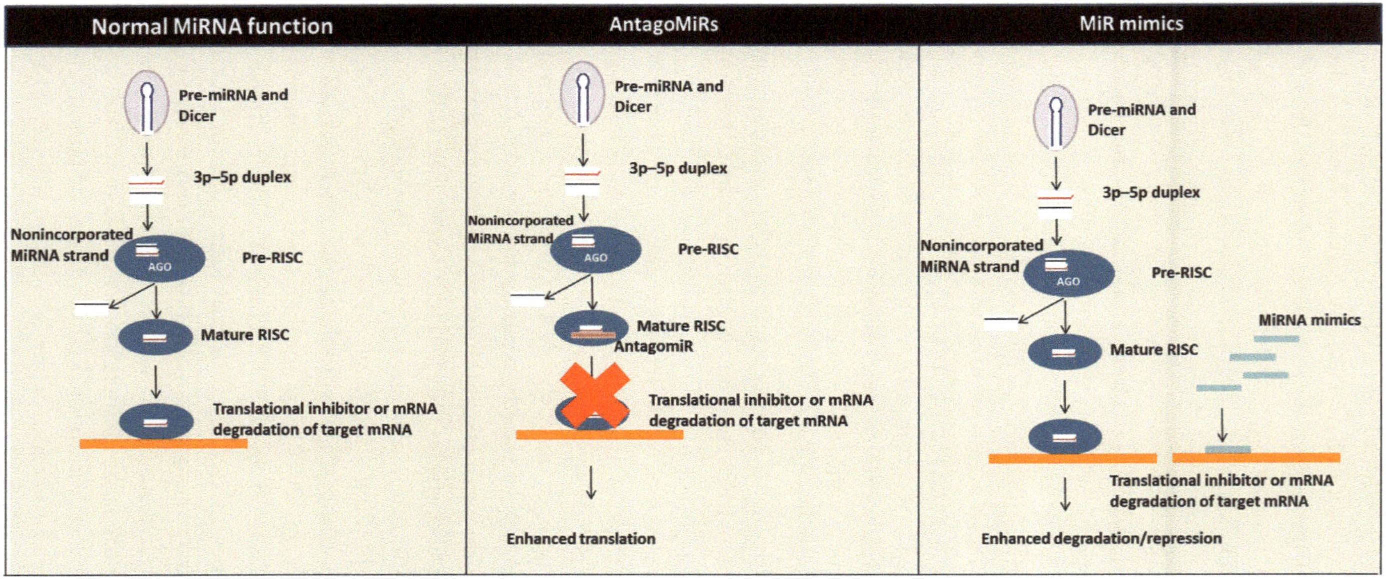

Fig. 4 Mechanism of action of antagomiRs and MiRNA mimics. AntagomiRs (MiRNA inhibitors) are thought to bind the mature MiRNAs preventing them from binding to their target mRNA(s); this results in elevated levels of the target protein(s). The addition of MiRNAs mimics result in more MiRNA being present and targeting the mRNA(s), therefore more degradation or repression of translation of the target mRNA.

on the efficiency and uptake of mimics and antagomiRs. Following systemic MiRNA delivery, a high accumulation of the MiRNAs in the kidney, spleen, liver, and urine has been shown (van Rooij and Kauppinen, 2014), suggesting that the levels of oligonucleotides that are needed to have the effect in other tissues can be extremely high. Usually great knowledge of the target tissue architecture and microenvironment is also needed to ensure efficient delivery. However, more recently, changes to the modifications of the antagomiRs and MiRNA mimics have improved with methyl- and cholesterol modifications to the mimics or antagomiRs, and the practice of targeting pri- and pre-miRNA transcript is also in use (Kloosterman et al., 2007). This has been demonstrated through successfully transfecting muscle with MiRNAs through tail vein injections of 2 mg/kg body weight (Beavers et al., 2015) of mimics and antagomiRs with no obvious side effects in other tissues, although this may be MiRNA dependent. This shows that therapeutic manipulation of MiRNA levels may be feasible in skeletal muscle.

4. CONCLUSIONS

The key role of MiRNAs in muscle development and maintaining muscle homeostasis in adulthood has been clearly described in the sections above. Moreover, dysregulation of MiRNA expression occurs in numerous muscle-wasting diseases, and there is a vast amount of evidence that this contributes to the resulting phenotype in skeletal muscle-wasting diseases.

Several questions with regards to the extent of MiRNA role in muscle disorders remain to be answered. It is still not clear whether the main function of MiRNAs in muscle is to fine-tune the physiological changes occurring during development, adulthood or disease, or whether MiRNAs are the driving force of these changes. It is also possible that changes in MiRNA expression and function in muscle in different conditions are a part of the compensatory mechanism aiming at preserving the homeostasis of skeletal muscle.

Changes in MiRNA expression in muscle during aging and disease have been described by many. However, there is a strikingly low level of overlap between the sets of MiRNAs differentially expressed in various diseases between the different studies. It remains to be established whether this is simply attributed to the limitations of the platforms used, or the experimental models used, or whether indeed there is little overlap in the differential expression of sets of MiRNAs in different diseases and model organisms.

Despite the result of muscular diseases being muscle wasting and atrophy, the lack of overlap of global MiRNA expression in muscular diseases is likely to be due to the differences in the mechanisms that contribute to the individual diseases. If there is truly no overlap in MiRNA expression, it is unlikely that a "one treatment fits all" approach would be able to treat changes in MiRNA expression in muscular diseases and individual diseases would require their own distinct treatments.

It is also not clear whether the same MiRNA plays a similar role in different contexts—studies suggesting this and the opposite exist. Nevertheless, functional studies of MiRNAs in various models of muscle wasting may provide a clearer answer with regards to the questions earlier.

The enormity of the role of MiRNAs in muscle wasting is yet to be fully understood; however, there is great therapeutic potential of MiRNAs as targets and as treatments to attenuate muscle wasting and future functional studies may provide better understanding of the potential MiRNA-based therapies against muscle wasting.

ACKNOWLEDGMENTS

The authors' work is supported by the Biotechnology and Biological Sciences Research Council (BBSRC; BB/L021668/1), the MRC and Arthritis Research UK as part of the MRC—Arthritis Research UK Centre for Integrated Research into Musculoskeletal Aging (CIMA), and the Wellcome Trust Institutional Strategic Support Fund (097826/Z/11/A). Due to space constraints, the authors would like to apologies to fellow colleagues whose work was not cited in this review.

Conflict of interest: The authors declare no conflict of interest regarding the publication of this chapter.

REFERENCES

Addison, O., Drummond, M.J., LaStayo, P.C., Dibble, L.E., Wende, A.R., McClain, D.A., et al., 2014. Intramuscular fat and inflammation differ in older adults: the impact of frailty and inactivity. J. Nutr. Health Aging 18, 532–538.

Aleman, H., Esparza, J., Ramirez, F.A., Astiazaran, H., Payette, H., 2011. Longitudinal evidence on the association between interleukin-6 and C-reactive protein with the loss of total appendicular skeletal muscle in free-living older men and women. Age Ageing 40, 469–475.

Alexander, M.S., Kawahara, G., Motohashi, N., Casar, J.C., Eisenberg, I., Myers, J.A., et al., 2013. MicroRNA-199a is induced in dystrophic muscle and affects WNT signaling, cell proliferation, and myogenic differentiation. Cell Death Differ. 20, 1194–1208.

Anderson, C., Catoe, H., Werner, R., 2006. MIR-206 regulates connexin43 expression during skeletal muscle development. Nucleic Acids Res. 34, 5863–5871.

Ansved, T., Larsson, L., 1990. Quantitative and qualitative morphological properties of the soleus motor nerve and the L5 ventral root in young and old rats. Relation to the number of soleus muscle fibers. J. Neurol. Sci. 96, 269–282.

Aoi, W., Naito, Y., Mizushima, K., Takanami, Y., Kawai, Y., Ichikawa, H., et al., 2010. The microRNA miR-696 regulates PGC-1α in mouse skeletal muscle in response to physical activity. Am. J. Physiol. Endocrinol. Metab. 298, E799–E806.

Ardite, E., Perdiguero, E., Vidal, B., Gutarra, S., Serrano, A.L., Muñoz-Cánoves, P., 2012. PAI-1-regulated miR-21 defines a novel age-associated fibrogenic pathway in muscular dystrophy. J. Cell Biol. 196, 163–175.

Auerbach, A., Vyas, G., Li, A., Halushka, M., Witwer, K., 2016. Uptake of dietary milk miRNAs by adult humans: a validation study. F1000Res. 5, 721.

Bagge, A., Clausen, T.R., Larsen, S., Ladefoged, M., Rosenstierne, M.W., Larsen, L., et al., 2012. MicroRNA-29a is up-regulated in beta-cells by glucose and decreases glucose-stimulated insulin secretion. Biochem. Biophys. Res. Commun. 426, 266–272.

Baier, S.R., Nguyen, C., Xie, F., Wood, J.R., Zempleni, J., 2014. MicroRNAs are absorbed in biologically meaningful amounts from nutritionally relevant doses of cow milk and affect gene expression in peripheral blood mononuclear cells, HEK-293 kidney cell cultures, and mouse livers. J. Nutr. 144, 1495–1500.

Baltimore, D., Boldin, M.P., O'Connell, R.M., Rao, D.S., Taganov, K.D., 2008. MicroRNAs: new regulators of immune cell development and function. Nat. Immunol. 9, 839–845.

Balzano, F., Deiana, M., Dei Giudici, S., Oggiano, A., Baralla, A., Pasella, S., et al., 2015. miRNA stability in frozen plasma samples. Molecules 20, 19030–19040.

Barreiro, E., 2016. The role of MicroRNAs in COPD muscle dysfunction and mass loss: implications on the clinic. Expert Rev. Respir. Med. 10, 1011–1022.

Bartel, D.P., 2004. MicroRNAs: genomics, biogenesis, mechanism, and function. Cell 116, 281–297.

Bartel, D.P., 2009. MicroRNAs: target recognition and regulatory functions. Cell 136, 215–233.

Bautmans, I., Onyema, O., Van Puyvelde, K., Pleck, S., Mets, T., 2011. Grip work estimation during sustained maximal contraction: validity and relationship with dependency and inflammation in elderly persons. J. Nutr. Health Aging 15, 731–736.

Beavers, K.R., Nelson, C.E., Duvall, C.L., 2015. miRNA inhibition in tissue engineering and regenerative medicine. Adv. Drug Deliv. Rev. 88, 123–137.

Bennett, C.F., Swayze, E.E., 2010. RNA targeting therapeutics: molecular mechanisms of antisense oligonucleotides as a therapeutic platform. Annu. Rev. Pharmacol. Toxicol. 50, 259–293.

Bhaumik, D., Scott, G.K., Schokrpur, S., Patil, C.K., Orjalo, A.V., Rodier, F., et al., 2009. MicroRNAs miR-146a/b negatively modulate the senescence-associated inflammatory mediators IL-6 and IL-8. Aging 1, 402–411.

Blijham, P.J., Schelhaas, H.J., Ter Laak, H.J., van Engelen, B.G., Zwarts, M.J., 2007. Early diagnosis of ALS: the search for signs of denervation in clinically normal muscles. J. Neurol. Sci. 263, 154–157.

Bloch, S.A., Donaldson, A.V., Lewis, A., Banya, W.A., Polkey, M.I., Griffiths, M.J., et al., 2015. MiR-181a: a potential biomarker of acute muscle wasting following elective high-risk cardiothoracic surgery. Crit. Care 19, 147.

Boettger, T., Wüst, S., Nolte, H., Braun, T., 2014. The miR-206/133b cluster is dispensable for development, survival and regeneration of skeletal muscle. Skelet. Muscle 4, 23.

Bohan, A., Peter, J.B., 1975. Polymyositis and dermatomyositis (first of two parts). N. Engl. J. Med. 292, 344–347.

Borchert, G.M., Lanier, W., Davidson, B.L., 2006. RNA polymerase III transcribes human microRNAs. Nat. Struct. Mol. Biol. 13, 1097–1101.

Boutz, P.L., Chawla, G., Stoilov, P., Black, D.L., 2007. MicroRNAs regulate the expression of the alternative splicing factor nPTB during muscle development. Genes Dev. 21, 71–84.

Bowerman, M., Salsac, C., Coque, E., Eiselt, E., Deschaumes, R.G., Brodovitch, A., et al., 2015. Tweak regulates astrogliosis, microgliosis and skeletal muscle atrophy in a mouse model of amyotrophic lateral sclerosis. Hum. Mol. Genet. 24, 3440–3456.
Brack, A., Conboy, M., Roy, S., Lee, M., Kuo, C., Keller, C., et al., 2007. Increased Wnt signaling during aging alters muscle stem cell fate and increases fibrosis. Science 317, 807–810.
Broome, C.S., Kayani, A.C., Palomero, J., Dillmann, W.H., Mestril, R., Jackson, M.J., et al., 2006. Effect of lifelong overexpression of HSP70 in skeletal muscle on age-related oxidative stress and adaptation after nondamaging contractile activity. FASEB J. 20, 1549–1551.
Bruneteau, G., Simonet, T., Bauche, S., Mandjee, N., Malfatti, E., Girard, E., et al., 2013. Muscle histone deacetylase 4 upregulation in amyotrophic lateral sclerosis: potential role in reinnervation ability and disease progression. Brain 136, 2359–2368.
Bua, E., Johnson, J., Herbst, A., Delong, B., McKenzie, D., Salamat, S., et al., 2006. Mitochondrial DNA-deletion mutations accumulate intracellularly to detrimental levels in aged human skeletal muscle fibers. Am. J. Hum. Genet. 79, 469–480.
Buckingham, M., Bajard, L., Chang, T., Daubas, P., Hadchouel, J., Meilhac, S., et al., 2003. The formation of skeletal muscle: from somite to limb. J. Anat. 202, 59–68.
Cacchiarelli, D., Legnini, I., Martone, J., Cazzella, V., D'Amico, A., Bertini, E., et al., 2011. miRNAs as serum biomarkers for Duchenne muscular dystrophy. EMBO Mol. Med. 3, 258–265.
Camera, D.M., Ong, J.N., Coffey, V.G., Hawley, J.A., 2016. Selective modulation of MicroRNA expression with protein ingestion following concurrent resistance and endurance exercise in human skeletal muscle. Front. Physiol. 7, 87.
Cerletti, M., Jang, Y.C., Finley, L.W.S., Haigis, M.C., Wagers, A.J., 2012. Short-term calorie restriction enhances skeletal muscle stem cell function. Cell Stem Cell 10, 515–519.
Chen, J.-F., Mandel, E.M., Thomson, J.M., Wu, Q., Callis, T.E., Hammond, S.M., et al., 2006. The role of microRNA-1 and microRNA-133 in skeletal muscle proliferation and differentiation. Nat. Genet. 38, 228–233.
Chen, X., Ba, Y., Ma, L., Cai, X., Yin, Y., Wang, K., et al., 2008. Characterization of microRNAs in serum: a novel class of biomarkers for diagnosis of cancer and other diseases. Cell Res. 18, 997–1006.
Chen, J.F., Tao, Y., Li, J., Deng, Z., Yan, Z., Xiao, X., et al., 2010. microRNA-1 and microRNA-206 regulate skeletal muscle satellite cell proliferation and differentiation by repressing Pax7. J. Cell Biol. 190, 867–879.
Chen, Y., Gelfond, J., McManus, L.M., Shireman, P.K., 2011. Temporal microRNA expression during in vitro myogenic progenitor cell proliferation and differentiation: regulation of proliferation by miR-682. Physiol. Genomics 43, 621–630.
Cheng, T.C., Wallace, M.C., Merlie, J.P., Olson, E.N., 1993. Separable regulatory elements governing myogenin transcription in mouse embryogenesis. Science 261, 215–218.
Cheng, C.S., Ran, L., Bursac, N., Kraus, W.E., Truskey, G.A., 2016. Cell density and joint microRNA-133a and microRNA-696 inhibition enhance differentiation and contractile function of engineered human skeletal muscle tissues. Tissue Eng. Part A 22, 573–583.
Cheung, T.H., Quach, N.L., Charville, G.W., Liu, L., Park, L., Edalati, A., et al., 2012. Maintenance of muscle stem cell quiescence by microRNA-489. Nature 482, 524–528.
Choi, M.-C., Ryu, S., Hao, R., Wang, B., Kapur, M., Fan, C.-M., et al., 2014. HDAC4 promotes Pax7-dependent satellite cell activation and muscle regeneration. EMBO Rep. 15, 1175–1183.
Chondrogiannia, N., Petropoulosb, I., Franceschic, C., Friguetb, B., Gonosa, E., 2000. Fibroblast cultures from healthy centenarians have an active proteasome. Exp. Gerontol. 35, 721–728.

Cohen, T.J., Waddell, D.S., Barrientos, T., Lu, Z., Feng, G., Cox, G.A., et al., 2007. The histone deacetylase HDAC4 connects neural activity to muscle transcriptional reprogramming. J. Biol. Chem. 282, 33752–33759.

Colman, R.J., Beasley, T.M., Allison, D.B., Weindruch, R., 2008. Attenuation of sarcopenia by dietary restriction in rhesus monkeys. J. Gerontol. A Biol. Sci. Med. Sci. 63, 556–559.

Conboy, I.M., Conboy, M.J., Wagers, A.J., Girma, E.R., Weissman, I.L., Rando, T.A., 2005. Rejuvenation of aged progenitor cells by exposure to a young systemic environment. Nature 433, 760–764.

Crist, C.G., Montarras, D., Pallafacchina, G., Rocancourt, D., Cumano, A., Conway, S.J., et al., 2009. Muscle stem cell behavior is modified by microRNA-27 regulation of Pax3 expression. Proc. Natl. Acad. Sci. U.S.A. 106, 13383–13387.

Crist, C.G., Montarras, D., Buckingham, M., 2012. Muscle satellite cells are primed for myogenesis but maintain quiescence with sequestration of Myf5 mRNA targeted by microRNA-31 in mRNP granules. Cell Stem Cell 11, 118–126.

Cuervo, A., Dice, J., 2000. Age-related decline in chaperone-mediated autophagy. J. Biol. Chem. 275, 31505–31513.

Cunningham, J.T., Rodgers, J.T., Arlow, D.H., Vazquez, F., Mootha, V.K., Puigserver, P., 2007. mTOR controls mitochondrial oxidative function through a YY1-PGC-1alpha transcriptional complex. Nature 450, 736–740.

Cuthbertson, D., Smith, K., Babraj, J., Leese, G., Waddell, T., Atherton, P., et al., 2005. Anabolic signaling deficits underlie amino acid resistance of wasting, aging muscle. FASEB J. 19, 422–424.

de Matos-Neto, E.M., Lima, J.D.C.C., de Pereira, W.O., Figuerêdo, R.G., Riccardi, D.M., Radloff, K., et al., 2015. Systemic inflammation in cachexia—is tumor cytokine expression profile the culprit? Front. Immunol. 6, 629.

Dey, B.K., Gagan, J., Dutta, A., 2011. miR-206 and -486 induce myoblast differentiation by downregulating Pax7. Mol. Cell. Biol. 31, 203–214.

Dey, B.K., Gagan, J., Yan, Z., Dutta, A., 2012. miR-26a is required for skeletal muscle differentiation and regeneration in mice. Genes Dev. 26, 2180–2191.

Dey, B.K., Pfeifer, K., Dutta, A., 2014. The H19 long noncoding RNA gives rise to microRNAs miR-675-3p and miR-675-5p to promote skeletal muscle differentiation and regeneration. Genes Dev. 28, 491–501.

Di Filippo, E.S., Mancinelli, R., Pietrangelo, T., La Rovere, R.M.L., Quattrocelli, M., Sampaolesi, M., et al., 2016. Myomir dysregulation and reactive oxygen species in aged human satellite cells. Biochem. Biophys. Res. Commun. 473, 462–470.

Dogra, C., Changotra, H., Mohan, S., Kumar, A., 2006. Tumor necrosis factor-like weak inducer of apoptosis inhibits skeletal myogenesis through sustained activation of nuclear factor-kappaB and degradation of MyoD protein. J. Biol. Chem. 281, 10327–10336.

Donaldson, A., Natanek, S.A., Lewis, A., Man, W.D.-C., Hopkinson, N.S., Polkey, M.I., et al., 2013. Increased skeletal muscle-specific microRNA in the blood of patients with COPD. Thorax 68, 1140–1149.

Drummond, M.J., McCarthy, J.J., Fry, C.S., Esser, K.A., Rasmussen, B.B., 2008. Aging differentially affects human skeletal muscle microRNA expression at rest and after an anabolic stimulus of resistance exercise and essential amino acids. Am. J. Physiol. Endocrinol. Metab. 295, E1333–E1340.

Drummond, M.J., McCarthy, J.J., Sinha, M., Spratt, H.M., Volpi, E., Esser, K.A., et al., 2011. Aging and microRNA expression in human skeletal muscle: a microarray and bioinformatics analysis. Physiol. Genomics 43, 595–603.

Duarte, F.V., Palmeira, C.M., Rolo, A.P., 2014. The role of microRNAs in mitochondria: small players acting wide. Genes 5, 865–886.

Dupuis, L., Gonzalez de Aguilar, J.L., Oudart, H., de Tapia, M., Barbeito, L., Loeffler, J.P., 2004. Mitochondria in amyotrophic lateral sclerosis: a trigger and a target. Neurodegener. Dis. 1, 245–254.
Eisenberg, I., Eran, A., Nishino, I., Moggio, M., Lamperti, C., Amato, A.A., et al., 2007. Distinctive patterns of microRNA expression in primary muscular disorders. Proc. Natl. Acad. Sci. U.S.A. 104, 17016–17021.
Epstein, J.A., Lam, P., Jepeal, L., Maas, R.L., Shapiro, D.N., 1995. Pax3 inhibits myogenic differentiation of cultured myoblast cells. J. Biol. Chem. 270, 11719–11722.
Evans, W.J., Conley, K.E., Cress, M.E., Jubrias, S.A., Esselman, P.C., Odderson, I.R., 1995. From muscle properties to human performance, using magnetic resonance. J. Gerontol. A Biol. Sci. Med. Sci. 50A, 35–40.
Fagiolo, U., Cossarizza, A., Scala, E., Fanales-Belasio, E., Ortolani, C., Cozzi, E., et al., 1993. Increased cytokine production in mononuclear cells of healthy elderly people. Eur. J. Immunol. 23, 2375–2378.
Feng, Y., Niu, L.L., Wei, W., Zhang, W.Y., Li, X.Y., Cao, J.H., et al., 2013. A feedback circuit between miR-133 and the ERK1/2 pathway involving an exquisite mechanism for regulating myoblast proliferation and differentiation. Cell Death Dis. 4, e934.
Gagan, J., Dey, B.K., Layer, R., Yan, Z., Dutta, A., 2011. MicroRNA-378 targets the myogenic repressor MyoR during myoblast differentiation. J. Biol. Chem. 286, 19431–19438.
Galimov, A., Merry, T.L., Luca, E., Rushing, E.J., Mizbani, A., Turcekova, K., et al., 2016. MicroRNA-29a in adult muscle stem cells controls skeletal muscle regeneration during injury and exercise downstream of fibroblast growth factor-2. Stem Cells 34, 768–780.
Gantier, M.P., McCoy, C.E., Rusinova, I., Saulep, D., Wang, D., Xu, D., et al., 2011. Analysis of microRNA turnover in mammalian cells following Dicer1 ablation. Nucleic Acids Res. 39, 5692–5703.
Georgantas, R.W., Streicher, K., Greenberg, S.A., Greenlees, L.M., Zhu, W., Brohawn, P.Z., et al., 2014. Inhibition of myogenic microRNAs 1, 133, and 206 by inflammatory cytokines links inflammation and muscle degeneration in adult inflammatory myopathies. Arthritis Rheumatol. 66, 1022–1033.
Giordani, L., Sandona, M., Rotini, A., Puri, P.L., Consalvi, S., Saccone, V., 2014. Muscle-specific microRNAs as biomarkers of Duchenne muscular dystrophy progression and response to therapies. Rare Dis. 2, e974969.
Gladman, M., Zinman, L., 2015. The economic impact of amyotrophic lateral sclerosis: a systematic review. Expert Rev. Pharmacoecon. Outcomes Res. 15, 439–450.
Gladman, M., Dharamshi, C., Zinman, L., 2014. Economic burden of amyotrophic lateral sclerosis: a Canadian study of out-of-pocket expenses. Amyotroph. Lateral Scler. Frontotemporal Degener. 15, 426–432.
Goljanek-Whysall, K., Sweetman, D., Abu-Elmagd, M., Chapnik, E., Dalmay, T., Hornstein, E., et al., 2011. MicroRNA regulation of the paired-box transcription factor Pax3 confers robustness to developmental timing of myogenesis. Proc. Natl. Acad. Sci. U.S.A. 108, 11936–11941.
Goljanek-Whysall, K., Sweetman, D., Münsterberg Andrea, E., 2012a. microRNAs in skeletal muscle differentiation and disease. Clin. Sci. 123, 611–625.
Goljanek-Whysall, K., Pais, H., Rathjen, T., Sweetman, D., Dalmay, T., Münsterberg, A., 2012b. Regulation of multiple target genes by miR-1 and miR-206 is pivotal for C2C12 myoblast differentiation. J. Cell Sci. 125, 3590–3600.
Goljanek-Whysall, K., Mok, G.F., Fahad Alrefaei, A., Kennerley, N., Wheeler, G.N., Münsterberg, A., 2014. myomiR-dependent switching of BAF60 variant incorporation into Brg1 chromatin remodeling complexes during embryo myogenesis. Development 141, 3378–3387.

Goljanek-Whysall, K., Iwanejko, L.A., Vasilaki, A., Pekovic-Vaughan, V., McDonagh, B., 2016. Ageing in relation to skeletal muscle dysfunction: redox homoeostasis to regulation of gene expression. Mamm. Genome 27, 341–357.

Grassi, B., Cerretelli, P., Narici, M.V., Marconi, C., 1991. Peak anaerobic power in master athletes. Eur. J. Appl. Physiol. 62, 394–399.

Greco, S., De Simone, M., Colussi, C., Zaccagnini, G., Fasanaro, P., Pescatori, M., et al., 2009. Common micro-RNA signature in skeletal muscle damage and regeneration induced by Duchenne muscular dystrophy and acute ischemia. FASEB J. 23, 3335–3346.

Guo, J., Zhao, W., Zhan, S., Li, L., Zhong, T., Wang, L., et al., 2016. Identification and expression profiling of miRNAome in goat longissimus dorsi muscle from prenatal stages to a neonatal stage. PLoS One 11, e0165764.

Hamrick, M.W., Herberg, S., Arounleut, P., He, H.-Z., Shiver, A., Qi, R.-Q., et al., 2010. The adipokine leptin increases skeletal muscle mass and significantly alters skeletal muscle miRNA expression profile in aged mice. Biochem. Biophys. Res. Commun. 400, 379–383.

Hennessy, E., Clynes, M., Jeppesen, P.B., O'Driscoll, L., 2010. Identification of microRNAs with a role in glucose stimulated insulin secretion by expression profiling of MIN6 cells. Biochem. Biophys. Res. Commun. 396, 457–462.

Hildyard, J.C., Wells, D.J., 2016. Investigating synthetic oligonucleotide targeting of Mir31 in Duchenne muscular dystrophy. PLoS Curr. 8. http://dx.doi.org/10.1371/currents.md.99d88e72634387639707601b237467d7.

Hitachi, K., Tsuchida, K., 2013. Role of microRNAs in skeletal muscle hypertrophy. Front. Physiol. 4, 408.

Hönig, J., Mizíková, I., Nardiello, C., Seeger, W., Morty, R., 2016. Optimising the delivery of antagomiRs to the neonatal mouse lung. Eur. Respir. J. 48.

Hu, J., Kong, M., Ye, Y., Hong, S., Cheng, L., Jiang, L., 2014a. Serum miR-206 and other muscle-specific microRNAs as non-invasive biomarkers for Duchenne muscular dystrophy. J. Neurochem. 129, 877–883.

Hu, Z., Klein, J.D., Mitch, W.E., Zhang, L., Martinez, I., Wang, X.H., 2014b. MicroRNA-29 induces cellular senescence in aging muscle through multiple signaling pathways. Aging (Albany NY) 6, 160–175.

Huang, M.B., Xu, H., Xie, S.J., Zhou, H., Qu, L.H., 2011. Insulin-like growth factor-1 receptor is regulated by microRNA-133 during skeletal myogenesis. PLoS One 6, e29173.

Huang, Z., Chen, X., Yu, B., He, J., Chen, D., 2012. MicroRNA-27a promotes myoblast proliferation by targeting myostatin. Biochem. Biophys. Res. Commun. 423, 265–269.

Hudson, M.B., Woodworth-Hobbs, M.E., Zheng, B., Rahnert, J.A., Blount, M.A., Gooch, J.L., et al., 2014. miR-23a is decreased during muscle atrophy by a mechanism that includes calcineurin signaling and exosome-mediated export. Am. J. Physiol. Cell Physiol. 306, C551–C558.

Hwang, H.W., Wentzel, E.A., Mendell, J.T., 2007. A hexanucleotide element directs microRNA nuclear import. Science 315, 97–100.

Israeli, D., Poupiot, J., Amor, F., Charton, K., Lostal, W., Jeanson-Leh, L., et al., 2016. Circulating miRNAs are generic and versatile therapeutic monitoring biomarkers in muscular dystrophies. Sci. Rep. 6, 28097.

Iwasaki, H., Imamura, T., Morino, K., Shimosato, T., Tawa, M., Ugi, S., et al., 2015. MicroRNA-494 plays a role in fiber type-specific skeletal myogenesis in human induced pluripotent stem cells. Biochem. Biophys. Res. Commun. 468, 208–213.

Janssen, H.L.A., Reesink, H.W., Lawitz, E.J., Zeuzem, S., Rodriguez-Torres, M., Patel, K., et al., 2013. Treatment of HCV infection by targeting MicroRNA. N. Engl. J. Med. 368, 1685–1694.

Ji, X., Wu, B., Fan, J., Han, R., Luo, C., Wang, T., et al., 2015. The anti-fibrotic effects and mechanisms of MicroRNA-486-5p in pulmonary fibrosis. Sci. Rep. 5, 14131.

Jiang, L.Q., Franck, N., Egan, B., Sjögren, R.J.O., Katayama, M., Duque-Guimaraes, D., et al., 2013. Autocrine role of interleukin-13 on skeletal muscle glucose metabolism in type 2 diabetic patients involves microRNA let-7. Am. J. Physiol. Endocrinol. Metab. 305, E1359–E1366.

Johnston, A.J., Murphy, K.T., Jenkinson, L., Laine, D., Emmrich, K., Faou, P., et al., 2015. Targeting of Fn14 prevents cancer-induced cachexia and prolongs survival. Cell 162, 1365–1378.

Kayani, A.C., Close, G.L., Jackson, M.J., McArdle, A., 2008. Prolonged treadmill training increases HSP70 in skeletal muscle but does not affect age-related functional deficits. Am. J. Physiol. Regul. Integr. Comp. Physiol. 294, R568–R576.

Kazantsev, A.G., Thompson, L.M., 2008. Therapeutic application of histone deacetylase inhibitors for central nervous system disorders. Nat. Rev. Drug Discov. 7, 854–868.

Khodadoust, M.S., Verhaegen, M., Kappes, F., Riveiro-Falkenbach, E., Cigudosa, J.C., Kim, D.S., et al., 2009. Melanoma proliferation and chemoresistance controlled by the DEK oncogene. Cancer Res. 69, 6405–6413.

Kim, H.K., Lee, Y.S., Sivaprasad, U., Malhotra, A., Dutta, A., 2006. Muscle-specific microRNA miR-206 promotes muscle differentiation. J. Cell Biol. 174, 677–687.

Kim, J.Y., Park, Y.-K., Lee, K.-P., Lee, S.-M., Kang, T.-W., Kim, H.-J., et al., 2014a. Genome-wide profiling of the microRNA-mRNA regulatory network in skeletal muscle with aging. Aging (Albany NY) 6, 524–544.

Kim, M., Yi, S.A., Lee, H., Bang, S.Y., Park, E.K., Lee, M.G., et al., 2014b. Reversine induces multipotency of lineage-committed cells through epigenetic silencing of miR-133a. Biochem. Biophys. Res. Commun. 445, 255–262.

Kim, N., Yoo, J.J., Atala, A., Lee, S.J., 2016. Combination of small RNAs for skeletal muscle regeneration. FASEB J. 30, 1198–1206.

Klingler, W., Jurkat-Rott, K., Lehmann-Horn, F., Schleip, R., 2012. The role of fibrosis in Duchenne muscular dystrophy. Acta Mycol. 31, 184–195.

Kloosterman, W.P., Lagendijk, A.K., Ketting, R.F., Moulton, J.D., Plasterk, R.H.A., 2007. Targeted inhibition of miRNA maturation with morpholinos reveals a role for miR-375 in pancreatic islet development. PLoS Biol. 5, e203.

Koning, M., Werker, P.M.N., van Luyn, M.J.A., Krenning, G., Harmsen, M.C., 2012. A global downregulation of microRNAs occurs in human quiescent satellite cells during myogenesis. Differentiation 84, 314–321.

Koutsoulidou, A., Mastroyiannopoulos, N.P., Furling, D., Uney, J.B., Phylactou, L.A., 2011. Expression of miR-1, miR-133a, miR-133b and miR-206 increases during development of human skeletal muscle. BMC Dev. Biol. 11, 34.

Koutsoulidou, A., Kyriakides, T.C., Papadimas, G.K., Christou, Y., Kararizou, E., Papanicolaou, E.Z., et al., 2015. Elevated muscle-specific miRNAs in serum of myotonic dystrophy patients relate to muscle disease progress. PLoS One 10, e0125341.

Krol, J., Loedige, I., Filipowicz, W., 2010. The widespread regulation of microRNA biogenesis, function and decay. Nat. Rev. Genet. 11, 597–610.

Krutzfeldt, J., Rajewsky, N., Braich, R., Rajeev, K.G., Tuschl, T., Manoharan, M., et al., 2005. Silencing of microRNAs in vivo with 'antagomirs'. Nature 438, 685–689.

Krutzfeldt, J., Poy, M.N., Stoffel, M., 2006. Strategies to determine the biological function of microRNAs. Nat. Genet. 38 (Suppl), S14–S19.

Kumar, D., Shadrach, J.L., Wagers, A.J., Lassar, A.B., 2009. Id3 is a direct transcriptional target of Pax7 in quiescent satellite cells. Mol. Biol. Cell 20, 3170–3177.

Larkindale, J., Yang, W., Hogan, P.F., Simon, C.J., Zhang, Y., Jain, A., et al., 2014. Cost of illness for neuromuscular diseases in the United States. Muscle Nerve 49, 431–438.

Laubier, J., Castille, J., Le Guillou, S., Le Provost, F., 2015. No effect of an elevated miR-30b level in mouse milk on its level in pup tissues. RNA Biol. 12, 26–29.

Lee, R.C., Feinbaum, R.L., Ambros, V., 1993. The C. elegans heterochronic gene lin-4 encodes small RNAs with antisense complementarity to lin-14. Cell 75, 843–854.

Lee, H., Jee, Y., Hong, K., Hwang, G.S., Chun, K.-H., 2013. MicroRNA-494, upregulated by tumor necrosis factor-α, desensitizes insulin effect in C(2)C(12) muscle cells. PLoS One 8, e83471.

Lee, K.-P., Shin, Y.J., Panda, A.C., Abdelmohsen, K., Kim, J.Y., Lee, S.-M., et al., 2015. miR-431 promotes differentiation and regeneration of old skeletal muscle by targeting Smad4. Genes Dev. 29, 1605–1617.

Levinger, I., Howlett, K.F., Peake, J., Garnham, A., Hare, D.L., Jerums, G., et al., 2010. Akt, AS160, metabolic risk factors and aerobic fitness in middle-aged women. Exerc. Immunol. Rev. 16, 98–104.

Lexell, J., Taylor, C.C., Sjöström, M., 1988. What is the cause of the ageing atrophy? Total number, size and proportion of different fiber types studied in whole vastus lateralis muscle from 15- to 83-year-old men. J. Neurol. Sci. 84, 275–294.

Li, Z., Rana, T.M., 2014b. Therapeutic targeting of microRNAs: current status and future challenges. Nat. Rev. Drug Discov. 13, 622–638.

Li, Z., Yang, C.S., Nakashima, K., Rana, T.M., 2011. Small RNA-mediated regulation of iPS cell generation. EMBO J. 30, 823–834.

Li, P., Jiao, J., Gao, G., Prabhakar, B.S., 2012. Control of mitochondrial activity by miRNAs. J. Cell. Biochem. 113, 1104–1110.

Li, Y., Li, Z., Zhou, S., Wen, J., Geng, B., Yang, J., et al., 2013. Genome-wide analysis of human microRNA stability. Biomed. Res. Int. 2013, 368975.

Li, X., Li, Y., Zhao, L., Zhang, D., Yao, X., Zhang, H., et al., 2014a. Circulating muscle-specific miRNAs in Duchenne muscular dystrophy patients. Mol. Ther. Nucleic. Acids 3, e177.

Li, X., Kong, D., Chen, H., Liu, S., Hu, H., Wu, T., et al., 2016. miR-155 acts as an anti-inflammatory factor in atherosclerosis-associated foam cell formation by repressing calcium-regulated heat stable protein 1. Sci. Rep. 6, 21789.

Liang, H., Ward, W.F., 2006. PGC-1alpha: a key regulator of energy metabolism. Adv. Physiol. Educ. 30, 145–151.

Lightfoot, A.P., McCormick, R., Nye, G.A., McArdle, A., 2014. Mechanisms of skeletal muscle ageing; avenues for therapeutic intervention. Curr. Opin. Pharmacol. 16, 116–121.

Liu, N., Bezprozvannaya, S., Williams, A.H., Qi, X., Richardson, J.A., Bassel-Duby, R., et al., 2008. microRNA-133a regulates cardiomyocyte proliferation and suppresses smooth muscle gene expression in the heart. Genes Dev. 22, 3242–3254.

Liu, N., Williams, A.H., Maxeiner, J.M., Bezprozvannaya, S., Shelton, J.M., Richardson, J.A., et al., 2012. microRNA-206 promotes skeletal muscle regeneration and delays progression of Duchenne muscular dystrophy in mice. J. Clin. Invest. 122, 2054–2065.

Lopez-Bastida, J., Perestelo-Perez, L., Monton-Alvarez, F., Serrano-Aguilar, P., Alfonso-Sanchez, J.L., 2009. Social economic costs and health-related quality of life in patients with amyotrophic lateral sclerosis in Spain. Amyotroph. Lateral Scler. 10, 237–243.

Lu, J., McKinsey, T.A., Zhang, C.-L., Olson, E.N., 2000. Regulation of skeletal myogenesis by association of the MEF2 transcription factor with class II histone deacetylases. Mol. Cell 6, 233–244.

Lund, E., Guttinger, S., Calado, A., Dahlberg, J.E., Kutay, U., 2004. Nuclear export of microRNA precursors. Science 303, 95–98.

Maciotta, S., Meregalli, M., Cassinelli, L., Parolini, D., Farini, A., Fraro, G.D., et al., 2012. Hmgb3 is regulated by microRNA-206 during muscle regeneration. PLoS One 7, e43464.

Maciotta, S., Meregalli, M., Torrente, Y., 2013. The involvement of microRNAs in neurodegenerative diseases. Front. Cell. Neurosci. 7, 265.

Mauro, A., 1961. Satellite cell of skeletal muscle fibers. J. Biophys. Biochem. Cytol. 9, 493–495.
McCarthy, J.J., Esser, K.A., Andrade, F.H., 2007. MicroRNA-206 is overexpressed in the diaphragm but not the hindlimb muscle of mdx mouse. Am. J. Physiol. Cell Physiol. 293, C451–C457.
McDonagh, B., 2016. Editorial: redox regulation in skeletal muscle aging and exercise. Front. Physiol. 7, 5.
McDonagh, B., Sakellariou, G.K., Smith, N.T., Brownridge, P., Jackson, M.J., 2014. Differential cysteine labeling and global label-free proteomics reveals an altered metabolic state in skeletal muscle aging. J. Proteome Res. 13, 5008–5021.
McDonagh, B., Sakellariou, G.K., Smith, N.T., Brownridge, P., Jackson, M.J., 2015. Redox proteomic analysis of the gastrocnemius muscle from adult and old mice. Data Brief 4, 344–348.
McDonagh, B., Scullion, S.M., Vasilaki, A., Pollock, N., McArdle, A., Jackson, M.J., 2016. Ageing-induced changes in the redox status of peripheral motor nerves imply an effect on redox signalling rather than oxidative damage. Free Radic. Biol. Med. 94, 27–35.
Mercken, E.M., Majounie, E., Ding, J., Guo, R., Kim, J., Bernier, M., et al., 2013. Age-associated miRNA alterations in skeletal muscle from rhesus monkeys reversed by caloric restriction. Aging (Albany NY) 5, 692–703.
Mielcarek, M., Landles, C., Weiss, A., Bradaia, A., Seredenina, T., Inuabasi, L., et al., 2013. HDAC4 reduction: a novel therapeutic strategy to target cytoplasmic huntingtin and ameliorate neurodegeneration. PLoS Biol. 11, e1001717.
Miska, E.A., Karlsson, C., Langley, E., Nielsen, S.J., Pines, J., Kouzarides, T., 1999. HDAC4 deacetylase associates with and represses the MEF2 transcription factor. EMBO J. 18, 5099–5107.
Mitchell, P.S., Parkin, R.K., Kroh, E.M., Fritz, B.R., Wyman, S.K., Pogosova-Agadjanyan, E.L., et al., 2008. Circulating microRNAs as stable blood-based markers for cancer detection. Proc. Natl. Acad. Sci. 105, 10513–10518.
Mohamed, J.S., Hajira, A., Pardo, P.S., Boriek, A.M., 2014. MicroRNA-149 inhibits PARP-2 and promotes mitochondrial biogenesis via SIRT-1/PGC-1alpha network in skeletal muscle. Diabetes 63, 1546–1559.
Moylan, J.S., Reid, M.B., 2007. Oxidative stress, chronic disease, and muscle wasting. Muscle Nerve 35, 411–429.
Munsterberg, A.E., Kitajewski, J., Bumcrot, D.A., McMahon, A.P., Lassar, A.B., 1995. Combinatorial signaling by Sonic hedgehog and Wnt family members induces myogenic bHLH gene expression in the somite. Genes Dev. 9, 2911–2922.
Naguibneva, I., Ameyar-Zazoua, M., Polesskaya, A., Ait-Si-Ali, S., Groisman, R., Souidi, M., et al., 2006. The microRNA miR-181 targets the homeobox protein Hox-A11 during mammalian myoblast differentiation. Nat. Cell Biol. 8, 278–284.
Nakasa, T., Ishikawa, M., Shi, M., Shibuya, H., Adachi, N., Ochi, M., 2010. Acceleration of muscle regeneration by local injection of muscle-specific microRNAs in rat skeletal muscle injury model. J. Cell. Mol. Med. 14, 2495–2505.
Nie, M., Liu, J., Yang, Q., Seok, H.Y., Hu, X., Deng, Z.L., et al., 2016a. MicroRNA-155 facilitates skeletal muscle regeneration by balancing pro- and anti-inflammatory macrophages. Cell Death Dis. 7, e2261.
Nie, Y., Sato, Y., Wang, C., Yue, F., Kuang, S., Gavin, T.P., 2016b. Impaired exercise tolerance, mitochondrial biogenesis, and muscle fiber maintenance in miR-133a-deficient mice. FASEB J. 30, 3745–3758.
Nielsen, S., Scheele, C., Yfanti, C., Akerstrom, T., Nielsen, A.R., Pedersen, B.K., et al., 2010. Muscle specific microRNAs are regulated by endurance exercise in human skeletal muscle. J. Physiol. 588, 4029–4037.
Nonaka, I., 1998. Animal models of muscular dystrophies. Lab. Anim. Sci. 48, 8–17.

Norman, K., Stobaus, N., Kulka, K., Schulzke, J., 2014. Effect of inflammation on handgrip strength in the non-critically ill is independent from age, gender and body composition. Eur. J. Clin. Nutr. 68, 155–158.
O'Rourke, J.R., Georges, S.A., Seay, H.R., Tapscott, S.J., McManus, M.T., Goldhamer, D.J., et al., 2007. Essential role for dicer during skeletal muscle development. Dev. Biol. 311, 359–368.
Palomero, J., Vasilaki, A., Pye, D., McArdle, A., Jackson, M.J., 2013. Aging increases the oxidation of dichlorohydrofluorescein in single isolated skeletal muscle fibers at rest, but not during contractions. Am. J. Physiol. Regul. Integr. Comp. Physiol. 305, R351–R358.
Panguluri, S.K., Bhatnagar, S., Kumar, A., McCarthy, J.J., Srivastava, A.K., Cooper, N.G., et al., 2010. Genomic profiling of messenger RNAs and microRNAs reveals potential mechanisms of TWEAK-induced skeletal muscle wasting in mice. PLoS One 5, e8760.
Parker, M.H., 2015. The altered fate of aging satellite cells is determined by signaling and epigenetic changes. Front. Genet. 6, 59.
Paul, R., Polkey, M., Kemp, P., Griffiths, M., 2015. S116 GDF-15, the miR-542 cluster and miR-422a are associated with muscle wasting in intensive care unit acquired paresis. Thorax 70, A66–A67.
Pedersen, M., Bruunsgaard, H., Weis, N., Hendel, H., Andreassen, B., Eldrup, E., et al., 2003. Circulating levels of TNF-alpha and IL-6-relation to truncal fat mass and muscle mass in healthy elderly individuals and in patients with type-2 diabetes. Mech. Ageing Dev. 124, 495–502.
Percival, J.M., Siegel, M.P., Knowels, G., Marcinek, D.J., 2013. Defects in mitochondrial localization and ATP synthesis in the mdx mouse model of Duchenne muscular dystrophy are not alleviated by PDE5 inhibition. Hum. Mol. Genet. 22, 153–167.
Ramaswamy, K.S., Palmer, M.L., van der Meulen, J.H., Renoux, A., Kostrominova, T.Y., Michele, D.E., et al., 2011. Lateral transmission of force is impaired in skeletal muscles of dystrophic mice and very old rats. J. Physiol. 589, 1195–1208.
Rao, P.K., Missiaglia, E., Shields, L., Hyde, G., Yuan, B., Shepherd, C.J., et al., 2010. Distinct roles for miR-1 and miR-133a in the proliferation and differentiation of rhabdomyosarcoma cells. FASEB J. 24, 3427–3437.
Reinhart, B.J., Slack, F.J., Basson, M., Pasquinelli, A.E., Bettinger, J.C., Rougvie, A.E., et al., 2000. The 21-nucleotide let-7 RNA regulates developmental timing in Caenorhabditis elegans. Nature 403, 901–906.
Remáková, M., Škoda, M., Faustová, M., Pleštilová, L., Mann, H., Vencovský, J., et al., 2013. THU0065 expression profiling of extracellular serum micrornas in patients with idiopathic inflammatory myopathy. Ann. Rheum. Dis. 72, A185.
Rennie, M.J., 2009. Anabolic resistance in critically ill patients. Crit. Care Med. 37, S398–S399.
Ruby, J.G., Jan, C.H., Bartel, D.P., 2007. Intronic microRNA precursors that bypass Drosha processing. Nature 448, 83–86.
Russell, A.P., Wada, S., Vergani, L., Hock, M.B., Lamon, S., Leger, B., et al., 2013. Disruption of skeletal muscle mitochondrial network genes and miRNAs in amyotrophic lateral sclerosis. Neurobiol. Dis. 49, 107–117.
Sakellariou, G.K., Vasilaki, A., Palomero, J., Kayani, A., Zibrik, L., McArdle, A., et al., 2013. Studies of mitochondrial and nonmitochondrial sources implicate nicotinamide adenine dinucleotide phosphate oxidase(s) in the increased skeletal muscle superoxide generation that occurs during contractile activity. Antioxid. Redox Signal. 18, 603–621.
Sakellariou, G.K., Davis, C.S., Shi, Y., Ivannikov, M.V., Zhang, Y., Vasilaki, A., et al., 2014. Neuron-specific expression of CuZnSOD prevents the loss of muscle mass and function that occurs in homozygous CuZnSOD-knockout mice. FASEB J. 28, 1666–1681.

Schepelmann, K., Winter, Y., Spottke, A.E., Claus, D., Grothe, C., Schroder, R., et al., 2010. Socioeconomic burden of amyotrophic lateral sclerosis, myasthenia gravis and facioscapulohumeral muscular dystrophy. J. Neurol. 257, 15–23.

Sempere, L.F., Freemantle, S., Pitha-Rowe, I., Moss, E., Dmitrovsky, E., Ambros, V., 2004. Expression profiling of mammalian microRNAs uncovers a subset of brain-expressed microRNAs with possible roles in murine and human neuronal differentiation. Genome Biol. 5, R13.

Shinde, S., Bhadra, U., 2015. A complex genome-microRNA interplay in human mitochondria. Biomed. Res. Int. 2015, 206382.

Short, K., Bigelow, M., Kahl, J., Singh, R., Coenen-Schimke, J., 2005. Decline in skeletal muscle mitochondrial function with aging in humans. PNAS 102, 5618–5623.

Small, E.M., O'Rourke, J.R., Moresi, V., Sutherland, L.B., McAnally, J., Gerard, R.D., et al., 2010. Regulation of PI3-kinase/Akt signaling by muscle-enriched microRNA-486. Proc. Natl. Acad. Sci. U.S.A. 107, 4218–4223.

Snow, J.W., Hale, A.E., Isaacs, S.K., Baggish, A.L., Chan, S.Y., 2013. Ineffective delivery of diet-derived microRNAs to recipient animal organisms. RNA Biol. 10, 1107–1116.

Soares, R.J., Cagnin, S., Chemello, F., Silvestrin, M., Musaro, A., De Pitta, C., et al., 2014. Involvement of microRNAs in the regulation of muscle wasting during catabolic conditions. J. Biol. Chem. 289, 21909–21925.

Song, M.Y., Ruts, E., Kim, J., Janumala, I., Heymsfield, S., Gallagher, D., 2004. Sarcopenia and increased adipose tissue infiltration of muscle in elderly African American women. Am. J. Clin. Nutr. 79, 874–880.

Sørensen, S.S., Nygaard, A.-B., Christensen, T., 2016. miRNA expression profiles in cerebrospinal fluid and blood of patients with Alzheimer's disease and other types of dementia—an exploratory study. Transl. Neurodegener. 5, 6.

Soriano-Arroquia, A., McCormick, R., Molloy, A.P., McArdle, A., Goljanek-Whysall, K., 2016a. Age-related changes in miR-143-3p:Igfbp5 interactions affect muscle regeneration. Aging Cell 15, 361–369.

Soriano-Arroquia, A., House L, Tregilgas, L., Canty-Laird, E., Goljanek-Whysall, K., 2016b. The functional consequences of age-related changes in microRNA expression in skeletal muscle. Biogerontology 17, 641–654.

Sousa-Victor, P., Gutarra, S., Garcia-Prat, L., Rodriguez-Ubreva, J., Ortet, L., Ruiz-Bonilla, V., et al., 2014. Geriatric muscle stem cells switch reversible quiescence into senescence. Nature 506, 316–321.

Swayze, E.E., Siwkowski, A.M., Wancewicz, E.V., Migawa, M.T., Wyrzykiewicz, T.K., Hung, G., et al., 2007. Antisense oligonucleotides containing locked nucleic acid improve potency but cause significant hepatotoxicity in animals. Nucleic Acids Res. 35, 687–700.

Sweetman, D., Goljanek, K., Rathjen, T., Oustanina, S., Braun, T., Dalmay, T., et al., 2008. Specific requirements of MRFs for the expression of muscle specific microRNAs, miR-1, miR-206 and miR-133. Dev. Biol. 321.

Tajrishi, M.M., Sato, S., Shin, J., Zheng, T.S., Burkly, L.C., Kumar, A., 2014. The TWEAK–Fn14 dyad is involved in age-associated pathological changes in skeletal muscle. Biochem. Biophys. Res. Commun. 446, 1219–1224.

Tanaka, S., Terada, K., Nohno, T., 2011. Canonical Wnt signaling is involved in switching from cell proliferation to myogenic differentiation of mouse myoblast cells. J. Mol. Signal 6, 12.

Thalacker-Mercer, A.E., Dell'Italia, L.J., Cui, X., Cross, J.M., Bamman, M.M., 2010. Differential genomic responses in old vs. young humans despite similar levels of modest muscle damage after resistance loading. Physiol. Genomics 40, 141–149.

Tiainen, K., Hurme, M., Hervonen, A., Luukkaala, T., Jylha, M., 2010. Inflammatory markers and physical performance among nonagenarians. J. Gerontol. A Biol. Sci. Med. Sci. 65, 658–663.

Title, A.C., Denzler, R., Stoffel, M., 2015. Uptake and function studies of maternal milk-derived microRNAs. J. Biol. Chem. 290, 23680–23691.
Tomlinson, B.E., Irving, D., 1977. The numbers of limb motor neurons in the human lumbosacral cord throughout life. J. Neurol. Sci. 34, 213–219.
Townley-Tilson, W.H., Callis, T.E., Wang, D., 2010. MicroRNAs 1, 133, and 206: critical factors of skeletal and cardiac muscle development, function, and disease. Int. J. Biochem. Cell Biol. 42.
Valdez, G., Heyer, M.P., Feng, G., Sanes, J.R., 2014. The role of muscle microRNAs in repairing the neuromuscular junction. PLoS One 9, e93140.
van der Steen, I., van den Berg, J.P., Buskens, E., Lindeman, E., van den Berg, L.H., 2009. The costs of amyotrophic lateral sclerosis, according to type of care. Amyotroph. Lateral Scler. 10, 27–34.
van Rooij, E., Kauppinen, S., 2014. Development of microRNA therapeutics is coming of age. EMBO Mol. Med. 6, 851–864.
van Rooij, E., Sutherland, L.B., Qi, X., Richardson, J.A., Hill, J., Olson, E.N., 2007. Control of stress-dependent cardiac growth and gene expression by a microRNA. Science 316, 575–579.
van Rooij, E., Quiat, D., Johnson, B.A., Sutherland, L.B., Qi, X., Richardson, J.A., et al., 2009. A family of microRNAs encoded by myosin genes governs myosin expression and muscle performance. Dev. Cell 17, 662–673.
Vasilaki, A., Mansouri, A., Van Remmen, H., van der Meulen, J.H., Larkin, L., Richardson, A.G., et al., 2006. Free radical generation by skeletal muscle of adult and old mice: effect of contractile activity. Aging Cell 5, 109–117.
Vasilaki, A., Pollock, N., Giakoumaki, I., Goljanek-Whysall, K., Sakellariou, G.K., Pearson, T., et al., 2016. The effect of lengthening contractions on neuromuscular junction structure in adult and old mice. Age (Dordr.) 38, 259–272.
Vaughan, D.W., 1992. Effects of advancing age on peripheral nerve regeneration. J. Comp. Neurol. 323, 219–237.
Vignier, N., Amor, F., Fogel, P., Duvallet, A., Poupiot, J., Charrier, S., et al., 2013. Distinctive serum miRNA profile in mouse models of striated muscular pathologies. PLoS One 8, e55281.
von Haehling, S., Anker, S.D., 2014. Prevalence, incidence and clinical impact of cachexia: facts and numbers-update 2014. J. Cachex. Sarcopenia Muscle 5, 261–263.
Wada, S., Kato, Y., Okutsu, M., Miyaki, S., Suzuki, K., Yan, Z., et al., 2011. Translational suppression of atrophic regulators by microRNA-23a integrates resistance to skeletal muscle atrophy. J. Biol. Chem. 286, 38456–38465.
Wakabayashi, K., Mori, F., Kakita, A., Takahashi, H., Utsumi, J., Sasaki, H., 2014. Analysis of microRNA from archived formalin-fixed paraffin-embedded specimens of amyotrophic lateral sclerosis. Acta Neuropathol. Commun. 2, 173.
Wei, W., Zhang, W., Bai, J., Zhang, H., Zhao, Y., Li, X., et al., 2016a. The NF-κB-modulated microRNAs miR-195 and miR-497 inhibit myoblast proliferation by targeting Igf1r, Insr and cyclin genes. J. Cell Sci. 129, 39–50.
Wei, Y., Tao, X., Xu, H., Chen, Y., Zhu, L., Tang, G., et al., 2016b. Role of miR-181a-5p and endoplasmic reticulum stress in the regulation of myogenic differentiation. Gene 592, 60–70.
Welle, S., Thornton, C., Jozefowicz, R., Statt, M., 1993. Myofibrillar protein synthesis in young and old men. Am. J. Physiol. 264, 693–698.
Wenz, T., Rossi, S.G., Rotundo, R.L., Spiegelman, B.M., Moraes, C.T., 2009. Increased muscle PGC-1α expression protects from sarcopenia and metabolic disease during aging. Proc. Natl. Acad. Sci. U.S.A. 106, 20405–20410.
Wightman, B., Ha, I., Ruvkun, G., 1993. Posttranscriptional regulation of the heterochronic gene lin-14 by lin-4 mediates temporal pattern formation in C. elegans. Cell 75, 855–862.

Williams, A.H., Valdez, G., Moresi, V., Qi, X., McAnally, J., Elliott, J.L., et al., 2009a. MicroRNA-206 delays ALS progression and promotes regeneration of neuromuscular synapses in mice. Science 326.

Williams, A.H., Liu, N., van Rooij, E., Olson, E.N., 2009b. MicroRNA control of muscle development and disease. Curr. Opin. Cell Biol. 21, 461–469.

Williams, A.H., Valdez, G., Moresi, V., Qi, X., McAnally, J., Elliott, J.L., et al., 2009c. MicroRNA-206 delays ALS progression and promotes regeneration of neuromuscular synapses in mice. Science 326, 1549–1554.

Winbanks, C.E., Beyer, C., Hagg, A., Qian, H., Sepulveda, P.V., Gregorevic, P., 2013. miR-206 represses hypertrophy of myogenic cells but not muscle fibers via inhibition of HDAC4. PLoS One 8, e73589.

Witwer, K.W., 2014. Diet-responsive mammalian miRNAs are likely endogenous. J. Nutr. 144, 1880–1881.

Wong, C.F., Tellam, R.L., 2008. MicroRNA-26a targets the histone methyltransferase enhancer of Zeste homolog 2 during myogenesis. J. Biol. Chem. 283, 9836–9843.

Wu, R., Li, H., Zhai, L., Zou, X., Meng, J., Zhong, R., et al., 2015. MicroRNA-431 accelerates muscle regeneration and ameliorates muscular dystrophy by targeting Pax7 in mice. Nat. Commun. 6, 7713.

Xie, W., Li, Z., Li, M., Xu, N., Zhang, Y., 2013. miR-181a and inflammation: miRNA homeostasis response to inflammatory stimuli in vivo. Biochem. Biophys. Res. Commun. 430, 647–652.

Yang, T., Chen, X.L., Huang, Z.Q., Wen, W.X., Xu, M., Chen, D.W., et al., 2014. MicroRNA-27a promotes porcine myoblast proliferation by downregulating myostatin expression. Animal 8, 1867–1872.

Yu, Z., Hecht, N.B., 2008. The DNA/RNA-binding protein, translin, binds microRNA122a and increases its in vivo stability. J. Androl. 29, 572–579.

Zacharewicz, E., Della Gatta, P., Reynolds, J., Garnham, A., Crowley, T., Russell, A.P., et al., 2014. Identification of microRNAs linked to regulators of muscle protein synthesis and regeneration in young and old skeletal muscle. PLoS One 9, e114009.

Zanotti, S., Gibertini, S., Curcio, M., Savadori, P., Pasanisi, B., Morandi, L., et al., 2015. Opposing roles of miR-21 and miR-29 in the progression of fibrosis in Duchenne muscular dystrophy. Biochim. Biophys. Acta 1852, 1451–1464.

CHAPTER SEVEN

MicroRNAs in Neurodegenerative Diseases

Sean Quinlan*, Aidan Kenny*, Miguel Medina†,‡, Tobias Engel*, Eva M. Jimenez-Mateos*,[1]

*Royal College of Surgeons in Ireland, Dublin, Ireland
†CIBERNED, Network Center for Biomedical Research in Neurodegenerative Diseases, Madrid, Spain
‡CIEN Foundation, Queen Sofia Foundation Alzheimer Center, Madrid, Spain
[1]Corresponding author: e-mail address: evajimenez@rcsi.ie

Contents

Abstract

MicroRNAs are small noncoding RNAs that regulate gene expression at the posttranscriptional level in a sequence-specific manner. After their discovery in 1993, mounting data have provided compelling evidence for their causal involvement in several human diseases, such as cancer and disorders of the brain.

International Review of Cell and Molecular Biology, Volume 334
ISSN 1937-6448
http://dx.doi.org/10.1016/bs.ircmb.2017.04.002

MicroRNAs have been described as the main regulator of homeostasis in neurons, and their dysregulation results in pathological conditions in the brain. In this review, we will focus on the role of MicroRNAs as novel drug targets and biomarkers of the three main neurodegenerative disorders (Alzheimer's disease, Parkinson's diseases, and Huntington's disease) and their role in other neurological disorders including traumatic brain injury and *status epilepticus*.

1. MicroRNAs

MicroRNAs (MiRNAs) are 18–25 nucleotides long, small, noncoding RNAs which regulate gene expression at the posttranscriptional level. The first MiRNA was discovered in *Caenorhabditis elegans* in 1993 (Lee et al., 1993; Wightman et al., 1993). However, the term MiRNA was not used until 2001, when MiRNAs were recognized as a unique class of biological regulators to control posttranscriptional protein expression (Ambros, 2008; Lee and Ambros, 2001). By June 2014, miRBase-21 (http://www.mirbase.org/) had annotated 2588 mature MiRNAs, which have been shown to be involved in several human diseases, such as cancer, immune-related diseases, and neurodegenerative disorders. To date, MiRNAs have been described as key regulators of cellular homeostasis in health and disease in all tissues and species analyzed. MiRNA expression is regulated temporally and spatially, being expressed in different cell types and at different developmental stages. Therefore, it comes as no surprise that an aberrant expression of MiRNAs has been implicated in numerous pathologies and disorders.

1.1 MiRNA Biogenesis Pathway and Mechanisms of Action

Mammalian genes encoding MiRNAs are generally expressed from polycistronic genes, with several MiRNAs under the control of the same promoter region and transcribed as a single unit (He and Hannon, 2004). MiRNA transcribing genes are localized within intronic regions of protein-coding or noncoding protein genes, or as independent transcriptional units. The majority of intragenic MiRNAs are localized antisense to the neighboring genes and are transcribed independently of adjacent genetic units (He and Hannon, 2004).

The components of the MiRNA biogenesis pathway and processing machinery are well conserved across species (Winter et al., 2009). The primary MiRNA (pri-miRNA) is cut by an RNase III domain-containing protein termed Drosha, along with the RNA-binding protein encoded by the DIGeorge Syndrome Critical Region Gene 8 (DGCR8)/Pasha (Winter et al., 2009). This processor complex binds to the lower stem region of the MiRNA's self-complementary region in the hairpin structure, i.e., DGCR8 binds to the RNA molecule, while Drosha cleaves the structure forming the precursor MiRNA (pre-miRNA) (Winter et al., 2009). The pre-miRNA is exported out of the nucleus into the cytoplasm through nuclear pore complexes via the action of Ras-related nuclear protein-guanosine triphosphate and the nuclear export protein Exportin-5 (Winter et al., 2009). In the cytoplasm, the pre-miRNA is cleaved by the RNase III protein Dicer forming a mature MiRNA duplex. In neurons, the formation of the mature MiRNA occurs in specific intracellular compartments. Pre-miRNAs can travel through neuronal dendrites to dendritic spines to be cleaved by Dicer after increased neuronal activity (Lugli et al., 2005). After the activation of glutamate or GABA receptors, influx of calcium into neurons activates calpain. Subsequently, activated calpain cleaves Dicer into an active 72 kDa form (Lugli et al., 2005). This active form allows MiRNAs to be generated rapidly in the correct subcellular compartment, dendritic, axonal, or cellular body. How MiRNAs translocate into dendrites remains unknown. However, the DEAH-box helicase DHX36 has been associated with the terminal loop of the precursor of MiR-134 which allows its translocation into the dendritic compartment (Bicker et al., 2013; Sambandan et al., 2017).

The MiRNA duplexes subsequently separate and the active strand is bound and stabilized by the Argonaute 2 protein (Ago-2) and then shuttled into the RNA-induced silencing complex (RISC) (Winter et al., 2009). The MiRNA directs the RISC complex to its complementary mRNA sequence, thereby silencing it. This has become the most commonly known and understood process in the biogenesis of MiRNAs.

Recently, a new mechanism has been identified in neurons, termed target-mediated MiRNA decay (TMMD). TMMD has been described as the main pathway for the regulation of MiRNA levels in neurons. This long-lasting binding of the MiRNA to the target gene will eventually lead to the degradation of the MiRNA (de la Mata et al., 2015).

Additionally, noncanonical pathways have also been described. One such example is that of miRtrons–MiRNAs that are generated from introns during RNA splicing. MiRNA generated in this fashion is exported immediately into the cytoplasm without the need to be cleaved by Drosha. Other novel biogenesis pathways include the cleavage of MiR-451 precursor via

Ago-2 instead of Dicer to generate the mature MiRNA, skipping the processing step via Dicer (Winter et al., 2009).

Once in the RISC complex, MiRNA will bind to the 3′ untranslated region (UTR) of the target mRNA (Adlakha and Saini, 2014). In neurons, MiRNAs bind to mRNA along a 7–8 nucleotide long "seed" sequence; however, in nonneuronal cells, full complimentary binding occurs (Winter et al., 2009). The incomplete complementary base pair matching in neurons results in repression of translation via halting off the translational machinery's ability to read the mRNA transcript, without affecting levels of the target mRNA (Winter et al., 2009). In contrast, a full complimentary seed region binding will induce degradation of the target mRNA, believed to be accomplished through destabilization and depolyadenylation of the target mRNA (Winter et al., 2009).

1.2 MiRNAs in the Brain

Genetic tools have demonstrated the crucial role MiRNAs play in brain development. Knockdown studies of proteins involved in different enzymatic steps in MiRNA biogenesis processes have allowed researchers to analyze in detail the role MiRNAs carry out in the brain (Tang et al., 2007; Wang et al., 2007). One of the first studies performed used a mutated version of Dicer, unable to bind to the MiRNA precursor, in zebrafish. This resulted in morphological defects in early brain patterning and morphogenesis (Giraldez et al., 2005). These abnormalities were rescued by the reintroduction of the MiR-430 family—suggesting MiRNAs playing a key role in the regulation of brain development (Giraldez et al., 2005).

Over the following years, several approaches have been applied to delete or overexpress different proteins involved in the biogenesis pathway of MiRNAs in mammals. For example, the conditional deletion of the Dgcr8 gene in cortical neurons during embryogenesis resulted in reduced soma size of neurons and less dendrites per neurons leading eventually to ataxia, tremors, and microcephaly (Babiarz et al., 2011). In contrast, a more subtle phenotype was found when Drosha was removed from hippocampal neurons from day 1 after birth. These Drosha-deficient mice presented microcephaly and reduced life span (Babiarz et al., 2011).

Supporting the role of MiRNAs during development, early lethality during embryogenesis and severe malformations have been reported in Dicer knockout mice (Bernstein et al., 2003). To avoid lethality during embryogenesis and to elucidate the role MiRNAs play in the CNS, several conditional transgenic mice have been developed over the last decade. Deletion of Dicer from hippocampal neurons from day 1 after birth resulted in larger ventricles and reduced hippocampal volume due to neuronal death in the hippocampus (Babiarz et al., 2011). Several other transgenic mice have been developed to delete Dicer in early adulthood. Interestingly, mice with a conditional deletion of Dicer during adulthood presented a similar phenotype including neuronal death, memory loss, tremors, and spontaneous seizure activity. This phenotype was independent of the subareas of the brain affected (cortex, hippocampus, cerebellum) and cell types targeted (astrocytes, oligodendrocytes, or neurons) (Harfe et al., 2005; Kawase-Koga et al., 2009; Konopka et al., 2010; Shin et al., 2009; Tao et al., 2011; Zhang et al., 2015).

Similar to results obtained using Dicer knockout mice, deficiency in Ago-2 also resulted in lethality in mice during embryogenesis, demonstrating the importance of MiRNAs during development. However, in contrast to inducible Dicer transgenic mice, deletion of Ago-2 in mature neurons had no effect on neuronal survival and morphology. Compensatory effects and cross functionality between the Ago family members, of which four members are expressed in the brain (Ago-1–4), may be a possible explanation for the observed discrepancy (Schaefer et al., 2010).

2. NEURODEGENERATIVE DISEASES

Neurodegenerative diseases are age-related disorders characterized by a progressive functional degeneration and/or death of neurons in the nervous system and can be the result of hereditary or sporadic causes (with no apparent family history or environmental impact) (Grasso et al., 2015). Neurodegenerative diseases are associated with distinct pathological changes in the brain including the formation of extracellular or intracellular protein aggregates. This may lead to the dysfunction of specific subareas of the brain or population of neurons resulting in movement or mental disorders with the associated behavior deficits.

Even though disease etiology differs between brain diseases, overlap of psychological symptoms is a common feature in neurodegenerative diseases, including depression, dementia, and anxiety. However, variation in the progression of the diseases (clinical and/or pathological features) may vary between patients of the same condition (Armstrong et al., 2005). The most common and representative, neurodegenerative diseases include

Alzheimer's disease, Parkinson's disease, and Huntington's disease (Grasso et al., 2015).

The study of the pathological changes leading to neurodegenerative disorders has been the focus of intense research over the last decade; however, the molecular events involved in these disorders remain elusive, and only a small number of pathological pathways have been associated with the etiology of neurodegenerative diseases. This includes the accumulation of misfolded proteins, failure of axonal transport, mitochondrial damage, oxidative injury, changes in channel dynamics, and altered autophagy (Brunden et al., 2014; Egorova et al., 2015; Hensley and Harris-White, 2015; Mena et al., 2015; Nixon and Yang, 2012; Vidal et al., 2014).

To date, there is no cure for neurodegenerative diseases and current medication merely treats the symptoms without providing any disease-modifying effects. Critically, once symptoms manifest, neuronal damage or loss is usually already in place and treatment will be exclusively palliative. Therefore, the identification of early diagnostics for neurodegenerative disorders is essential for an effective therapy or disease-modifying treatment.

2.1 Biomarkers in Neurodegenerative Diseases

Biomarkers are biochemical or molecular elements expressed by cells and tissues which can be found in biofluids (typically blood, urine, or cerebrospinal fluid (CSF)). In neurodegenerative diseases, identifying reliable and stable biomarkers will help to: (1) accurately diagnose diseases and/or support a differential diagnosis of diseases with similar symptomatology and (2) predict diseases progression or treatment response.

Neurodegenerative diseases are mechanistically complex, involving different molecular pathways and targets, and, to date, standard methodologies have failed to clarify and understand the underlying mechanism. The possibility to identify specific biomarkers for neurodegenerative diseases such as Alzheimer's, Parkinson's, and Huntington's disease would lead to a more accurate diagnosis and a better treatment of these disorders. In the past, biomarkers from blood, plasma, serum, and CSF have been proposed. The most used biomarkers in AD have been Aβ42, total Tau and phospho-Tau in CSF, while α-synuclein is the most promising biomarker for PD.

MiRNAs have been identified as ideal biomarkers; they are stable and easily detectable in biofluids, including plasma and serum, CSF and urine. Circulating cell-free MiRNAs are associated to protein complexes (Argonaute protein family or lipoprotein) and encapsulated in vesicles (microparticles, exosomes, or apoptotic bodies). This association to proteins and vesicles protects MiRNAs from being degraded by RNases in the extracellular environment. Furthermore, circulating MiRNAs are stable and resistant to low and high pH levels, and freeze–thaw cycles, making them ideal biomarkers for pathological conditions.

In this review, we will focus on the role MiRNAs play as key modulators during disease pathogenesis and their potential as diagnostics tools in neurodegenerative diseases.

3. ALZHEIMER'S DISEASES

Alzheimer's disease (AD) is a progressive fatal neurodegenerative disease first described by Alzheimer (1906). AD is the most prevalent neurodegenerative diseases in the developed world with over 46 million cases worldwide, primarily affecting individuals over the age of 65 (Querfurth and LaFerla, 2010; Reitz and Mayeux, 2014). AD is the most common cause of dementia in the elderly and results in a gradual decline of cognitive function affecting memory formation, behavioral changes, and language skills (Querfurth and LaFerla, 2010).

The two main histopathological hallmarks of AD are senile plaques, constituted by extracellular deposit of amyloid-beta (Aβ) and neurofibrillary tangles (NFTs), made up mainly of intracellular aggregates of hyperphosphorylated Tau protein (Hardy and Selkoe, 2002).

The amyloid precursor protein (APP) is a type I integral membrane protein ubiquitously expressed with no known function (Hardy and Selkoe, 2002). Under physiological conditions, APP is primarily cleaved by the α-secretase (ADAM10) and γ-secretase, to generate a nonamyloidogenic soluble peptide. However, under pathological conditions, APP is cleaved by β-secretases (particularly BACE1) generating amyloidogenic neurotoxic Aβ peptide (Hardy and Selkoe, 2002). It is believed that an increase in APP production and/or an imbalance of α- and β-secretase activity plays a key role in the buildup of Aβ and the development of plaques contributing to the progression of the amyloid cascade, with α- and β-secretases being competitors for the cleavage of the APP protein, however, with opposing results in Aβ production (Hardy and Selkoe, 2002). NFT pathology involves the abnormal phosphorylation of the microtubule-associated protein (MAP) Tau. Hyperphosphorylation of Tau results in misfolding and aggregation of the protein resulting in the formation of NFTs. NFT aggregates strongly correlate with the amount of neuronal cell death and synapse loss (Farias et al., 2011; Spillantini and Goedert, 2013).

Two different forms of AD have been described: sporadic AD, which accounts for over 95% of the cases, and familial or early-onset AD, which accounts for the remaining 5% of cases. Sporadic AD has been associated with age, lifestyle, diet, type 2 diabetes, stroke, and the apolipoprotein allele E4. In contrast, early-onset AD is mainly associated with mutations in genes directly involved in the production of Aβ (APP, PSEN-1, PSEN-2) (Bertram et al., 2010; Guerreiro and Hardy, 2014).

3.1 Current Treatment for Alzheimer's Disease

The main treatment for AD is supportive care from family and other caregivers. Quality of life for patients with dementia is improved in a known environment, which meets their needs. Therefore, family caregivers need to learn how to manage the progression of the illness and the specificities of AD patients (Scheltens et al., 2016).

Presently, the only treatment options available for AD are acetylcholinesterase inhibitors (e.g., donepezil, rivastigmine, galantamine) and NMDA-type glutamate receptor antagonists (memantine); both of which are minimally effective and only symptomatic (Scheltens et al., 2016). Acetylcholinesterase inhibitors stabilize cognitive performance during the first year of treatment, and memantine provides some benefits in some patients with moderate to severe dementia (Scheltens et al., 2016). Consequently, there is an urgent need to identify new disease-modifying treatments that will stop or delay AD pathology. Clinical trials have mainly focused on anti-amyloid approaches, including immunization strategies, secretase inhibitors, and anti-Aβ aggregation drugs. Several large phase III trials have been published with disappointing results testing antibodies or drugs targeting the formation of Aβ showing no improvement of clinical symptoms (Scheltens et al., 2016).

3.2 Alzheimer's Disease and MicroRNAs

Several studies have investigated the role of MiRNAs in AD, in particular their role in the formation of extracellular amyloid plaques and NFTs. Eight MiRNA families have been consistently shown to be deregulated in experimental models and/or human samples of AD including MiR-29, MiR-15 MiR-107, MiR-146, MiR-9, MiR-101, MiR-106, and the cluster

Table 1 Selected Deregulated MicroRNAs in Tissue From Patients or Animal Models of AD and Target Genes

MicroRNA	Tissue	Specie	Target Genes
MiR-29a/b/c	Hippocampus	Mouse and human	BACE1, Bim, Bmf, Hrk, Puma
MiR-15	CSF	Patients	ERK1
MiR-15	Brain	Patients	ERK1
MiR-46a/b	Cortex and hippocampus	Patients	Il1b
MiR-9	Cortex and hippocampus	Mouse and patients	Tau
MiR-101	Cortex and hippocampus	Mouse and patients	Tau
MiR-106	Cortex and hippocampus	Mouse and patients	β-Amyloid

MiR-212/132 (for a detailed description of MiR-212/132, see Section 6.1; Table 1).

The MiR-29 family consists of three mature members including MiR-29a, MiR-29b, and MiR-29c, expressed from two different genes. Under normal conditions, MiR-29 family members are expressed from 2 weeks of age onward in mice with levels increasing over time, suggesting a critical role in neuronal homeostasis and maintenance of neurons during aging (Hebert et al., 2009, 2010; Ling et al., 2011; Smirnova et al., 2005). MiR-29 has been reported to be downregulated in the brain of animal models of the diseases as well as in AD patients (Cogswell et al., 2008; Geekiyanage and Chan, 2011; Hebert et al., 2008; Nunez-Iglesias et al., 2010; Shioya et al., 2010). Emerging evidences suggest that MiR-29b is an inhibitor of neuronal apoptosis via targeting a family of proapoptotic regulators including Bim, Bmf, Hrk, and Puma (Kole et al., 2011). In in vitro models, MiR-29a has been reported to target the β-secretase BACE1 (a key regulator of APP production), and MiR-29a overexpression leads to reduced levels of BACE1 in mice, suggesting a role in AD pathology (Hebert et al., 2008).

MiR-15 has also been shown to be downregulated in experimental models of AD and in the CSF of AD patients (Hebert et al., 2010). One of the most promising targets of MiR-15 in AD is the kinase ERK1.

In AD, downregulation of MiR-15 increases ERK1 levels, resulting in the hyperphosphorylation of Tau, the main component of NFTs (Hebert et al., 2010). Another extensive studied MiRNA in AD is MiR-107. Interestingly, MiR-107 has the same seed region as MiR-15; however, in contrast to MiR-15, MiR-107 has been shown to target BACE1, a key component of the formation of Aβ.

One of the most fascinating MiRNA families in AD is the MiR-146 family with both members of the family dysregulated in AD patient brain. Whereas MiR-146a has been shown to be upregulated, MiR-146b has been shown to be downregulated (Cogswell et al., 2008; Cui et al., 2010; Lukiw et al., 2008; Sethi and Lukiw, 2009). The mechanism underlying this discrepancy in expression, which is unique to AD, is unknown. A possible explanation might be that even though both MiRNAs are activated under inflammatory conditions, only MiR-146a increases are associated with the proinflammatory cytokine interleukin-1β (IL-1β). IL-1β release is an early pathological event in AD, and therefore, the observed differences in the regulation of MiR-146a and MiR-146b may be in part due to the increased IL-1β levels during early pathology stages of AD (Lukiw, 2012).

Other MiRNAs implicated in AD include MiR-9, MiR-101, and MiR-106, which have been mainly observed to be downregulated in experimental models of AD and brain tissue from patients. MiR-9 and MiR-101 have been associated with Tau pathology, and MiR-106 has been associated with the production of Aβ plaques (Hebert et al., 2010).

3.3 MiRNAs as Biomarkers of Alzheimer's Disease

MiRNAs as biomarkers of AD have been extensively studied using blood, serum, plasma, and CSF. In one of the first studies, Schipper and colleagues isolated blood mononuclear cells from 16 AD patients and 16 controls using microarrays. Here, the authors observed only two deregulated MiRNAs, MiR-34a and MiR-181b, with both MiRNAs being upregulated (Schipper et al., 2007). In a later study, small RNA-Seq in blood samples from 48 AD patients and 22 control identified 27 deregulated MiRNAs in AD patients (Satoh et al., 2015). 13 MiRNAs were upregulated, including MiR-26b-3p, MiR-30c-5p, MiR-30d-5p, MiR-148b-5p, MiR-151a-3p, and MiR-186-5p; and 14 MiRNAs were downregulated, including four members of the let-7 family, MiR-15a-5p, MiR-17-3p, MiR-29b-3p, and MiR-144-5p (Satoh et al., 2015).

Using serum or plasma, several studies have shown numerous differentially regulated MiRNAs in AD patients when compared to healthy subjects. In one of the first studies profiling MiRNAs from serum, MiR-137, MiR-181c, MiR-9, MiR-19a, and MiR-29b were found downregulated in AD patients (Geekiyanage et al., 2012). In a separate study, MiR-125b, MiR-23a, and MiR-26b were found exclusively downregulated in AD compared to noninflammatory neurological diseases, inflammatory neurological diseases, and dementia patients (Galimberti et al., 2014). A similar regulation of these three MiRNAs was found in the CSF of AD patients. Interestingly, levels of MiR-26b negatively correlated with levels of Tau and P-Tau in the CSF of these AD patients (Galimberti et al., 2014). In parallel to this study, Tan and colleagues profiled MiRNAs from serum samples from AD patients compared to control. Here, the authors found MiR-125b, MiR-181c, and MiR-9 to be deregulated in AD patients, with MiR-125b showing the highest diagnostic potential for AD (Tan et al., 2014).

In a major effort to identify MiRNAs as biomarker for AD, RNA sequencing was carried out in 50 AD patients and 50 matched controls. In a second step, 14 putative candidates were validated in 158 AD patients and 155 matched controls. Among these, only six MiRNAs have been validated including MiR-483-3p, MiR-342-3p, MiR-98-5p, MiR-191-5p, MiR-855-5p, and let-7d-5p with MiR-342-3p showing the highest diagnostic accuracy for AD (Kumar et al., 2013). Interestingly, the diagnostic accuracy was increased if MiR-342-3p was isolated from exosome, showing that exosomal purified MiRNAs may strengthen the diagnostic accuracy for AD (Lugli et al., 2015).

In the same year, Dong and colleagues profiled MiRNAs using RNA sequencing from serum from AD and patients suffering from mild cognitive impairment (MCI), vascular dementia (VD), and healthy controls. MiR-31, MiR-93, MiR-143, and MiR-146a were upregulated in AD patients compared to controls; however, no MiRNA was able to discriminate between AD and MCI or VD patients (Dong et al., 2015).

To reduce the bias from the peripheral tissue in blood, plasma, and serum, two different studies have profiled MiRNAs using CSF. In a first study, microarray analysis was carried out using CSF from six AD patients and six matched controls. MiR-9, MiR-125b, MiR-146a, and MiR-155 were overexpressed in AD patients (Alexandrov et al., 2012). In a second study, MiRNAs from CSF samples from 22 AD patients and 28 healthy controls were profiled using the Open Array. Seven MiRNAs were found upregulated including MiR-146a, and eight MiRNAs were downregulated.

ROC analysis was carried out in the 15 MiRNAs with MiR-146a presenting the highest diagnostic accuracy (Denk et al., 2015).

Finally, in a recent study, MiRNA profiling was carried out using CSF from 10 AD patients and 10 patients with other dementias using a qPCR-based panel from Exiqon. Here, three MiRNAs, MiR-199b-5p, MiR-22-5p, and MiR-206, were detected more often in AD patients than in patients with other dementias, demonstrating its accuracy to differentiate between AD and other types of dementias (Sorensen et al., 2016).

4. PARKINSON'S DISEASE

Parkinson disease (PD) is a progressive neurodegenerative disorders with one of the highest prevalence, affecting approximately 1% of the population over 55 years of age. PD symptoms include bradykinesia (slowness of movements), rigidity, resting tremor, and posture instability (Majidinia et al., 2016; Wu et al., 2011). PD patients may develop other secondary non-motor symptoms overtime, such as cognitive impairment, depression, sleep disorders, and dementia (Chaudhuri and Odin, 2010; Poewe, 2008).

The main pathological hallmark of PD is the formation of Lewy bodies, cytoplasmic inclusions of α-synuclein–ubiquitin complex, which cannot be degraded by the proteasome (Ibanez et al., 2004). PD is characterized by the degeneration of dopaminergic cells, possibly due to an abnormal deposition of α-synuclein within the *substantia nigra* (Papapetropoulos and McCorquodale, 2007; Zheng et al., 2010). As the disease advances, other regions of the brain are affected, including cingulate gyrus, amygdala, and higher cortical regions, causing psychosis and dementia (Hamza et al., 2010). However, the molecular mechanisms of PD remain largely unknown.

The etiology of PD is a complex combination of genetic causes, environmental causes, and brain aging. Several genes have been associated with PD including mutations in α-synuclein (SNCA), leucine-rich repeat kinase 2 (LRRK2), oncogene DJ1-, and PTEN-induced putative kinase 1 (PINK1). SNCA and LRRK2 are mainly associated with late-onset PD, and DJ1 and PTEN are associated with early-onset PD (Coppede, 2012).

4.1 Current Treatments for Parkinson's Disease

Two different types of treatments are currently used in patients suffering from PD including pharmacology-based and nonpharmacology-based treatments.

The majority of tremor symptoms in PD patients are thought to be produced by a lack of dopamine in the *substantia nigra*. Therefore, to diminish motor symptoms, drug treatments are focused on increasing levels of dopamine. Levodopa (L-Dopa) has been the most widely used treatment for PD over the last 30 years. Once taken, L-Dopa is metabolized to dopamine, thereby restoring normal dopamine levels in the *substantia nigra*. Other treatments to increase levels of dopamine include inhibition of catechol-*O*-methyltransferase enzyme (COMT, which degrades dopamine), such as Tolcapone, which is generally used in parallel with levodopa to increase the efficiency of the treatment.

Other dopamine agonists in the clinic include bromocriptine, pergolide, and pramipexole. Compared to levodopa, dopamine agonists may delay motor complications, but are less effective in controlling symptoms; nevertheless, they are sufficiently effective to control symptoms during the initial years of PD. Thus, dopamine agonists are usually used in early stages of the disease and levodopa in the late phases of the disease.

The number one nonpharmaceutical treatment for PD is surgery. However, surgery is generally only used in advanced PD cases when patients have become intolerant to medication and drugs are ineffective suppressing symptoms. Surgery for PD is divided into lesion and deep brain stimulation (DBS) with the main targeted areas being the thalamus, globus pallidus, and subthalamic nucleus.

4.2 MicroRNAs in Parkinson's Disease

MiRNAs have been associated in the pathogenesis of PD in two different ways, regulating the survival of dopaminergic neurons and regulating the activity and expression of α-synuclein, main component of Lewy bodies aggregates (Table 2).

Table 2 Deregulated MicroRNAs in Tissue From Patients or Animal Models of PD and Target Genes

MicroRNA	Tissue	Specie	Target Genes
MiR-133b	Cortex and substantia nigra	Patients, Aphakia, and 6-OHDA mouse models	Not identified
MiR-205	Cortex	Patients	LRRK2
MiR-34b	Cortex	Patients	Not identified (affects mitochondria dysfunction)

Evidence for a direct or indirect function of MiRNAs in the pathophysiology of PD is now accumulating. The first study to show a role of MiRNAs in the maintenance of midbrain dopaminergic neurons was reported by Kim et al. (2007). Deletion of Dicer in embryonic-stem cell cultures resulted in a selective loss of dopaminergic neurons (Kim et al., 2007). Similar results were found in a mouse model lacking Dicer in dopaminergic neurons. These mice presented a progressive loss of midbrain dopaminergic neurons and their nigrostriatal axonal processes (Kim et al., 2007).

In the same study, MiRNA profiling from midbrain, cerebellum, and cerebral cortex samples from PD patients and healthy control found eight dysregulated MiRNAs exclusive to the midbrain and not in the cerebellum and cortex (Kim et al., 2007). Among these MiRNAs, MiR-133b was specifically reduced in PD patients compared to control, and two PD mouse models (Aphakia mice and 6-OHDA-treated mice) (Kim et al., 2007). Furthermore, overexpression of MiR-133b rescued the PD phenotype in in vitro models, suggesting a causal involvement of MiR-133b in the pathogenesis of PD (Kim et al., 2007).

The LRRK2 gene locus mutations have been described as a major genetic risk factor for PD. Cho and colleagues showed MiR-205 being able to target LRRK2 and LRRK2-targeting MiR-205 was elevated in the cortex of patients with sporadic PD, while levels of LRRK2 were increased in the same patients. Overexpressing MiR-205 decreased LRRK2 protein levels in HEK293 cells and primary neurons. Similar results were found when MiR-205 was inhibited resulting in increased LRRK2 protein levels. Finally, overexpressing MiR-205 in a transgenic mouse model of PD reversed aspects of the PD phenotype (neuronal damage and delayed in behavior symptoms) (Cho et al., 2013). Similar to MiR-205, let-7 and MiR-184 have been described as regulators of homeostasis in dopaminergic neurons. MiRNAs let-7 and MiR-184 have also been shown to regulate the expression of LRRK2 (Gehrke et al., 2010).

Finally, Minones-Moyano and colleagues profiled MiRNAs from brain tissue from 11 PD patients and 6 healthy volunteers. The MiR-34b/c cluster was found to be downregulated in advanced stages of PD. The downregulation of MiR-34b/c was associated with a malfunction of mitochondria and neuronal death (Minones-Moyano et al., 2011).

Many other MiRNAs have been proposed to be involved in the pathology of PD such as MiR-153 and MiR-7 which bind directly to the 3′ UTR of α-synuclein in vitro (Doxakis, 2010; Junn et al., 2009). MiR-433 has been related to the regulation of FGF20, a transcription factor which regulates the expression of α-synuclein (Junn et al., 2009).

4.3 Circulating MicroRNAs in Parkinson's Diseases

In one of the first studies investigating the role of MiRNAs as biomarkers for brain diseases, a MiRNA profile was carried out in blood from treated and untreated PD patients and healthy controls (Margis et al., 2011). MiR-1, MiR-22*, and MiR-29a were found to be decreased in the plasma of non-treated PD patients when compared to healthy controls (Margis et al., 2011). MiR-16-2*, MiR-26a-2*, and MiR-30a were found higher in untreated patients (Margis et al., 2011). Interestingly, all these six MiRNAs have been associated directly or indirectly with the regulation of α-synuclein or dopamine transport (Margis et al., 2011).

In a separate study, a MiRNA expression profile was carried out in leukocytes obtained from PD patients before and after DBS using a combination of exon microarray and small RNA-sequencing approaches. In this study, Soreq and colleagues identified five MiRNAs (MiR-4293, MiR-378c, MiR-18b, MiR-20c, and MiR-1249) which were dysregulated after DBS.

More recently, 384 plasma MiRNAs were analyzed from three independent pools of PD patients and controls. Seven MiRNAs were significantly overexpressed: MiR-181c, MiR-331-5p, MiR-193a-3p, MiR-196b, MiR-454, MiR-125a-3p, and MiR-137. In a second step, MiRNAs were validated in plasma using 31 patients and 25 control samples. Here, MiR-331-5p was found significantly higher in PD patients compared to control (20-fold increase) (Cardo et al., 2013).

In 2014, three different studies analyzed MiRNA levels in serum and/or CSF samples of PD patients. In the first study, four MiRNAs were found downregulated in two different families carrying the LRRK2-G2019S mutation and in patients suffering from idiopathic PD: MiR-29a/b and MiR-19a/b. Results were validated by individual PCRs in 65 idiopathic PC and 65 controls (Botta-Orfila et al., 2014). In the same year, Burgos and colleagues performed RNA sequencing from 67 PD patients and 78 controls from serum and CSF (Burgos et al., 2014). 17 MiRNAs were dysregulated in the CSF of PD patients, including the previously PD-associated MiRNAs, MiR-128, MiR-132, and MiR-212 (Burgos et al., 2014). In contrast, only five MiRNAs were differentially expressed in PD serum samples compared to control (Burgos et al., 2014). Vallelunga and colleagues found nine differentially expressed MiRNAs in PD serum samples, including MiR-30c and MiR-148b (Vallelunga et al., 2014).

In a recent study, Marques and colleagues profiled MiRNAs from CSF from PD patients and controls without neurological conditions. From the differentially regulated MiRNAs, 10 MiRNAs were analyzed based on previous publications and target genes, MiR-19a, MiR-19b, MiR-24, MiR-30c, MiR-34b, MiR-133b, MiR-148b, and MiR-205. Only two MiRNAs validated the screening profile, MiR-24 was downregulated and MiR-205 was upregulated in PD compared to control. However, a combination of MiR-19a, MiR-19b, MiR-24, MiR-30c, MiR-34b, MiR-133b, and MiR-205 increased the accuracy for the diagnosis of PD (Marques et al., 2016).

5. HUNTINGTON'S DISEASE

Huntington's disease (HD) was first mentioned in 1872 (Huntington, 1872). In Huntington's original paper, HD was described as a hereditary chorea with a tendency toward insanity and suicide, with clear manifestations at midlife. However, the gene location was not discovered until 1983 (Gusella et al., 1983), and it was not until 1993 that the *huntingtin* gene was described as the main cause of HD (Eddy et al., 2016; MacDonald et al., 1993).

HD is a genetic neurodegenerative disease caused by an abnormal CAG nucleotide expansion in the gene encoding *huntingtin*, giving rise to a lack-of-function protein. Mutant huntingtin accumulates in neurons, causing cell toxicity and dysfunction of neurons throughout the brain as the disease progresses. The neuropathological alterations are found widely throughout the brain, but the main affected areas are in the striatum (with loss of the medium spiny neurons) and the cerebral cortex (Landles and Bates, 2004).

HD manifests with motor and behavioral/mental symptoms, including involuntary choreiform movements (characterized by unwanted contractions of the facial muscles, trunk, and extremities), progressive cognitive impairment (such as loss of working memory and attention deficits), and neuropsychiatric symptoms (e.g., depression and psychosis) (Eddy et al., 2016).

The mean age of disease onset is between 30 and 50 years, with the mean duration between 17 and 20 years until death. The clinical onset of the disease is determined by motor symptoms, although mental symptoms are apparent 10 years before of the conventional motor onset (Duff et al., 2007).

5.1 Current Pharmacological Treatments for Huntington's Diseases

There is no cure for HD, current treatments do not stop or prevent the progression of the disease, and they are mainly focused on improving daily functioning by reducing severity of symptoms.

The loss of medium spiny neurons in the striatum results in abnormal neurotransmission of the dopamine, glutamate, and GABA systems. Dopaminergic dysfunction has been observed in presymptomatic and symptomatic cases; however, the loss of striatal dopamine receptors has only been observed in manifested cases of HD, the main focus of pharmacotherapy in HD (Kaur et al., 2016).

Generally, treatment is started when chorea interferes with daily functioning at work and at home which includes disturbances of balance, falls, and physical injury. In a Europe-wide assessment, 39.3% of all registered HD patients receive dopamine-depleting drugs, antipsychotics, or antiparkinsonian medication to treat chorea symptoms (Kaur et al., 2016).

The main drugs for the treatment of HD target the dopamine systems, including dopamine receptor antagonists, tiapride and clozapine; and the dopamine-depleting drug, tetrabenazine. Tetrabenazine is the only drug approved by the FDA for the treatment of chorea since 2008; it is a reversible dopamine-depleting drug, which has been shown to reduce chorea severity by 23.5% in HD patients; however, no improvement of the total motor score has been observed (Kaur et al., 2016).

5.2 MicroRNAs in Huntington's Disease

Several studies have suggested an involvement of MiRNAs in the pathogenesis of HD. While wild-type *huntingtin* is associated with neuroprotective effects, mutant *huntingtin* is neurotoxic (Savas et al., 2008). Mutant *huntingtin* interacts with two members of the MiRNA biogenesis pathway, Ago-1 and Ago-2, and inhibits the formation of the RISC complex, thus reducing the formation of MiRNAs (Savas et al., 2008). Furthermore, the transcription factor REST and coREST are sequestered by wild-type huntingtin. However, under pathological conditions, REST and coREST can be transported into the nucleus to induce the abnormal expression of several genes including MiR-9, the MiR-124 family, and MiR-132 (Johnson and Buckley, 2009; Table 3).

In one of the first studies, Marti and colleagues profiled MiRNAs from cortex from HD patients and controls using RNA sequencing. Among the

Table 3 Selected Deregulated MicroRNAs in Tissue From Patients or Animal Models of HD and Target Genes

MicroRNA	Tissue	Specie	Target Genes
MiR-22	Striatum	Mouse primary neurons. YAC128 and R6/2 mouse model	CREB
MiR-9	Striatum	YAC128 and R6/2 mouse model	Not identified
MiR-124	Striatum	R6/2 mouse model and induced HD rat model	Not identified
MiR-132	Striatum	YAC128 and R6/2 mouse model	Not identified

380 MiRNAs expressed in the cortex, MiR-100, MiR-106b, MiR-148b, MiR-151, MiR-222, MiR-382, and MiR-383 were significantly dysregulated in HD patient brain (Marti et al., 2010).

The role of MiRNAs in HD was analyzed using transgenic and induced animal models. Lee and colleagues performed MiRNA microarrays with brain tissue from two transgenic HD mouse models (YAC128 and R6/2, which express the human mutant huntingtin protein) and one induced HD rat model (by intracerebral of injection of the neurotoxin 3-nitropropionic acid) at different stages of the diseases (Lee et al., 2011). In early diseases stages, the main response was an upregulation of MiRNAs (21.3% of MiRNAs were upregulated, and 5.9% were downregulated) (Lee et al., 2011). In contrast, in the chronic stage of the disease, the main response was a downregulation of MiRNAs. In the YAC128 mouse model, 4.1% of MiRNAs were upregulated and 30.2% were downregulated; similar results were found in the R6/2 mouse model, 20.8% of MiRNAs were downregulated, and only 6.5% were upregulated (Lee et al., 2011). Nine MiRNAs were commonly regulated in both models, including MiR-22, MiR-29c, MiR-128, MiR-132, MiR-138, MiR-218, MiR-222, MiR-344, and MiR-674* (Lee et al., 2011). However, different MiRNAs were upregulated in the chronic stages. Finally, as downregulation was a common strong response in both chronic transgenic animal models, the biogenesis pathway of MiRNAs was analyzed. In the chronic stage on the YAC128 mouse model, Dicer was found to be downregulated, though none of the other proteins (Drosha, Dgcr8, or Exp5) was regulated in this transgenic mouse model. In contrast, in the R6/2 mouse model, only Drosha was downregulated, with no changes in the other proteins of the biogenesis pathway (Lee et al., 2011).

To identify potential MiRNA candidates in HD, Johnson and colleagues analyzed MiRNAs which are under the control of REST (a transcription factor extensively studied in HD). Similar to results by Lee and colleagues, MiR-29a, MiR-124a, MiR-132, and MiR-135b were downregulated in the R6/2 mouse model. However, from the four MiRNAs, only MiR-132 was found downregulated in human HD brains (Johnson et al., 2008).

In a separate study, MiRNAs were profiled using prefrontal cortex samples from 26 HD patients, 2 presymptomatic HD patients, and 36 controls using next-generation sequencing. In this study, 938 MiRNAs were reliably quantified, and 75 were significantly differentially expressed in HD vs control, 46 MiRNAs were upregulated, and 29 MiRNAs were downregulated (Hoss et al., 2015a). The most dysregulated MiRNAs were MiR-10b-5p, MiR-196a-5p, MiR-615-3p, MiR-10b-3p, and MiR-196-5p. Surprisingly, both arms of the precursor sequences were differentially expressed for eight MiRNAs, including MiR-10b, MiR-129, MiR-1298, MiR-142, MiR-144, MiR-148a, MiR-302a, and MiR-486 (Hoss et al., 2015a). Finally, levels of MiRNAs were correlated with clinical aspects, including Vonsattel grade of neuropathological involvement. Nine MiRNAs were correlated with neuropathological involvement (MiR-10b-5p, MiR-196-5p, MiR-10b-3p, MiR-196b-5p, MiR-302a-3p, MiR-200c-3p, MiR-4488, MiR-4449, and MiR-663b), with MiR-10b-5p showing the highest correlation to predict HD status and age of onset (Hoss et al., 2015a).

More recently, the functional role of MiR-22 has been analyzed in HD. Levels of MiR-22 are downregulated in the brain of experimental models of HD. Suggesting a neuroprotective role in HD, overexpression of MiR-22 in primary striatal and cortical neurons protected against a neurotoxic stimulus (Jovicic et al., 2013). This neuroprotective effect was associated with the MiR-22 target gene, REST, transcription factor strongly linked to HD (Jovicic et al., 2013).

5.3 Circulating MicroRNAs in Huntington's Disease

Much effort has been undertaken to date to identify reliable biomarkers of HD to predict the age of disease onset. Increased levels of MiR-10b-5p and MiR-486-5p have been found to be upregulated in the plasma of HD patients ($n=26$) compared to healthy volunteers ($n=8$). Both MiRNAs have also been found upregulated in the brain of HD patients (Hoss et al., 2015b). In an additional analysis, no differences were found in levels of MiR-10b-5p between asymptomatic *huntingtin* gene carriers and symptomatic HD patients. However, levels of MiR-10b-5p correlated with disease onset age (Hoss et al., 2015b).

More recently, a MiRNA microarray profile was carried out using blood from HD patients and healthy controls. Surprisingly, MiR-34b was the only MiRNA found to be elevated in plasma from premanifest HD patients when compared to control. Interestingly, MiR-34b has been shown to block the toxicity of mutant *huntingtin* in vitro (Gaughwin et al., 2011).

Finally, 752 MiRNAs were profiled using 15 HD patients at different disease stages and 7 healthy volunteers. 390 MiRNAs were detected across samples from which 168 were differently expressed. Among these, 139 MiRNAs were upregulated and 29 MiRNAs were downregulated. MiR-132-3p, MiR-22-5p, and MiR-223 were found dysregulated in HD patients, and MiR-122-5p and MiR-100-5p were associated to specific disease stages (Diez-Planelles et al., 2016).

6. COMMON DEREGULATED MicroRNAs IN NEURODEGENERATIVE DISEASES

Two MiRNAs in particular, MiR-132 and MiR-22, have been extensively studied in the pathogenesis of AD, PD, and HD in one or several studies. The next section will give an overview of the potential role of both MiRNAs in the maintenance of neurons, neuroinflammation, and their role under pathological conditions.

6.1 MicroRNA-132

MiRNA-132 is a member of the MiR-212/132 family, with both MiRNAs are expressed from the same cluster and having identical seed regions, suggesting that they might share the same mRNA targets (Tognini and Pizzorusso, 2012; Wanet et al., 2012). MiR-132 and MiR-212 are both expressed in the brain; however, MiR-212 has also been shown to be expressed in the heart and mammary glands (Ucar et al., 2010). This difference in tissue expression is most likely due to posttranscriptional regulatory mechanisms.

In neuronal cells, the MiR-212/132 cluster is regulated by CREB with the MiR-212/132 promoter region containing at least four different CREB-binding sites (Remenyi et al., 2010; Vo et al., 2005).

In particular, MiR-132 has been shown to be regulated by neuronal activity. It has been demonstrated on several occasions that primary and precursor levels of MiR-132 are increased during long-term potentiation (LTP) in the rat adult dentate gyrus (Wibrand et al., 2010). Subsequently, using small RNA sequencing, Remenyi found that MiR-132 and MiR-212 were the most upregulated MiRNA upon BDNF stimulation of primary cortical mouse neurons (Remenyi et al., 2010). In contrast, others studies have found that the stimulation of neurons resulted in an increase of mature MiR-132 levels, but not MiR-212, which was possibly due either to a preferential processing of the pre-miR-132 by Dicer or to differences in the stability of MiR-132 and MiR-212.

6.1.1 MicroRNA-132 in Normal Brain Function

MiR-132 has been shown to regulate neurite outgrowth, dendrite morphology, and spine plasticity in hippocampal neurons via the inhibition of the GTPse-activating protein p250GAP (Vo et al., 2005; Wibrand et al., 2010). Inhibition of p250GAP activates the RAC1–PAK actin remodeling pathway and thereby regulates activity-dependent spine plasticity in hippocampal neurons (Impey et al., 2010).

Additionally, MiR-132 has been reported to target methyl CpG-binding protein 2 (MeCP2), a known regulator of neuronal maturation, synapse formation, as well as dendrite and axon branching (Klein et al., 2007). Furthermore, it has been proposed that MeCP2 regulates BDNF, which triggers the induction of MiR-212/132 expression (Klein et al., 2007).

Levels of MiR-132 increase progressively during postnatal development until adulthood in naïve mice, indicating that MiR-132 might play a critical role in activity-dependent synaptic plasticity and the formation of neuronal circuits (Nudelman et al., 2010). Consistently, upregulation of MiR-132 led to an increase in mature dendritic spines, while downregulation to immature dendritic spines (Mellios et al., 2011).

6.1.2 MicroRNA-132 in Neurodegenerative Diseases

Emerging evidence links MiR-132 to the pathogenesis of AD. Downregulation of MiR-132 has been repeatedly shown in the hippocampus and prefrontal cortex of AD patients (Cogswell et al., 2008; Lau et al., 2013; Tan et al., 2013; Wong et al., 2013). MiR-132 has also been shown to be downregulated in patients suffering from progressive supranuclear palsy, another tauopathy (Smith et al., 2011). Supporting the role of MiR-132 in AD, several genes involved in AD are MiR-132 targets, including phosphatase and tensin homolog (PTEN), E1A-binding protein p300 (P300), forkhead box O1a and O3a (FOXO1a and FOXO3a) (Smith et al., 2011).

The role of MiR-132 in PD has been controversial. Gillardon et al., using microfluidic chips, found reduced levels of MiR-132 in a well-established mouse model of PD (α-synuclein A30P transgenic model) (Gillardon et al., 2008). In contrast, Lungu and colleagues showed a significant upregulation of MiR-132 in mesencephalic brain areas in a rat model of PD using microarray and qRT-PCR (Lungu et al., 2013). Moreover, in a separate study, the levels of the MiR-132 target gene nuclear receptor related 1 protein (NURR1) were inversely correlated with MiR-132 levels (Yang et al., 2012).

Similar to AD, MiR-132 has been shown to be downregulated in mouse models of HD and in patients suffering from HD (Johnson et al., 2008; Lee et al., 2011). The downregulation of MiR-132 in HD has been associated with an increase of the inhibitory transcription factor RE1-silencing transcription factor (REST), which is known to inhibit neuronal-specific genes in nonneuronal cells under normal conditions (Johnson et al., 2008).

6.1.3 MicroRNA-132 in Other Brain Disorders

In addition to its role in neurodegenerative disorders, MiR-132 has been involved in other neurological conditions, characterized by learning disabilities, such as Rett syndrome (RTT) and Fragile X syndrome (FXS) (Jin et al., 2004). RTT and FXS are both X-linked neurodevelopmental disorders (Jin et al., 2004; Na and Monteggia, 2011). RTT is associated with a loss of function and/or mutations in the MeCP2 gene. MeCP2 has been linked to MiR-132, and a genome-wide analysis of MiRNA expression in a mouse model of schizophrenia showed downregulation of MiR-132 and MiR-212 (Na and Monteggia, 2011). Interestingly, duplication of the MeCP2 gene also resulted in RTT syndrome (Na and Monteggia, 2011).

FXS is caused by loss of FMRP. FMRP interacts with several proteins of the MiRNA biogenesis pathways (Dicer and Ago-1) and with specific mature MiRNAs, including MiR-132. Interestingly, FMRP is required for the spine regulatory function of MiR-132 (Jin et al., 2004).

Furthermore, MiR-132 is altered in other neuropsychiatric disorders, including schizophrenia and bipolar disorders. In schizophrenia, the analysis of MiR-132 showed contradictory results. Kim and colleagues profiled MiRNAs of cortex samples from 35 schizophrenic patients and 35 controls using high-throughput qPCR TLDA cards. This study showed an increase of MiR-132 in schizophrenic patients when compared to controls (Kim et al., 2010). However, by using a MiRNA microarray profile, Miller and colleagues showed a downregulation of MiR-132 in 35 schizophrenic patients compared to 35 controls (Miller et al., 2012). Recently, MiR-132 has been suggested as a potential biomarker for schizophrenia with levels of MiR-132 decreased in plasma of untreated schizophrenic patients when compared to controls. Levels were restored to normal after treatment with risperidone (Yu et al., 2015b).

In line with its role as modulator of neuronal activity, MiR-132 has been found to be upregulated after *status epilepticus*, and inhibition of MiR-132 in vivo protects the brain against *status epilepticus*-induced neuronal death (Jimenez-Mateos et al., 2011).

In conclusion, these studies show the relevance of MiR-132 in brain disorders regulating inflammation and neuronal activity, and suggest MiR-132 as a key regulator of neuronal homeostasis and function.

6.2 MicroRNA-22

The MiR-22 gene was initially cloned from HeLa cells in 2002 from the chromosome 17p13 (Lagos-Quintana et al., 2002). MiR-22 is an evolutionary-conserved gene which, in contrast to the majority of MiRNAs, is expressed in a single genetic unit. After its discovery, MiR-22 has been associated with diverse biological processes including epigenetic modifications, embryonic development, and skeletal metabolism (Houbaviy et al., 2003; Kim et al., 2015; Muinos-Gimeno et al., 2011). Over the past decade, MiR-22 has been extensively studied in cancer where it acts as tumor suppressor and oncogene, depending on the cell type. However, the mechanism underlying this dual effect has not been elucidated yet. In contrast, MiR-22 has been shown to be always neuroprotective in brain disorders including *status epilepticus* (Jimenez-Mateos et al., 2015), cerebral ischemia (Yu et al., 2015a), HD (Jovicic et al., 2013), AD (Hebert et al., 2008), and traumatic brain injury (Ma et al., 2016), demonstrating its importance in neuronal function and homeostasis.

Additionally, MiR-22 has been described as a regulator of inflammation via targeting the histone deacetylase protein 4 (HDAC4), thereby changing the expression of numerous genes, in particular the expression of members of the interleukin family (Lu et al., 2015). Supporting a role as inflammatory modulator, MiR-22 targets the purinergic P2X7 receptor in the brain, a well-known neuroinflammatory mediator (Jimenez-Mateos et al., 2015).

6.2.1 MicroRNA-22 in the Brain

Originally, MiR-22 was associated with neuronal migration and differentiation. During embryogenesis, MiR-22 regulates radial migration in the cortex and differentiation of neurons. During cortical formation, MiR-22 is expressed mainly in the intermediate level throughout the cortical wall, blocking MiR-22 impaired the entry of neurons into the cortical plate, and, accordingly, overexpression of MiR-22 facilitates migration of neurons into the cortical plate (Volvert et al., 2014). Furthermore, inhibition of MiR-22 during embryogenesis results in an accumulation of multipolar neurons at the expense of bipolar neurons (typical neurons in the cortex). By targeting the coREST/REST transcriptional repressor complex, MiR-22 regulates the expression of doublecortin, a well-known MAP that contributes to neuronal differentiation and migration (Volvert et al., 2014).

The role of MiR-22 in adult brain function has been demonstrated recently. MiR-22 has been shown to regulate the memory-related long-term facilitation pathway associated with serotonin. In an *Aplysia* model, serotonin triggers activation of MAPK/Erk pathway, resulting in downregulation of MiR-22. The downregulation of MiR-22 results in upregulation of the cytoplasmic polyadenylating element-binding protein (CPEB). CPEB proteins regulate the extension of the poly-A tail in mRNAs and as a consequence protein production. In memory formation during LTP, serotonin triggers the downregulation of MiR-22, resulting in an increase of CPEB proteins and changes in the protein production in the synapsis, promoting thereby the maintenance of learning-related long-term synaptic plasticity (Fiumara et al., 2015).

Recent studies have now shed light on how the expression of MiR-22 is regulated in the brain. Several transcription factors have been found to bind to the MiR-22 promoter including P53 and specificity protein 1 (SP1). In a recent publication, we have shown by using primary hippocampal neurons that MiR-22 expression is regulated via SP1 and calcium. Under basal conditions levels of MiR-22 are low; however, a small increase in neuronal activity leads to an increase in MiR-22 levels via SP1. In contrast, strongly increased neuronal activity with the associated increases in intracellular calcium concentrations prevents SP1 binding to the MiR-22 promoter blocking MiR-22 increases (Engel et al., 2017).

6.2.2 MicroRNA-22 in Neurodegenerative Diseases

MiRNA-22 has been consistently linked to neurodegenerative diseases. Levels of MiR-22 are downregulated in HD (Packer et al., 2008) and

AD patient's brain (Hebert et al., 2008) and in from PD patients. Additionally, MiR-22 has been linked to neurodegeneration and aging and the overexpression of MiR-22 protects primary hippocampal neurons against senescence (Jovicic et al., 2013). The mechanism underlying the neuroprotective effects of MiR-22 in neurons has not been yet elucidated; however, several evidences have been found in HD and AD.

According with a potential role of MiR-22 in HD, Jovicic and colleagues analyzed the role of MiR-22 in an in vitro model of HD. In rat striatal and cortical neurons, the authors demonstrated the neuroprotective effects of MiR-22 against *huntingtin* (htt)-induced neurotoxicity; in fact, overexpression of MiR-22 reduced the number of htt protein aggregates by targeting proteins linked to HD pathogenesis including Rcor1, Rgs2, and HDAC4 (Jovicic et al., 2013). MiR-22 has been shown to be consistently downregulated in plasma from experimental mouse models of HD and HD patients (Diez-Planelles et al., 2016). However, no correlation was found between MiR-22 levels and clinical outcomes (Diez-Planelles et al., 2016).

In AD, the functional role of MiR-22 has been hardly analyzed. MiR-22 has been found to be downregulated in sporadic AD patients' brain; however, a potential functional mechanism during AD pathology has not been analyzed to date (Hebert et al., 2008). In contrast, its role as a biomarker of AD has been extensively studied. Levels of MiR-22 have been proposed as differential diagnostic tool between AD and other types of dementias. Levels of MiR-22 were undetectable in CSF samples in patients suffering from dementia; however, MiR-22 was present in 6 out of 10 AD patients; however, a possible correlation with disease stages was not confirmed (Sorensen et al., 2016). The observed discrepancy in the levels of MiR-22 in the brain and CSF has not been elucidated yet. However, neurodegenerative disorders are associated with disturbances in the blood–brain barrier. The higher levels of MiR-22 in the CSF of AD patients may be related to infiltration from the blood, where MiR-22 is highly expressed (Sorensen et al., 2016).

Different results from the AD patients were found when MiRNA profiling of plasma samples of PD patients was carried out. Levels of MiR-22 were found downregulated in the plasma of PD patients. MiR-22 levels returned to normal levels after patients were treated with levodopa (Margis et al., 2011).

Supporting these results, overexpression of MiR-22 in PC12 cells treated with 6-hydroxydopamine (in vitro model of PD) promoted cell survival and proliferation. In this model, the protective effects of MiR-22 were associated with the regulation of the transient receptor potential melastatin 7 (TRPM7) (Yang et al., 2016).

6.2.3 MiR-22 in Other Neurological Diseases

MiR-22 has also been analyzed in other neurological conditions, including *status epilepticus* and amyloid lateral sclerosis (ALS). To determine the role MiR-22 plays in seizure-induced damage, we used a unilateral model of *status epilepticus* where seizures are triggered by an intraamygdala injection of kainic acid. These results in neuronal death in the ipsilateral hippocampus, while the contralateral hippocampus, also recruited during seizures, is spared from damage (Jimenez-Mateos et al., 2015). To elucidate why neurons survived in the contralateral hippocampus, MiRNAs were coimmunoprecipitated from the RISC complex in the damaged and undamaged hippocampus. MiR-22 was the most upregulated MiRNA in the undamaged hippocampus (Jimenez-Mateos et al., 2015). Furthermore, by targeting the purinergic receptor P2X7, MiR-22 reduces neuroinflammation and neuronal death in the brain (Jimenez-Mateos et al., 2015). A similar mechanism has been described using and in vitro model of ALS expressing a mutated version of superoxide dismutase 1 (SOD1). MiR-22 was upregulated in microglia primary cell cultures expressing mutated SOD1 when compared to microglia expressing normal SOD1. As seen during *status epilepticus*, MiR-22 upregulation resulted in a downregulation of P2X7 (Parisi et al., 2013).

Supporting a neuroprotective role of MiR-22, the overexpression of MiR-22 was protective in an in vitro model of TBI (by scratch injury) protecting cortical neurons from dying (Ma et al., 2016).

Interestingly, in a recent study, exosomal MiR-22 was found downregulated in patients suffering from migraine when compared to healthy controls. These results were confirmed using circulating blood monocytes from the same individuals. Furthermore, a combination of four MiRNAs, including MiR-22, showed a high accuracy for the diagnosis of migraine (Tafuri et al., 2015).

7. FUTURE PERSPECTIVES AND CONCLUSION

MiRNAs have been shown to be key regulators of neuronal survival in animal models, but their potential as drug targets have still to be proven. MiRNAs are master regulators of gene expression; one single MiRNA can regulate hundreds of genes. Neurological diseases are complex conditions, which involves several processes, such as inflammation, hyperexcitability, and neuronal death. Thus, the multitargeting characteristics of MiRNAs make them ideal targets for the development of novel therapeutics for neurological conditions. To date, the inhibition of MiRNAs has demonstrated their utility in experimental models and the intracerebral delivery of MiRNAs has provided beneficial effects in animal models of TBI and epilepsy. However, this approach is unlikely to be used in patients and to date no study has shown the efficiency of targeting MiRNAs in neurological disorders via systemic injections.

On the other hand, MiRNAs are promising biomarkers for brain conditions. MiRNAs are stable in extracellular biofluids where they can be easily detected. Without doubt future multicenter studies will significantly advance the potential of MiRNAs as reliable biomarkers for the differential diagnosis of numerous brain diseases including the neurodegenerative diseases Alzheimer's, Parkinson's, and Huntington's.

ACKNOWLEDGMENTS

This work was supported by funding from Science Foundation Ireland (13/SIRG/2114 to E. M.J.-M. and 12/COEN/18 and 13/SIRG/2098 to T.E.), from the Health Research Board Ireland (HRA-POR-2015-1243 to T.E.), and from Spanish Ministry of Economy, Industry and Competitiveness and Carlos III Institute (SAF2016-78603-R to M.M.).

REFERENCES

Adlakha, Y.K., Saini, N., 2014. Brain microRNAs and insights into biological functions and therapeutic potential of brain enriched miRNA-128. Mol. Cancer 13, 33.

Alexandrov, P.N., Dua, P., Hill, J.M., Bhattacharjee, S., Zhao, Y., Lukiw, W.J., 2012. microRNA (miRNA) speciation in Alzheimer's disease (AD) cerebrospinal fluid (CSF) and extracellular fluid (ECF). Int. J. Biochem. Mol. Biol. 3, 365–373.

Alzheimer, A., 1906. Über einen eigenartigen schweren Erkrankungsprozeβ der Hirnrincle. Neurol. Central 25, 1134.

Ambros, V., 2008. The evolution of our thinking about microRNAs. Nat. Med. 14, 1036–1040.

Armstrong, R.A., Lantos, P.L., Cairns, N.J., 2005. Overlap between neurodegenerative disorders. Neuropathology 25, 111–124.

Babiarz, J.E., Hsu, R., Melton, C., Thomas, M., Ullian, E.M., Blelloch, R., 2011. A role for noncanonical microRNAs in the mammalian brain revealed by phenotypic differences in Dgcr8 versus Dicer1 knockouts and small RNA sequencing. RNA 17, 1489–1501.

Bernstein, E., Kim, S.Y., Carmell, M.A., Murchison, E.P., Alcorn, H., Li, M.Z., Mills, A.A., Elledge, S.J., Anderson, K.V., Hannon, G.J., 2003. Dicer is essential for mouse development. Nat. Genet. 35, 215–217.

Bertram, L., Lill, C.M., Tanzi, R.E., 2010. The genetics of Alzheimer disease: back to the future. Neuron 68, 270–281.

Bicker, S., Khudayberdiev, S., Weiss, K., Zocher, K., Baumeister, S., Schratt, G., 2013. The DEAH-box helicase DHX36 mediates dendritic localization of the neuronal precursor-microRNA-134. Genes Dev. 27, 991–996.

Botta-Orfila, T., Morato, X., Compta, Y., Lozano, J.J., Falgas, N., Valldeoriola, F., Pont-Sunyer, C., Vilas, D., Mengual, L., Fernandez, M., Molinuevo, J.L., Antonell, A., Marti, M.J., Fernandez-Santiago, R., Ezquerra, M., 2014. Identification of blood serum micro-RNAs associated with idiopathic and LRRK2 Parkinson's disease. J. Neurosci. Res. 92, 1071–1077.

Brunden, K.R., Trojanowski, J.Q., Smith 3RD, A.B., Lee, V.M., Ballatore, C., 2014. Microtubule-stabilizing agents as potential therapeutics for neurodegenerative disease. Bioorg. Med. Chem. 22, 5040–5049.

Burgos, K., Malenica, I., Metpally, R., Courtright, A., Rakela, B., Beach, T., Shill, H., Adler, C., Sabbagh, M., Villa, S., Tembe, W., Craig, D., Van Keuren-Jensen, K., 2014. Profiles of extracellular miRNA in cerebrospinal fluid and serum from patients with Alzheimer's and Parkinson's diseases correlate with disease status and features of pathology. PLoS One e94839, 9.

Cardo, L.F., Coto, E., de Mena, L., Ribacoba, R., Moris, G., Menendez, M., Alvarez, V., 2013. Profile of microRNAs in the plasma of Parkinson's disease patients and healthy controls. J. Neurol. 260, 1420–1422.

Chaudhuri, K.R., Odin, P., 2010. The challenge of non-motor symptoms in Parkinson's disease. Prog. Brain Res. 184, 325–341.

Cho, H.J., Liu, G., Jin, S.M., Parisiadou, L., Xie, C., Yu, J., Sun, L., Ma, B., Ding, J., Vancraenenbroeck, R., Lobbestael, E., Baekelandt, V., Taymans, J.M., He, P., Troncoso, J.C., Shen, Y., Cai, H., 2013. MicroRNA-205 regulates the expression of Parkinson's disease-related leucine-rich repeat kinase 2 protein. Hum. Mol. Genet. 22, 608–620.

Cogswell, J.P., Ward, J., Taylor, I.A., Waters, M., Shi, Y., Cannon, B., Kelnar, K., Kemppainen, J., Brown, D., Chen, C., Prinjha, R.K., Richardson, J.C., Saunders, A.M., Roses, A.D., Richards, C.A., 2008. Identification of miRNA changes in Alzheimer's disease brain and CSF yields putative biomarkers and insights into disease pathways. J. Alzheimers Dis. 14, 27–41.

Coppede, F., 2012. Genetics and epigenetics of Parkinson's disease. ScientificWorldJournal 2012, 489830.

Cui, J.G., Li, Y.Y., Zhao, Y., Bhattacharjee, S., Lukiw, W.J., 2010. Differential regulation of interleukin-1 receptor-associated kinase-1 (IRAK-1) and IRAK-2 by microRNA-146a and NF-kappaB in stressed human astroglial cells and in Alzheimer disease. J. Biol. Chem. 285, 38951–38960.

de la Mata, M., Gaidatzis, D., Vitanescu, M., Stadler, M.B., Wentzel, C., Scheiffele, P., Filipowicz, W., Grosshans, H., 2015. Potent degradation of neuronal miRNAs induced by highly complementary targets. EMBO Rep. 16, 500–511.

Denk, J., Boelmans, K., Siegismund, C., Lassner, D., Arlt, S., Jahn, H., 2015. MicroRNA profiling of CSF reveals potential biomarkers to detect Alzheimer's disease. PLoS One 10, e0126423.

Diez-Planelles, C., Sanchez-Lozano, P., Crespo, M.C., Gil-Zamorano, J., Ribacoba, R., Gonzalez, N., Suarez, E., Martinez-Descals, A., Martinez-Camblor, P., Alvarez, V., Martin-Hernandez, R., Huerta-Ruiz, I., Gonzalez-Garcia, I., Cosgaya, J.M., Visioli, F., Davalos, A., Iglesias-Gutierrez, E., Tomas-Zapico, C., 2016. Circulating microRNAs in Huntington's disease: emerging mediators in metabolic impairment. Pharmacol. Res. 108, 102–110.

Dong, H., Li, J., Huang, L., Chen, X., Li, D., Wang, T., Hu, C., Xu, J., Zhang, C., Zen, K., Xiao, S., Yan, Q., Wang, C., Zhang, C.Y., 2015. Serum microRNA profiles serve as novel biomarkers for the diagnosis of Alzheimer's disease. Dis. Markers 2015, 625659.

Doxakis, E., 2010. Post-transcriptional regulation of alpha-synuclein expression by mir-7 and mir-153. J. Biol. Chem. 285, 12726–12734.

Duff, K., Paulsen, J.S., Beglinger, L.J., Langbehn, D.R., Stout, J.C., Predict, H.D., Investigators of the Huntington Study Group, 2007. Psychiatric symptoms in Huntington's disease before diagnosis: the predict-HD study. Biol. Psychiatry 62, 1341–1346.

Eddy, C.M., Parkinson, E.G., Rickards, H.E., 2016. Changes in mental state and behaviour in Huntington's disease. Lancet Psychiatry 3, 1079–1086.

Egorova, P., Popugaeva, E., Bezprozvanny, I., 2015. Disturbed calcium signaling in spinocerebellar ataxias and Alzheimer's disease. Semin. Cell Dev. Biol. 40, 127–133.

Engel, T., Brennan, G.P., Sanz-Rodriguez, A., Alves, M., Beamer, E., Watters, O., Henshall, D.C., Jimenez-Mateos, E.M., 2017. A calcium-sensitive feed-forward loop regulating the expression of the ATP-gated purinergic P2X7 receptor via specificity protein 1 and microRNA-22. Biochim. Biophys. Acta 1864, 255–266.

Farias, G., Cornejo, A., Jimenez, J., Guzman, L., Maccioni, R.B., 2011. Mechanisms of tau self-aggregation and neurotoxicity. Curr. Alzheimer Res. 8, 608–614.

Fiumara, F., Rajasethupathy, P., Antonov, I., Kosmidis, S., Sossin, W.S., Kandel, E.R., 2015. MicroRNA-22 gates long-term heterosynaptic plasticity in Aplysia through presynaptic regulation of CPEB and downstream targets. Cell Rep. 11, 1866–1875.

Galimberti, D., Villa, C., Fenoglio, C., Serpente, M., Ghezzi, L., Cioffi, S.M., Arighi, A., Fumagalli, G., Scarpini, E., 2014. Circulating miRNAs as potential biomarkers in Alzheimer's disease. J. Alzheimers Dis. 42, 1261–1267.

Gaughwin, P.M., Ciesla, M., Lahiri, N., Tabrizi, S.J., Brundin, P., Bjorkqvist, M., 2011. Hsa-miR-34b is a plasma-stable microRNA that is elevated in pre-manifest Huntington's disease. Hum. Mol. Genet. 20, 2225–2237.

Geekiyanage, H., Chan, C., 2011. MicroRNA-137/181c regulates serine palmitoyltransferase and in turn amyloid beta, novel targets in sporadic Alzheimer's disease. J. Neurosci. 31, 14820–14830.

Geekiyanage, H., Jicha, G.A., Nelson, P.T., Chan, C., 2012. Blood serum miRNA: non-invasive biomarkers for Alzheimer's disease. Exp. Neurol. 235, 491–496.

Gehrke, S., Imai, Y., Sokol, N., Lu, B., 2010. Pathogenic LRRK2 negatively regulates microRNA-mediated translational repression. Nature 466, 637–641.

Gillardon, F., Mack, M., Rist, W., Schnack, C., Lenter, M., Hildebrandt, T., Hengerer, B., 2008. MicroRNA and proteome expression profiling in early-symptomatic alpha-synuclein(A30P)-transgenic mice. Proteomics Clin. Appl. 2, 697–705.

Giraldez, A.J., Cinalli, R.M., Glasner, M.E., Enright, A.J., Thomson, J.M., Baskerville, S., Hammond, S.M., Bartel, D.P., Schier, A.F., 2005. MicroRNAs regulate brain morphogenesis in zebrafish. Science 308, 833–838.

Grasso, M., Piscopo, P., Crestini, A., Confaloni, A., Denti, M.A., 2015. Circulating microRNAs in neurodegenerative diseases. EXS 106, 151–169.

Guerreiro, R., Hardy, J., 2014. Genetics of Alzheimer's disease. Neurotherapeutics 11, 732–737.

Gusella, J.F., Wexler, N.S., Conneally, P.M., Naylor, S.L., Anderson, M.A., Tanzi, R.E., Watkins, P.C., Ottina, K., Wallace, M.R., Sakaguchi, A.Y., et al., 1983. A polymorphic DNA marker genetically linked to Huntington's disease. Nature 306, 234–238.

Hamza, T.H., Zabetian, C.P., Tenesa, A., Laederach, A., Montimurro, J., Yearout, D., Kay, D.M., Doheny, K.F., Paschall, J., Pugh, E., Kusel, V.I., Collura, R., Roberts, J., Griffith, A., Samii, A., Scott, W.K., Nutt, J., Factor, S.A., Payami, H., 2010. Common genetic variation in the HLA region is associated with late-onset sporadic Parkinson's disease. Nat. Genet. 42, 781–785.

Hardy, J., Selkoe, D.J., 2002. The amyloid hypothesis of Alzheimer's disease: progress and problems on the road to therapeutics. Science 297, 353–356.

Harfe, B.D., McManus, M.T., Mansfield, J.H., Hornstein, E., Tabin, C.J., 2005. The RNaseIII enzyme Dicer is required for morphogenesis but not patterning of the vertebrate limb. Proc. Natl. Acad. Sci. U.S.A. 102, 10898–10903.

He, L., Hannon, G.J., 2004. MicroRNAs: small RNAs with a big role in gene regulation. Nat. Rev. Genet. 5, 522–531.

Hebert, S.S., Horre, K., Nicolai, L., Papadopoulou, A.S., Mandemakers, W., Silahtaroglu, A.N., Kauppinen, S., Delacourte, A., de Strooper, B., 2008. Loss of microRNA cluster miR-29a/b-1 in sporadic Alzheimer's disease correlates with increased BACE1/beta-secretase expression. Proc. Natl. Acad. Sci. U.S.A. 105, 6415–6420.

Hebert, S.S., Horre, K., Nicolai, L., Bergmans, B., Papadopoulou, A.S., Delacourte, A., de Strooper, B., 2009. MicroRNA regulation of Alzheimer's amyloid precursor protein expression. Neurobiol. Dis. 33, 422–428.

Hebert, S.S., Papadopoulou, A.S., Smith, P., Galas, M.C., Planel, E., Silahtaroglu, A.N., Sergeant, N., Buee, L., de Strooper, B., 2010. Genetic ablation of Dicer in adult forebrain neurons results in abnormal tau hyperphosphorylation and neurodegeneration. Hum. Mol. Genet. 19, 3959–3969.

Hensley, K., Harris-White, M.E., 2015. Redox regulation of autophagy in healthy brain and neurodegeneration. Neurobiol. Dis. 84, 50–59.

Hoss, A.G., Labadorf, A., Latourelle, J.C., Kartha, V.K., Hadzi, T.C., Gusella, J.F., Macdonald, M.E., Chen, J.F., Akbarian, S., Weng, Z., Vonsattel, J.P., Myers, R.H., 2015a. miR-10b-5p expression in Huntington's disease brain relates to age of onset and the extent of striatal involvement. BMC Med. Genomics 8, 10.

Hoss, A.G., Lagomarsino, V.N., Frank, S., Hadzi, T.C., Myers, R.H., Latourelle, J.C., 2015b. Study of plasma-derived miRNAs mimic differences in Huntington's disease brain. Mov. Disord. 30, 1961–1964.

Houbaviy, H.B., Murray, M.F., Sharp, P.A., 2003. Embryonic stem cell-specific microRNAs. Dev. Cell 5, 351–358.

Huntington, G., 1872. On chorea. Med. Surg. Rep. 26, 317–321.

Ibanez, P., Bonnet, A.M., Debarges, B., Lohmann, E., Tison, F., Pollak, P., Agid, Y., Durr, A., Brice, A., 2004. Causal relation between alpha-synuclein gene duplication and familial Parkinson's disease. Lancet 364, 1169–1171.

Impey, S., Davare, M., Lesiak, A., Fortin, D., Ando, H., Varlamova, O., Obrietan, K., Soderling, T.R., Goodman, R.H., Wayman, G.A., 2010. An activity-induced microRNA controls dendritic spine formation by regulating Rac1-PAK signaling. Mol. Cell. Neurosci. 43, 146–156.

Jimenez-Mateos, E.M., Bray, I., Sanz-Rodriguez, A., Engel, T., McKiernan, R.C., Mouri, G., Tanaka, K., Sano, T., Saugstad, J.A., Simon, R.P., Stallings, R.L., Henshall, D.C., 2011. miRNA expression profile after status epilepticus and hippocampal neuroprotection by targeting miR-132. Am. J. Pathol. 179, 2519–2532.

Jimenez-Mateos, E.M., Arribas-Blazquez, M., Sanz-Rodriguez, A., Concannon, C., Olivos-Ore, L.A., Reschke, C.R., Mooney, C.M., Mooney, C., Lugara, E., Morgan, J., Langa, E., Jimenez-Pacheco, A., Silva, L.F., Mesuret, G., Boison, D., Miras-Portugal, M.T., Letavic, M., Artalejo, A.R., Bhattacharya, A., Diaz-Hernandez, M., Henshall, D.C., Engel, T., 2015. MicroRNA targeting of the P2X7 purinoceptor opposes a contralateral epileptogenic focus in the hippocampus. Sci. Rep. 5, 17486.

Jin, P., Zarnescu, D.C., Ceman, S., Nakamoto, M., Mowrey, J., Jongens, T.A., Nelson, D.L., Moses, K., Warren, S.T., 2004. Biochemical and genetic interaction between the fragile X mental retardation protein and the microRNA pathway. Nat. Neurosci. 7, 113–117.

Johnson, R., Buckley, N.J., 2009. Gene dysregulation in Huntington's disease: REST, microRNAs and beyond. Neuromolecular Med. 11, 183–199.

Johnson, R., Zuccato, C., Belyaev, N.D., Guest, D.J., Cattaneo, E., Buckley, N.J., 2008. A microRNA-based gene dysregulation pathway in Huntington's disease. Neurobiol. Dis. 29, 438–445.

Jovicic, A., Zaldivar Jolissaint, J.F., Moser, R., Silva Santos Mde, F., Luthi-Carter, R., 2013. MicroRNA-22 (miR-22) overexpression is neuroprotective via general anti-apoptotic effects and may also target specific Huntington's disease-related mechanisms. PLoS One 8, e54222.

Junn, E., Lee, K.W., Jeong, B.S., Chan, T.W., Im, J.Y., Mouradian, M.M., 2009. Repression of alpha-synuclein expression and toxicity by microRNA-7. Proc. Natl. Acad. Sci. U.S.A. 106, 13052–13057.

Kaur, N., Kumar, P., Jamwal, S., Deshmukh, R., Gauttam, V., 2016. Tetrabenazine: spotlight on drug review. Ann. Neurosci. 23, 176–185.

Kawase-Koga, Y., Otaegi, G., Sun, T., 2009. Different timings of Dicer deletion affect neurogenesis and gliogenesis in the developing mouse central nervous system. Dev. Dyn. 238, 2800–2812.

Kim, J., Inoue, K., Ishii, J., Vanti, W.B., Voronov, S.V., Murchison, E., Hannon, G., Abeliovich, A., 2007. A microRNA feedback circuit in midbrain dopamine neurons. Science 317, 1220–1224.

Kim, A.H., Reimers, M., Maher, B., Williamson, V., McMichael, O., Mcclay, J.L., van den Oord, E.J., Riley, B.P., Kendler, K.S., Vladimirov, V.I., 2010. MicroRNA expression profiling in the prefrontal cortex of individuals affected with schizophrenia and bipolar disorders. Schizophr. Res. 124, 183–191.

Kim, B., Kim, M.K., Kim, S.W., Kim, K.M., Kim, H.S., An, H.J., Kim, J.O., Choi, T.K., Kim, N.K., Lee, S.H., 2015. Association of human microRNAs miR-22 and miR-491 polymorphisms with panic disorder with or without agoraphobia in a Korean population. J. Affect. Disord. 188, 118–126.

Klein, M.E., Lioy, D.T., Ma, L., Impey, S., Mandel, G., Goodman, R.H., 2007. Homeostatic regulation of MeCP2 expression by a CREB-induced microRNA. Nat. Neurosci. 10, 1513–1514.

Kole, A.J., Swahari, V., Hammond, S.M., Deshmukh, M., 2011. miR-29b is activated during neuronal maturation and targets BH3-only genes to restrict apoptosis. Genes Dev. 25, 125–130.

Konopka, W., Kiryk, A., Novak, M., Herwerth, M., Parkitna, J.R., Wawrzyniak, M., Kowarsch, A., Michaluk, P., Dzwonek, J., Arnsperger, T., Wilczynski, G., Merkenschlager, M., Theis, F.J., Kohr, G., Kaczmarek, L., Schutz, G., 2010. MicroRNA loss enhances learning and memory in mice. J. Neurosci. 30, 14835–14842.

Kumar, P., Dezso, Z., Mackenzie, C., Oestreicher, J., Agoulnik, S., Byrne, M., Bernier, F., Yanagimachi, M., Aoshima, K., Oda, Y., 2013. Circulating miRNA biomarkers for Alzheimer's disease. PLoS One 8, e69807.

Lagos-Quintana, M., Rauhut, R., Yalcin, A., Meyer, J., Lendeckel, W., Tuschl, T., 2002. Identification of tissue-specific microRNAs from mouse. Curr. Biol. 12 (9), 735.

Landles, C., Bates, G.P., 2004. Huntingtin and the molecular pathogenesis of Huntington's disease. Fourth in molecular medicine review series. EMBO Rep. 5, 958–963.

Lau, P., Bossers, K., Janky, R., Salta, E., Frigerio, C.S., Barbash, S., Rothman, R., Sierksma, A.S., Thathiah, A., Greenberg, D., Papadopoulou, A.S., Achsel, T., Ayoubi, T., Soreq, H., Verhaagen, J., Swaab, D.F., Aerts, S., de Strooper, B., 2013. Alteration of the microRNA network during the progression of Alzheimer's disease. EMBO Mol. Med. 5, 1613–1634.

Lee, R.C., Ambros, V., 2001. An extensive class of small RNAs in Caenorhabditis elegans. Science 294, 862–864.

Lee, R.C., Feinbaum, R.L., Ambros, V., 1993. The C. elegans heterochronic gene lin-4 encodes small RNAs with antisense complementarity to lin-14. Cell 75, 843–854.

Lee, S.T., Chu, K., Im, W.S., Yoon, H.J., Im, J.Y., Park, J.E., Park, K.H., Jung, K.H., Lee, S.K., Kim, M., Roh, J.K., 2011. Altered microRNA regulation in Huntington's disease models. Exp. Neurol. 227, 172–179.

Ling, K.H., Brautigan, P.J., Hahn, C.N., Daish, T., Rayner, J.R., Cheah, P.S., Raison, J.M., Piltz, S., Mann, J.R., Mattiske, D.M., Thomas, P.Q., Adelson, D.L., Scott, H.S., 2011. Deep sequencing analysis of the developing mouse brain reveals a novel microRNA. BMC Genomics 12, 176.

Lu, W., You, R., Yuan, X., Yang, T., Samuel, E.L., Marcano, D.C., Sikkema, W.K., Tour, J.M., Rodriguez, A., Kheradmand, F., Corry, D.B., 2015. The microRNA miR-22 inhibits the histone deacetylase HDAC4 to promote T(H)17 cell-dependent emphysema. Nat. Immunol. 16, 1185–1194.

Lugli, G., Larson, J., Martone, M.E., Jones, Y., Smalheiser, N.R., 2005. Dicer and eIF2c are enriched at postsynaptic densities in adult mouse brain and are modified by neuronal activity in a calpain-dependent manner. J. Neurochem. 94, 896–905.

Lugli, G., Cohen, A.M., Bennett, D.A., Shah, R.C., Fields, C.J., Hernandez, A.G., Smalheiser, N.R., 2015. Plasma exosomal miRNAs in persons with and without Alzheimer disease: altered expression and prospects for biomarkers. PLoS One 10, e0139233.

Lukiw, W.J., 2012. NF-small ka, cyrillicB-regulated micro RNAs (miRNAs) in primary human brain cells. Exp. Neurol. 235, 484–490.

Lukiw, W.J., Zhao, Y., Cui, J.G., 2008. An NF-kappaB-sensitive micro RNA-146a-mediated inflammatory circuit in Alzheimer disease and in stressed human brain cells. J. Biol. Chem. 283, 31315–31322.

Lungu, G., Stoica, G., Ambrus, A., 2013. MicroRNA profiling and the role of microRNA-132 in neurodegeneration using a rat model. Neurosci. Lett. 553, 153–158.

Ma, J., Shui, S., Han, X., Guo, D., Li, T., Yan, L., 2016. MicroRNA-22 attenuates neuronal cell apoptosis in a cell model of traumatic brain injury. Am. J. Transl. Res. 8, 1895–1902.

MacDonald, M.E., Ambrose, C.M., Duyao, M.P., Myers, R.H., Lin, C., Srinidhi, L., Barnes, G., Taylor, S.A., James, M., Groot, N., MacFarlane, H., Jenkins, B., 1993. A novel gene containing a trinucleotide repeat that is expanded and unstable on Huntington's disease chromosomes. The Huntington's Disease Collaborative Research Group. Cell 72, 971–983.

Majidinia, M., Mihanfar, A., Rahbarghazi, R., Nourazarian, A., Bagca, B., Avci, C.B., 2016. The roles of non-coding RNAs in Parkinson's disease. Mol. Biol. Rep. 43, 1193–1204.

Margis, R., Margis, R., Rieder, C.R., 2011. Identification of blood microRNAs associated to Parkinsonis disease. J. Biotechnol. 152, 96–101.

Marques, T.M., Kuiperij, H.B., Bruinsma, I.B., van Rumund, A., Aerts, M.B., Esselink, R.A., Bloem, B.R., Verbeek, M.M., 2016. MicroRNAs in cerebrospinal fluid as potential biomarkers for Parkinson's disease and multiple system atrophy. Mol. Neurobiol. (Date of publication: Nov 14, 2016).

Marti, E., Pantano, L., Banez-Coronel, M., Llorens, F., Minones-Moyano, E., Porta, S., Sumoy, L., Ferrer, I., Estivill, X., 2010. A myriad of miRNA variants in control and Huntington's disease brain regions detected by massively parallel sequencing. Nucleic Acids Res. 38, 7219–7235.

Mellios, N., Sugihara, H., Castro, J., Banerjee, A., Le, C., Kumar, A., Crawford, B., Strathmann, J., Tropea, D., Levine, S.S., Edbauer, D., Sur, M., 2011. miR-132, an experience-dependent microRNA, is essential for visual cortex plasticity. Nat. Neurosci. 14, 1240–1242.

Mena, N.P., Urrutia, P.J., Lourido, F., Carrasco, C.M., Nunez, M.T., 2015. Mitochondrial iron homeostasis and its dysfunctions in neurodegenerative disorders. Mitochondrion 21, 92–105.

Miller, B.H., Zeier, Z., Xi, L., Lanz, T.A., Deng, S., Strathmann, J., Willoughby, D., Kenny, P.J., Elsworth, J.D., Lawrence, M.S., Roth, R.H., Edbauer, D., Kleiman, R.J., Wahlestedt, C., 2012. MicroRNA-132 dysregulation in schizophrenia has implications for both neurodevelopment and adult brain function. Proc. Natl. Acad. Sci. U.S.A. 109, 3125–3130.
Minones-Moyano, E., Porta, S., Escaramis, G., Rabionet, R., Iraola, S., Kagerbauer, B., Espinosa-Parrilla, Y., Ferrer, I., Estivill, X., Marti, E., 2011. MicroRNA profiling of Parkinson's disease brains identifies early downregulation of miR-34b/c which modulate mitochondrial function. Hum. Mol. Genet. 20, 3067–3078.
Muinos-Gimeno, M., Espinosa-Parrilla, Y., Guidi, M., Kagerbauer, B., Sipila, T., Maron, E., Pettai, K., Kananen, L., Navines, R., Martin-Santos, R., Gratacos, M., Metspalu, A., Hovatta, I., Estivill, X., 2011. Human microRNAs miR-22, miR-138-2, miR-148a, and miR-488 are associated with panic disorder and regulate several anxiety candidate genes and related pathways. Biol. Psychiatry 69, 526–533.
NA, E.S., Monteggia, L.M., 2011. The role of MeCP2 in CNS development and function. Horm. Behav. 59, 364–368.
Nixon, R.A., Yang, D.S., 2012. Autophagy and neuronal cell death in neurological disorders. Cold Spring Harb. Perspect. Biol. 43 (1), 38–45.
Nudelman, A.S., Dirocco, D.P., Lambert, T.J., Garelick, M.G., Le, J., Nathanson, N.M., Storm, D.R., 2010. Neuronal activity rapidly induces transcription of the CREB-regulated microRNA-132, in vivo. Hippocampus 20, 492–498.
Nunez-Iglesias, J., Liu, C.C., Morgan, T.E., Finch, C.E., Zhou, X.J., 2010. Joint genome-wide profiling of miRNA and mRNA expression in Alzheimer's disease cortex reveals altered miRNA regulation. PLoS One 5, e8898.
Packer, A.N., Xing, Y., Harper, S.Q., Jones, L., Davidson, B.L., 2008. The bifunctional microRNA miR-9/miR-9* regulates REST and CoREST and is downregulated in Huntington's disease. J. Neurosci. 28, 14341–14346.
Papapetropoulos, S., McCorquodale, D., 2007. Gene-expression profiling in Parkinson's disease: discovery of valid biomarkers, molecular targets and biochemical pathways. Future Neurol. 2, 29–38.
Parisi, C., Arisi, I., D'Ambrosi, N., Storti, A.E., Brandi, R., D'Onofrio, M., Volonte, C., 2013. Dysregulated microRNAs in amyotrophic lateral sclerosis microglia modulate genes linked to neuroinflammation. Cell Death Dis. 4, e959.
Poewe, W., 2008. Non-motor symptoms in Parkinson's disease. Eur. J. Neurol. 15 (Suppl. 1), 14–20.
Querfurth, H.W., Laferla, F.M., 2010. Alzheimer's disease. N. Engl. J. Med. 362, 329–344.
Reitz, C., Mayeux, R., 2014. Alzheimer disease: epidemiology, diagnostic criteria, risk factors and biomarkers. Biochem. Pharmacol. 88, 640–651.
Remenyi, J., Hunter, C.J., Cole, C., Ando, H., Impey, S., Monk, C.E., Martin, K.J., Barton, G.J., Hutvagner, G., Arthur, J.S., 2010. Regulation of the miR-212/132 locus by MSK1 and CREB in response to neurotrophins. Biochem. J. 428, 281–291.
Sambandan, S., Akbalik, G., Kochen, L., Rinne, J., Kahlstatt, J., Glock, C., Tushev, G., Alvarez-Castelao, B., Heckel, A., Schuman, E.M., 2017. Activity-dependent spatially localized miRNA maturation in neuronal dendrites. Science 355 (6325), 634–637.
Satoh, J., Kino, Y., Niida, S., 2015. MicroRNA-seq data analysis pipeline to identify blood biomarkers for Alzheimer's disease from public data. Biomark Insights 10, 21–31.
Savas, J.N., Makusky, A., Ottosen, S., Baillat, D., Then, F., Krainc, D., Shiekhattar, R., Markey, S.P., Tanese, N., 2008. Huntington's disease protein contributes to RNA-mediated gene silencing through association with Argonaute and P bodies. Proc. Natl. Acad. Sci. U.S.A. 105, 10820–10825.
Schaefer, A., Im, H.I., Veno, M.T., Fowler, C.D., Min, A., Intrator, A., Kjems, J., Kenny, P.J., O'carroll, D., Greengard, P., 2010. Argonaute 2 in dopamine 2 receptor-expressing neurons regulates cocaine addiction. J. Exp. Med. 207, 1843–1851.

Scheltens, P., Blennow, K., Breteler, M.M., de Strooper, B., Frisoni, G.B., Salloway, S., Van der Flier, W.M., 2016. Alzheimer's disease. Lancet 388, 505–517.
Schipper, H.M., Maes, O.C., Chertkow, H.M., Wang, E., 2007. MicroRNA expression in Alzheimer blood mononuclear cells. Gene Regul. Syst. Bio. 1, 263–274.
Sethi, P., Lukiw, W.J., 2009. Micro-RNA abundance and stability in human brain: specific alterations in Alzheimer's disease temporal lobe neocortex. Neurosci. Lett. 459, 100–104.
Shin, D., Shin, J.Y., McManus, M.T., Ptacek, L.J., Fu, Y.H., 2009. Dicer ablation in oligodendrocytes provokes neuronal impairment in mice. Ann. Neurol. 66, 843–857.
Shioya, M., Obayashi, S., Tabunoki, H., Arima, K., Saito, Y., Ishida, T., Satoh, J., 2010. Aberrant microRNA expression in the brains of neurodegenerative diseases: miR-29a decreased in Alzheimer disease brains targets neurone navigator 3. Neuropathol. Appl. Neurobiol. 36, 320–330.
Smirnova, L., Grafe, A., Seiler, A., Schumacher, S., Nitsch, R., Wulczyn, F.G., 2005. Regulation of miRNA expression during neural cell specification. Eur. J. Neurosci. 21, 1469–1477.
Smith, P.Y., Delay, C., Girard, J., Papon, M.A., Planel, E., Sergeant, N., Buee, L., Hebert, S.S., 2011. MicroRNA-132 loss is associated with tau exon 10 inclusion in progressive supranuclear palsy. Hum. Mol. Genet. 20, 4016–4024.
Sorensen, S.S., Nygaard, A.B., Christensen, T., 2016. miRNA expression profiles in cerebrospinal fluid and blood of patients with Alzheimer's disease and other types of dementia—an exploratory study. Transl. Neurodegener. 5, 6.
Spillantini, M.G., Goedert, M., 2013. Tau pathology and neurodegeneration. Lancet Neurol. 12, 609–622.
Tafuri, E., Santovito, D., de Nardis, V., Marcantonio, P., Paganelli, C., Affaitati, G., Bucci, M., Mezzetti, A., Giamberardino, M.A., Cipollone, F., 2015. MicroRNA profiling in migraine without aura: pilot study. Ann. Med. 47, 468–473.
Tan, L., Yu, J.T., Hu, N., Tan, L., 2013. Non-coding RNAs in Alzheimer's disease. Mol. Neurobiol. 47, 382–393.
Tan, L., Yu, J.T., Tan, M.S., Liu, Q.Y., Wang, H.F., Zhang, W., Jiang, T., Tan, L., 2014. Genome-wide serum microRNA expression profiling identifies serum biomarkers for Alzheimer's disease. J. Alzheimers Dis. 40, 1017–1027.
Tang, F., Kaneda, M., O'carroll, D., Hajkova, P., Barton, S.C., Sun, Y.A., Lee, C., Tarakhovsky, A., Lao, K., Surani, M.A., 2007. Maternal microRNAs are essential for mouse zygotic development. Genes Dev. 21, 644–648.
Tao, J., Wu, H., Lin, Q., Wei, W., Lu, X.H., Cantle, J.P., Ao, Y., Olsen, R.W., Yang, X.W., Mody, I., Sofroniew, M.V., Sun, Y.E., 2011. Deletion of astroglial Dicer causes non-cell-autonomous neuronal dysfunction and degeneration. J. Neurosci. 31, 8306–8319.
Tognini, P., Pizzorusso, T., 2012. MicroRNA212/132 family: molecular transducer of neuronal function and plasticity. Int. J. Biochem. Cell Biol. 44, 6–10.
Ucar, A., Vafaizadeh, V., Jarry, H., Fiedler, J., Klemmt, P.A., Thum, T., Groner, B., Chowdhury, K., 2010. miR-212 and miR-132 are required for epithelial stromal interactions necessary for mouse mammary gland development. Nat. Genet. 42, 1101–1108.
Vallelunga, A., Ragusa, M., Di Mauro, S., Iannitti, T., Pilleri, M., Biundo, R., Weis, L., Di Pietro, C., De Iuliis, A., Nicoletti, A., Zappia, M., Purrello, M., Antonini, A., 2014. Identification of circulating microRNAs for the differential diagnosis of Parkinson's disease and multiple system atrophy. Front. Cell. Neurosci. 8, 156.
Vidal, R.L., Matus, S., Bargsted, L., Hetz, C., 2014. Targeting autophagy in neurodegenerative diseases. Trends Pharmacol. Sci. 35, 583–591.

Vo, N., Klein, M.E., Varlamova, O., Keller, D.M., Yamamoto, T., Goodman, R.H., Impey, S., 2005. A cAMP-response element binding protein-induced microRNA regulates neuronal morphogenesis. Proc. Natl. Acad. Sci. U.S.A. 102, 16426–16431.

Volvert, M.L., Prevot, P.P., Close, P., Laguesse, S., Pirotte, S., Hemphill, J., Rogister, F., Kruzy, N., Sacheli, R., Moonen, G., Deiters, A., Merkenschlager, M., Chariot, A., Malgrange, B., Godin, J.D., Nguyen, L., 2014. MicroRNA targeting of CoREST controls polarization of migrating cortical neurons. Cell Rep. 7, 1168–1183.

Wanet, A., Tacheny, A., Arnould, T., Renard, P., 2012. miR-212/132 expression and functions: within and beyond the neuronal compartment. Nucleic Acids Res. 40, 4742–4753.

Wang, Y., Medvid, R., Melton, C., Jaenisch, R., Blelloch, R., 2007. DGCR8 is essential for microRNA biogenesis and silencing of embryonic stem cell self-renewal. Nat. Genet. 39, 380–385.

Wibrand, K., Panja, D., Tiron, A., Ofte, M.L., Skaftnesmo, K.O., Lee, C.S., Pena, J.T., Tuschl, T., Bramham, C.R., 2010. Differential regulation of mature and precursor microRNA expression by NMDA and metabotropic glutamate receptor activation during LTP in the adult dentate gyrus in vivo. Eur. J. Neurosci. 31, 636–645.

Wightman, B., Ha, I., Ruvkun, G., 1993. Posttranscriptional regulation of the heterochronic gene lin-14 by lin-4 mediates temporal pattern formation in C. elegans. Cell 75, 855–862.

Winter, J., Jung, S., Keller, S., Gregory, R.I., Diederichs, S., 2009. Many roads to maturity: microRNA biogenesis pathways and their regulation. Nat. Cell Biol. 11, 228–234.

Wong, H.K., Veremeyko, T., Patel, N., Lemere, C.A., Walsh, D.M., Esau, C., Vanderburg, C., Krichevsky, A.M., 2013. De-repression of FOXO3a death axis by microRNA-132 and -212 causes neuronal apoptosis in Alzheimer's disease. Hum. Mol. Genet. 22, 3077–3092.

Wu, Y., Le, W., Jankovic, J., 2011. Preclinical biomarkers of Parkinson disease. Arch. Neurol. 68, 22–30.

Yang, D., Li, T., Wang, Y., Tang, Y., Cui, H., Tang, Y., Zhang, X., Chen, D., Shen, N., Le, W., 2012. miR-132 regulates the differentiation of dopamine neurons by directly targeting Nurr1 expression. J. Cell Sci. 125, 1673–1682.

Yang, C.P., Zhang, Z.H., Zhang, L.H., Rui, H.C., 2016. Neuroprotective role of microRNA-22 in a 6-hydroxydopamine-induced cell model of Parkinson's disease via regulation of its target gene TRPM7. J. Mol. Neurosci. 60, 445–452.

Yu, H., Wu, M., Zhao, P., Huang, Y., Wang, W., Yin, W., 2015a. Neuroprotective effects of viral overexpression of microRNA-22 in rat and cell models of cerebral ischemia-reperfusion injury. J. Cell. Biochem. 116, 233–241.

Yu, H.C., Wu, J., Zhang, H.X., Zhang, G.L., Sui, J., Tong, W.W., Zhang, X.Y., Nie, L.L., Duan, J.H., Zhang, L.R., Lv, L.X., 2015b. Alterations of miR-132 are novel diagnostic biomarkers in peripheral blood of schizophrenia patients. Prog. Neuropsychopharmacol. Biol. Psychiatry 63, 23–29.

Zhang, C., Ge, X., Liu, Q., Jiang, M., Li, M.W., Li, H., 2015. MicroRNA-mediated non-cell-autonomous regulation of cortical radial glial transformation revealed by a Dicer1 knockout mouse model. Glia 63, 860–876.

Zheng, B., Liao, Z., Locascio, J.J., Lesniak, K.A., Roderick, S.S., Watt, M.L., Eklund, A.C., Zhang-James, Y., Kim, P.D., Hauser, M.A., Grunblatt, E., Moran, L.B., Mandel, S.A., Riederer, P., Miller, R.M., Federoff, H.J., Wullner, U., Papapetropoulos, S., Youdim, M.B., Cantuti-Castelvetri, I., Young, A.B., Vance, J.M., Davis, R.L., Hedreen, J.C., Adler, C.H., Beach, T.G., Graeber, M.B., Middleton, F.A., Rochet, J.C., Scherzer, C.R., Global PD Gene Expression (GPEX) Consortium, 2010. PGC-1alpha, a potential therapeutic target for early intervention in Parkinson's disease. Sci. Transl. Med. 2, 52ra73.